Second International Conference on Mechanical and Aeronautical Engineering (ICMAE 2016)

IOP Conference Series: Materials Science and Engineering
Volume 187

Hong Kong, China
28 – 30 December 2016

Editors:

Ian McAndrew
Andrew Rae

ISBN: 978-1-5108-3842-0
ISSN: 1757-8981

Printed from e-media with permission by:

Curran Associates, Inc.
57 Morehouse Lane
Red Hook, NY 12571

CURRAN ASSOCIATES INC.
proceedings
.com

Some format issues inherent in the e-media version may also appear in this print version.

Printed by Curran Associates, Inc. (2017)

Institute of Physics
Dirac House, Temple Back
Bristol BS1 6BE UK

Phone: 44 1 17 929 7481
Fax: 44 1 17 920 0979

techtracking@iop.org

Additional copies of this publication are available from:

Curran Associates, Inc.
57 Morehouse Lane
Red Hook, NY 12571 USA
Phone: 845-758-0400
Fax: 845-758-2633
Email: curran@proceedings.com
Web: www.proceedings.com

Table of contents

Volume 187

2016 Second International Conference on Mechanical and Aeronautical Engineering

(ICMAE 2016)28–30 December 2016, Hong Kong

Accepted papers received: 16 March 2017Published online: 23 March 2017

Preface

Aeronautical Science

Aeronautical Systems

Mechanical Science

Mechanical Systems

IOP Conf. Series: Materials Science and Engineering **187** (2017) 011001 doi:10.1088/1757-899X/187/1/011001

PREFACE

ICMAE 2016 was held in Hong Kong during December 28-30, 2016. ICMAE 2016 was organized by Hong Kong Society of Mechanical Engineers (HKSME). The conference provides a useful and wide platform both for display the latest research and for exchange of research results and thoughts in broad Mechanical and Aeronautical Engineering. The participants of the conference were from almost every part of the world, with background of either academia or industry, even well-known enterprise. The success and prosperity of the conference is reflected high level of the papers received.

The proceedings are a compilation of the accepted papers and represent an interesting outcome of the conference. This proceeding covers 4 chapters: Aeronautical Science, Aeronautical Systems, Mechanical Science and Mechanical Systems.

We would like to acknowledge all of those who supported ICMAE 2016. Each individual and institutional help were very important for the success of this conference. Especially we would like to thank the organizing committee for their valuable advices in the organization and helpful peer review of the papers.

We sincerely hope that ICMAE 2016 will be a forum for excellent discussions that will put forward new ideas and promote collaborative researches. We are sure that the proceedings will serve as an important research source of references and the knowledge, which will lead to not only scientific and engineering progress but also other new products and processes.

Prof. Ian McAndrew, Embry Riddle Aeronautical University, UK
Prof. Andrew Rae, University of the Highlands and Islands, UK

COMMITTEES

Advisory Committee Chairs

Prof. Andrew Rae, University of the Highlands and Islands, UK
Prof. Ian McAndrew, Embry Riddle Aeronautical University, UK

Conference Committee Chairs

Prof. WONG Chee How, Nanyang Technological University, Singapore
Prof. Christopher Y. H. CHAO, Hong Kong University of Science and Technology, Hong Kong
Prof. Dan Zhang,York University, Canada
Prof. Cheng Li, The Hong Kong Polytechnic University, Hong Kong

Program Committee Chairs

Prof. ERASMO CARRERA, Politecnico di Torino, Italy
Prof. Matthew Parkinson,Pennsylvania State University, USA

International Technical Committee

Prof. Musa bin Mailah, Universiti Teknologi Malaysia, Malaysia
Prof. Chee-Meng Chew, National University of Singapore
Debra T. Bourdeau, Embry Riddle Aeronautical University, UK
Dr.David Scott Cross, Embry Riddle Aeronautical University, UK
Dr. Michele Ferlauto, Politecnico di Torino, Italy
Prof. Lin Shih-Chieh, National Tsing Hua University, Taiwan
Prof. Do, Hyungrok,Seoul National University, Korea
Prof. Yu-Ren Wu,National Central University, Taiwan
Prof. Nasir Hayat, University of Engineering and Technology, Pakistan
Prof. Ugur GUVEN, University of petroleum & energy studies, USA
Dr. TRAN Anh Tuan, Nanyang Technology University, Singapore
Dr. VINOTHKUMAR SIVALINGAM, Anna University, India
Dr. Farzad Ismail, Universiti Sains Malaysia, Malaysia
Dr. Eltayeb Eljack, King AbdulAziz University, Saudi Arabia
Dr. P.Sivaprakash, A.S.L.Pauls College of Engineering and Technology, India
Elena Navarro, Embry Riddle Aeronautical University, USA
Dr. Satish S. Chinchanikar, Department of Mechanical Engineering, VIIT, India
Dr. Redouane Zitoune, Clément Ader Institute of Toulouse University, France
Dr. Anis Hor, Clément Ader Institute of ISAE - SUPAERO, Toulouse, France
Dr. Vijayan Krishnaraj, PSG College of Coimbatore, India

IOP Conf. Series: Materials Science and Engineering **187** (2017) 011003 doi:10.1088/1757-899X/187/1/011003

Peer review statement

All papers published in this volume of *IOP Conference Series: Materials Science and Engineering* have been peer reviewed through processes administered by the proceedings Editors. Reviews were conducted by expert referees to the professional and scientific standards expected of a proceedings journal published by IOP Publishing.

IOP Conf. Series: Materials Science and Engineering **187** (2017) 012001 doi:10.1088/1757-899X/187/1/012001

Numerical Study of a Three Dimensional Interaction between two bow Shock Waves and the Aerodynamic Heating on a Wedge Shaped Nose Cone

N Wu[1,*], J H Wang[1] and L Shen[1]

[1]Department of Thermal Science and Energy Engineering, School of Engineering Science, University of Science and Technology of China, Hefei 230027, Anhui, China

[*]nanwu@mail.ustc.edu.cn

Abstract. This paper presents a numerical investigation on the three-dimensional interaction between two bow shock waves in two environments, i.e. ground high-enthalpy wind tunnel test and real space flight, using Fluent 15.0. The first bow shock wave, also called induced shock wave, which is generated by the leading edge of a hypersonic vehicle. The other bow shock wave can be deemed objective shock wave, which is generated by the cowl clip of hypersonic inlet, and in this paper the inlet is represented by a wedge shaped nose cone. The interaction performances including flow field structures, aerodynamic pressure and heating are analyzed and compared between the ground test and the real space flight. Through the analysis and comparison, we can find the following important phenomena: 1) Three-dimensional complicated flow structures appear in both cases, but only in the real space flight condition, a local two-dimensional type IV interaction appears; 2) The heat flux and pressure in the interaction region are much larger than those in the no-interaction region in both cases, but the peak values of the heat flux and pressure in real space flight are smaller than those in ground test. 3) The interaction region on the objective surface are different in the two cases, and there is a peak value displacement of 3 mm along the stagnation line.

1. Introduction

Shock-shock interactions in hypersonic flight may cause extremely high aerodynamic heating, which will lead to the failure of thermal protection systems. For an SSTO (Single-Stage to Orbit) vehicle, the maximum heat flux of the cowl leading edge reaches up to 579MW/m^2 due to a type IV shock-shock interaction [1]. Besides the heating problem, the research on the flow field structures of shock/shock interactions is also an interesting problem. Edney [2] conducted a series of experiments on the two-dimensional interaction between a bow shock wave and a planar oblique shock wave, and classified the interactions into six types, among which the type IV induces the most intense heating around the interaction region.

After Edney [2], many further studies have been conducted on shock-shock interactions both in experimental and numerical aspects [3-6]. However, most of the investigations are limited within two-dimensional problems, mainly focusing on the two-dimensional type IV interaction or the three-dimensional interaction between a planar oblique and a bow shock wave. In real situations, the interaction between the bow shock generated by vehicle front and the bow shock of inlet cowl lip shows three-dimensional effects. Therefore, it is significant to understand the real three-dimensional performances.

Before a real hypersonic flight, to validate a new thermal protection system, ground high-enthalpy wind tunnel tests are necessary. In the ground test, stagnation temperature is usually seen an important factor. To create the same stagnation temperature to real space flight, one has to use a lower Mach number but higher static temperature and pressure in ground high-enthalpy wind tunnel testes. Therefore, differences exist between the two environments. However, a few literatures mentioned the differences before this work, especially with the consideration of shock/shock interactions. To understand the differences of the two cases better, numerical simulation as an effective tool is used in this investigation. The aim is to investigate the flow field and the aerodynamic heating caused by the three-dimensional interaction, and to reveal the differences between the ground high-enthalpy wind tunnel test and the real space hypersonic flight.

2. Physical model

To simulate the typical three-dimensional interaction between the two bow shock waves generated by vehicle front and inlet cowl clip, the simplified physical model used in this paper is shown in figure1. It consists of two parts, a shock wave generator with a 10mm radius and a compression angle of 23° to freestream direction, an objective nose cone with a radius of 5mm and a cone angle of 12° and an extension length of 80mm. The model is symmetrical and has a 2mm gap in horizontal direction.

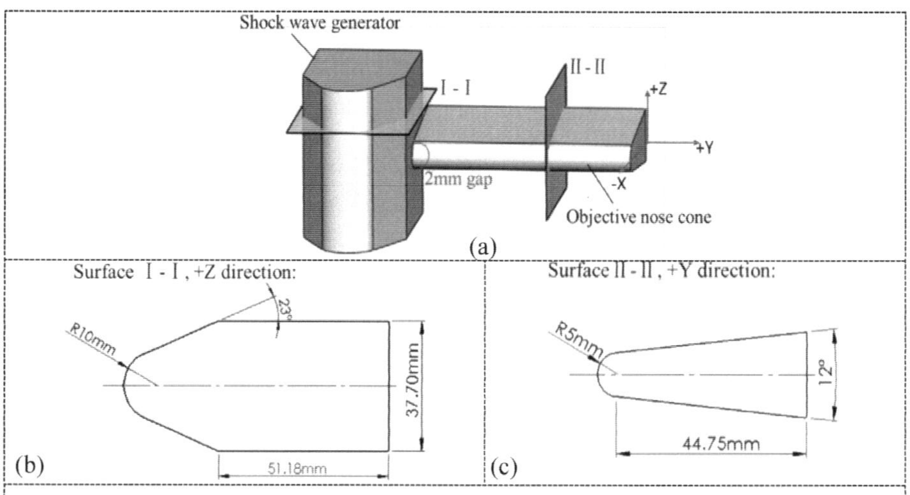

Figure 1. The diagram of geometric model and size.

3. Numerical method

3.1. Mesh setup

The mesh is generated by ICEM, and the computational domain is extended by ten times of the size of the model. Using the symmetrical condition, half of the model is built in order to save computation time. Here, non-uniform structured grid is used, and the grid near the wall is locally refined. After the validation of grid independence and mesh adaption, the grid strategy with about 4.5 million elements is used considering the balance of computational load and accuracy.

3.2. Solution strategies

Steady state assumption and Spalart-Allmaras turbulent model are used in the following simulations with Fluent 15.0 as the CFD solver. The differential equations are discretized using second–order upwind scheme. The calculations are performed using density-based, coupled implicit method. The gas parameters are calculated by "Ideal gas model" and Sutherland formula. The inlet boundary condition of main flow is set by "pressure far-field" and at the exit by "pressure outlet". The wall of the model is set as adiabatic with no slip condition.

3.3. Calculation conditions
At the same stagnation temperature 2400K, the calculation conditions in ground test and real space flight are listed in Table 1, and the atmosphere parameters are computed by the formulas presented by the United States in 1976 [7].

Table 1. Calculation conditions in two cases.

Parameters	Mach number	Altitude	Total temperature	Static Temperature	Static Pressure
Values	4.2	Ground	2400K	530.035K	15030Pa
	7	26.5km	2400K	222.034K	2304Pa

4. Results and discussion
Using the parameters shown in Table 1, numerical simulations are carried out in the two cases, and the results including flow field, aerodynamic pressure and heating are analyzed and compared.

4.1. Flow field
Figure 2 shows Mach number contours in a horizontal section passing through the stagnation line and four vertical sections in ground test, among which section D represents the no-interaction region. In section A and B, as shown in figure 2b and 2c, we can see the shock location is far away from the stagnation point. In section C, as shown in figure 2d, there is a discontinuity between the objective nose cone and the objective shock wave. This phenomenon is caused by the induced bow shock wave, which divides the flow behind the shock wave into two direction flows, along the free stream and the span wise flow of the objective nose cone. The same three-dimensional complicated flow structures can also be seen in space flight condition.

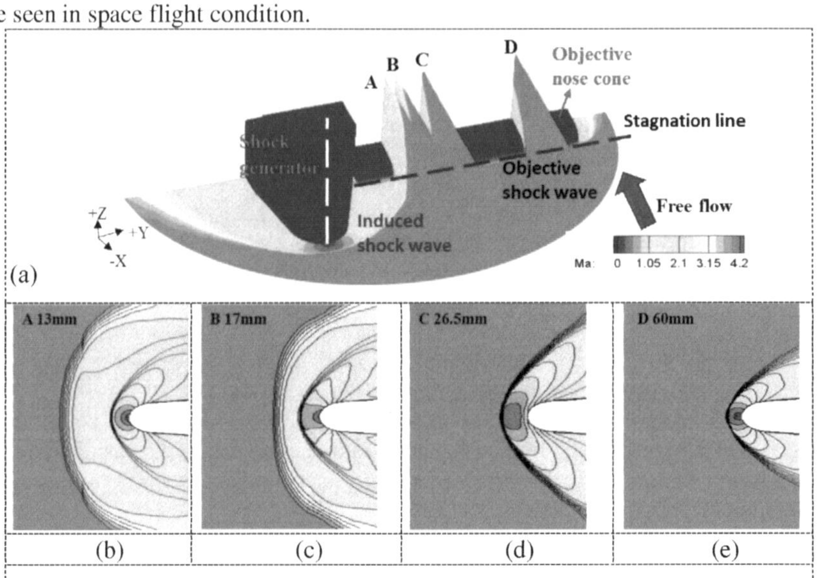

Figure 2. Mach number contours in vertical sections in ground test.

Figure 3 further exhibits the flow field in a horizontal section (z=0) in detail both in the ground test and real space flight. From this figure, we can find that: 1) In the two cases, near the nose cone surface, there is a shear layer, which is generated by the interaction of the induced shock wave with the objective shock wave. In this shear layer, a local boundary layer separation and reattachment exist. 2) Only in the real space flight condition, a local supersonic region (Ma>1) appears inside the boundary layer, as shown in figure 3b. Now the interaction type is called as type IV shock wave by Edney [1]. The difference in flow field can be explained by the fact that: at the same stagnation temperature, the Mach number in the ground test is smaller than and the induced shock intensity is lower than that in the real high space, and therefore in the ground test there is no supersonic region, as shown in figure 3a.

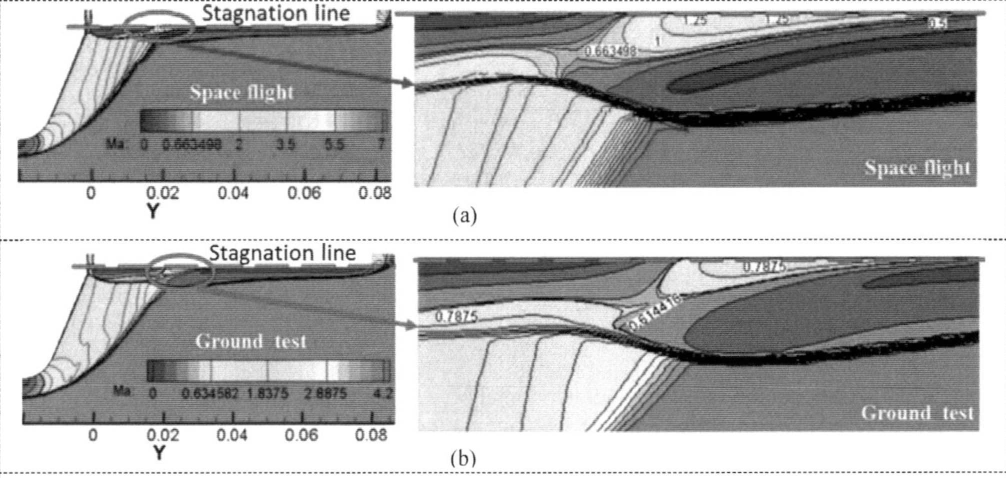

Figure 3. Mach number contours in horizontal section (z=0) in ground wind tunnel test and real space flight.

4.2. Aerodynamic pressure load

Figure 4a shows the pressure distribution in the stagnation region of the objective nose cone, and figure 4b further displays the pressure distribution along the stagnation line (the dashed red line shown in figure 4a).

From the analysis and comparison of the figures, we can find the following three interesting phenomena: 1) Pressure increases in the interaction region in both cases. 2) For the ground test, the peak value in the interaction region is 0.63MPa, about 1.8 times of the no-interaction region and 42 times of the inlet pressure (0.015MPa). However the peak value for the space flight is about 0.28 MPa, which is much lower than that in the ground test. 3) The peak value is at Y=18mm along the stagnation line in the ground test, and Y=15mm in the space flight. There is a difference of 3mm along the stagnation line in y direction between the two cases. The reason is that to achieve the same stagnation temperature 2400K, Mach number is 4.2 in the ground test, which is much smaller than Ma=7 in the space flight. Accordingly, the angle of the Mach cone in the ground test is larger and the interaction region moves farther along the extension direction.

The above phenomena indicates that in active thermal protection systems, to cool the stagnation area or nose cone, for the real space flight, the driving force to inject cooling media is much lower than that in the ground test, and the driving force in the shock-shock interaction region is larger than that of the no-interaction region in both cases.

IOP Conf. Series: Materials Science and Engineering **187** (2017) 012001 doi:10.1088/1757-899X/187/1/012001

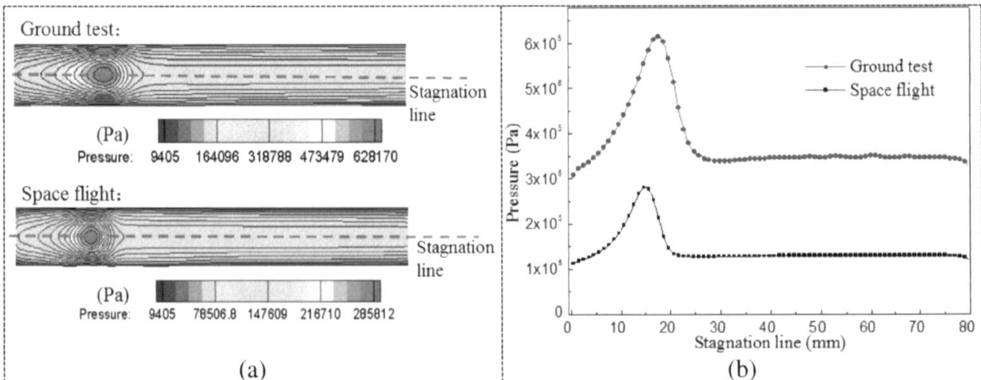

Figure 4. Pressure distribution of the nose cone along the stagnation line in ground wind tunnel experiment and real space flight.

4.3. Aerodynamic heating

Here, the adiabatic temperature boundary condition is replaced using a constant temperature of 280K and the aerodynamic heating on the objective nose cone is investigated. Figure 5 shows the heat flux distribution along the stagnation line, the dashed red line shown in figure 4a.

Through figure 5, one can find the following phenomena: 1) Heat flux increases in the interaction region in both cases. 2) The peak heat flux in the ground test is much larger than that in the space flight. For the ground test, the peak heat flux reaches to 17.76MW/m^2, which is about 1.6 times of the no-interaction region. While for the real space flight, the maximum heat flux is close to 11.94 MW/m^2, about 1.93 times of the no-interaction region. This can be explained by the fact that though the stagnation temperature and the wall temperature are set the same in the two cases, the Reynold number of the free flow in the ground test is much larger, and the local heat transfer in the interaction region is more intense. 3) Comparing the heat flux distribution with the pressure distribution as shown in figure 4, we can find that they have the same location of the peak value in the two cases.

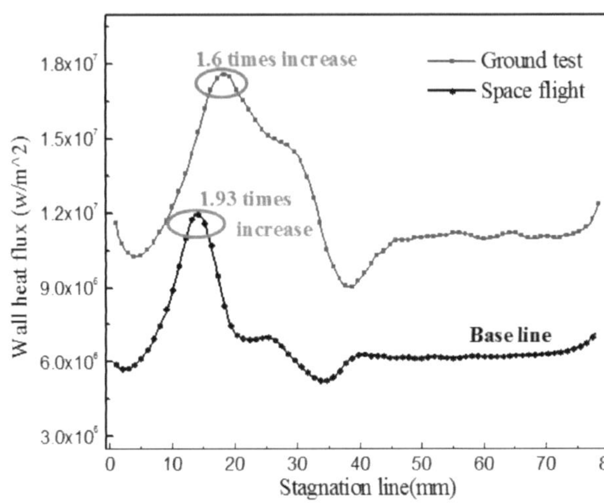

Figure 5. Wall heat flux distribution along the stagnation line in ground test.

5. Conclusions

A numerical investigation is carried out to study the three-dimensional interaction of two bow shock waves at two conditions, ground high-enthalpy wind tunnel test and real space flight. The differences of flow field and aerodynamic characteristics between the two cases are analyzed and compared. Through the analysis and comparison, the following conclusions can be drawn:

1) Three-dimensional complicated flow structures appear in both cases, but only in the real space flight a local two-dimensional type IV shock wave appears in the horizontal section.
2) At the same stagnation temperature, the pressure and heat flux in the interaction region are much larger than those in the no-interaction region in both cases, but the peak values in the real space flight are much smaller than those in the ground test, which means that the thermal environment is more severe in ground tests.
3) At the same stagnation temperature, the peak values of the heat flux and pressure occupy the same location in both cases. But the interaction regions are different in the two cases, and there is a peak value displacement of 3 mm along the stagnation line. It's a good guidance for the design of thermal protection systems.

6. References

[1] Glass D E 2008 Ceramic matrix composite (CMC) thermal protection systems (TPS) and ho structures for hypersonic vehicles *15th AIAA Space Planes and Hypersonic Systems and Technologies Conf.* (Dayton) pp 1-36
[2] Edney B 1968 Anomalous heat-transfer and pressure distribution on blunt bodies at hypersonic speeds in the presence of an impinging Shock(Sweden:Stockholm) *FFA* **No 115**
[3] Duquesne 1998 Numerical investigation of a three dimensional turbulent shock/shock interaction *AIAA 36th Aerospace Sciences Meeting & Exhibit* (NV.: Reno)
[4] Nowak R J and Holden M S 1990 Shock/Shock Interference on a Transpiration Cooled Hemispherical Model *AIAA 21st Fluid Dynamics, Plasma Dynamics and Lasers Conf.* (WA: Seattle)**Vol 1**
[5] Wieting A R and Holden M S 1989 Experimental Study on Shock Wave Interference Heating on a Cylindrical Leading Edge at Mach 6 and 8 *AIAA Journal* **Vol. 27** pp1557-1565
[6] Miyaji K 1999 Numerical analysis of three-dimensional shock/shock interactions and the aerodynamic heating *37th Aerospace Sciences Meeting and Exhibit* (NV:Reno)
[7] Lin J and Wang J H 2014 Numerical Investigation on Aero-Thermodynamic Characteristics of Wedge Shaped Nose Cone of Near Space Supersonic Flight *In Applied Mechanics and Materials* **vol 541** pp 608-612

Acknowledgements

This work is supported by the National Natural Science Foundation of China (No. 51376168) and the 3rd Institute of China Aerospace Science and Industry Corp.

IOP Conf. Series: Materials Science and Engineering **187** (2017) 012002 doi:10.1088/1757-899X/187/1/012002

Research on Turbofan Engine Model above Idle State Based on NARX Modeling Approach

Bing Yu[1,2]**, Wenjun Shu**[1,2]

[1]Jiangsu Province Key Laboratory of Aerospace Power Systems ,College of Energy and Power Engineering, Nanjing University of Aeronautics & Astronautics, Nanjing 210016, China
[2]Advanced Aero Engine Collaborative Innovation Center, Nanjing University of Aeronautics & Astronautics, Nanjing 210016, China

Corresponding author:Bing Yu(Yb203@nuaa.edu.cn)

Abstract. The nonlinear model for turbofan engine above idle state based on NARX is studied. Above all, the data sets for the JT9D engine from existing model are obtained via simulation. Then, a nonlinear modeling scheme based on NARX is proposed and several models with different parameters are built according to the former data sets. Finally, the simulations have been taken to verify the precise and dynamic performance the models, the results show that the NARX model can well reflect the dynamics characteristic of the turbofan engine with high accuracy.

1. Introduction

The modern aircraft engine is a complicated thermomechanical system; therefore, it is a challenging work to modeling it with high accuracy and real time characteristic. In general, there are two basic methods for modeling the system: the theory modeling and the identification modeling [1-2]. Among the existing modeling methods, the nonlinear autoregressive exogenous (NARX) modeling approach can well capture the dynamics of many nonlinear dynamical systems [3-4]. The NARX is a kind of identification technology based on time series [5], this modeling approach has been employed to the modeling work of gas turbine engine by several researchers. For instance, a nonlinear autoregressive exogenous model of the PGTA10B1 General Electric-Nuovo Pignone gas turbine operating in isolated and non-isolated conditions was reported by M.Basso et al. [6]. The NARX models of a heavy-duty single-shaft gas turbine(GT) were developed and validated by Hamid Asgari et al. [7]. In this article, the modeling of the JT9D engine is chosen as the plant, and the data for modeling are based on the component level turbofan model of JT9D [8]. By refining the identification parameters, several NARX models are built, and the mean errors of the best one are lower than 0.01, which can well reflect the dynamics characteristic of the turbofan engine with high accuracy.

2. The specification of the turbofan engine

The identification modeling of nonlinear system should be based on actual flight or test data sets in practical environment. Due to the restriction of our research work, it hard to get the real experimental data, thus the component level model of a turbofan engine introduced by the NASA is employed as the plant and the data sets are obtained from its simulation. Then, the data sets are applied to develop the

NARX model of the turbofan engine. The main specifications of the turbofan engine employed in this article are illustrated in table 1. The component level model consists eight parts, which are respectively the inlet, fan, low pressure compressor, high pressure compressor, burner, high pressure turbine, low pressure turbine and nozzle. And its input signals are the flight height H, Mach number Ma and fuel-air ratio FAR, meanwhile, the output signals are low pressure rotation speed n_L and high pressure rotation speed n_H.

Table 1. The main specifications of the turbofan engine.

Take-off thrust(kN)	Air flow rate(kg/s)	Bypass ratio	Total compression ratio	Thrust-weight ratio	High pressure rotation speed (rpm)
222.4	721.2	5.1	23.5	5.63	7800

3. NARX model

The NARX model is based on the extension of the autoregressive exogenous (ARX) model [9-11], and its mathematical structure could be described as equation (1):

$$y(t) = f[y(t-1), \dots, y(t-m_a), u(t-m_k), \dots, u(t-m_k-m_b-1)] \quad (1)$$

Where $y(t)$ is the output of the NARX model, $u(t)$ is the input of the NARX model, and f is a nonlinear function based on a finite number of input and output data; m_a, m_b and m_k are number of past output terms, number of past input terms and delay from input to the output in terms of the number of samples, respectively. The NARX model structure is shown as figure1:

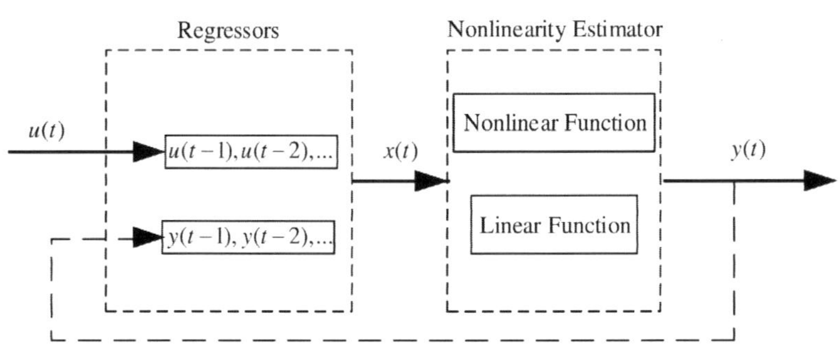

Figure 1. The NARX model structure.

As presented in figure 1, the NARX model structure is consisted of two stages which are the regressors block and the nonlinearity estimator block. In this article, the wavelet network is used as the nonlinear function in the nonlinearity estimator [12-13].

4. NARX model identification

4.1. The NARX model identification process

Figure 2 presents the NARX model identification process. The input and output data sets obtained through running the turbofan engine component level model are divided into two groups, which are respectively employed to develop and validate the NARX mode.

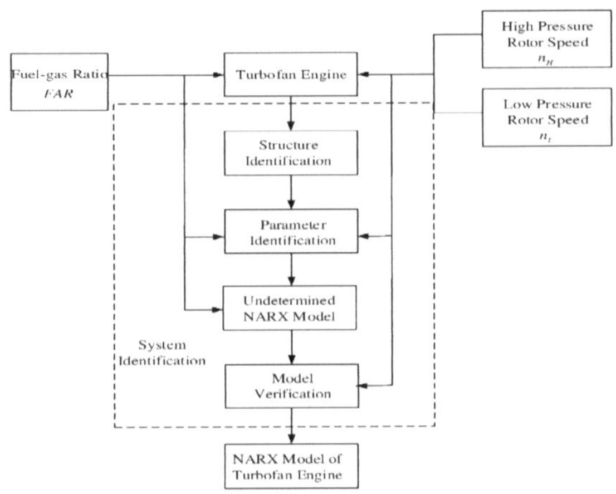

Figure 2. The NARX model identification process.

4.2. The NARX model structure and parameters identification

As shown in figure 1, both the input values and past output values are applied in the regressors block of the NARX model. Therefore, according to the equation (1), the following expressions can be obtained:

$$n_H(t) = f[n_H(t-1), n_H(t-2),\cdots, n_H(t-n_{a11}), n_L(t-1), n_L(t-2),\cdots, n_L(t-n_{a22}), FAR(t-n_{k11}),\cdots, FAR(t-n_{k11}-n_{b11}+1)] \quad (2)$$

$$n_L(t) = f[n_H(t-1), n_H(t-2),\cdots, n_H(t-n_{a21}), n_L(t-1), n_L(t-2),\cdots, n_L(t-n_{a22}), FAR(t-n_{k21}),\cdots, FAR(t-n_{k21}-n_{b21}+1)] \quad (3)$$

Where n_L and n_H are low and high pressure rotation speed of the NARX model, respectively; FAR is the fuel-air ratio of the NARX model; n_a, n_b and n_k can be expressed as follows:

$$n_a = \begin{bmatrix} n_{a11} & n_{a12} \\ n_{a21} & n_{a22} \end{bmatrix},\ n_b = \begin{bmatrix} n_{b11} \\ n_{b21} \end{bmatrix},\ n_k = \begin{bmatrix} n_{k11} \\ n_{k21} \end{bmatrix} \quad (4)$$

Based on the NARX model structure of the turbofan engine and the limitation of computing ability, n_a, n_b and n_k are limited within the range of [1, 5].

4.3. The input and output data sets acquisition

In view of the turbofan engine component level model mentioned above, the input data change scope is set to make the engine work in idling and above in this article. The range of the FAR are from 0.008 to 0.023, and the flight height H is 0m and Mach number Ma is 0. In the simulation, the integral method based on the limited white noise is applied to simulate the input data sets (that is FAR), and the limits on the integral are FAR ranges; Then, when FAR is imported into the turbofan engine component level model, the output data sets (that are n_L and n_H) can be obtained. The acquisition time is 200 seconds, and the step time is 0.04 seconds. The data sets in the first 80 seconds are utilized to develop the NARX model, and the rest of data sets are applied to validate the model.

4.4. The NARX model verification

The NARX model is employed to identify the nonlinear system, which can select different input and output terms and the delay, and then a series of NARX model can be determined. Therefore, the Best Fit as evaluation standard is employed to obtain the best NARX model from the determined model.

The Degree of fitting between the output of the turbofan engine component level model and the output of the determined NARX model is calculated, and the numerical most one has the highest accuracy.

$$f_{FIT} = \left(1 - \frac{|y-\tilde{y}|}{|y-\bar{y}|}\right) \times 100\% \quad (5)$$

Where f_{FIT} is the degree of fitting, y is the output value of the turbofan engine component level model, that are low and high pressure rotation speed; $\tilde{y}$ is the output value of the NARX model; $\bar{y}$ is the mean value of y.

The best NARX model named Best-name is required to meet two factors: (1) Both of the Degree of fitting of high pressure rotation speed n_L and the Degree of fitting of high pressure rotation speed n_H are more than 95%; (2) On the basis of meeting (1), the sum of the Degree of fitting of n_L and n_H is highest. With the value of n_a, n_b and n_k changed and the Degree of fitting of the determined NARX model calculated, the best-model can be obtained. Table 2 shows results of the best-model and three other NARX models.

Table 2. The degree of fitting for the best-model and three other NARX models.

MODEL	n_a	n_b	n_k	n_HFIT%	n_LFIT%
Best-model	[2 1; 2 1]	[1; 1]	[1; 1]	96.43	96.95
M1	[3 1; 1 1]	[2; 3]	[2; 3]	80.83	89.49
M2	[4 2; 5 4]	[5; 4]	[2; 4]	59.52	68.50
M3	[2 2; 2 2]	[2; 4]	[3; 4]	36.46	46.03

Figure 3 reveals variations of the rotation speed for the NARX model and component level model when the input data are same. NARX model output curves are consistent with component level model output curves, and the degree of fitting for n_L and n_H are 96.43% and 96.95%. The results show that the NARX model can be successfully applied to modeling of the turbofan engine component level model, and it meets the requirements for further research.

For further comparison the variations, the error Δn is defined:

$$\Delta n = \sqrt{\left(\frac{n_L - n_L'}{n_L}\right)^2 + \left(\frac{n_H - n_H'}{n_H}\right)^2} \quad (6)$$

Where n and n' are the output value of component level model and the NARX model at the same time, respectively. And figure 4 shows the errors tendency. And the errors of the Best-model, the mean values of Δn are 0.008, are smallest and being within the acceptable range.

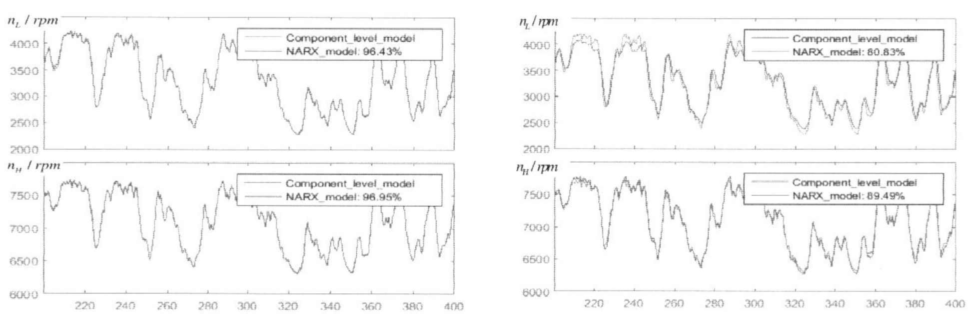

(a) Best-model (b) M2

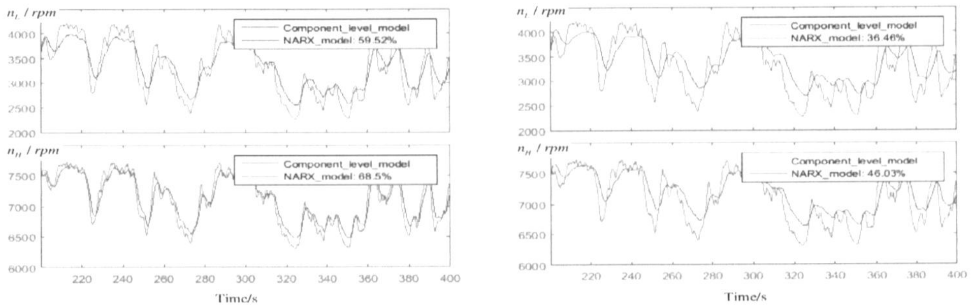

(c) M3 (d) M4

Figure 3. Comparison between component level model and the NARX model.

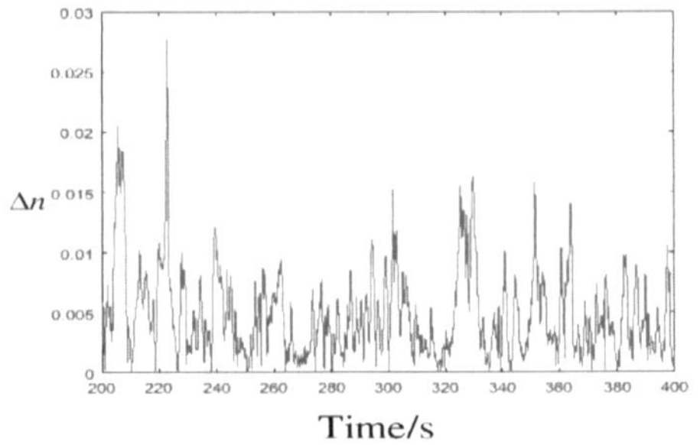

(a) Best-model

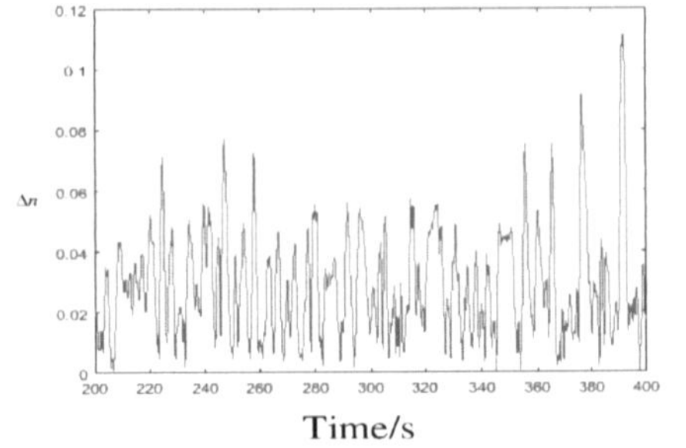

(b) M2

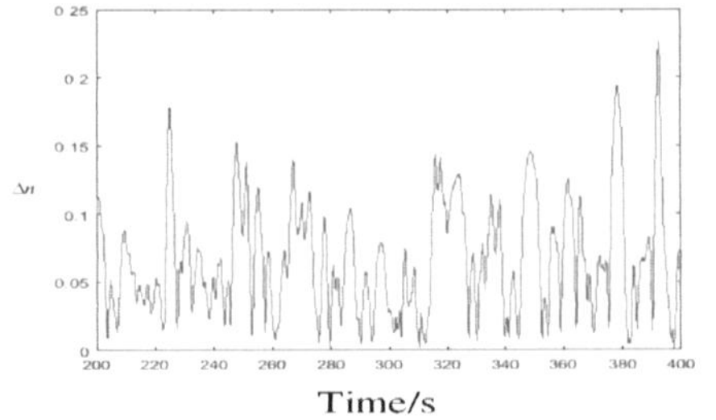

(c) M3

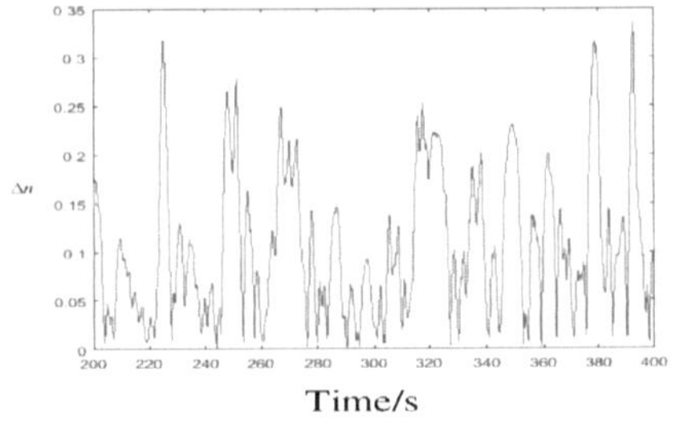

(d) M4

Figure 4. The errors of the output data between component level and the NARX model.

5. Conclusion

The NARX model structure and its identification process are introduced in the article. the nonlinear modeling scheme based on NARX is proposed and several models with different parameters are built according to the former data sets. Finally, the simulations have been taken to verify the precise and dynamic performance the models, the results show that the NARX model can well reflect the dynamics characteristic of the turbofan engine with high accuracy.

6. References

[1] Zheng Wei. Identification Modeling Method and Application of Photovoltaic Grid-Connected Inverter[D]. Chongqing: Chongqing University, 2014.

[2] Niu Peifeng, Zhang Xiulin, Peng Ce, et al. Process Control System[M]. Bei Jing: Publishing House of Electronics Industry, 2011.

[3] Ting L I, Xia L, Zhang X H, et al. CPS robust control strategy research based on NARX neural network pre-sentient algorithms and fuzzy logic controller[J]. Dianli Xitong Baohu Yu

Kongzhi/power System Protection & Control, 2012, 40(14).
[4] Leontaritis I J, Billings S A. Input-output parametric models for non-linear systems Part I: deterministic non-linear systems[J]. International Journal of Control, 1985, 41(41):303-328.
[5] Arain M A, Ayala H V H, Ansari M A. Nonlinear System Identification Using Neural Network[J]. Emerging Trends & Applications in Information Communication Technologies, 2012, 51(6):1191-1214.
[6] Basso M, Giarre L, Groppi S, et al. NARX models of an industrial power plant gas turbine[J]. IEEE Transactions on Control Systems Technology, 2005, 13(4):599-604.
[7] Asgari H, Chen X Q, Morini M, et al. NARX models for simulation of the start-up operation of a single-shaft gas turbine[J]. Applied Thermal Engineering, 2015, 93:368-376.
[8] Chapman J W, May R D, Lavelle T M, et al. Toolbox for the Modeling and Analysis of Thermodynamic Systems (T-MATS) User's Guide[J]. 2014.
[9] Selami Beyhan, Musa Alci. Fuzzy functions based ARX model and new fuzzy basis function models for nonlinear system identification[J]. Applied Soft Computing, 2010, 10(2):439-444.
[10] Billings S A, Leontaritis I J, Billings S A, et al. Parameter estimation tech-niques for nonlinear systems[J]. 1982, 18(2):299 - 305.
[11] Radmaneshfar E, Karrari M. A New Method for Structure Detection of Nonlinear ARX model: ANOVA_BSD[J]. World Congress on Engineering, 2007, 2165(1).
[12] Liu Xia. The Modeling and Application of the Wavelet Neural Network on Complicated Nonlinear System[D]. Daqing Petroleum Institute, 2005.
[13] Zheng L, Liu S, Xie X. Modified Wavelet Neural Network Based Nonlinear Model Identification for Photovoltaic Generation System[J]. Power System Technology, 2011, 35(10):159-164.

Acknowledgements
This work is supported by the Natural Science Foundation of Jiangsu Province (No. BK20140820), the Fundamental Research Funds for the Central Universities (No. NJ20160037), and Natural Science Foundation of China (51406083).

Experimental Investigation on Effect of Fin Height on Microscale Heat Transfer and Fluid Flow for Macro Scale Industrial Applications

K X Cheng[1], A L Goh[1], M Hadi[1], K T Ooi[1]

[1] School of Mechanical and Aerospace Engineering, Nanyang Technological University, 50 Nanyang Avenue, Singapore 639798.

Abstract. Microchannel for macro geometry application is gaining popularity particularly in aerospace, biomedical and photovoltaic. A novel method of employing microchannel in macro geometry at lower cost using conventional machining methods has been developed. A solid cylinder on outer diameter 19.4 mm is placed concentrically into a copper pipe of inner diameter 20 mm, forming an annular microchannel with 300 μm gap. This study takes a step further by introducing surface profile of different heights on the surface of solid cylinder and investigating the effect on two main design objectives- increasing heat removal capability at same pumping power and reducing pumping power for the same heat removal duty. Four surface profiles -parallel fins as well as fins with height of 0.1, 0.2 and 0.3 mm, were investigated experimentally at constant heat flux at Reynolds number from 690 to 4600. The amount of fluid in the microchannel, channel length of 30 mm, bifurcating angle of 75 degrees and mean hydraulic diameter of 600 μm are kept as constant parameters. A plain insert is used as benchmark for comparison of enhancement. In this study, insert with fins of 0.3 mm attains the highest enhancement of 43 percent increment in heat transfer as compared to plain insert using the same pumping power. While keeping the heat removal duty constant, the same insert is able to perform the duty using less than 50 percent the pumping power required by the plain insert at low Reynolds numbers.

1. Introduction

Single-phase heat transfer and fluid flow in microscale channels have been an active research regime since 1981. Tuckerman and Pease [1] first implemented microscale heat sink for silicon integrated circuit. Their work has then spurred numerous works on microchannel which can be categorised into design and implementation, fundamental understanding and classification and practical implementation with enhancement for the next few decades, as reported by Kandlikar [2]. Goodling [3], Sobhan and Garimella [4], Palm [5] and Mohammed [6] have done excellent reviews on this topic.

Kandlikar [7] summarised the works up to 2012 and highlighted the research needs for the implementation of microchannel in heat exchangers. Kong and Ooi [8] made a paradigm shift in the implementation of microscale heat transfer in macro geometry in 2013. An annular channel with a microscale gap of 300 μm was created by inserting a solid cylinder of mean diameter 19.4 mm concentrically into another hollow cylinder of inner diameter 20 mm. The research demonstrated the feasibility of the implementation of microscale heat transfer in macro scale system at a relatively low cost as the system was manufactured through conventional machining processes, instead of the complex and costly microfabrication techniques.

Goh and Ooi [9] further improved the thermal performance of the system by implementing surface profile on the inner solid cylinder. The nature-inspired inverted fish scale profile enhanced the overall thermal performance, citing re-initialisation of boundary layers, flow recirculation and higher turbulence intensity as the possible reasons. Nonetheless, less attention was paid on the fluid flow in order to attempt the high pressure drop woes.

Other studies looked into reducing pressure drop in conventional microchannels. Lorenzini and Kandlikar [10] studied numerically the effect of introducing fins of variable density along the length of the microchannel. The work draw attention to pressure drop issue on top of thermal performance of the channel by considering the coefficient of performance. Wang et al. [11] examined tree and leaf-like network in terms of thermal and hydrodynamic performance. Their work highlighted the role of bifurcation in recovering the local pressure drop.

Hence this project takes previous work done by Goh and Ooi [9] to a step further by implementing fins on the surface of the inner cylinder to reduce the overall pressure drop. This study hypothesises that flow bifurcation aids in recovering local pressure while the fins keep the flow within the developing region along the length of the microchannel. This project experimentally investigates three different fin heights and the effect on the thermal and hydrodynamic performance, using a plain profile as the reference. The heat transfer area, total amount of heat supply, average channel gap size, channel length and degree of bifurcation are kept constant in this study.

2. Experimental setup and procedures

2.1. Experimental setup

Figure 1 shows the experimental setup used in this study. The chiller, bypass loop, filter, test loop, measuring devices, test module and data acquisition system are the key components of the setup.

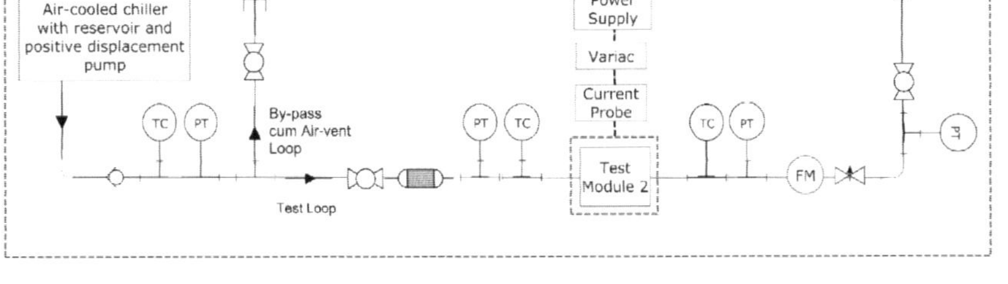

Figure 1: Experimental test bed

The working fluid is water and it is stored in the reservoir. It is recirculated and cooled to the desired inlet temperature and pumped into the test loop by a positive displacement pump integrated within the chiller. The volumetric flow of the water is controlled by a bypass loop and a needle valve located downstream of the test module. The air-vent loop serves to remove air bubbles trapped in the test loop.

The water is first channeled through a 40-micron filter to remove solid particles before entering the test loop. The flow conditions including the fluid temperature and pressure are monitored by a T-type thermocouples and pressure transducers respectively, along the test loop. The heat generated by the test module is removed by water which will be channeled back to the reservoir thereafter for recirculation.

2.2. Experimental test module

Figure 2 and **Figure 3** illustrate sectional view and side view of the test module respectively. An insert of a nominal diameter of 19.4 mm is placed concentrically into the copper block with an inner diameter of 20 mm, forming an annular channel of 0.6 mm width. Essentially, a 300 μm microchannel is formed within the test module as highlighted in **Figure 3**.

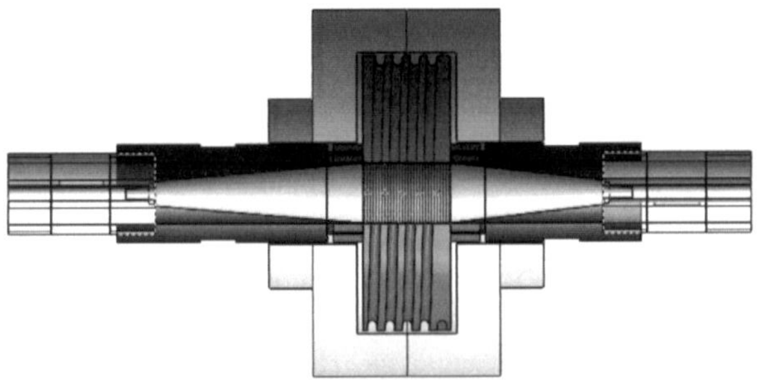

Figure 2: Sectional view of the test module

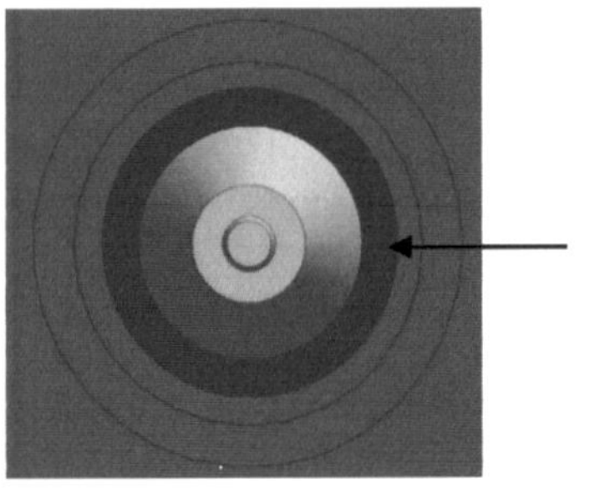

Figure 3: Side view of test module

2.3. Insert Design

A total of five inserts are investigated experimentally in this study, namely plain, parallel-fin, fin_1, fin_2 and fin_3 inserts. **Figure 4**, **Figure 5** and **Figure 6** illustrate plain, parallel-fin and fin_1 insert respectively. Fin_1, fin_2 and fin_3 have 75 degree bifurcating fins on the surface of the insert with three different heights of 0.1 mm, 0.2 mm and 0.3 mm. The height of the fin is depicted in **Figure 7**. The parallel-fin insert has a fin height of 0.2 mm. All the inserts have a nominal diameter of 19.4 mm in the heating region to ensure constant amount of fluid in the microchannel.

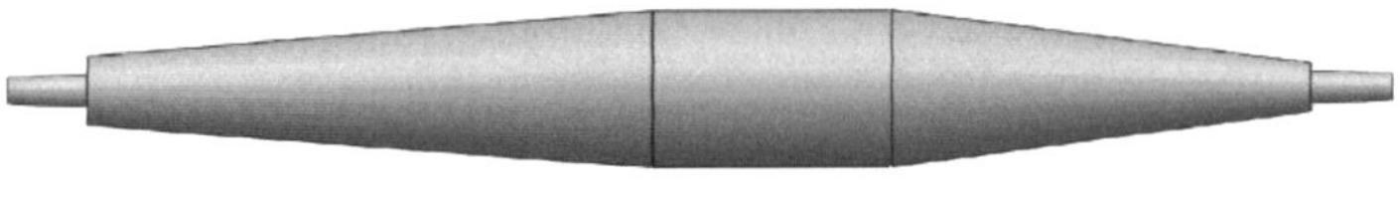

Figure 4: Plain insert

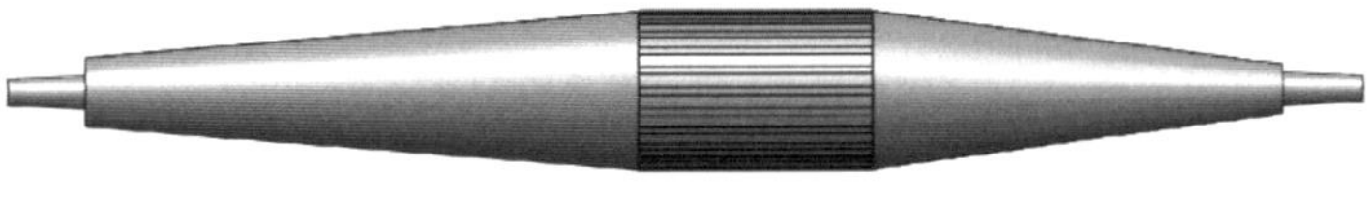

Figure 5: Parallel-fin insert

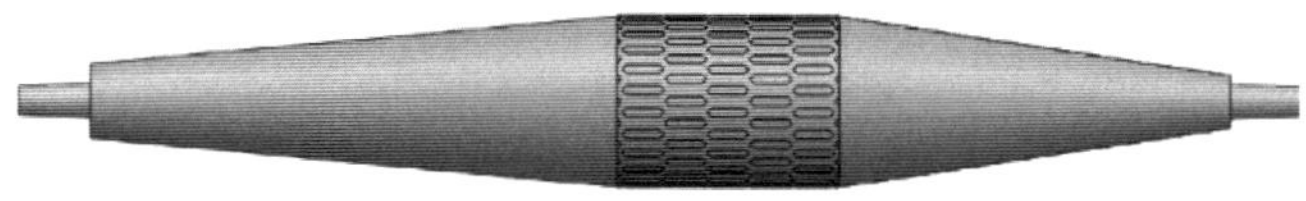

Figure 6: cfin_1 insert

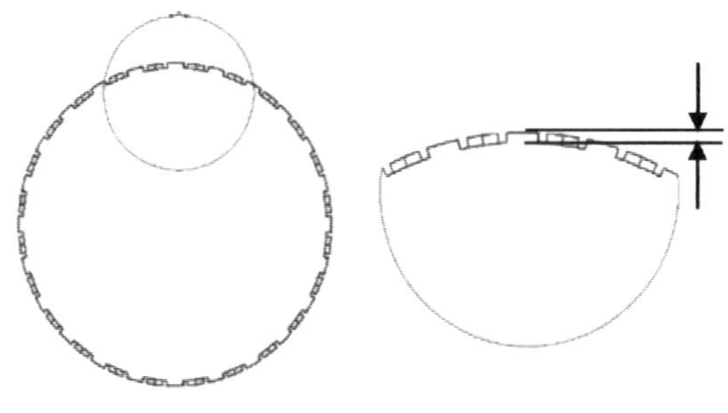

(a) (b)

Figure 7: (a) Side view of the portion of the insert within the heating region; (b) magnified view of the portion highlighted by circle

2.4. Measuring devices

J- and T-type thermocouples are used to measure water temperature in the experimental setup. A total of five T-type thermocouples are used. Two of them are installed before and after the test module while the other three are installed along the flow loop for monitoring purposes. A total of twelve J-type thermocouples are incorporated into the test module. Five of these thermocouples are installed on each side of the test module, 2.5 mm radially away from the wall of the microchannel to acquire readings for the computation of wall temperature. The other two are placed 5 mm from the outer surface of the test module, sending signal to data acquisition system and circuit breaker for safety reasons.

WIKA model A-10 pressure transducers are used to monitor the pressure in the system, giving current signals of 4 to 20 mA which correspond to pressure readings of 0 to 6 barg. Two of the pressure transducers are installed before and after the test module while the other three are positioned along the flow loop for monitoring purposes. The flow rate is measured by a Coriolis mass flow meter.

2.5. Data reduction

The data acquired during the 10-minute steady state condition is processed to obtain thermal and hydrodynamic parameters.

The heat transfer coefficient, h is obtained from Newton's law of cooling as shown in Equation (1).

$$h = \frac{\dot{Q}}{A\left(T_{w,m} - T_{f,m}\right)} \quad (1)$$

The rate of heat transferred to the fluid, $\dot{Q}$ is deduced from the inlet and outlet fluid temperatures as shown in Equation (2). The mass flow rate, $\dot{m}$ is the product of temperature-dependent density and volumetric flow rate. All the fluid properties are obtained from National Institute of Standards and Technology (NIST) Reference Fluid and Thermodynamic and Transport Properties Database (REFPROP).

$$\dot{Q} = \dot{m}c_{p,m}\left(T_{out} - T_{in}\right) \quad (2)$$

The heat transfer area is kept constant at 1885 mm^2. The fluid temperature, is computed by taking the average of the inlet and outlet fluid temperature, as shown in Equation (3).

$$T_{f,m} = \frac{T_{in} + T_{out}}{2} \tag{3}$$

The wall temperature, on the other hand, is deduced from J-type thermocouples. The thermal resistance of the copper block in the radial direction is taken into account to get the wall temperature as depicted in Equation (4).

$$T_{w,i} = T_{r,i} - q\frac{\ln(d_r / d_w)}{2\pi L k_c} \tag{4}$$

The wall temperature is therefore assumed to be the average of the ten points along the wall of the copper block, according to Equation (5).

$$T_{w,m} = \left(\sum_{i=1}^{10} T_{w,i}\right) / 10 \tag{5}$$

Equation (6) and (7) are used to compute Nusselt number and Reynolds number respectively.

$$Nu = \frac{hD_h}{k_f} \tag{6}$$

$$\mathrm{Re} = \frac{\dot{m}D_h}{A_c \mu} \tag{7}$$

The pressure drop is deduced from the pressure difference between the inlet and outlet of the test module as shown in Equation (8).

$$\Delta P = P_{in} - P_{out} \tag{8}$$

Performance evaluation criteria for two main objectives of the design: increased heat duty for equal pumping power and reduced pumping power for constant heat duty, were introduced by Webb [12-14]. The subscript s stands for smooth surface or plain insert for our case. Equation (9) is the objective function for increased heat duty at the same pumping power while Equation (10) evaluates the performance aimed at reducing pumping power for the same heat duty.

$$\frac{UA}{U_s A_s} = \frac{Nu / Nu_s}{(\Delta P / \Delta P_s)^{1/3}} \tag{9}$$

$$\frac{P}{P_s} = \frac{\Delta P / \Delta P_s}{(Nu / Nu_s)^3} \tag{10}$$

2.6. Experimental operating conditions

The study was conducted for Reynolds number from 690 to 4600 with a constant heat supply of 500 W. The heat supply area was kept constant at 18.85 cm^2.

3. Results and Discussion

Figure 8 shows the convective heat transfer coefficient for all the inserts. The highest heat transfer coefficient obtained for the study is 45.0 $kW/m^2 \cdot K$ using fin_3 insert at a Reynolds number of 4563. This is equivalent to a Nusselt number of 42.46.

It is obvious from **Figure 8** that all the inserts with fins achieve higher heat transfer coefficients as compared to plain insert. Fin_3 yields the highest heat transfer coefficients at all Reynolds numbers, This is probably due to the lowest minimum gap size between the fin and the wall despite the constant nominal diameter. Nonetheless, fin_3 results in higher pressure drop at all flow rates as depicted in **Figure 9**. Although the pressure drop is relatively high at higher Reynolds numbers especially for fin_3 insert, the pumping requirement can be easily met by a commercially readily available pump.

IOP Conf. Series: Materials Science and Engineering **187** (2017) 012003 doi:10.1088/1757-899X/187/1/012003

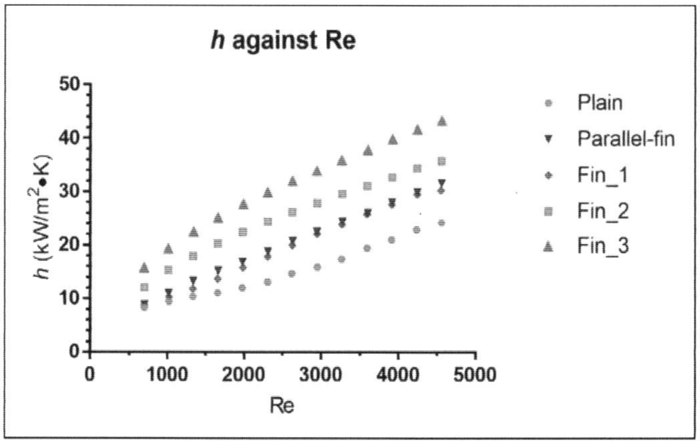

Figure 8: Heat transfer coefficient against Reynolds number

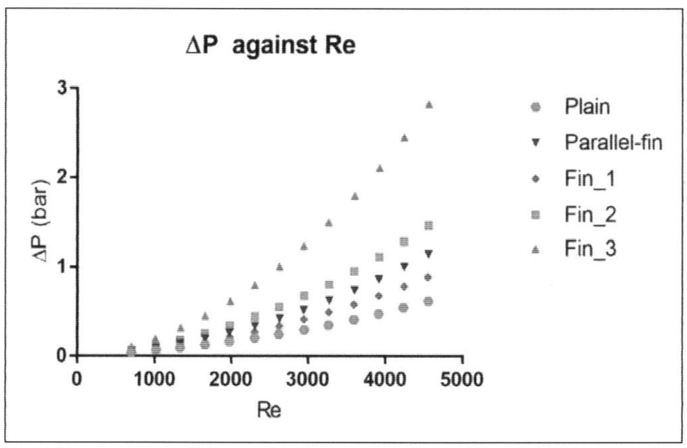

Figure 9: Pressure drop against Reynolds number

Figure 10 compares the performance of all the inserts against the plain insert, aiming to incearse the heat removal capability at the same pumping power. The highest index achieved is 1.48, and it is by fin_3 insert at Reynolds number from 1660 to 2000. On the other hand, fin_2 depicts similar performance index particularly at high Reynolds number. At low Reynolds number, fin_3 slightly out performs fin_2 insert. This is likely due to smaller minimum gaps formed by fin_3 might have initiated higher turbulence as compared to those of fin_2. Notably, the transition Reynolds number is around 2000, which is close enough to the critical Reynolds number for conventionally sized channels of 2300. Interestingly, both parallel-fin and fin_1 inserts have similar performance index despite having differing surface profile. The re-initilisation of boundary layers along the length of the channel by the discontinuous surface profile on fin_1 insert might have compensated for the larger minimum gaps as compared to those of the parallel-fin insert which has a higher fin height at 0.2 mm. It is obvious that- channels with a higher fin height show higher heat removal capability especially at low Reynolds numbers.

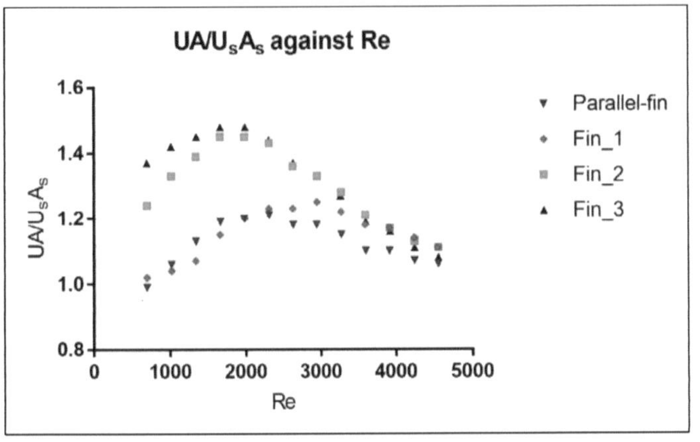

Figure 10: UA/U_sA_s against Reynolds number

Figure 11 shows the plot of performance index which aims at reducing pumping power but keeping heat removal capability constant. Although all the inserts have similar low pressure drop at low Reynolds as shown in **Figure 9**, this phenomenon is accompanied by low heat removal ability for parallel-fin and fin_1 inserts. Therefore, both inserts have a relatively high performance index. The fin_2 and fin_3 account for the slightly higher pressure drop at low Reynolds number by much higher thermal performance, requiring less than 50 percent of the pumping power as compared to the plain insert to achieve the same heat removal duty. Similar to the performance index for heat removal capability, all the inserts have close index for pumping power requirement at high Reynolds number. However, it is obvious that all the inserts have an index less than 1, implying that the surface profiles actually lower the pumping power for the same heat removal capability. The results show that both performance indices for heat duty and pumping power are the most ideal near to Reynolds number of 2000.

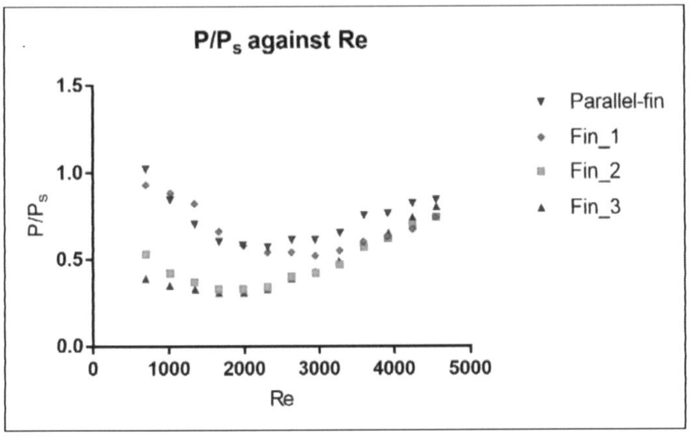

Figure 11:P/P_s against Reynolds number

4. Conclusion

a. This study demonstrates the feasibility of enhancing the thermal and hydrodynamic performance of microchannel in macro geometry using surface profiles. It shows that bifurcating fins perform better than parallel fins both in heat transfer capability and pressure drop.

b. For the same pumping power, a higher fin enhances heat transfer more than the shorter fin, particularly at low Reynolds numbers. The highest enhancement of 43 % in heat transfer as compared to plain insert is achieved by fin_3 at a Reynolds number close to 2000.
c. Considering the same amount of heat being removed, all the surface profiles require less pumping power as compared to plain insert. Both fin_2 and fin_3 need less than half the pumping power required by plain insert to remove the same amount of heat at low Reynolds numbers.

The findings from this study can be used a reference for further studies on the microchannel enhancement in macro geometries as fin height plays significant role both in terms of heat transfer and pressure drop issue.

5. References

[1] Tuckerman, D.B. and R. Pease, *High-performance heat sinking for VLSI.* IEEE Electron device letters, 1981. **2**(5): p. 126-129.
[2] Kandlikar, S.G., History, advances, and challenges in liquid flow and flow boiling heat transfer in microchannels: a critical review. Journal of Heat Transfer, 2012. **134**(3): p. 034001.
[3] Goodling, J.S. Microchannel heat exchangers: A review. in SPIE's 1993 International Symposium on Optics, Imaging, and Instrumentation. 1993. International Society for Optics and Photonics.
[4] Sobhan, C.B. and S.V. Garimella, *A comparative analysis of studies on heat transfer and fluid flow in microchannels.* Microscale Thermophysical Engineering, 2001. **5**(4): p. 293-311.
[5] Palm, B., *Heat transfer in microchannels.* Microscale Thermophysical Engineering, 2001. **5**(3): p. 155-175.
[6] Mohammed, H., et al., Heat transfer and fluid flow characteristics in microchannels heat exchanger using nanofluids: a review. Renewable and Sustainable Energy Reviews, 2011. **15**(3): p. 1502-1512.
[7] Kandlikar, S.G., et al., *Heat transfer in microchannels—2012 status and research needs.* Journal of Heat Transfer, 2013. **135**(9): p. 091001.
[8] Kong, K.S. and K.T. Ooi, A numerical and experimental investigation on microscale heat transfer effect in the combined entry region in macro geometries. International Journal of Thermal Sciences, 2013. **68**: p. 8-19.
[9] Goh, A.L. and K.T. Ooi, *Nature-inspired Inverted Fish Scale microscale passages for enhanced heat transfer.* International Journal of Thermal Sciences, 2016. **106**: p. 18-31.
[10] Lorenzini-Gutierrez, D. and S.G. Kandlikar, Variable Fin Density Flow Channels for Effective Cooling and Mitigation of Temperature Nonuniformity in Three-Dimensional Integrated Circuits. Journal of Electronic Packaging, 2014. **136**(2): p. 021007-021007.
[11] Wang, X.-Q., et al., *Flow and thermal characteristics of offset branching network.* International Journal of Thermal Sciences, 2010. **49**(2): p. 272-280.
[12] Webb, R., Performance evaluation criteria for use of enhanced heat transfer surfaces in heat exchanger design. International Journal of Heat and Mass Transfer, 1981. **24**(4): p. 715-726.
[13] Webb, R. and M. Scott, A parametric analysis of the performance of internally finned tubes for heat exchanger application. Journal of Heat Transfer, 1980. **102**(1): p. 38-43.
[14] Gee, D.L. and R. Webb, *Forced convection heat transfer in helically rib-roughened tubes.* International Journal of Heat and Mass Transfer, 1980. **23**(8): p. 1127-1136.

IOP Conf. Series: Materials Science and Engineering **187** (2017) 012004 doi:10.1088/1757-899X/187/1/012004

Global Sliding Mode Control for the Bank-to-Turn of Hypersonic Glide Vehicle

J Zhang[1,*], Y F Yu[1], P P Yan[1], Y H Fan[1] and X W Guo[2]

[1]School of Astronautics, Northwestern Polytechnical University, Xi'an 710072, China
[2]Xi'an Aerospace Propulsion Institute, Xi'an 710199, China

[*]ypp328@163.com

Abstract. The technology of Bank-to-Turn has been recognized as an attractive direction due to their significance for the control of hypersonic glide vehicle. Strong coupling existing among pitch, yaw and roll channel was a great challenge for banking to turn, and thus a novel global sliding mode controller was designed for hypersonic glider in this paper. Considering the coupling among channels as interference, we can use invariance principle of sliding mode motion to realize the decoupling control. The global sliding mode control system could eliminate the stage of reaching, which can lead to the realization of whole systematic process decoupling control. When the global sliding mode factor was designed, a minimum norm pole assignment method of the sliding mode matrix was introduced to improve the robustness of the system. The method of continuity of symbolic function was adopted to overcome the chatter, which furtherly modify the transient performance of the system. The simulation results show that this method has good performance of three channel decoupling control and guidance command tracking. And it can meet the requirements of the dynamic performance of the system.

1. Introduction

Hypersonic vehicle has received more and more attentions of military powers because of its high speed, short reaction time, big combat radius, and good for hiding and strong penetration ability [1-6]. At present, the hypersonic vehicle has been a hot topic in the aerospace area [7, 8]. Especially the hypersonic boost-glide vehicle, whose technical bottleneck is easy to break, has become the first choice for the development of hypersonic weapon in military powers [9, [10].

Plane-symmetry configuration is always used in the hypersonic glide vehicle, which determines the application of the Bank-to-Turn control in the lateral maneuvers [11]. A typical flight trajectory of hypersonic glide vehicle is shown in Figure 1.However, when banking to turn, the aircraft has a bigger roll rate in the roll channel. And obvious kinematic and aerodynamic coupling [12] among pitch, yaw and roll channel are also existed. The design of the control system also will be more complex. In addition, due to the characteristics of hypersonic vehicle, coupling and model uncertainty problems are more prominent.

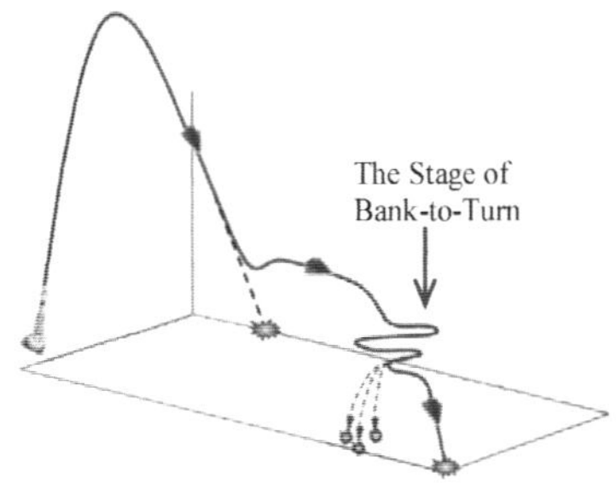

Figure 1 flight trajectory of hypersonic glide vehicle

Considering coupling and model uncertainty problems the hypersonic glide vehicle has when banking to turn, the domestic and foreign scholars put forward the trajectory linearization method, dynamic inverse method and adaptive control method etc as solutions[13]. For example, in Ref [13], according to the singular perturbation theory, a attitude control system of the aircraft is designed by using the trajectory linearization method. The designed controller has good control performance, but with the increase of altitude, the flight control ability has declined. Ref [15] presents a design of a nonlinear adaptive dynamic inversion controller for a Generic Hypersonic Vehicle. In Ref [16], a decoupling controller based on nonlinear dynamic inversion and full state feedback is designed, but the design method of dynamic inverse control law requires accurate mathematical model, and the robustness of the system is poor. On the basis of multi input and multi output feature model, the robustness of the nonlinear strong coupling uncertainty system is verified by the adaptive control method in Ref [19], but in the present study, the measurement error of the attitude angle is not considered as well as the delay and error of the actuator, the research is limited to the feasibility of this method. Ref [20] introduces an adaptive control design which aims to improve tracking and stability of the generic hypersonic vehicle. In Ref [21], a robust controller was designed for the longitudinal dynamics of a hypersonic aircraft. Ref [23] and [24] show that sliding mode control is an effective robustness control method for strong coupling and model uncertainty problems. A sliding mode variable structure control law is designed for hypersonic glide vehicle in reference [25], but only when in the sliding mode motion, the system will be in the state of decoupling and resisting parameter distortion.

In order to overcome the shortcomings of the general sliding mode control design, this paper proposes a method of global sliding mode control to design a high speed glide vehicle with a bank-to-turn controller. By introducing a global sliding mode factor in the sliding hyperplane, the designed control law is allowed to make the system remain on the switching hyperplane from the beginning, which can achieve the whole process decoupling control of the system. In the design of the whole sliding mode factor, a minimum norm pole assignment method of the sliding mode matrix is introduced. In order to overcome the chattering of the system, the method of continuity symbolic function is adopted, which improves the transient performance of the system. Finally, the effectiveness of the proposed method is verified by the simulation of a hypersonic glide vehicle.

2. Mathematical model of hypersonic glide vehicle with sliding mode control

The pitch, yaw and roll channel equation with BTT control can be expressed as[26]:

$$\begin{cases} \dot{\alpha} = \omega_z - a_{34}\alpha - a_{35}\delta_z \\ \dot{\beta} = \omega_y + \alpha_0\omega_x - b_{34}\beta - b_{37}\delta_y \\ \dot{\omega}_x = -b_{14}\beta - b_{11}\omega_x - b_{12}\omega_y - b_{18}\delta_x - b_{17}\delta_y \\ \dot{\omega}_y = -b_{24}\beta + b_{21}\omega_x - b_{22}\omega_y - b_{23}\delta_x - b_{27}\delta_y \\ \dot{\omega}_z = -a_{24}\alpha - a_{22}\omega_z - a_{25}\delta_z \end{cases} \tag{1}$$

Where, α,β are attack and sideslip angle of the system, $\omega_x,\omega_y,\omega_z$ are roll, yaw and pitch rate, $\delta_x,\delta_y,\delta_z$ are roll, yaw and pitch rudder angle, α_0 is constant balance angle of attack. $\alpha_0\omega_x$ is the kinematic coupling term and $-b_{14}\beta$ 、 $b_{21}\omega_x - b_{23}\delta_x$ 、 $-b_{12}\omega_y - b_{17}\delta_x$ are aerodynamic coupling terms. The coefficients of these coupling terms are comparatively large and can not be ignored easily. The calculation formulas of the dynamic coefficient of the formula (1) are as followed:

$$a_{22} = -m_z^{\bar{\omega}_z} qSL^2/(J_z V) , \quad a_{24} = -m_z^{\alpha} qSL/J_z$$
$$a_{25} = -m_z^{\delta_z} qSL/J_z , \quad a_{34} = -C_y^{\alpha} qS/(mV)$$
$$a_{35} = -C_y^{\delta_z} qS/(mV) , \quad b_{11} = -m_x^{\bar{\omega}_x} qSL^2/(J_x V)$$
$$b_{18} = -m_x^{\delta_x} qSL/J_x , \quad b_{12} = -m_x^{\bar{\omega}_y} qSL^2/(J_x V)$$
$$b_{14} = -m_x^{\beta} qSL/J_x , \quad b_{17} = -m_x^{\delta_y} qSL/J_x$$
$$b_{22} = -m_y^{\bar{\omega}_y} qSL^2/(J_y V) , \quad b_{24} = -m_y^{\beta} qSL/J_y$$
$$b_{27} = -m_y^{\delta_y} qSL/J_y , \quad b_{34} = -C_Z^{\beta} qS/(mV)$$
$$b_{37} = -C_Z^{\delta_y} qS/(mV), \quad b_{21} = m_y^{\bar{\omega}_x} qSL^2/(J_y V)$$
$$b_{23} = -m_y^{\delta_x} qSL/J_y$$

Where m is the mass of the vehicle, J_x,J_y,J_z are the moment of inertia, q is the dynamic pressure, V is the speed, S is the reference area, and L is the reference length.

Write the roll angle of the vehicle as γ, then

$$\dot{\gamma} = \omega_x \tag{2}$$

Let's select state variable as $Z = [\alpha,\beta,\gamma,\omega_x,\omega_y,\omega_z]'$, considering the formula (1) and (2), the mathematical model of the state equation of a hypersonic glide vehicle can be derived as follows:

$$\dot{Z} = A_z \cdot Z + B_z \cdot U \tag{3}$$

Where

$$A_z = \begin{bmatrix} -a_{34} & 0 & 0 & 0 & 0 & 1 \\ 0 & -b_{34} & 0 & \alpha_0 & 1 & 0 \\ 0 & 0 & 0 & 1 & 0 & 0 \\ 0 & -b_{14} & 0 & -b_{11} & -b_{12} & 0 \\ 0 & -b_{24} & 0 & b_{21} & -b_{22} & 0 \\ -a_{24} & 0 & 0 & 0 & 0 & -a_{22} \end{bmatrix}$$

$$B_z = \begin{bmatrix} 0 & 0 & -a_{35} \\ 0 & -b_{37} & 0 \\ 0 & 0 & 0 \\ -b_{18} & -b_{17} & 0 \\ -b_{23} & -b_{27} & 0 \\ 0 & 0 & -a_{25} \end{bmatrix}, \quad U = \begin{bmatrix} \delta_x \\ \delta_y \\ \delta_z \end{bmatrix}$$

When designing the control system of a hypersonic glide vehicle, we hope the roll angle γ of the roll channel can quickly track the roll angle command signal and the pitch channel can also track the overload signal n_{yc} fast, in the meantime keep the sideslip angle as 0. Because on every feature point of the flight process, the overload command signal n_{yc} and the angle of attack command signal α_c have the one-to-one corresponding relation, the purpose of the controller design can also be viewed as the angle of attack α tracking angle of attack command signal α_c, the roll angle tracking roll angle command signal γ_c, and at the same time keeping the sideslip angle command signal as 0.

The global sliding mode control is to adjust the state of the controlled system from nonzero to zero, which achieves the function of the state regulator. Therefore, in order to transform the regulator problem into the tracking problem, the error model is introduced.

$$e_\alpha = \alpha - \alpha_c,\ e_\beta = \beta - \beta_c,\ e_\gamma = \gamma - \gamma_c \tag{4}$$

Then

$$\dot{e}_\alpha = \dot{\alpha} - \dot{\alpha}_c,\ \dot{e}_\beta = \dot{\beta} - \dot{\beta}_c,\ \dot{e}_\gamma = \dot{\gamma} - \dot{\gamma}_c \tag{5}$$

Note state variable $Z_e = [e_\alpha, e_\beta, e_\gamma, \omega_x, \omega_y, \omega_z]^{\mathrm{T}}$. Considering the formula (3) and (5), we can obtain:

$$\dot{Z}_e = A_z \cdot Z_e + B_z \cdot U + D_z \tag{6}$$

Where

$$D_z = [-a_{34} - \dot{\alpha}_c, -b_{34} - \dot{\beta}_c, -\dot{\gamma}_c, -b_{14}\beta_c, -b_{24}\beta_c, -a_{24}\alpha_c]^{\mathrm{T}}$$

The formula (6) represents the error model of the controlled object, which keep the angle of attack, sideslip angle and roll angle in expectations, that is, control the error to 0. Further, in order to facilitate the design of variable structure controller, the error model of the formula (6) is needed to be transformed into a simple standard type of variable structural control by linear transformation.

$$X = P^{-1} \cdot Z_e \tag{7}$$

Then

$$\dot{X} = P^{-1}A_e P \cdot X + P^{-1}B_e \cdot U + P^{-1}D_e \tag{8}$$

Note $P^{-1}A_e P = A$, $P^{-1}B_e = B$, $P^{-1}D_e = D$, then formula (8) can be written as followed:

$$\dot{X} = A \cdot X + B \cdot U + D \tag{9}$$

Where $B = [0 \quad B_2]^{\mathrm{T}}$, U is a $m \times 1$ matrix and B_2 is a $m \times m$ matrix. The equation can be written in the following form:

$$\begin{bmatrix} \dot{X}_1 \\ \dot{X}_2 \end{bmatrix} = \begin{bmatrix} A_{11} & A_{12} \\ A_{21} & A_{22} \end{bmatrix} \cdot \begin{bmatrix} X_1 \\ X_2 \end{bmatrix} + \begin{bmatrix} 0 \\ B_2 \end{bmatrix} \cdot U + D \tag{10}$$

The mathematical model of the hypersonic gliding vehicle with sliding mode variable structure control is obtained.

3. Design of global sliding mode controller for hypersonic glide vehicle

The formula (9) presents the state equation of the multivariable system without considering parameter perturbation. If the parameter perturbation is taken into account, the general multivariable uncertain system can be expressed as

$$\dot{X}(t) = (A + \Delta A) \cdot X(t) + (B + \Delta B) \cdot U(t) + D \cdot f(t) \tag{11}$$

Where, $X \in \mathrm{R}^n$ is the state variable, $U \in \mathrm{R}^m$ is the control vector and $f(t) \in \mathrm{R}^l$ is the external disturbance; A , B and D have the corresponding dimension. ΔA and ΔB are the parameter perturbation matrix.

According to the system described, we chose the sliding hyper plane as:

$$S(X,t) = CX - CE(t)X(0) \tag{12}$$

Where,

$$E(t) = \begin{bmatrix} E_1(t) & 0 \\ 0 & E_2(t) \end{bmatrix}$$

$$E_1(t) = diag\left[\exp(-\beta_1 t) \quad \cdots \quad \exp(-\beta_{n-m} t)\right]$$

$$E_2(t) = diag\left[\exp(-\beta_{n-m+1} t) \quad \cdots \quad \exp(-\beta_n t)\right]$$

And $\mathrm{Re}(\beta_i) > 0, i = 1, \cdots n$ is the sliding mode parameter to be designed.

3.1. Design of modal parameter matrix C

Here we use pole assignment method to design matrix C .

The system that has become a simple standard type of variable structure control, can be written as follows:

$$\begin{cases} \dot{X}_1 = A_{11}X_1 + A_{12}X_2 + D_1 \\ \dot{X}_2 = A_{21}X_1 + A_{22}X_2 + B_2U + D_2 \\ S = C_1X_1 + C_2X_2 - C_1E_1(t)X_1(0) - C_2E_2(t)X_2(0) \end{cases} \tag{13}$$

Where $S = 0$, we can get:

$$X_2 = -KX_1 + KE_1(t)X_1(0) + E_2(t)X_2(0) \tag{14}$$

Where $K = C_2^{-1}C_1$. Substitute X_2 into the differential equation about X_1 in equation (13):

$$\dot{X}_1 = (A_{11} - A_{12}K)X_1 \tag{15}$$

Because (A, B) is completely controllable, (A_{11}, A_{12}) is also completely controllable. Choose the right K, to make sure that the characteristic root λ_i of $(A_{11} - A_{12}K)$ is the expected characteristic root, so that the system has good dynamic quality in the sliding phase.Then,

$$C = C_2[K \quad I_m] \tag{16}$$

Formula (16) is the sliding mode parameter matrix we need. Where C_2 is usually taken as a unit matrix.

Because it is a multiple input multiple output system, there are many solutions can configure the system to poles where required. In order to avoid the norm of matrix C is too large, resulting in transient control value of variable structure control become too large, referring to [27], a minimum norm pole assignment method for a sliding mode matrix is introduced, so that multiple solution problem in pole assignment can be transformed into a least squares problem:

$$\min_{\lambda(A_{11} - A_{12}K) = \lambda^*} (\max_i \sum_{j=1}^{n} K_{ij}) \tag{17}$$

3.2. Design of sliding mode motion parameters

The system is in a global sliding mode motion state, that is, the system always has $S = 0$, then

$$C_1(X_1 - E_1(t)X_1(0)) + C_2(X_2 - E_2(t)X_2(0)) = 0 \tag{18}$$

From formula (15), we can know that the sliding mode motion of the system is expected to be

$$X_1 = \tilde{E}_1(t)X_1(0)) \tag{19}$$

So choose $-\beta_i$ as the desired pole, that is

$$\beta_i = -\lambda_i(A_{11} - A_{12}K), \quad (i=1, \cdots, n-m) \tag{20}$$

Considering formula (18) and (19), the desired motion characteristics of the system state X_2 can be

$$X_2 = \tilde{E}_2(t)X_2(0)) \tag{21}$$

Similarly, we choose β_i and $i = (n-m+1, \cdots, n)$ according to the requirements of the dynamic performance of the system, so as to make X_2 achieve the desired motion characteristics.

3.3. Design of global sliding mode variable structure control law

The task of variable structure control law is to keep the state in the sliding mode, that is the designed $U(t)$ meet the accessibility condition.

$$S^T\dot{S} < 0 \tag{22}$$

The variable structure control law is constructed as follows:

$$U(t) = -g(t)(CB)^{-1}\operatorname{sgn}(S) \tag{23}$$

Where $g(t) > 0$ is the designed variable structure control coefficient. Substitute formula (23) into $S^T\dot{S}$:

$$\begin{aligned}S^T\dot{S} &= -g(t)S^T\operatorname{sgn}(S) - g(t)S^T C\Delta B(CB)^{-1}\operatorname{sgn}(S) \\ &+S^T[CAX + C\Delta AX + CDf + CHE(t)X_0] \\ &\le -g(t)\|S^T\|[1-\psi_b\|C\|\cdot\|(CB)^{-1}\|]+ \\ &\|S^T\|[(\|CA\|+\psi_a\|C\|)\|X\|+\|CD\|\psi_f \\ &+\beta_{\max}\exp(-\beta_{\min}t)\|CX(0)\|]\end{aligned} \tag{24}$$

Where $H = diag[\beta_1,\cdots,\beta_n]$, ψ_a, ψ_b, ψ_f are known positive constants, and meet $\|\Delta A\|\le\psi_a$, $\|\Delta B\|\le\psi_b$, $\|f\|\le\psi_f$. Simultaneous formula (22) and formula (24) we can get:

$$\begin{aligned}g(t) > [1-\psi_b\|C\|\cdot\|(CB)^{-1}\|]^{-1}\cdot[(\|CA\| \\ +\psi_a\|C\|)\|X\|+\|CD\|\psi_f \\ +\beta_{\max}\exp(-\beta_{\min}t)\|CX(0)\|]\end{aligned} \tag{25}$$

Formula (25) can be written as:

$$g(t) = (1-a_4)^{-1}\cdot[a_1\|X\|+a_2+a_3\exp(-\beta_{\min}t)]+\varepsilon \tag{26}$$

Where, ε is a small positive number, the remaining coefficients are:

$$\begin{aligned}a_1 &= \|CA\|+\psi_a\|C\| \\ a_2 &= \|CD\|\psi_f \\ a_3 &= \beta_{\max}\|CX(0)\| \\ a_4 &= \psi_b\|C\|\cdot\|(CB)^{-1}\|\end{aligned} \tag{27}$$

Substitute $g(t)$ from formula (26) into formula (23), so that the variable structure control law satisfying the condition of the sliding mode accessibility is obtained.

In addition, the flutter is a common phenomenon in sliding mode control system. In order to eliminate high frequency flutter, a simple and practical method is used, that is replace sgn(S) in formula (23) by M(s).

$$\begin{aligned}M(S) &= [m(s_1) \ \cdots \ m(s_m)]^T \\ m(s_i) &= \frac{s_i}{|s_i|+\sigma_i}, \quad i=1,\cdots,m\end{aligned} \tag{28}$$

Where, δ_i is a small positive constant.

4. Simulation results and analysis

Suppose a hypersonic glide vehicle is at 40 Km, Ma=6 and the cruise balance angle of attack α_0=5, the aircraft's kinetic coefficients in the cruise state are as follows:

$$\begin{aligned}&a_{22}=0.0614\ ,\quad a_{24}=-37.4079,\ a_{25}=36.5729 \\ &a_{34}=0.0648\ ,\quad a_{35}=0.0056\ ,\quad b_{11}=0.0528 \\ &b_{18}=100.237\ ,\quad b_{12}=0.0078\ ,\quad b_{14}=123.5126 \\ &b_{17}=27.8024\ ,\quad b_{22}=0.0118\ ,\quad b_{24}=7.3013 \\ &b_{27}=6.6308\ ,\quad b_{34}=0.0121\ ,\quad b_{37}=0.0011, \\ &b_{21}=0.0008\ ,\quad b_{23}=6.2823\end{aligned}$$

Substitute these kinetic coefficients into formula (3), the mathematical model of a hypersonic glide vehicle in the form of state equation is obtained.

Select a linear transformationa matrix P^{-1} as:

$$P^{-1}=\begin{bmatrix} 1 & 0.7049 & 0 & 0 & -0.0002 & -0.0002 \\ 1 & 0.7049 & 1 & 0 & -0.0002 & -0.0002 \\ 1 & 0 & 1 & 0 & 0 & -0.0002 \\ 0 & 0 & 0 & 1 & 0 & 0 \\ 0 & 0 & 0 & 0 & 1 & 0 \\ 0 & 0 & 0 & 0 & 0 & 1 \end{bmatrix}$$

Substitute P^{-1} into formula (8), so that the mathematical model of aircraft can be changed into a simple standard type of variable structure control.

Suppose the initial angle of attack α_0 of the vehicle is 5° , the sideslip angle has a initial deviation of 1° and the roll angle is 0. Give the angle of attack command α_c =10° , the sideslip angle command β_c =0 and roll angle command γ_c =45° .

A global sliding mode controller for hypersonic glide vehicle is designed in this paper. In order to ensure the control system step response rise time is less than 1 second, when using the pole assignment method, we can set the pole to {-5} and calculate the variable structure control law of the formula (23).

To validate the robustness of controller, the uncertainties of the model are considered as $|\Delta m| \le 0.03$, $|\Delta S| \le 0.03$, $|\Delta J_x| \le 0.05$, $|\Delta J_y| \le 0.05$, $|\Delta J_z| \le 0.10$, $|\Delta m_x| \le 0.10$, $|\Delta m_y| \le 0.10$, $|\Delta m_z| \le 0.10$, $|\Delta q| \le 0.05$ .The simulation results as shown in Figure 2 to Figure 7:

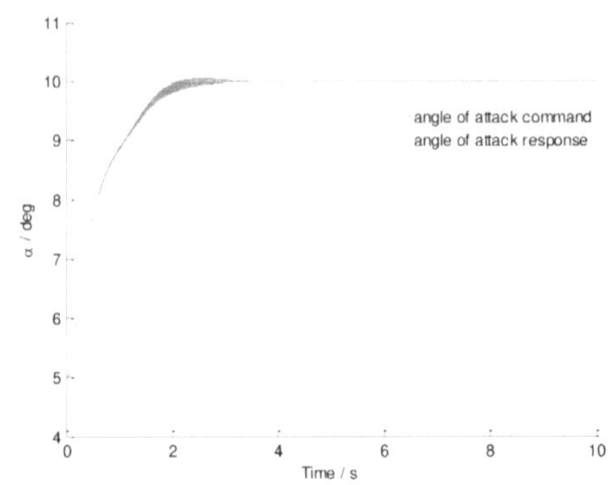

Figure 2
Response of Angle of attack

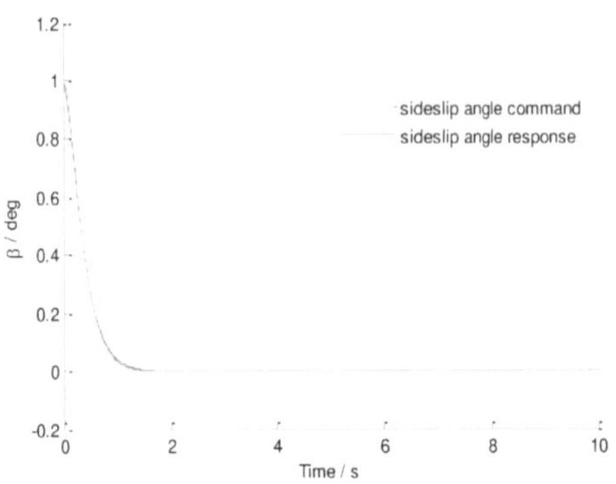

Figure 3
Response of sideslip angle

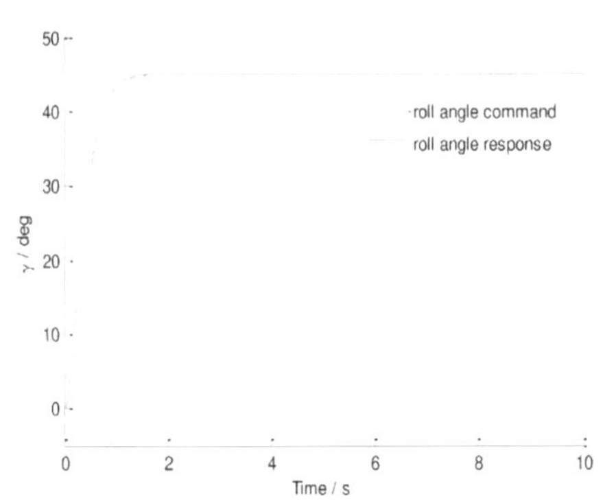

Figure 4
Response of roll angle

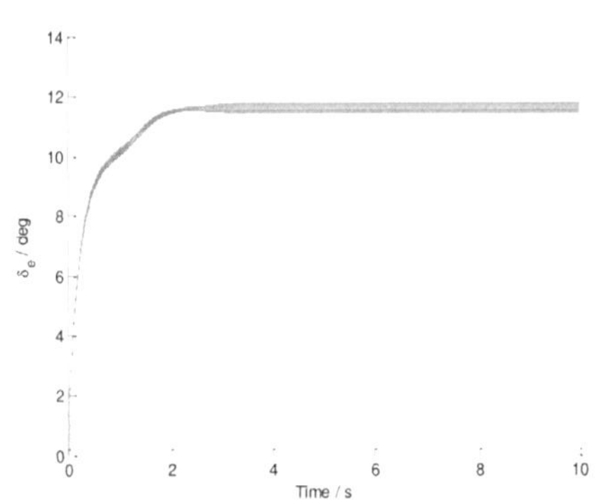

Figure 5
Results of elevator angle

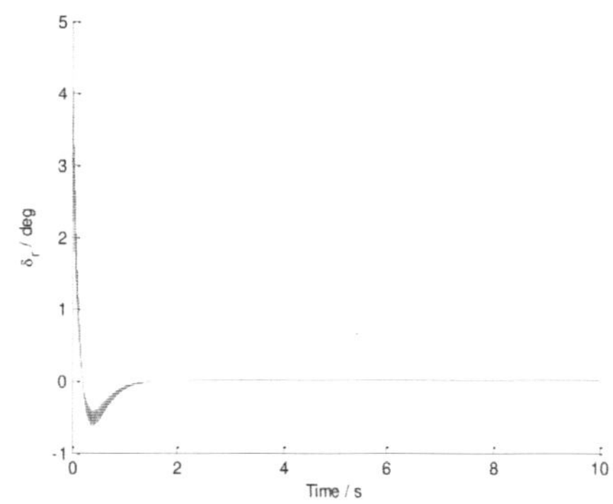

Figure 6
Results of rudder angle

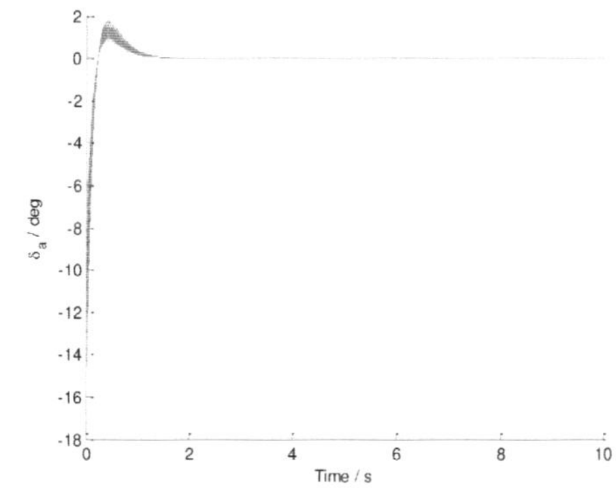

Figure 7
Results of aileron angle

The simulation results show that the angle of attack and roll angle can track angle and roll angle command signal well and the system keeps the sideslip angle as 0. The designed controller can be applied to the BTT control of hypersonic glide vehicle.

5. Conclusions

In response to the phenomenon of the high coupling of the hypersonic glide vehicle in BTT control, a global sliding mode variable structure controller is designed. The coupling among channels is considered as a kind of interference. By using the invariance principle of sliding mode motion state, we can realize decoupling control. At the same time, the global sliding mode control system can eliminate the stage of reaching, which can enable the system to be in a state of sliding mode motion at the beginning, and realize the whole process decoupling control of the system. The given simulation example in this paper illustrates the effectiveness of the proposed method.

6. References

[1] Charania, A. C., and Bradford, J. 2013. Mission effectiveness tradeoffs for future military space access vehicles. *Intervention,6*(1), 1749-1774.
[2] Walker, S., Sherk, J., Shell, D., Schena, R., Bergmann, J., and Gladbach, J. 2013. The DARPA/AF Falcon Program: The Hypersonic Technology Vehicle #2 (HTV-2) Flight Demonstration Phase. *Aiaa International Space Planes and Hypersonic Systems and Technologies Conference*.
[3] Walker, S., and Rodgers, F. 2005. Falcon Hypersonic Technology Overview. *Aiaa/cira, International Space Planes and Hypersonics Systems and Technologies Conference*.
[4] AIAA. 2000. Hypersonic vehicle control surface development.
[5] Wiese, D. P., Annaswamy, A. M., Muse, J. A., Bolender, M. A., and Lavretsky, E. 2015. Adaptive output feedback based on closed-loop reference models for hypersonic vehicles. *Journal of Guidance Control & Dynamics, 38*(12), 1-12.
[6] Rehman, O. U., Fidan, B., and Petersen, I. 2013. Minimax LQR Control Design for a Hypersonic Flight Vehicle. *Aiaa/dlr/dglr International Space Planes and Hypersonic Systems and Technologies Conference*.
[7] Johnson, E. N., Calise, A. J., Curry, M. D., Mease, K. D., and Corban, J. E. 2015. Adaptive guidance and control for autonomous hypersonic vehicles. *Georgia Institute of Technology, 29*(3), 725-737.
[8] Fan, Y. H., Yan, P. P., Wang, F., and Xu, H. Y. 2016. Discrete sliding mode control for hypersonic cruise missile. Discrete Dynamics in Nature & Society, 2016(1), 1-9.

[9] Qu, C.Y. 2015, Analysis on the development of the boost glide weapon of United States, *Tactical Missile Technology*, 04:1-4+52.
[10] Acton, J. M. 2015. Hypersonic boost-glide weapons. *Science & Global Security, 23*(3), 191-219.
[11] Williams, D. E., Friedland, B., and Madiwale, A. N. 1986. Modern control theory for design of autopilots for bank-to-turn missiles. *Journal of Guidance Control & Dynamics, 10*(4), 1130-1136.
[12] Li, Q. 2015, Study on Reentry Guidance and Control Method for Hypersonic Glide Vehicle, *Beijing Institute of Technology*.
[13] Chen, X.Q., Hou, Z. X., Liu, J.X. 2012, Analysis and controller design of bank-to-turn system for hypersonic gliding vehicle, *Journal of National University of Defense Technology*, 03:17-23.
[14] Xu, B., Fan, Y., and Zhang, S. 2015. Minimal-learning-parameter technique based adaptive neural control of hypersonic flight dynamics without back-stepping. Neurocomputing, 164(C), 201-209.
[15] Rollins, E., Valasek, J., Muse, J. A., and Bolender, M. A. 2013. Nonlinear Adaptive Dynamic Inversion Applied to a Generic Hypersonic Vehicle.*Aiaa Guidance, Navigation, and Control*.
[16] An, X. Y., Wang, X. H. 2010, Analysis and controller design of bank-to-turn system for hypersonic gliding vehicle, *Journal of System Simulation*, S1: 107-110.
[17] Weerdt, E., Kampen, E. J., Gemert, D., Chu, Q. P., and Mulder, J. A. 2013. *Adaptive Nonlinear Dynamic Inversion for Spacecraft Attitude Control with Fuel Sloshing. AIAA Guidance, Navigation and Control Conference and Exhibit*.
[18] AIAA. 2007. Robust Nonlinear Dynamic Inversion Control for a Hypersonic Cruise Vehicle. *AIAA Guidance, Navigation and Control Conference and Exhibit*.
[19] Gong, Y. L., and Hong-Xin, W. U. 2010. Characteristic model-based adaptive attitude control for hypersonic vehicle. *Journal of Astronautics*.
[20] Creagh, M., Kearney, M., and Beasley, P. 2011. Adaptive Control for a Hypersonic Glider using Parameter Feedback from System Identification.*AIAA Guidance, Navigation, and Control Conference*(Vol.1, pp.532-549).
[21] Fan, Y., Yang, J., and Zhang, Y. 2007. Robust control of hypersonic aircraft. Proceedings of SPIE - The International Society for Optical Engineering, 6795, 67955A-67955A-6.
[22] Marrison, C. I., and Stengel, R. F. 1998. Design of robust control systems for a hypersonic aircraft. *Journal of Guidance Control and Dynamics, 21*(1), 58-63.
[23] Chang, Y., Wang, X., Chun, H. E., and Liu, X. 2015. Study on adaptive variable structure control law for hypersonic gliding vehicle. *Aerospace Control*.
[24] Li, Y. J., Zhang, K. 2005. Adaptive control theory and its application, *Northwestern Polytechnical University Press*, p 179.
[25] Chang, Y., Wang, X., Chun, H. E., and Liu, X. 2015. Study on adaptive variable structure control law for hypersonic gliding vehicle. *Aerospace Control*.
[26] Lin, D.F., Wang, H., Wang, J., Fan, J.F. 2012, Autopilot design and guidance law analysis of tactical missile, *Beijing Institute of Technology*, p10-40.
[27] Li, S., Zhang, X. G., Xiang, J. W. 2000, Pole assignment method of output feedback with minimized norm in vibration control system design, *Chinese Journal of Aeronautics*, 05:446-449.

Acknowledgments
This paper is supported by Fundamental Research Funds for the Central University under Grant 3102015BJ008, Aeronautical Science Foundation of China under Grant No. 2015ZA53003 and this work is also supported by the Special Science Research Foundation of Doctor Subject and Higher education under Grant No.20136102120012.

Research on Time-series Modeling and Filtering Methods for MEMS Gyroscope Random Drift Error

Xiao Yi Wang* **and Xiu Yun Meng**

School of Aerospace Engineering, Beijing Institute of Technology, Zhongguancun South Street, Haidian District, 100081, CHINA

*18813126829@sina.cn

Abstract. The precision of MEMS gyroscope is reduced by random drift error. This paper applied time series analysis to model random drift error of MEMS gyroscope. Based on the model established, Kalman filter was employed to compensate for the error. To overcome the disadvantages of conventional Kalman filter, Sage-Husa adaptive filtering algorithm was utilized to improve the accuracy of filtering results and the orthogonal property of innovation in the process of filtering was utilized to deal with outliers. The results showed that, compared with conventional Kalman filter, the modified filter can not only enhance filter accuracy, but also resist to outliers and this assured the stability of filtering thus improving the performance of gyroscopes.

1. Introduction

With the rapid development of micro-electromechanical systems (MEMS) technology, MEMS gyroscope has been widely used in low-cost inertial navigation systems. It has small sizes, low cost, and light weights, but its application in high-precision navigation systems is limited due to comparatively low precision [1]. The random drift errors of MEMS gyroscopes accumulate with data integration in the inertial navigation algorithm, which will lead to large navigation errors. Therefore, effective modeling and compensation to random drift error signals of MEMS gyroscopes has been paid increasing attention to improve the measurement accuracy.

ARMA model is a dynamic linear model in random process that is currently widely used in system analysis, forecasting, identification and control. Compared with neural network and wavelet analysis, it is suitable for online real-time estimations in simple and low-cost systems. Therefore, time series analysis was used to model the random drift errors of MEMS gyroscopes in this paper.

In the filtering process, outliers will lead to serious deviations of the best estimations from the true state values of the system. Kalman filter is a linear estimator based on minimum mean square errors.

It is an effective way to use Kalman filter to suppress the random noise of gyroscopes in terms of precision enhancement. However, its filtering accuracy depends on the accuracy of the mathematical models and the statistical properties of the noise, and the presence of outliers affects the stability of the filter. Therefore, an anti-outlier adaptive Kalman filtering method was used and the simulation results demonstrated the effectiveness and feasibility of the method.

2. The time series model of MEMS gyroscope random drift error

In an autoregressive integrated moving average model, the future value of a variable is assumed to be a linear function of several past observations and random errors [2]. The general form of time series model ARMA(p,q) is expressed as follows :

$$\hat{X}_k = \varphi_1\hat{X}_{k-1} + \varphi_2\hat{X}_{k-2} + \cdots + \varphi_p\hat{X}_{k-p} + a_k - \theta_1 a_{k-1} - \theta_2 a_{k-2} - \cdots - \theta_q a_{k-q} \tag{1}$$

Where the $\varphi_i (i = 1, 2, ...p)$ and $\theta_j (j = 1, 2, ...q)$ are the coefficients of auto-regression and moving average respectively, which can be determined by the least-squares estimation method; p and q are integers and referred to as orders of the model; a_k is the random error at time period t . If q = 0, then (1) becomes a model of AR (p); When p = 0, the model reduces to a model of MA (q).The values of p and q can be determined by Akaike information criterion (AIC) [3].

Place a Micro Inertial Measurement Unit (MIMU) on a level surface, and collect original output signals, which are shown in Figure 1.

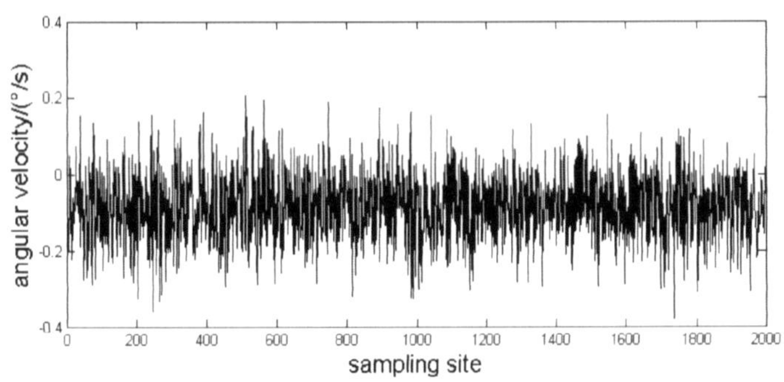

Figure 1. The static output of a MEMS gyroscope

Only when the random series have the characteristics of zero mean, stationarity, and normality, time-series model can be established [4].So in order to build time-series model of drift error, the data collected need to be preprocessed. After remove outliers, constant and trends component, the remaining data upon testing fulfill the requirement of timing model. The model parameters and AIC values obtained by calculation are shown in table 1.

Table 1. The model parameters and AIC values of random drift error

	AR(1)	AR(2)	AR(3)	ARMA(1,1)	ARMA(2,1)
φ_1	0.2134	0.2178	0.2182	0.1085	- 0.2219
φ_2	—	- 0.0207	- 0.0240	—	0.0721
φ_3	—		0.0149	—	—
θ_1	—	—	—	0.1098	0.4403
AIC	-2.1085	-2.1089	-2.1092	-2.1090	- 2.1091

The AIC values show no significant difference in different model. Considered the real-time demand for a low cost system, AR(1) model is chosen to establish the time-series model, which is expressed as:

$$x_k = \phi x_{k-1} + w_k \tag{2}$$

where x_k is the output of the timing model, $\phi = 0.218$, White noise $w_k \sim NID(0, \sigma_w^{\ 2})$.

3. The improved algorithm of Kalman filter

3.1 The establishment of filtering model

When process noise is colored noise and observation noise is white noise, the method of amplifying state variables is used to establish state space model. Set $X_k = [r_k, x_k]^T$, r_k is the real angular rate of gyroscope, x_k is the output of the timing model. Then state space model is given as

$$\begin{bmatrix} r_k \\ x_k \end{bmatrix} = \begin{bmatrix} 1 & 1 \\ 0 & \phi \end{bmatrix} \begin{bmatrix} r_{k-1} \\ x_{k-1} \end{bmatrix} + \begin{bmatrix} 0 \\ 1 \end{bmatrix} w_{k-1} \tag{3}$$

$$Z_k = [1, 0] \begin{bmatrix} r_k \\ x_k \end{bmatrix} + v_k \tag{4}$$

That is,

$$X_k = AX_{k-1} + \Gamma W_k \tag{5}$$

$$Z_k = HX_k + V_i \tag{6}$$

Where $A = \begin{bmatrix} 1 & 1 \\ 0 & \phi \end{bmatrix}$, $\Gamma = \begin{bmatrix} 0 & 1 \end{bmatrix}^T$, $W_{k-1} = w_{k-1}$, $V_{k-1} = v_{k-1}$, $H = [1, 0]$. The process noise w_k and observation noise v_k are uncorrelated, zero mean. Their variance values are Q_k and R_k respectively. Q_k is set as the residuals covariance of AR(1). R_k is set as the covariance of actual measurement data.

3.2 Adaptive Kalman filter

When affected by the working environment, the statistical characteristics of noise may change. The conventional kalman filter needs to be improved as the performance of Kalman filter depends on the accuracy of the model established [5]. An adaptive Kalman filter, using Sage-Husa filtering algorithm to adjust the statistical characteristics of process noise and measurement noise on-line is described as follows [6] [7]:

$$\hat{X}_{k|k-1} = A\hat{X}_{k-1} + q_{k-1} \tag{7}$$

$$P_{k|k-1} = AP_{K-1}A' + \Gamma_{k|k-1}Q_{k-1}\Gamma_{k|k-1}' \tag{8}$$

$$K_k = P_{k|k-1}H_k'\left[H_kP_{k|k-1}H_k' + R_{k-1}\right]^{-1} \tag{9}$$

$$e_k = Z_k - \hat{Z}_k = Z_k - H_k\hat{X}_{k|k-1} \tag{10}$$

$$\hat{X}_k = \hat{X}_{k|k-1} + K_k(e_k - r_{k-1}) \tag{11}$$

$$P_k = \left[I - K_kH_k\right]P_{k|k-1} \tag{12}$$

The time varying noise estimator is described as follows:

$$d_{k-1} = (1-b)/(1-b\wedge k) \tag{13}$$

$$r_k = (1-d_{k-1})r_{k-1} + d_{k-1}\left[Z_k - H_k\hat{X}_{(k|k-1)}\right] \tag{14}$$

$$q_k = (1-d_{k-1})q_{k-1} + d_{k-1}\left[\hat{X}_k - A\hat{X}_{k-1}\right] \tag{15}$$

$$R_k = (1\text{-}d_{k-1})R_{k-1} + d_{k-1}*[[Z_k - H_k\hat{X}_{(k|k-1)} - r_k][Z_k - H_k\hat{X}_{(k|k-1)} - r_k]^T - H_kP_{k|k-1}H_k^T] \tag{16}$$

$$Q_k = (1-d_{k-1})Q_{k-1} + d_{k-1}[K_k[Z_k - H_k\hat{X}_{(k|k-1)} - r_k][Z_k - H_k\hat{X}_{(k|k-1)} - r_k]^T K_k^T + P_k - AP_{k|k-1}A^T] \tag{17}$$

b is the forgetting factor and usually take 0.95 ~ 0.99.

3.3 Algorithm of anti-outliers

Outliers in measurements can result in divergence of filtering result, so it is desirable to improve the filter to have the effects of anti-outliers. During the Kalman filtering process, innovation e_k possess the characteristic of orthogonality [8]. Then according to equation (10),

$$E(Z_kZ_k^T) = E(\hat{Z}_k\hat{Z}_k^T) + E(e_ke_k^T) = H_k\hat{X}_{(k|k-1)}\hat{X}^T_{(k|k-1)}H_k^T + H_kP_{k|k-1}H_k' + R_{k-1} \tag{18}$$

Set $J = H_k\hat{X}_{(k|k-1)}\hat{X}^T_{(k|k-1)}H_k^T + H_kP_{k|k-1}H_k' + R_{k-1}$ and ε is supposed to be a disturbance caused by the calculation error and other factors. When observation Z_k satisfies $E(Z_kZ_k^T) \in [J-\varepsilon, J+\varepsilon]$, we can draw a conclusion that Z_k isn't an outlier; Conversely, it is. When outliers exit in measurements, orthogonality will be destroyed. An activation function is needed to keep the orthogonality of innovation after Z_k has been considered an outlier to assure the stability of filter, which can be defined as follows [9],

$$f(r)=\begin{cases} 1 & (m \geq r) \\ m/r & (m < r) \end{cases} \tag{19}$$

Where $m=\sqrt{J}, r=|Z_k|$, Then the state estimation equation (11) can be replaced by

$$\hat{X}_k = \hat{X}_{k|k-1} + K_k(f(r)Z_k - H_k\hat{X}_{k|k-1} - r_{k-1}) \tag{20}$$

4. Result

Take the static output of a MEMS gyroscope as shown in Figure 1 as observations of the filters. The results of filtering for random drift error are shown in Figure 2.

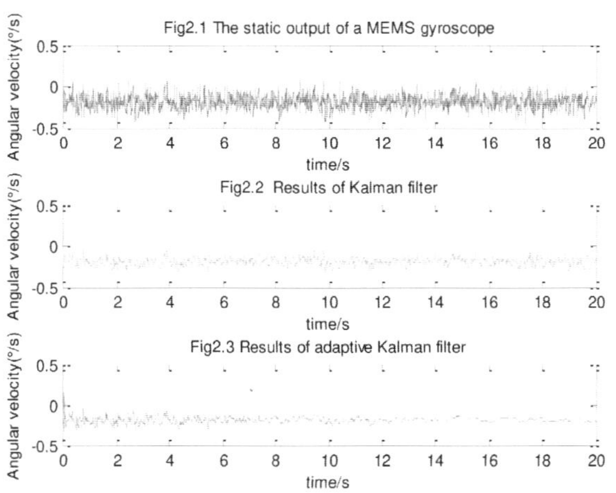

Figure 2. Results of static experiments

As can be seen from Figure 2, compared with conventional Kalman filter, filtering accuracy of adaptive Kalman filter has improved obviously. The mean and standard deviation values of noise before and after filtering are shown in table 2.

Table 2. Mean and standard deviation of noise before and after filtering

	mean（°/s）	Standard deviation（°/s）
random drift error	-0.184049	0.0863
After Kalman filter	-0.183968	0.048196
After adaptive Kalman filter	-0.184041	0.034918

The mean remains unchanged before and after filter because Kalman filter estimation is a method of unbiased estimation. The standard deviation of signals after adaptive Kalman filter is relatively smaller than after Kalman filter, which showed that improved kalman filter has a better filtering effect for random drift error.

For achieving the dynamic simulation and viewing the effect of eliminating outliers, add a sinusoidal signal and isolated, continuous outliers to the static output shown in Fig.1 as observations of the filters. The results of filtering are shown in Figure3.

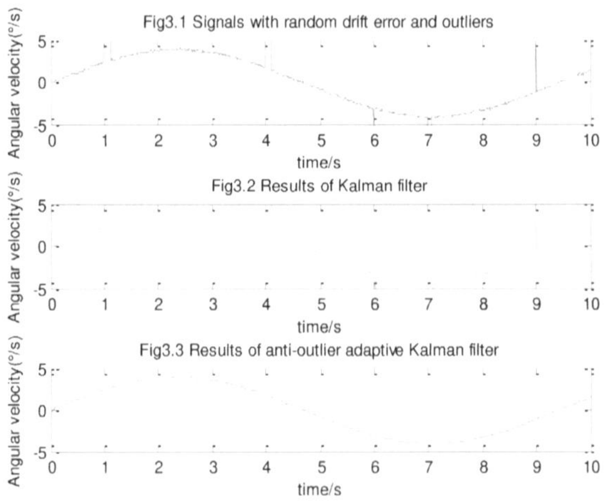

Figure 3. Results of dynamic simulation

As we can see from Figure 3, anti-outlier adaptive Kalman filter can eliminate outliers and decrease noise, which can improve the using precision of gyroscopes.

5. Conclusion

In this paper, enhancing the accuracy and stability of conventional Kalman filter for random drift error of MEMS gyroscope is the research topic, and the achievements are summarized as following: 1. According to time series analysis methods, the time series model of MEMS gyroscope has been established; 2. To improve the accuracy of the filter, an adaptive Kalman filter has been used; 3. An anti-outliers function for identifying outliers and an activation function as the weight to each measurement for modifying the orthogonal property of innovation sequence have been used to eliminate negative influences on performances of filter caused by outliers.

6. References

[1] Guo Y N, Han F, Du S H, Ma G F and Zhu L K 2015 Performance analysis of MEMS gyro and improvement using Kalman filter *Control Conference (CCC), 2015 34th Chinese* (New Jersey:IEEE) pp173–214

[2] Zhang G P 2003 Time series forecasting using a hybrid ARIMA and neural network model *Neurocomputing* **50** 159

[3] Zhao W C, Fu M Y, Zhang Q H and Deng Z H 2005 Investigation on modeling and filter

IOP Conf. Series: Materials Science and Engineering **187** (2017) 012005 doi:10.1088/1757-899X/187/1/012005

methods for MEMS IMU drift data *Journal of Chinese Inertial Technology* **6** 003

[4] Yang P P and Li Q 2009 Kalman Filtering of MEMS Gyro Based on Time-series Model *Electronic Measurement & Instruments* **09** 367

[5] Rizal K and Bagus E 2012 The discrete Kalman filtering approach for seismic signals deconvolution *AIP Conf. Proc.* **1454** 91

[6] Lv W and Wang Y D 2005 Application of Sage-Husa adaptive kalman filter on initial alignment SINS *Control Technology of Tactical Missile* **15** 6

[7] Zhang L M, Zhang X H and Chen Z Q 2013 Improvement of adaptive Kalman filtering algorithm and its application in SINS/GPS integrated navigation *J Southeast Univ* **43** 89

[8] Gong X L and Fang J C 2009 Application of Modified Kalman Filtering Restraining Outliers Based on Orthogonality of Innovation to POS *Acta Aeronautica et Astronautica Sinica* **30** 2348

[9] Liu H F, Yao Y, Lu D and Ma J 2003 Study for Outliers Based on Kalman filtering *Electric Machines and Control* **7** 40

IOP Conf. Series: Materials Science and Engineering **187** (2017) 012006 doi:10.1088/1757-899X/187/1/012006

A Modeling Method of Multiple Targets Assignment under Multiple UAVs' Cooperation

Q H Wang [1], G Wan [1], X F Cao [1] and L X Xie[1]

[1]Zhengzhou Institute of Surveying and Mapping, China

e-mail: dcn_ing@163.com

Abstract. Aiming at the multiple UAVs' cooperation in the complex environment, detailed analysis about targets assignment model is made in the paper. Firstly, three basic situations are discussed according to the quantitative relationship between the UAVs and the targets. Then in order to make the targets model more practical, the probability that the UAVs' damage is also taken into consideration. Following, basic particle swarm optimization algorithm is adopted to solve the model which has great performance in efficiency and convergence. Finally, three-dimensional environment is simulated to verify the model. Simulation results show that the model is practical and close to the actual environment.

1. Introduction

More and more UAVs are put into practical use with their performance improves. They complete a variety of tasks instead of people in dangerous and complex environment, such as targets search, targets attack and etc. In the face of complex tasks, it is necessary of UAVs' cooperation to improve the implementation efficiency. Then targets assignment is value of discussion. Targets assignment is reasonable planning between multiple targets and multiple UAVs based on UAVs' performance constraints, environment constraints, etc. Targets assignment helps to improve the efficiency of UAVs and the probability of tasks' completion. Multiple Targets assignment under multiple UAVs' cooperation is a kind of combinatorial optimization problem.

Targets assignment model is modelling in the specific of some environment. The common targets assignment models include mixed integer linear programming (MILP), network flow model (NFM), vehicle routing problem (VRP) and multiple traveling salesman problem (MTSP) [1-3].

Zengin [4] et al. researched on the problem of the cooperative control and the targets assignment around the target tracking. Bellingham [5] proposed the mixed integer linear programming aiming at the minimum time of tasks completion. Bertuccelli [6] et al. studied the effects of the dynamic uncertain environment on the multiple targets assignment modeling. In the case where UAVs is less than targets. Shaferman [7] et al. paid attention to the multiple tasks assignment which is especially suitable for the modeling under the environment of the city. However, the discussion is made under the situation where targets is more than targets. At present, there is no detailed discussion about the quantitative relationship between the UAVs and the targets in any literature.

For the problem that not all possibilities are discussed when modeling, and the corresponding factors are not comprehensive when constructing cost functions, three basic situations are discussed according to the quantitative relationship between the UAVs and the targets. The probability that UAVs' damage are taken into consideration when constructing cost function. Then Basic particle swarm optimization algorithm is adopted to solve the model. Finally, the three-dimensional environment is simulated to verify the model. The simulation results show that the model is practical and close to the actual environment. The particle swarm algorithm is easy to implement and has good performance in efficiency.

2. Targets assignment model

2.1. Cost function

It is necessary to solve the assignment problem between targets and UAVs in the multiple UAVs' cooperation. The basic principle is to minimize the overall cost in the premise of completing the tasks.

Numbered the UAVs $1,2,\cdots,U$ and Numbered the targets $1,2,\cdots,T$ where U and T respectively represent the quantities of the UAVs and the targets. The relationship can be described as $U>T$, $U=T$, $U<T$. It must be ensured that each UAV will be assigned to the target and each object must have the corresponding UAV. Assuming that all UAVs' and targets' initial positions are known before assignment, there are three factors we considered in flight cost function, including the total flight voyage of multiple UAVs, the total time of executing tasks and the probability of the UAVs' damage. The three factors are discussed as follows.

The total flight voyage of multiple UAVs

$$d_{sum}=\sum_{u,v=1}^{U}\sum_{i,j=1}^{T}d(u\sim v,i\sim j)x(u\sim v,i\sim j) \tag{1}$$

where $u\sim v$ are UAVs; $i\sim j$ are targets; $d(u\sim v,i\sim j)$ are flight voyage.

The total time of executing tasks

$$t_{sum}=\max\{\sum_{u,v=1}^{U}\sum_{i,j=1}^{T}t(u\sim v,i\sim j)x(u\sim v,i\sim j)\} \tag{2}$$

where $t(u\sim v,i\sim j)$ is the flight time of some UAV which satisfies

$$t(u\sim v,i\sim j)=\frac{d(u\sim v,i\sim j)}{v_{u\sim v}} \tag{3}$$

where $v_{u\sim v}$ is the velocity of UAVs.

The probability of the UAVs' damage

$$th_{sum}=\sum_{u,v=1}^{U}\sum_{i,j=1}^{T}p(u\sim v,i\sim j)x(u\sim v,i\sim j) \tag{4}$$

where $p(u\sim v,i\sim j)$ is the probability to the damage of some UAVs.

According to [8], here is the probability when the u-th UAV flied to the j-th target,

$$p_u=\begin{cases}\exp\left(-\dfrac{(R-r_{radar}{}^{\max}/2)^2}{2\left(\sqrt{2}r_{radar}{}^{\max}\right)^2}\right), R\le r_{radar}{}^{\max}\\ 0, R>r_{radar}{}^{\max}\end{cases} \tag{5}$$

where $r_{radar}{}^{\max}$ is the effective radius of radars, R is the distance between the UAVs and radars.

From (5), without the effective range of the radar, the probability of the UAVs' damage is 0. Within the effective range, the probability obeys Gauss distribution.

$x(u\sim v,i\sim j)$ is the decision variable whose value is 0 or 1. It represents the corresponding relationship between the UAVs and the targets and following is the three situations.

1) When $U=T$, the relationship between the UAVs and the targets is one to one, then $x(u\sim v,i\sim j)=x(u,i)$. In that case, $\forall u=1,2,\cdots,U$, $\sum_{i=1}^{m}x(u,i)=1$; $\forall i=1,2,\cdots,T$ $\sum_{u=1}^{n}x(u,i)=1$.

2) When $U > T$, each UAV has the only one corresponding target. It means that some target may be assigned to several UAVs, then $x(u \sim v, i \sim j) = x(u \sim v, i)$. In that case, For any UAV or UAV groups which means the several UAVs assigned to the same target, $\sum_{i=1}^{m} x(u \sim v, i) = 1$.

3) When $U < T$, each target has the only one corresponding UAV. It means that some UAV may be assigned several targets, then $x(u \sim v, i \sim j) = x(u, i \sim j)$. In that case, for any target or target groups which means the several targets assigned to the same UAV, $\sum_{i=1}^{m} x(u \sim v, i) = 1$.

Considering the above factors, the flight cost function is given as follows:

$$F(x) = w_1 \alpha_1 d_{\text{sum}} + w_2 \alpha_2 t_{\text{sum}} + w_3 th_{\text{sum}} \tag{6}$$

w_1, w_2, w_3 are the weight factors within (0,1) and $w_1 + w_2 + w_3 = 1$,. The weight factors are given by experts according to the specific tasks. α_1, α_2 are the ratio factors which make the total voyage cost, the total time cost and the threat cost remain the same magnitude level.

2.2. *Analysis of constraints*

There are constraints as follows,

1) Voyage constraints

The single voyage of each UAV is limited. It is influenced by the UAV's performance or the oil supplement. The voyage constraint can be expressed as

$$\sum_{i,j=1}^{T} d(u, i \sim j) x(u \sim v, i \sim j) \le D_u \tag{7}$$

where D_u is the maximum voyage of the u-th UAV, $u = 1, 2, \cdots, U$.

2) Flight time constraints

Similar to the voyage, the single flight time of each UAV is limited, which can be expressed as follows

$$\sum_{i,j=1}^{T} t(u, i \sim j) x(u \sim v, i \sim j) \le T_u \tag{8}$$

where T_u is the maximum flight time of the u-th UAV, $u = 1, 2, \cdots, U$.

3) Target execution order constraints

In most cases, the concept of target value is fuzzy. The commander or experts seldom provide the accurate value while they describe the target value with words like 'high' and 'low'. For convenience, the target value is divided into seven grades, 'especially high', 'very high', 'high', 'medium', 'low', 'very low', 'especially low'.

In the paper, the seven grades are transmit into quantitative numbers [7] within the range of [0, 10]. The correspondence is as follows, 'especially high=[9,10]', 'very high=[8,9]', 'high=[7,8]', 'medium=[5,7]', 'low=[3,5]', 'very low=[1,3]', 'especially low[0,1]'. This will provide a quantitative reference for the experts so that they can determine the value of the targets, which has an effect on the order of the target execution.

Except the target value which influences the execution order, some target may have the specific execution order. For example, if the i-th target must be executed after the j-th target, then the constraint can be expressed as

$t_j \ge t_i + \Delta t$

Where t_i represents the time when the i-th target is executed, the same as the t_j. Δt is the minimum interval between the t_i and t_j.

3. Solutions to the model

Particle swarm algorithm(PSO) is an intelligent optimization algorithm proposed by Dr. R.Eberhart and Dr. J.Kenedy in 1995, which stems from the simulation of birds flock's looking for food [9, 10]. It is a kind of calculation methods of artificial life. However, it is different from other evolution algorithms, which generates the optimal solution by the group cooperation mechanism instead of the group competition mechanism. PSO algorithm possesses the advantages of simplified, rather quick convergence speed, global optimization performance, and less controlling parameters, et al. PSO has attracted more and more attention recently.

In the paper, PSO is adopted to solve the model. Each particle represents a kind of solution to the targets assignment, which is denoted as $\bar{x} = < \bar{p}, \bar{v} > = <$ target assignment solution, velocity vector $>$, $\bar{p}, \bar{v} \in S \subset R^D$ (D-dimensional space), that is to say that the position and velocity of the i-th particle is respectively $\bar{p}_i = (p_{i1}, p_{i2}, \cdots, p_{id}, \cdots, p_{iD})$ and $\bar{v}_i = (v_{i1}, v_{i2}, \cdots, v_{id}, \cdots, v_{iD})$.

Step 1 (Initialization): $t = 1$, generate K particles $\bar{x}_i(t) = < \bar{p}_i(t), \bar{v}_i(t) >$ randomly to form the initial particle group.

Step 2 (Fitness value calculation): calculate the individual extremum which corresponding to the best fitness value of each particle, denoted as $pbest_i(t)$ and the optimal solution found by the whole particle swarm, called the global extremum $gbest_i(t)$.

Step 3 (Evolutionary learning): $t = t + 1$, update the positions and velocities according to (9)(10)

$$\begin{aligned} v_i(t+1) = v_i(t) + c_1 rand_1 \left[pbest_i(t) - p_i(t) \right] \\ + c_2 rand_2 \left[gbest_i(t) - p_i(t) \right] \end{aligned} \tag{9}$$

$$p_i(t+1) = p_i(t) + v_i(t+1) \tag{10}$$

where $v_i(t)$ is the velocity of the current particle, $rand_1$ and $rand_2$ are random numbers within [0,1]. c_1 and c_2 are learning factors (also called acceleration constant), which represent the memory learning and the motivation of learning from others. Recalculate $pbest_i(t)$ and $gbest_i(t)$.

Step 4 (Algorithm end) If $t > \max\_gen$ or the end conditions are satisfied, output $gbest(t)$, otherwise go to Step 3.

4. Simulations and results analysis

Target assignment model and solving algorithm are verified by MATLAB simulation. Firstly, the environment area where the UAVs and the targets exist is constructed.

Though the two-dimensional simulation is relatively easy [11], the description is not actual enough and will influence the effectiveness of the targets assignment. Therefore, the three-dimensional simulation is made. Three-dimensional environment is built based on the standard digital elevation model data, and the radar threat is taken into consideration at the same time. The related parameters are shown in appendix.

Table 1 and table 2 are the initial parameters of the UAVs.

Table 1. Parameters of UAVs when $U < T$

Number of the UAV	Three-dimensional coordinates（km）	The maximum voyage（km）	The longest flight time（min）
1	（75,20,13）	500	50
2	（110,30,8）	700	70
3	（100,25,9）	600	60
4	（45,50,12）	600	60

IOP Conf. Series: Materials Science and Engineering **187** (2017) 012006 doi:10.1088/1757-899X/187/1/012006

Table 2. Parameters of UAVs when $U = T$ and $U > T$

Number of the UAV	Three-dimensional coordinates（km）	The maximum voyage（km）	The longest flight time（min）
1	（75,20,13）	500	50
2	（110,30,8）	700	70
3	（100,25,9）	600	60
4	（45, 0,12）	600	60
5	（35,65,10）	650	65
6	（65,35,8）	610	61

Table 3 and table 4 are the initial parameters of the targets.

Table 3. Target parameters when $U > T$

Number of the UAV	Three-dimensional coordinates（km）	Targets value（0~10）	Order of targets execution
1	（70,190,8）	5	[2 1],[4 1],[1 3]
2	（120,170,3）	8	
3	（100,175,3）	2	
4	（160,180,3）	7	

Table 4. Target parameters when $U = T$ and $U < T$

Number of the UAV	Three-dimensional coordinates（km）	Targets value（0~10）	Order of targets execution
1	（70,190,8）	5	[3 4],[5 2],[6 4]
2	（65,110,8）	3	
3	（120,170,3）	8	
4	（100,175,3）	2	
5	（160,180,3）	7	
6	（80,195,5）	7	

200 Monte Carlo simulations have been carried out. Figure 1 is the three-dimensional simulation results when $U = T$. In the figure 1, sphere represents the radar threat in the flight environment, diamond sign indicates the starting position of the UAV and the asterisk sign is the positions of the targets. The lines between the UAVs and the targets represent the flight tracks. Figure 2 is the contour map of the simulations. The solutions of the targets assignment under multiple UAVs' cooperation are also shown in table 5.

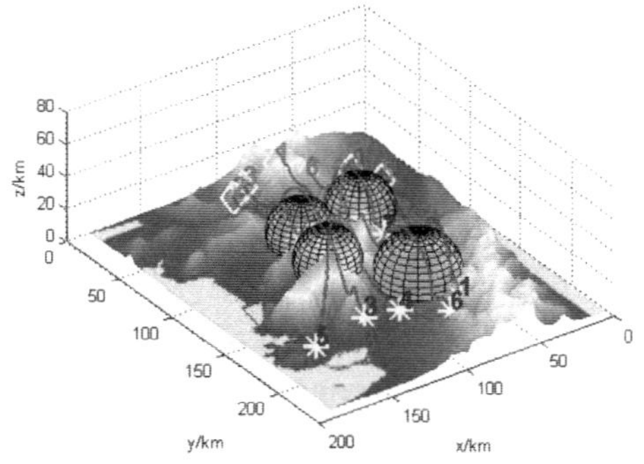

Figure 1. Three dimensional simulation results when $U = T$

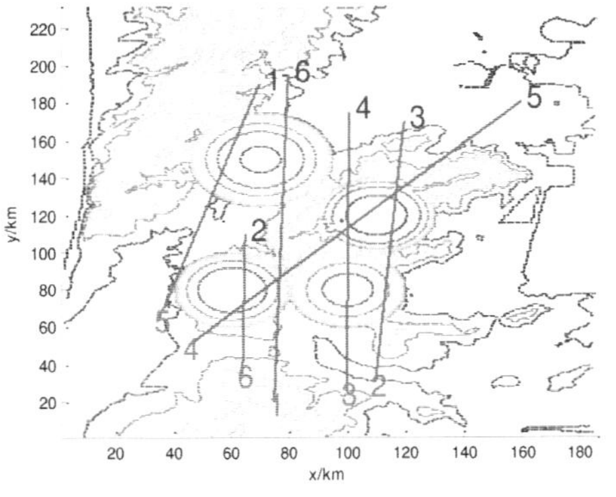

Figure 2. Contour map when $U = T$

Table 5. Targets assignment solutions when $U = T$

Number of the UAV	1	2	3	4	5	6
Number of the target	6	3	4	5	1	2

When $U > T$, the three-dimensional simulation results are shown in figure 3. The corresponding contour map is shown as figure 4. The targets assignment solutions are shown in table 6.

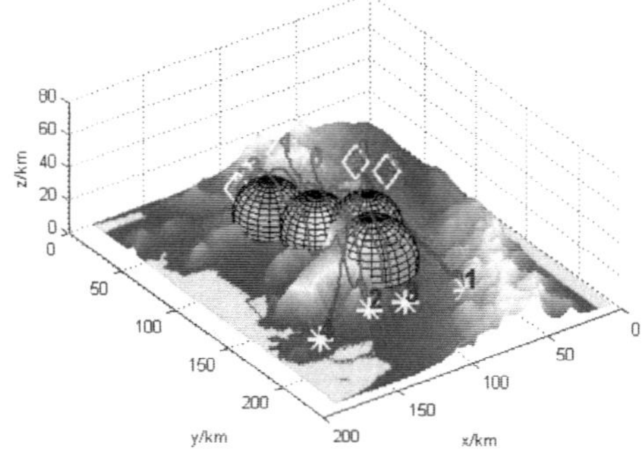

Figure 3. Three dimensional simulation results when $U > T$

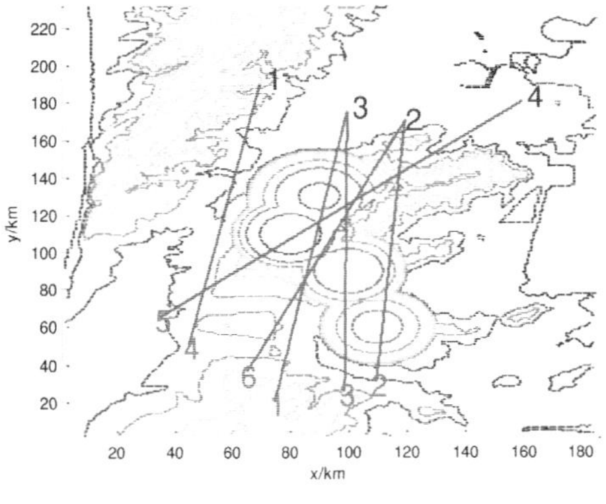

Figure 4. Contour map when $U > T$

Table 6. Targets assignment solutions when $U > T$

Number of the UAV	1	2	3	4	5	6
Number of the target	3	2	3	1	4	2

When $U < T$, the three-dimensional simulation results are shown in figure 5. The corresponding contour map is shown as figure 6. The targets assignment solutions are shown in table 7.

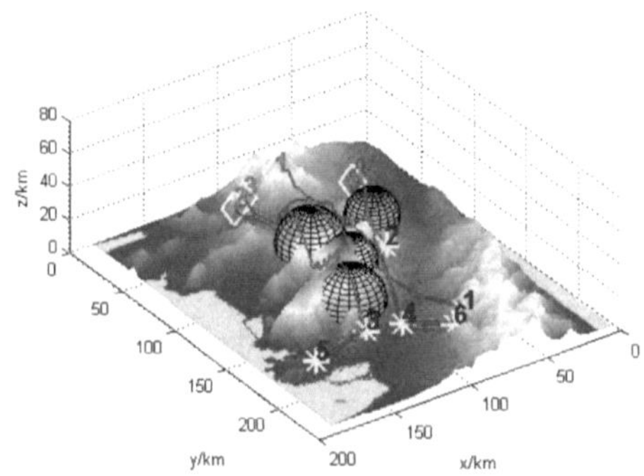

Figure 5. Three dimensional simulation results when $U < T$

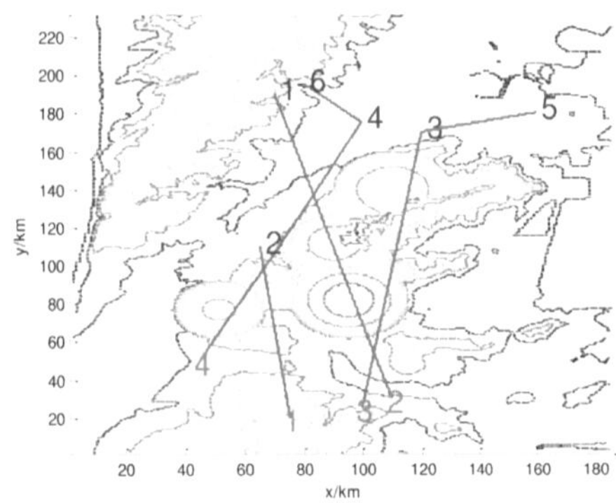

Figure 6. Contour map when $U < T$

Table 7. Targets assignment solutions when $U < T$

Number of the UAV	2	1	3	4	3	4
Number of the target	1	2	3	4	5	6

5. Conclusion

Multiple targets assignment of the multiple UAVs' cooperation is researched on in the paper. Firstly, the quantitative relationship between UAVs and targets is discussed in detail and the probability that the UAVs' damage is taken into consideration. Thus the model is more practical. Basic particle swarm optimization algorithm is adopted to solve the model which has great performance in efficiency. Finally, three-dimensional environment is simulated to verify the model. In the future, it is necessary to consider the effect of dynamic threat on the modeling. In addition, more intelligent optimization algorithms are in multi UAV cooperative application. The algorithm can also be combined with other intelligent optimization algorithms to further improve the performance.

6. References

[1] Griggs B J, Parnell G S and Lehmkuhl L J 1997 *Oper. Res.*, **45** 662-76.
[2] O'Rourke K P, Carlton W B, Bailey T G and Hill R R 2001 *Military Operations Research* **6** 5-30.
[3] Beard R W, Mclain T W and Goodrich M 2003 *IEEE Trans. Rob. Autom.* **18** 911-22.
[4] Zengin U and Dogan 2013 *In Proc. of AIAA Guidance, Navigation and Control Conference and Exhibit* (USA: Chicago) **12** p2006.
[5] Beilingham J, Tilterson M and Richards 2003 *Cooperative Control: Models, Applications and Algorithms*(US: Springer) p23.
[6] Bertuccelli L, Choi H L and Cho P 2009 *AIAA Guidance, Navigation, and Control Conference* (USA: American Institute of Aeronautics and Astronautics) **10** p2009.
[7] Shaferman, V and Shima T 2013 *In Proc. of AIAA Guidance, Navigation and Control Conference and Exhibit*(USA: Chicago), **10** p2009.
[8] Washburn A and Kress M *Combat Modelling* (US: Springer) p 39.
[9] Kennedy J and Eberhart R 1995 *In Proc. of International Conference on Neural Networks* **4** p1942.
[10] Eberhart R and Kennedy J 1995 *International Symposium on MICRO Machine and Human Science* p39.
[11] Bahar M R B, Bahar H B and Hashemzadeh F 2011 *International Conference on Automation and Computing IEEE* p75.

A 2D Simulation of the Flow Separation Control over a NACA0015 Airfoil Using a Synthetic Jet Actuator

M A Boukenkoul [1, 2, *], **F C Li** [1] **and M Aounallah** [3]

[1] School of Energy Science and Engineering, Harbin Institute of Technology, Harbin, CO 15001 China.
[2] Energy and Mechanical Engineering department, Sciences and technologic University of Tlemcen, Tlemcen, CO 13000 Algeria.
[3] Marine Engineering department, Sciences and Technologic University of Oran, Oran, CO 31000 Algeria.

*boukenkoulmedamine@hotmail.fr

Abstract. The present study aims to investigate numerically the flow control possibility using a synthetic jet actuation over a bi-dimensional NACA0015 airfoil manoeuvring at a highly turbulent flow ($8.9e10^5$ Reynolds to chord number). The 2-D flow behaviour was computed using the ANSYS Fluent commercial code. The so-called Reynolds Averaged Navier-Stocks (RANS) approach has been tested for one (Spalat-Allmaras S-A) and two (K-ε) transport equations for the turbulence modelling. Both present a weakness to predict the stall angle effectively. The S-A lift coefficient slope seems to be the closest to the experimental data. The synthetic jet control exhibits an extraordinary lift coefficient enhancement at high Angles Of Attack (AOA) but seems to be less obvious at low AOA, where the flow is still attached. A synthetic jet of a Strouhal ($St = 2$) and momentum ($C\mu$ of 0.56%), delays the stall onset from 15 to 19deg with enhancing the lift coefficient by 40%. The actual work has been enriched by studying the effect of the jet's frequency and momentum on the lift temporal signal. Also, the interaction between the mean flow and the synthetic jet structures topology was undertaken.

1. Introduction

Despite the huge advancement reached in the aero/hydrodynamic fields, the boundary layer instability remains one of the most challenging purposes facing scientists and engineers. Aiming usually to delay or eliminate the boundary layer separation, -unless otherwise for spoilers or speed brakes- in order to enhance the aerodynamic performances at the same wing body geometry. Postponing the separation onset can be traduced by a lift increase, drag reduction and a stall delay. These three latter lead to deal with huge issues in both civilian and military domains.

In the civilian aviation, the economic and environmental issues are more and more strict, reducing the drag goes together with a fuel consumption reduction and therefore more green flights. While, the military sector looks for higher air superiority and tactic ability. Both can be increased by enhancing the aircraft's manoeuvrability and shortening the take-off and landing required distance.

According to the energy expended to set a flow control, Gad-el Hal et al. [1] classified the control strategies into two main kinds. A *passive* control requiring any auxiliary power to be used (ex: vortex generators) and an *active* control which requires contrary to the first one an external power to be set (ex: suction/blowing actuators). Even if the named control strategies reached a high-efficiency level, they are usually fed from the engine compressed air, ex: the MiG21 can reduce his approach speed by 40 *km/h* using a continue blowing control. That requires a complex piping apparatus and constitutes an additional load for the engines. Also in the case of an engine failure, pilots may lose the lift force when the most needed.

Nowadays, research focus on how to enable continuous suction or blowing technics to save on the actuator energy by reducing the injected rates. Continuous control has become alternative or pulsed. At a first approximation, one can say that a synthetic jet is generated by a succession of blowing and suction phases. The mean jet resulting does not bring any additional fluid from where they named as a Zero Net Mass Flux ZNMF technic. Technically, we find plenty of solutions enabling the generation of this type of jet. Devices driven by piezoelectricity, electrostatic, thermal or acoustic can be small enough to be accommodated within the controlled body and have a localised action. However, they cannot provide a high-speed ejection, particularly when their membranes are activated at frequencies away from their resonance frequency. The thing that has delayed the use of this type of actuators compared to their counterpart based on a simple mechanical principle as carried out by Gilarranz *et al* [2].

It was found that the actuator geometry scale and the fluid characteristics are key points in the ring or pair vortex formation in a ZNMF. Several studies aim to parameterize its generation criteria. A number of dimensionless parameters governing the production and evolution of the SJA are carried out. Glezer *et al* [3], and Smith and Glezer [4] started by proposing two independent parameters. The first one relative to the blown fluid column size named the stroke length ratio. The second one based on the blowing stroke Reynolds number. Utturkar *et al* [5], and Holman *et al*[6] found that the ratio $\frac{Vi}{Vs}$ of the self-induced velocity in a vortex generation at the blowing stroke Vi and the mean velocity at the suction stroke Vs must be greater than a known value K (dependent to the ZNMF) to avoid the vortex reversibility to the actuator during the suction stroke. They proposed the following:

$$\frac{Vi}{Vs} \sim \frac{1}{St} = \frac{Re_{U_0}}{S^2} > K \tag{1}$$

Where, $S = (\frac{2\pi Re_{U_0}}{L})^{1/2}$ is the Stokes number and $St = \frac{2\pi f D_0}{U_0} = \frac{2\pi}{L}$ is the Strouhal number. Thereby the Strouhal number used in the present study found to be relative to the vortex shedding is based on the flow field and not on the actuator scale.

For a ZMNF interacting with a mean flow boundary layer -the case usually met in the flow control applications- the momentum amount released by the SJ to this latter is found to be exceptionally essential for the control efficiency. Different definitions of the momentum coefficient can be found in the literature McCormick [7], Amitay *et al* [8], Tuck and Soria [9] and Seifert *et al* [10].In the present study we take the momentum coefficient as defined in Gilarranz *et al* [2]'s work:

$$C_\mu = \frac{e(\rho\, {u_{max}}^2)_{jet}}{c(\rho\, U^2)_\infty} \tag{2}$$

Where, e is the slot diameter and c is the airfoil chord length.

In the present work, the commercial software ANSYS FLUENT is used to solve the steady Reynolds-Averaged Navier-Stokes (RANS) equation for a highly turbulent flow over the two-dimensional NACA0015 airfoil under two turbulent model closure (S-A) and (K-ε). Once the turbulence model was chosen, the synthetic jet control technic is computed and then compared to the experimental data. Further topology analyses are made concerning the mean flow-synthetic jet interaction.

2. Problem position

The stall phenomenon is more or less an abrupt loss of the lift force due to the huge flow separation. Even if the stall onset is usually encountered at reduced speeds, it can occur practically at any flow velocity and depends on other parameters as the flight configuration, the load factor and the meteorological parameters. Aiming to simulate the synthetic jet actuation ability to postpone the stall's occurrence, a 2D NACA 0015 was chosen because of its symmetry, making the validation easy at a zero Angle Of Attack AOA from one hand and its abundance in the literature, on the other.

The same experimental apparatus carried out by Gilarranz *et al* [11] was reproduced to facilitate the computing validation. The airfoil is an NACA 0015 with 375 *mm* chord-wise, dotted by a 2 *mm* diameter synthetic jet actuator SJA, fitted at 12% its chord. The Reynolds number *Re* of 8.96 10^5 is reached at a free stream velocity of $U = 35 m/s$.

3. Computation procedures

The flow behaviour over the airfoil is estimated by resolving the Reynolds Averaged Navier-Stokes (RANS) equations using the commercial software (ANSYS Fluent) based on the finite volume method. The flow is supposed to be steady, viscous and two-dimensional. We assume that the flow is incompressible as the velocities encountered are far below the sound's speed maintaining the air density constant. (Mach number $M \langle 0.3$). The energy equation is decoupled, as there is no heat transfer. As the air is a light gas, we can easily neglect the volume forces and the air weight. The RANS equations constitute an open system due to the new random variables resulting from the averaged quantities. This requires additive equations to solve the system. In the present work, two turbulence first order models –well described in the literature- were used for the system closure. The first one at one transport equation called Spalart-Allmaras S-A and the second one is at two transport equations called K-ε.

For an external flow as over an airfoil, it is beneficial to define the boundaries as far as possible from the body geometry to enhance the *far field* boundary condition accuracy. A *constant velocity* defines the *inlet* condition. The different AOA are obtained in each case by varying the inlet velocity components. The wing body is defined by its adherence and set as *wall* condition. The control parameters are introduced by setting a modulated speed as a *User Defined Function UDF* to the SJA section (2 *mm*) at 12% of the upper surface using a *moving wall* condition. The controlling flow will be normal to the airfoil chord, sinusoidal in time and constant in the space. The jet speed is defined as follow:

$$v(x, y) = (0\, , Umax. \sin(\omega t)) \tag{3}$$

While the flow is considered incompressible, as said before, it is more convenient to resolve the partial differential equations using the Fluent's coupled algorithm based on the pressure. The second order upwind scheme was used to interpolate the convection terms of momentum and turbulence. In which concern the pressure-velocity coupling, the algorithm SIMPLE is used. The resolution convergence can be acceptable for all the simulations, where all the residual quantities go below a 10^{-4} order with a stabilisation of the calculated aerodynamic coefficients.

4. Grid generation

After importing the airfoil geometry into the software Gambit, the computational domain is defined and filled using a quadrangular structured 2D 'C grid' ("*quad-map*" for Gambit) by imposing the same discretization for all segments facing. This for all the simulations carried out. One notes that the airfoil discretization is performed accordingly to the profile curvature. The boundary layer is generated by a front mounted to ensure a sufficiently fine grid near the walls where the higher pressure gradients are located. The blockings are so meshed by defining the number of elements and the geometric reasons at each face. Once the ratio height-width allows, the mesh becomes coarser to alleviate the computational requirements. To reach an optimal computational accuracy, three grids were tested for both turbulence models while looking out the grid-to-solution dependency and the value of **y+**.

Table 1. *y+* values obtained at a Re of 8.96 10^5 for both K-ε and S-A models for a 18 900, 26 300 and 37 400 quadrilateral cells.

	Coarse	Medium	Fine
S-A	83	24	5
Kε	70	28	9

Simulations were done for each two degrees from 0 till 22 deg AOA using the coarse grid for the $K\varepsilon$ and the fine grid for the S-A turbulence model which correspond respectively to the relatively high value 70 of $y+$ and the moderately low value of 5. For a zero AOA, both turbulent models present a perfect symmetry between the suction and pressure surfaces with a zero lift coefficient as expected from the NACA0015 bilateral symmetry independently to the grid.

Figure 1 shows how the upper surface Cp superposes perfectly on the lower surface one and highlights the high depression near the leading edge. Then, the softening of the depression along the chord with the airfoil thickness decrease. This applies to both turbulence models. Curves coincide on the whole airfoil except at the leading edge where the stop pressure is the exception. This difference is important in view of the simplicity of this test case. The problem is due probably to the fact that the mesh is not refined enough at the leading edge.

For validation purposes, the lift coefficient is plotted in the Figure 2 for both S-A and K-ε against the Gilarranz *et al* [11]'s experimental one.

For small AOA (lower than 7deg) the two models match perfectly with the experimental data. When the airflow starts separate (higher than 7deg AOA) both simulations experience a slight gap from the experimental values. This latter becomes more and more remarkable for the nearby stall AOA.

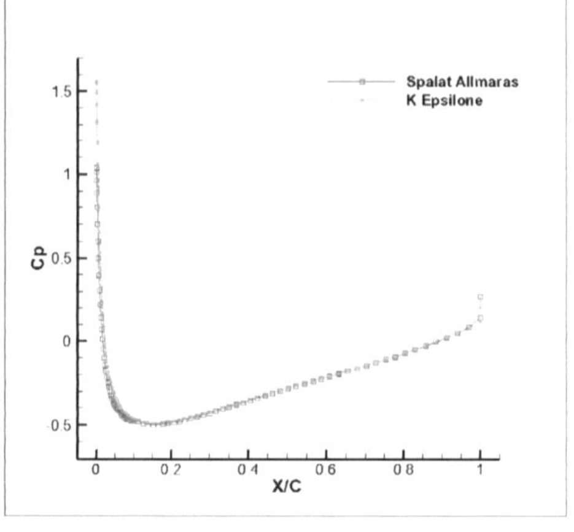

Figure 1. Pressure Coefficient Cp at zero AOA for both S-A and K-ε.

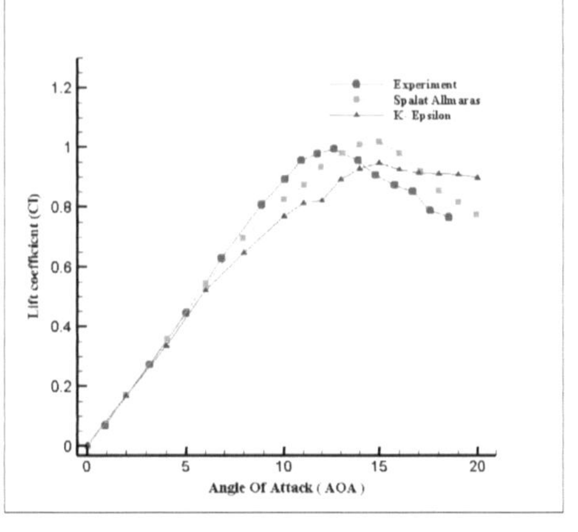

Figure 2. Lift coefficient validation, both S-A and K-ε plotted against the Gilarranz *et al* [11] data.

Table 2. Simulation validation parameters

Models	Stall AOA	Cl_{max}	$\frac{\alpha_{num}-\alpha_{exp}}{\alpha_{exp}}$	$\frac{Cl_{num}-Cl_{exp}}{Cl_{exp}}$	Slop Behaviour
Experiment	12.6	1.01			
S-A	15	1.03	1.9%	0.01	Identic
K-ε	18	0.96	4.9%	0.05	different

Table 2, summarises the different validation parameters, where α designs the AOA. Both models display a stall delay prediction with a module variation. These differences between the experimental data and the numerical simulations show that the current RANS methods cannot simulate the airfoil stall properly. However, in the present study, authors opt for the S-A model for all SJA controlled cases. Because it offers the lowest stall angle variation and a typical thick airfoil lift coefficient behaviour, contrary to the K-ε presenting a double jump behaviour, which usually characterises the thin airfoils. The pulsed slot of the SJA has been defined as a 2 *mm* skin segment at 12% of the chord. A new grid is generated by refining the mesh near the slot and conserving the previous y+ and distortion conditions.

5. Results

5.1. Aerodynamic coefficients

The SJA control strategy combines both suction and blowing control mechanisms. It against the flow separation by energising the boundary layer and therefore, overcomes the adverse pressure gradient. That can be realised at the different control cycles by adding a momentum to the weak boundary layer. At the suction cycle, the lower momentum is aspired into the actuator cavity shifting, therefore, the higher momentum closer to the airfoil surface. Then, at the blowing period, the actuator releases the same flow amount, now at higher momentum. In addition, the system frequency creates vortices cores leading to enhance the mixture region between the inner (weak) region and the outer (energised) region. This makes the boundary layer more turbulent and thereby, effectively energises it.

Figure 3 shows a ($St = 2$, $C\mu = 0.56\%$) SJA control effect on the NACA0015's lift coefficient. The flow control does not seem to bring any positive effect at the low AOA. Until 8 deg AOA, where the flow is mainly attached to the wing body, the two curves remain slightly superposed. For the higher AOA, when the separation is more obvious and affects more the flow patterns, the control strategy enhances the lift coefficient dramatically by energising the boundary layer and therefore delaying the separation occurrence. Under the present actuation parameterisation, the lift coefficient knows an extraordinary jump of about 40% from 1.01 to 1.44 for the airfoil without and with control respectively. This can be traduced practically by a higher load and action range. The stall onset is also postponed from 15deg to 19deg, giving thereby four more exploitable degrees allowing to higher manoeuvrability.

Once reaching the Cl_{max}, The two cases present different slide slope. The basic airfoil smoothly stalls as expected from a thick airfoil as the NACA0015. Here the flow separation starts from the trailing edge and overtakes the leading part when pitching to further AOA. In contrary, the controlled airfoil experiences a different behaviour. It abruptly stalls and loses his lift similarly to the leading edge stalls usually encountered for the medium aerofoils. These stalls are characterised by forming a bubble just after the leading edge. The size this latter does not vary too much with the AOA increase until it bursts under the action of the high adverse pressure leading to a more aggressive stall. The SJA control does not provide a significant effect on the lift enhancement for the AOA greater than 19 deg. This at least for the present parameterization.

In Figure 4 the drag coefficient is plotted against the lift one for both controlled and not cases. Two domains are easily distinguished. For the lower AOA, the drag force is mainly composed of a parasitic drag. The SJA makes the boundary layer more turbulent. It, therefore, procures a higher drag level without any consistent lift gain.

For higher AOA leading to a $Cl > 0.75$, the induced drag amount increases significantly for both airfoils. The flow separation leads to an additive drag clearly noticed for the uncontrolled airfoil. This extra drag supposed to result from the wake blockage and the vortex cores shedding can be largely reduced by delaying the flow separation onset, and therefore the SJA succeeds to reduce the drag coefficient dramatically. This can be explained by the fact that a laminar boundary layer induces a lower friction drag than a turbulent one. However, the latter resists more to the separation. Thus, for a separated laminar boundary layer, the viscous drag is much greater than if the boundary layer is turbulent and remained attached.

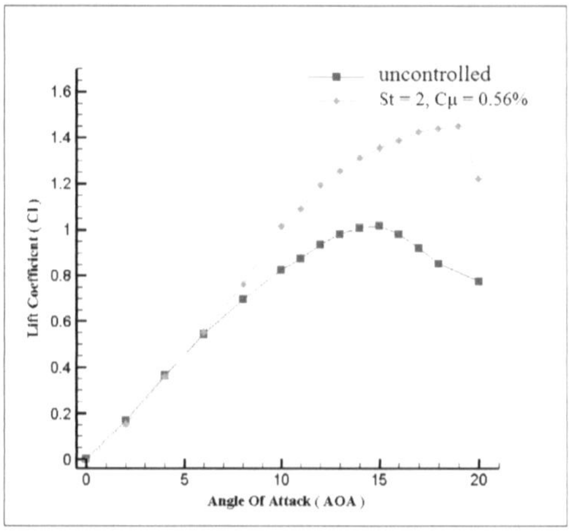

Figure 3. SJA effects on the lift coefficient at Re of 8.96 105.

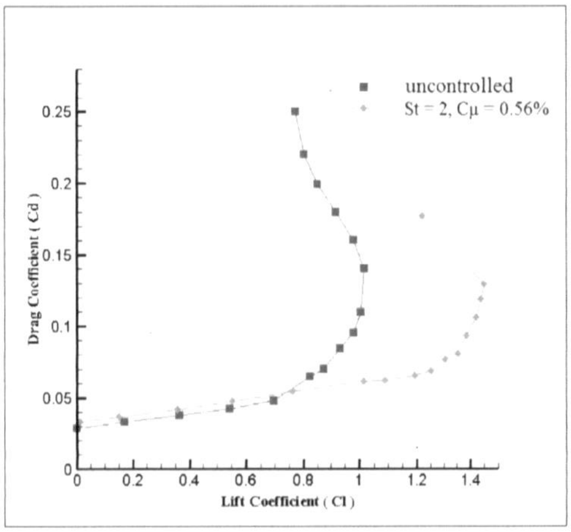

Figure 4. SJA effects on the aerodynamic performance, Drag coefficient vs. Lift coefficient.

5.2. Topological analysis of the flow created by the actuator

Figure 5a shows how the flow separates over the NACA0015 at 15deg AOA. The separation is clearly a trailing edge separation as presented by the streamlines and it causes the airfoil stall as noticed from the lift coefficient in the section above. Figure5b illustrates the ability of the current control to eliminate the flow separation all over the suction surface. The flow establishment delays then the stall occurrence and procures higher lift force with lower drag.

For better understanding of the control phenomenon, this section undertakes the interaction (free stream-synthetic jet) by limiting to the streamlines pattern at the different actuator phases focusing on the evolution of the votex topology.

Under the same control settings as the section above, at 15deg AOA Figures 6 show the near slot vortex development over one control cycle.

For, t = T / 8, the synthetic jet starts blowing in an attached boundary layer in the *y* direction as shown in (Figure 6a). The incoming flow enters in a collision with the synthetic blown detaching the boundary layer locally, and rolling a pretty strong vortex, located just downstream of the slot. The upstream counter-clockwise vortex expected to be noticed seems to be annulated by the boundary layer having a clockwise vorticity.

For, t = T / 4 as the synthetic jet intensifies until its maximum, the first vortex moves aft away from the slot, while growing as its size reaches approximately ten times the slot diameter 10 *e* of length, and 5 *e* in height releasing enough space for a second vortex generation. Upstream of the slot a strong shear zone is visible. The flow plunges to the wall just before being thrown up under the influence of the jet blowing giving birth to a new vortex structure as Figure 6b expresses.

In which concern the blow reducing phase (t = 3T / 8), as the border between the latter vortex and the jet still intense is strongly sheared and unstable, it creates a third vortex structure extending almost to the body skin. The last two vortices are weaker and smaller compared to first one generated (Figure 6c).

For t = T / 2, the vortex structures continue moving away from the slot. While the free stream is driving these structures downward, it tends to stretch them while still creeping to the wall (Figure 6d).

Contrary to the first half-period, the suction phase is manifested only by one behaviour. It is the place of the boundary layer reattachment near the slot (Figures 6 e & f) and the different vortex convection along the upper surface. The dominant element of this flow is the primary vortex, which is the largest and one whose life is the longer.

a) b)

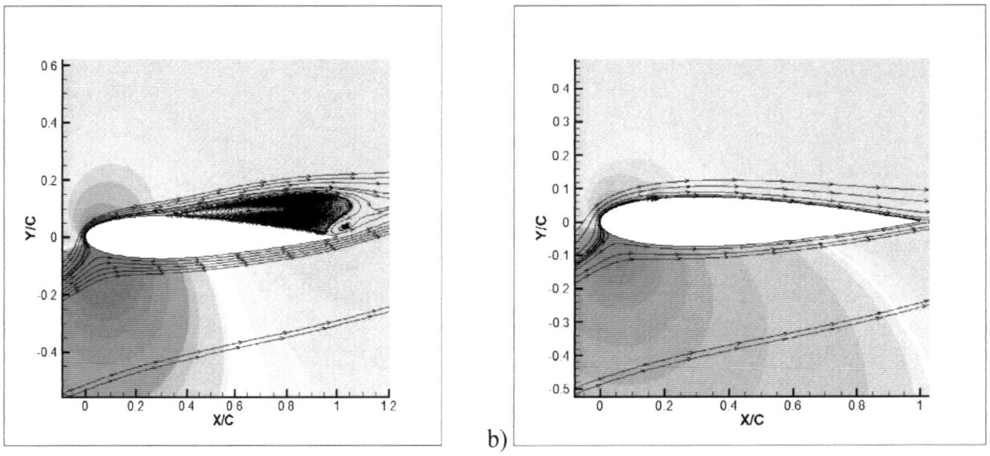

Figure 5. streamline pattern for a 15deg AOA flow: a) uncontrolled, b) a St = 2, Cμ = 0.56% SJA control

5.3. Synthetic jet control parameterisation

Synthetic jets are a parietal source of momentum to which the length and the time scales characteristics are controlled. This feature becomes remarkable in view to achieve a flow control since it becomes possible to combine the influence of two physical quantities, namely the time scale by an alternative excitation (*St*) and a spatial size by a momentum ejection rate ($C\mu$). It is common to delay the flow separation by simple boundary layer suction or momentum injection, but if we can provide an alternative jet having a frequency near those of the large unstable coherent structures. This may promote the boundary layer mixing and therefore increases the momentum amount added thereto, which can delay much more efficiently the separation phenomenon. Below will assessed the effect of the two parameters on the lift coefficient separately.

a) b)

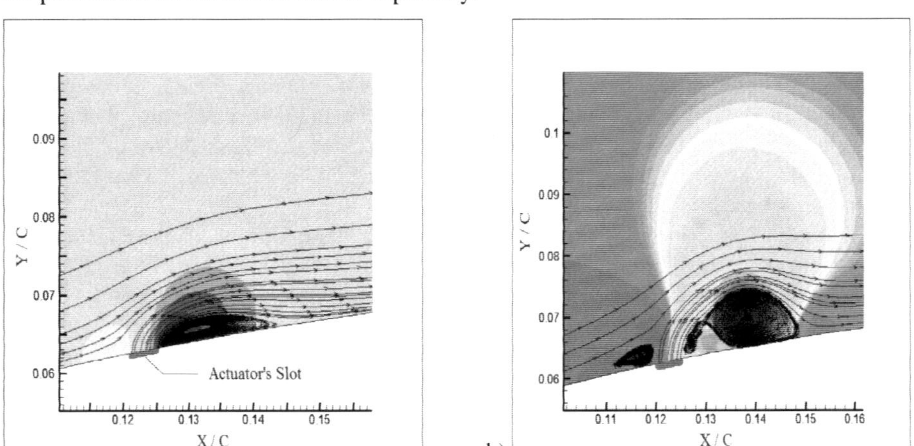

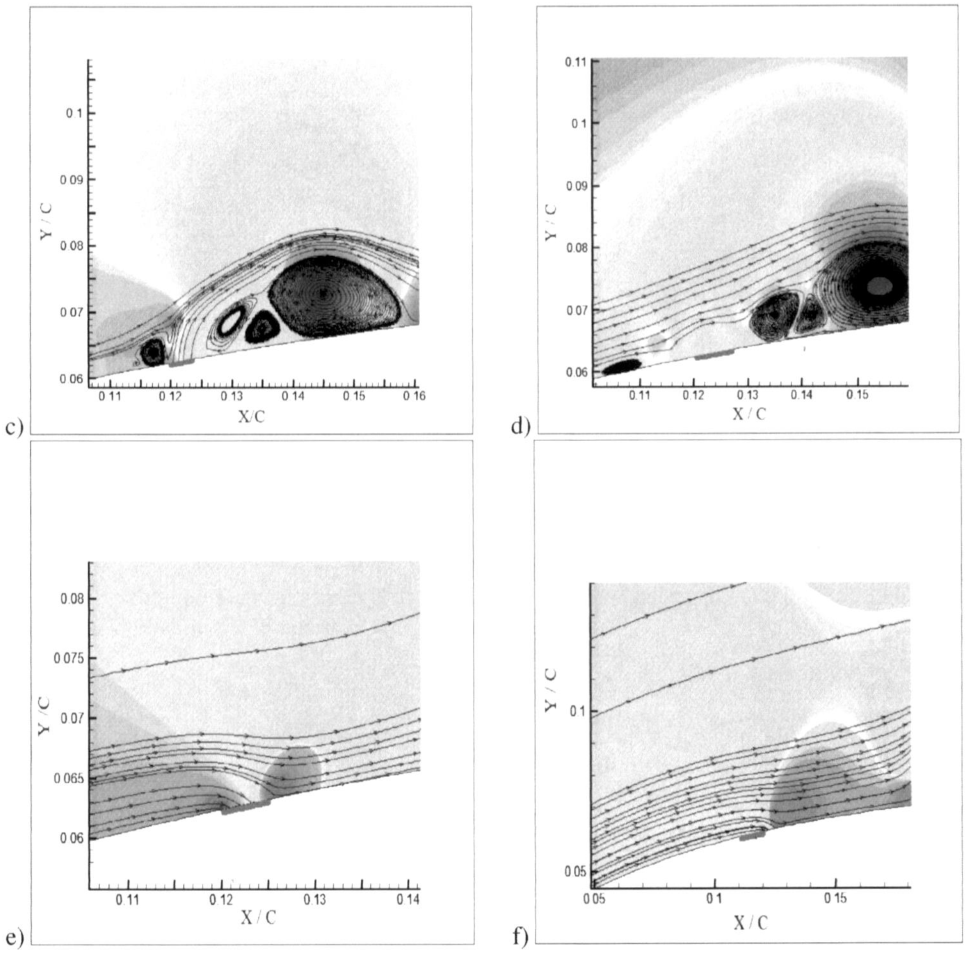

Figure 6. Formation of vortex structures at the different cycle phases: a) T/8, b) T/4, c) 3T/8, d) T/2, e) 3T/4, f) 1T.

5.3.1. Jet frequency effect. In order to distinguish the jet frequency effect on the control mechanism, various dimensionless excitation on *St* are introduced while holding the $C\mu$ at 0.56%.

Figure 7 shows the lift coefficient temporal reaction to four control frequencies *St*: 0, 0.2, 0.8 and 2. The lower control frequency seems incapable to fix the lift signal and other frequencies not related to the SJA appear as a mean flow coherent structures generation. Once the SJA reaches to lock the lift temporal as for a *St* of 0.8 and 2, the lift averaged value remains slightly the same while the signal amplitudes are proportionally shrunk according to the excitation frequency.

5.3.2. Jet momentum effect. Once the reduced operating frequency of the synthetic jet is fixed, the effect of its momentum is investigated. The blown momentum $C\mu$ is changed by varying the maximum jet amount (U_{max}). Three pic speeds U_{max} were tested: of *U, U/2* and *U/4* which correspond respectively to a $C\mu$ of 0.53%, 0.13%, 0.03%.

Figure 8 shows how the lift temporal demurs mainly random for a $C\mu$ of 0.03% where no frequency locking is noticed and the peak to peak amplitudes remains significant. This attests a control failure under this SJA parameterisation. Higher speed control U_{max} leads imperatively to a signal lock into the control excitation frequency. A proportional relation is to highlight between the temporal lift amplitudes and the momentum injected while the averaged lift remains invariant. The $C\mu$ of 0.13% which seems optimally controlling the present flow (15deg AOA) is not able to do the same at 19deg AOA where the 0.53% jet does. Higher AOA generate greater wake and larger structures, which require higher momentum to be locked.

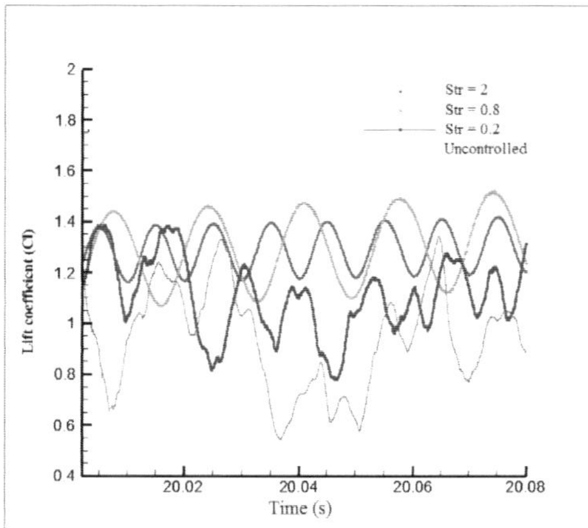

Figure 7. Lift coefficient temporal signal for different *St* excitation control.

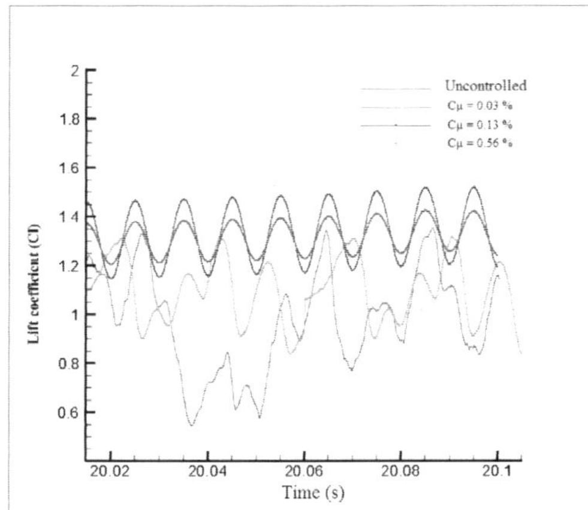

Figure 8. Lift coefficient temporal signal response to a various SJA momentum $C\mu$.

6. Conclusion

The interaction of a synthetic jet with a turbulent boundary layer "Re = 8, 96 10^5" developed over an NACA0015 was simulated over a range of 20 degrees of incidence and compared to a similar experimental apparatus data. The S-A turbulent model expressed a greater agreement with experimental data compared to the K-ε model. It provides a closer lift slope with a certain gap. These variations proportional to the AOA, probably result from the RANS limitations, including the loss of the turbulent information by averaging. The SJA control was able to expand spectacularly the flight domain, where it shifted the critical stall angle from 15deg to 19deg with a lift gain exceeding 40% accompanied by a significant drag reduction.

A control parameterization was based on the frequency and jet speed dimensionless respectively on *St* and $C\mu$. It concludes a proportional relationship between the jet momentum and the lift signal amplitudes. Contrary, the control frequencies react inversely with the temporal peak-to-peak values. Thus, it can be concluded in the current state of research that is quite difficult to design a unique grid of SJA capable of controlling the flow at different flight configurations. This is the consequence of the jet aerodynamic parameters dependence and also its position and orientation, the slot shape, the jet modulation, the *Re* and the *Mach* number and the airfoil geometry.

Perspectives are to enhance the SJA effectiveness furthermore. In this sense, the focus on a reactive smart control having the ability to vary the jet proprieties depending on the main flow state. It is well known from the bibliography that a boundary layer requires less momentum to be kept attached than to reattach from a separated state. Similarly for the frequencies required.

7. References

[1] Gad-el Hak, Mohamed & Bushnell, Dennis M 1991 Separation control: review. *Journal of Fluids Engineering* 1991; 113 (**1**), 5-30.

[2] Gilarranz J, Traub L, Rediniotis O. A new class of synthetic jet actuators—part I: design, fabrication and bench top characterization. *Journal of fluids engineering*. 2005; 127(**2**):367-76.

[3] Glezer A, Amitay M, Honohan AM. Aspects of low-and high-frequency actuation for aerodynamic flow control. *AIAA journal*. 2005; 43(**7**):1501-11.

[4] Smith BL, Glezer A. The formation and evolution of synthetic jets. *Physics of Fluids* (1994). 1998; 10(9):2281-97.

[5] Utturkar Y, Holman R, Mittal R, Carroll B, Sheplak M, Cattafesta L. A jet formation criterion for synthetic jet actuators. *AIAA paper*. 2003; 636:2003.

[6] Holman R, Utturkar Y, Mittal R, Smith BL, Cattafesta L. Formation criterion for synthetic jets. *AIAA journal*. 2005; 43(10):2110-6.

[7] McCormick D. Boundary layer separation control with directed synthetic jets. *AIAA paper*. 2000; 519:2000.

[8] Amitay M, Honohan A, Trautman M, Glezer A. Modification of the aerodynamic characteristics of bluff bodies using fluidic actuators. *AIAA paper*. 1997; 2004.

[9] Seifert A, Darabi A, Wyganski I. Delay of airfoil stall by periodic excitation. *Journal of Aircraft*. 1996; 33(4):691-8.

[10] Tuck A, Soria J, editors. Active flow control over a NACA 0015 airfoil using a ZNMF jet. *15th Australasian fluid mechanics conference*; 2004.

[11] Gilarranz J, Traub L, Rediniotis O. A new class of synthetic jet actuators—part II: application to flow separation control. *Journal of fluids engineering*. 2005; 127(2):377-87.

Numerical Investigation of the Low-Frequency Flow Oscillation over NACA-0012 Aerofoil

Yasir A. Elawad* **and Eltayeb M. Eljack**

Aeronautical Engineering Department, King AbdulAziz University, Jeddah, Saudi Arabia.

*Corresponding author's e-mail address: yasir.abuobaida@gmail.com

Abstract. The present study investigates the low-frequency flow oscillation phenomenon near stall conditions for NACA0012 aerofoil at Reynolds number of 9×10^4, Mach number of 0.4 and angle of attack of $11.0°$. The phenomenon is clearly captured in the time histories of aerodynamic coefficients. Statistical and spectral analysis are carried out for time histories of aerodynamic coefficients. Qualitative study is performed to investigate the unsteady behaviour of laminar separation bubble over one low-frequency cycle.

1. Introduction

At low Reynolds number a laminar flow separation is highly expected near the aerofoil leading edge due to strong adverse pressure gradient. At small angles of attack, the separated shear layer moves toward the aerofoil surface, where the flow is reattached to the surface. The region between the separation and reattachment locations is termed as laminar separation bubble (LSB). This bubble is classified as a short and a long bubble [1]. At higher angles of attack, the separated shear layer moves away from the surface, and the flow becomes fully separated, which is termed as bubble bursting. The leading edge stall occurs as a consequence of bubble bursting. The aerodynamic performance of the aerofoil is markedly affected, since the lift force decreases and the drag force increases. The unsteady behaviour of the LSB leads the flow to switch between stalling and non-stalling conditions, which is found to be in a quasi-periodic manner with a low frequency, and it is known as low frequency flow oscillation (LFO).

Zaman et al. [2] carried out the first study of the LFO, they found that LFO occurs if the Strouhal number (St) is less than 0.2, and when there is a trailing edge or a thin aerofoil stall. In Bragg et al. [3] study they captured the LFO using hot wire measurements in the aerofoil wake, in which they pointed out the relation between shear layer flapping and the flow oscillations. Furthermore, Broeren and Bragg [4] used conditionally averaged laser Doppler velocimetry (LDV) measurements, they showed the periodic turning between stalling and non-stalling conditions with an unsteady LSB, and the mechanism of the LSB interaction with the trailing edge separation. Rinoie and Takemura [5] studied an experiment of flow-field around a NACA-0012 aerofoil at $Re = 1.3 \times 10^5$. They captured the LFO phenomenon at $\alpha = 11.5°$, where the frequency of oscillation was 2 Hz. Tanaka [6] carried out a similar experiment of Rinoie and Takemura [5]. He used Particle Image Velocimetry (PIV) measurements, and he revealed that the amplitude of oscillations is greatest at $\alpha = 11.5°$, and the oscillation is quasi-periodic. Sandham [7] used a time accurate viscous-inviscid interaction method (VII) to study LFO phenomenon, he noticed that the dominant St of the oscillations is dependent on the aerofoil shape. Also he mentioned that the bubble bursting occurred as a consequence of potential flow/boundary-layer interactions, therefore, a simple model of boundary layer transition and turbulence could be used.

In addition, several numerical simulations were carried out to investigate the unsteadiness of LSB and the mechanism of LFO phenomenon. Almutairi and AlQadi [8] investigated the phenomenon of natural LFO using the same LES method used in [9]. They studied numerical simulations of flow around NACA-0012 aerofoil at $Re = 5 \times 10^4$, angles of attack around $\alpha = 9.30°$, and Mach number of $M_\infty = 0.4$. They found that the time history of lift coefficient exhibits a self-sustained natural LFO at $\alpha = 9.25°$ and $9.40°$ in a quasi-periodic manner. Almutairi et al. [10] performed LES for the flow over NACA-0012 aerofoil at $Re = 1.3 \times 10^5$ and $\alpha = 11.5°$, which is similar to the experiments in [5] and [6]. They showed the time histories of the aerodynamic coefficients in which the LFO phenomenon was featured in a quasi-periodic manner. They found that the dominant Strouhal number was 0.00826 which is comparable to that of the experiment [5].

2. Mathematical Model and Computational Setup

In the present study, LES is carried out to simulate the flow field around NACA-0012 aerofoil. The LES requires decomposition of the flow field in which the large scales are resolved and the small scales are modelled using sub-grid scale model (SGS). The mixed-time-scale (MTS) model is used as SGS model, this model was developed by Inagaki et al. [11]. The model is constructed without using any explicit wall-damping function to overcome the drawbacks of other SGS models. A fourth-order accurate central difference scheme is used for the spatial discretization of the internal points and a Carpenter fourth order accurate scheme, Carpenter et al. [12], is applied to treat points near the boundary Jones [13] and Jaber [14]. A fourth-order Runge-Kutta explicit scheme is used for the temporal discretization.

The spatial filter is employed to the flow-field for small scales without disturbing large scales in order to overcome grid to grid oscillations. Visbal and Rizzeta [15] found that in order to maintain numerical stability, it is convenient to use compact schemes of order equal or greater than the order of spatial discretization. Therefore, an explicit fourth order tri-diagonal filter is applied and the conservative flow field variables are filtered in x, y and then in z direction.

The domain and curvilinear coordinate system are illustrated in figure 1. A NACA-0012 aerofoil is oriented at an angle of attack of 11.0°. The dimensions $W = 5c$, $L_\eta = 7.3c$ and $L_\zeta = 0.5c$ refer to the wake length, C-grid radius, and the domain length in span-wise direction, respectively. Where c is the aerofoil chord.

Adequate grid resolution is required in order to resolve the large eddies in the flow-field. A C-grid of $832 \times 351 \times 167$ is constructed around the aerofoil in curvilinear coordinates ξ, η, and ζ, respectively. The wall viscous units (y^+, Δx^+ and Δz^+) are usually used to examine the grid resolution. The first grid point was allocated at $y_1^+ < 0.7$ in the wall-normal direction. The grid resolution in the stream-wise and span-wise directions are $\Delta x^+ < 13$ and $\Delta z^+ < 11$, respectively.

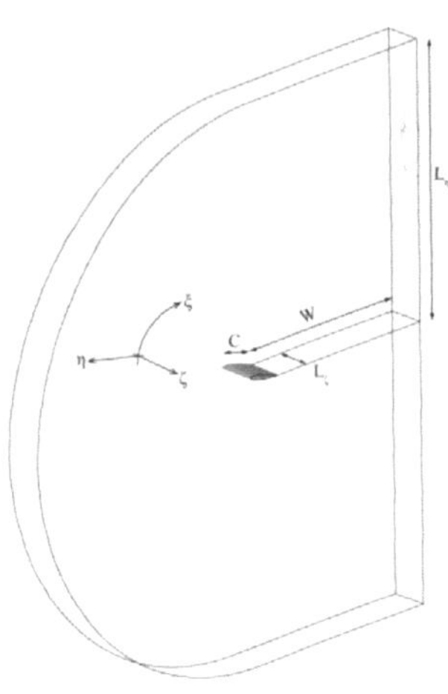

Figure 1. Domain

At the free-stream and far-field boundaries, the integral characteristic boundary condition [16] is applied, while at the downstream exit zonal characteristic boundary condition [17] is applied. An adiabatic, no slip condition is applied at the aerofoil surface. A periodic boundary condition is applied in the span-wise direction for each sub-step of the Runge-Kutta time steps.

3. Results and Discussion

In this study, LES is carried out for the flow-field over NACA-0012 aerofoil at Reynolds number of 9×10^4, Mach number of 0.4, and angle of attack of $11.0°$. The free-stream was set parallel to the horizontal axis. The time-step is set to 10^{-4} non-dimensional time units. No artificial forcing is applied, whilst a clean free-stream conditions are used. The LFO phenomenon is captured in the time history of lift coefficient as shown in figure 2. A natural self-sustained oscillation is exhibited at $\alpha = 11.0°$ in a quasi-periodic manner. The flow is oscillating around a mean of 0.7104 between the stalled and non-stalled conditions with a low frequency of 0.0394 and Strouhal number of 0.0075. The observed randomness and irregularity in the cycle are due to the unsteady nature of the transition process and the subsequent turbulent flow.

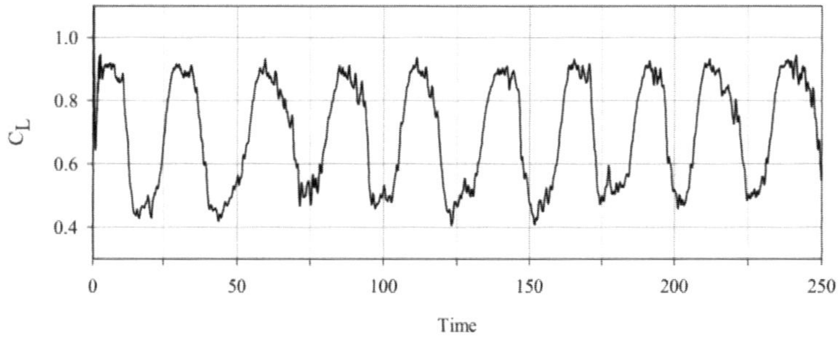

Figure 2. Time history of the lift coefficient (C_L).

The spectrum of the lift coefficient features the dominant frequency modes on the flow-field. The spectrum is estimated using the fast Fourier transform (FFT) algorithm. The lift spectrum is presented in figure 3. The lift spectrum features a low frequency mode, $St = 0.0075$, which is in the same order of magnitude as that of the experimental measurements [5].

The instantaneous iso-surface of Q-criterion ($Q = 200$) coloured by the total energy per unit mass (E) is illustrated in figure 4 for both attached and separated flow phases. Figure 4a shows the early transition process where the shear layer is attached. The shear layer is a little bit distorted and a two dimensional scales convert to a large three dimensional structures followed by a breakdown and consequently the flow becomes turbulent. The separated flow phase, figure 4b, shows a late transition process. The shear layer is detached from the surface towards the free-stream and the breakdown process of the two dimensional scales moves downstream.

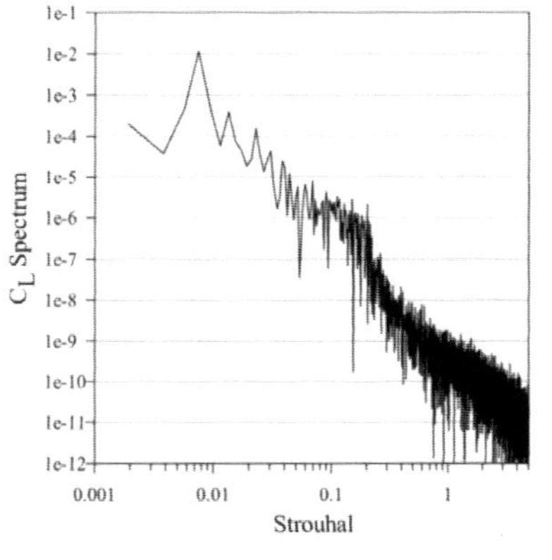

Figure 3. Lift coefficient spectrum.

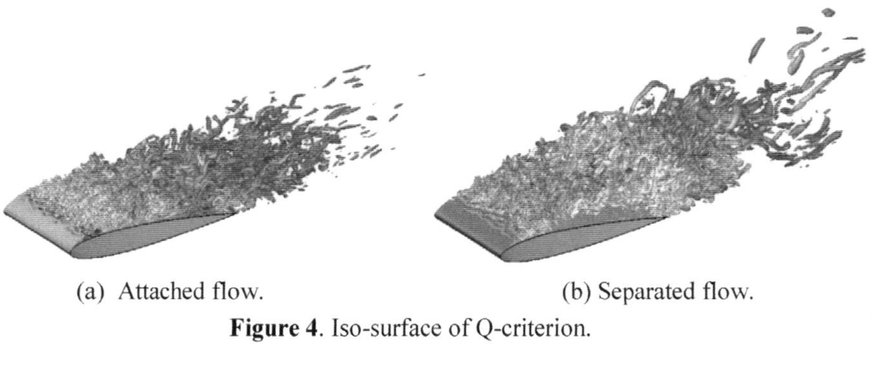

(a) Attached flow. (b) Separated flow.

Figure 4. Iso-surface of Q-criterion.

2D $x - y$ planes of data are sampled every 50 time-steps, non-dimensional sampling frequency of 204. The sampled data contains time and span-wise ensemble averaged velocity components, pressure, Reynolds stresses and density. The sampling started after the flow was fully developed and the flow became stationary in time. The starting sampling time was $t = 210$ non-dimensional time units and it ended at $t = 240$, covering one LFO cycle. The data was locally averaged over sixty subintervals and each subinterval consist of 5,000 time steps (0.5 non-dimensional time units).

Figure 5 shows streamlines for the complete LFO cycle. At the beginning of the cycle, $\phi = 18°$, a short bubble forms near the aerofoil leading-edge on the suction side. The short bubble grows in length and height in time as shown in figure 5b. The bubble continue to grow in size until it eventually burst and a fully separated flow is observed at $\phi = 138°$. The flow remains separated for $\phi = 198°$ and $258°$ and the bubble size varies with time. The flow reattaches to the aerofoil surface at $\phi = 318°$ and the bubble becomes identical to that observed at $\phi = 18°$. The cycle continue in a quasi-periodic manner. This behaviour is in agreement with Rinoie and Takemura [5] and Tanaka [6] observations in which they showed flow visualization. The behaviour also agrees very well with Almutairi et al. [10].

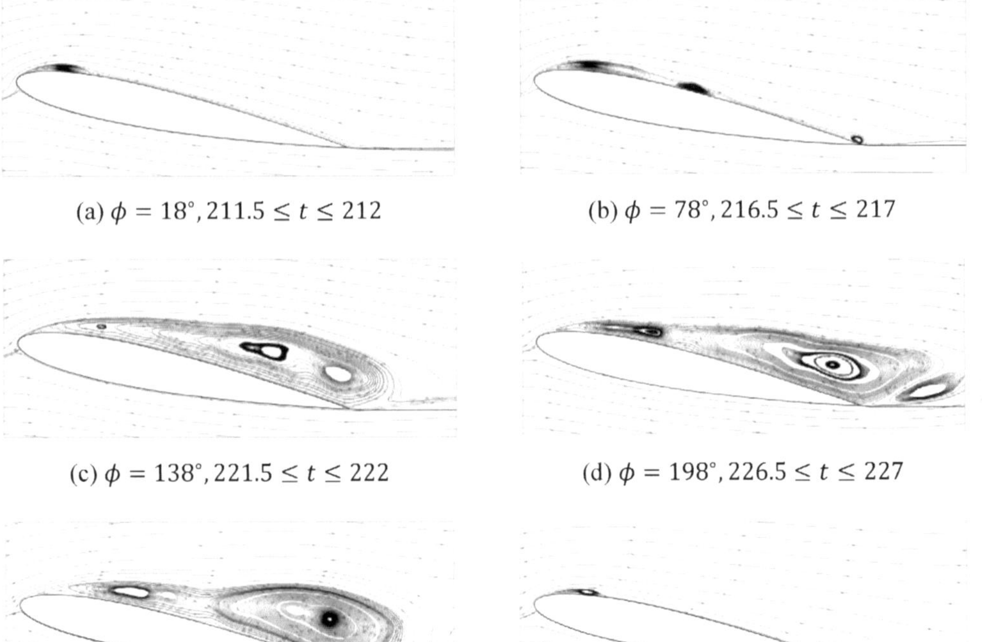

(a) $\phi = 18°, 211.5 \leq t \leq 212$

(b) $\phi = 78°, 216.5 \leq t \leq 217$

(c) $\phi = 138°, 221.5 \leq t \leq 222$

(d) $\phi = 198°, 226.5 \leq t \leq 227$

(e) $\phi = 258°, 231.5 \leq t \leq 232$

(f) $\phi = 318°, 236.5 \leq t \leq 237$

Figure 5. Streamlines of the flow field over one LFO

The averaged pressure coefficient (C_p) over the previously selected subintervals is plotted in figure 6. The pressure decreases drastically at the aerofoil leading-edge due to the curvature shape as the velocity tends to increase. A strong adverse pressure gradient is present near the aerofoil leading edge at the beginning of the cycle ($\phi = 18°$) which causes the flow separation, followed by a small declining and increment in the pressure for a small distance where the flow is reattached and the bubble is formed. Downstream the bubble there is a rapid increase in the pressure due to the turbulent flow. Further downstream C_p remains almost constant. A slight reduction of C_p with time is observed when the bubble burst (from $\phi = 48°$ to $\phi = 238°$). During the fully separated flow phase the C_p distribution is almost flat along the aerofoil suction side ($\phi = 258°$). The C_p increases at $\phi = 288°$ which is indicative that the flow has reattached. The C_p continue to increase until the short bubble is formed near the leading edge at $\phi = 348°$. The C_p behaviour is in agreement with the experimental work of [5] and [6] and the numerical work of [10].

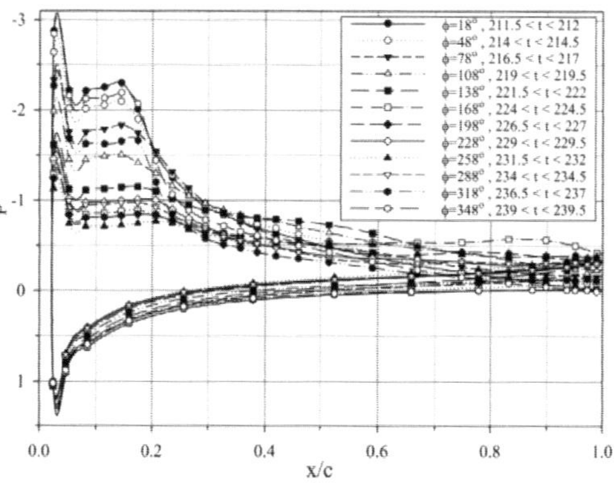

Figure 6. Pressure coefficient.

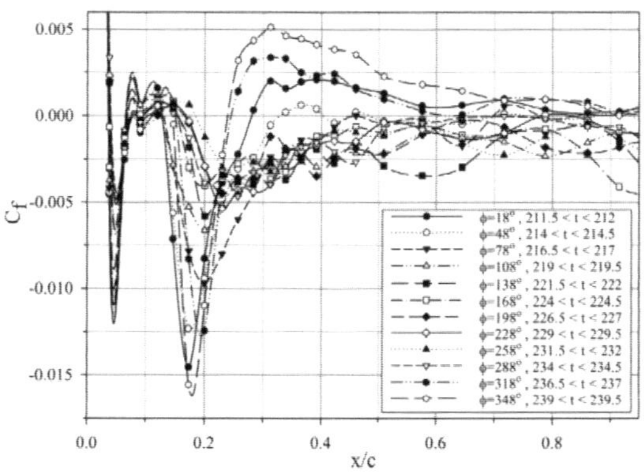

Figure 7. Skin friction coefficient.

The averaged friction coefficient C_F over the selected subintervals is shown in figure 7. Maximum values of C_F are observed at the leading-edge all over the cycle. The C_F is then dropped sharply near the leading-edge where the averaged C_F crossed the $x-$ axis. This is indicative that the C_F has negative values and the flow is reversed, i.e., separated. The C_F continued to decrease until it reached minimum values in the interval $0.04 \leq x/c \leq 0.05$, separation locations over the cycle. After that the C_F switched to positive values indicating flow reattachment locations and recirculation zone. The relatively small recirculation zone features a small counter-clockwise vortex as shown in figure 8. The C_F decreased again and reached a second negative peak in the interval $0.18 \leq x/c \leq 0.285$ followed by a sharp increase forming a larger recirculation zone constituting laminar separation bubble. The second recirculation region has a higher velocity gradient than the first recirculation region. At $\phi = 318°$ and $348°$ the flow became similar to that at $\phi = 18°$ and $48°$ where the flow is fully attached except for a short laminar separation bubble near the leading edge.

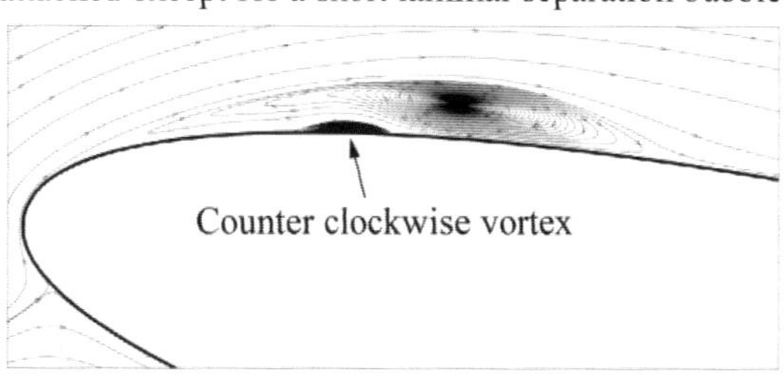

Figure 8. The counter-clockwise vortex.

4. Conclusion

In the current study, the low frequency flow oscillation phenomenon is captured using LES technique for the flow field over NACA-0012 aerofoil at Reynolds number of 9×10^4, Mach number of 0.4 and angle of attack of 11.0°. The time history of the lift coefficient exhibited the natural self-sustained LFO phenomenon, where the flow is switching between attached flow and separated flow phases. The lift coefficient spectrum showed that the dominant low frequency mode in the flow-field is at Strouhal number of 0.0075, which is in agreement with experimental measurements.

5. References

[1] PR Owen and L Klanfer. On the laminar boundary layer separation from the leading edge of a thin aerofoil. Citeseer, 1953.

[2] KBMQ Zaman, DJ McKinzie, and CL Rumsey. A natural low-frequency oscillation of the flow over an airfoil near stalling conditions. Journal of Fluid Mechanics, 202:403-442, 1989.

[3] BM Bragg, Douglas C Heinrich, FA Balow, and KBMQ Zaman. Flow oscillation over an airfoil near stall. AIAA journal, 34(1):199-201, 1996.

[4] AP Broeren and MB Bragg. Flowfield measurements over an airfoil during natural low-frequency oscillations near stall. AIAA journal, 37(1):130-132, 1999.

[5] K Rinoie and N Takemura. Oscillating behaviour of laminar separation bubble formed on an aerofoil near stall. Aero-nautical Journal, 108(1081):153-163, 2004.

[6] Hiroyuki Tanaka. Flow visualization and piv measurements of laminar separation bubble oscillating at low frequency on an airfoil near stall. In 24th International Congress of the Aeronautical Sciences, pages 1-15, 2004.

[7] ND Sandham. Transitional separation bubbles and unsteady aspects of aerofoil stall. Aeronautical Journal, 112(1133):395-404, 2008.

[8] Jaber H Almutairi and Ibraheem M AlQadi. Large-eddy simulation of natural low-frequency oscillations of separating-reattaching ow near stall conditions. AIAA journal, 51(4):981-991, 2013.

[9] Jaber H Almutairi, Lloyd E Jones, and Neil D Sandham. Intermittent bursting of a laminar separation bubble on an airfoil. AIAA journal, 48(2):414-426, 2010.

[10] J AlMutairi, I AlQadi, and E ElJack. Large eddy simulation of a naca-0012 airfoil near stall. In Direct and Large-Eddy Simulation IX, pages 389-395. Springer, 2015.

[11] Masahide Inagaki, Tsuguo Kondoh, and Yasutaka Nagano. A mixed-time-scale sgs model with fixed model-parameters for practical les. Journal of uids engineering, 127(1):1-13, 2005.

[12] Mark H Carpenter, Jan Nordstrom, and David Gottlieb. A stable and conservative interface treatment of arbitrary spatial accuracy. Journal of Computational Physics, 148(2):341-365, 1999.

[13] Lloyd Edward Jones. Numerical studies of the flow around an airfoil at low Reynolds number. PhD thesis, University of Southampton, 2008.

[14] Jaber Almutari. Large-eddy simulation of ow around an airfoil at low Reynolds number near stall. PhD thesis, University of Southampton, 2011.

[15] Miguel R Visbal and DP Rizzetta. Large-eddy simulation on curvilinear grids using compact di_erencing and _ltering schemes. Journal of Fluids Engineering, 124(4):836-847, 2002.

[16] HS Sandhu and ND Sandham. Boundary conditions for spatially growing compressible shear layers. Rept. QMW-EP-1100, Faculty of Engineering, Queen Mary and Westfield College, Univ. of London, London, 1994.

[17] Richard D Sandberg and Neil D Sandham. Nonreecting zonal characteristic boundary condition for direct numerical simulation of aerodynamic sound. AIAA journal, 44(2):402-405, 2006.

IOP Conf. Series: Materials Science and Engineering **187** (2017) 012009 doi:10.1088/1757-899X/187/1/012009

Laminar Flame Velocity and Temperature Exponent of Diluted DME-Air Mixture

Abdul Naseer Mohammed, Muzammil Anwar, Khalid A Juhany, Akram Mohammad

Department of Aeronautical Engineering, King Abdulaziz University, Jeddah, Saudi Arabia

akramnmania@gmail.com

Abstract. In this paper, the laminar flame velocity and temperature exponent diluted dimethyl ether (DME) air mixtures are reported. Laminar premixed mixture of DME-air with volumetric dilutions of carbon dioxides (CO_2) and nitrogen (N_2) are considered. Experiments were conducted using a preheated mesoscale high aspect-ratio diverging channel with inlet dimensions of 25 mm $\times$ 2 mm. In this method, flame velocities are extracted from planar flames that were stabilized near adiabatic conditions inside the channel. The flame velocities are then plotted against the ratio of mixture temperature and the initial reference temperature. A non-linear power law regression is observed suitable. This regression analysis gives the laminar flame velocity at the initial reference temperature and temperature exponent. Decrease in the laminar flame velocity and increase in temperature exponent is observed for CO_2 and N_2 diluted mixtures. The addition of CO_2 has profound influence when compared to N_2 addition on both flame velocity and temperature exponent. Numerical prediction of the similar mixture using a detailed reaction mechanism is obtained. The computational mechanism predicts higher magnitudes for laminar flame velocity and smaller magnitudes of temperature exponent compared to experimental data.

1. Introduction

With the recent progress in the implementation of alternative fuels, dimethyl ether (DME) has emanated as one of the prominent clean fuels [1]. It is the simplest of ethers with high reactivity due to higher cetane number ($\approx$60), low boiling point (-25 °C) and low auto-ignition temperature. It reduces the formation of particulates due to the presence of high oxygen content (about 35% by mass), no carbon-carbon bond and low C:H ratio which results in almost smoke-free combustion [1].

The flame velocity of DME-air/diluent mixtures has been reported by very few. Daly et al. [2] used the combustion bomb method to report the flame velocity of the DME-air mixture. Zhao et al. [3] used the stagnation flame burner method to calculate the flame velocity of the DME-air mixture with the help of PIV technique, the flame velocity obtained through the stagnation flame burning is slightly higher than the spherical bomb method as reported by Zhao et al. [3]. This might be due to the linear exploration methods been used for specifying the stretch effect. Qin and Ju [4] reported the flame velocity for the DME-air diluent mixture at elevated pressure by using a constant pressure bomb method and Schlieren system to capture the images and found to reasonably good when compared with the results of Daly et al. [2] and Zhao et al. [3] at ambient conditions. Recently Huang et al. [5] and De Vries et al. [6] measured the flame velocity of DME-air diluent mixtures by using a constant volume spherical bomb and a high-speed Schlieren system and found much lower than the flame velocity obtained through PIV technique.

The exercise of inert gasses as diluents for DME-air combustion elucidates the increasing array of experimental and computational studies on combustion characteristics such as flame propagation and extinction. The combustion of DME in the presence of these gasses was studied in connection with the EGR (exhaust gas recirculation) technique, which was intended to lower combustion instabilities and reduce NOx emissions and the rate of heat transfer in engines. Considering this growing interest in reliable values of laminar flame velocities for DME-air under various initial conditions of temperatures and dilutions becomes very essential. The dilution gasses result in a decrement of the flame velocity of fuel-air mixtures. However, studies in literature were limited to ambient temperature condition only and can worsen at elevated temperatures.

Considering the practical application of diluted DME-air mixtures at higher mixture temperature in exhaust gas recirculation conditions, the objective of this paper is to measure the laminar flame velocity of CO_2/N_2 diluted stoichiometric DME-air mixtures at elevated temperatures. A preheated mesoscale diverging channel method validated by Akram et al. [7] was used for measurements.

2. Experimental method

2.1. Experimental setup

Direct image of the experimental setup is shown in Fig. 1. A mesoscale channel with a 10° divergent

Figure 1. High aspect ratio diverging channel with external heater

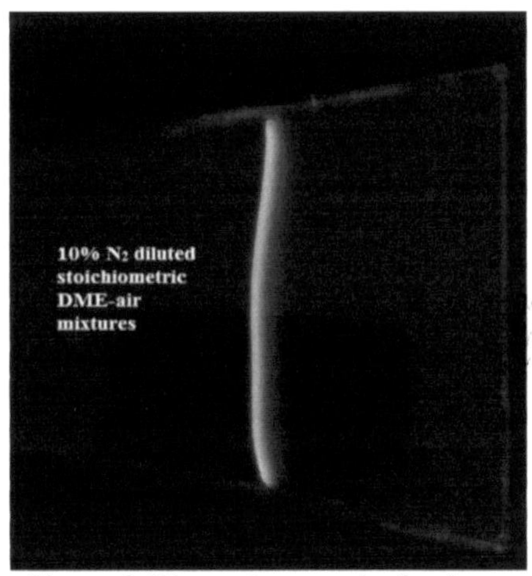

Figure 2. Direct photograph of a planar flame

angle, and inlet dimension of 25 mm × 2 mm is used as referred by [7]. The channel is made up of fused quartz. An external heater is used to preheat the channel and stabilize the flames at different initial temperatures. The temperature inside the channel due to this heating is noted apriory in both axial and transverse directions by supplying constant airflow through the channel. The temperature of the channel walls at the location of the stabilized flame is measured using a K-type thermocouple.

The diluted stoichiometric mixture is fed to the channel through electric mass flow controllers (MFCs). The fresh mixture is ignited at the outlet of the channel. The flame moves upstream and stabilizes at a location where the mixture velocity matches with its flame velocity at that particular temperature. Once stabilized, the flame remains stationary at the same location. Direct images of the stabilized flame are taken with a digital camera. One of such photographs is shown in Fig. 2 taken for a nitrogen-diluted stoichiometric DME-air mixture. These flames were planar in both crosswise and depth direction of the channel. Other details of the setup and method can be found in [7].

Measurement of flame position is also carried out extinguishing the flame for each case. The experiments are repeated to confirm the reproducibility of the method and data.

2.2. Computational method

The freely propagating steady adiabatic flames were simulated to compute the laminar flame velocity of the different mixtures. The PREMIX code [8] of CHEMKIN II package [9] was used for computations. Zhao et al. mechanism [10] is used for laminar flame velocity calculations of diluted mixtures. The thermodynamic and transport properties were taken from supplementary data available in ref. [10] Thermal and multi-component diffusion were considered. The upwind difference scheme is chosen for the study, which permits for the adaptive grid refinement. Zhao et al. [3] have reported some discrepancies in the computational values due to the degree of refinement of the numerical domain with grid adaptations. Accurate and the grid independent flame velocities were confirmed with very refined mesh using GRAD = 0.02 and CURV = 0.03.

3. Results and Discussion

3.1. Effect of temperature on laminar flame velocity

The change of laminar flame velocity for CO_2/N_2 diluted stoichiometric DME-air mixture with three different percentage dilution rates of Φ_d= 10, 20 and 30 % is observed from measurements. The experiments were carried over a range of temperatures (330-570 K). The initial unburned mixture temperature $T_{u,o}$ and pressure for the entire experiments are maintained at 300 K and 1 atm respectively. The magnitudes of laminar flame velocity and the temperature exponent of pure stoichiometric DME-air mixture were taken from Varghese et al. [11] for comparison purpose.

The laminar flame velocity variation for pure, 30 % N_2 and 30 % CO_2 diluted stoichiometric DME-air mixtures at elevated mixture temperatures is shown in Figs. 3 and 4 respectively. Symbols with continuous lines show the experimental data (power law fits) while dashed lines show the computational data obtained using Zhao et al. mechanism [10]. Experimental data are associated with their respective uncertainty calculated as per the method reported in [7] for this experimental method. The continuous line shows the power law correlation fitted on the measured experimental data to obtain the values of temperature exponent α and reference flame velocity S_{lo} at temperature $T_{u,o}$. The flame velocity of the diluted mixture increases with an increase in the temperature for both the

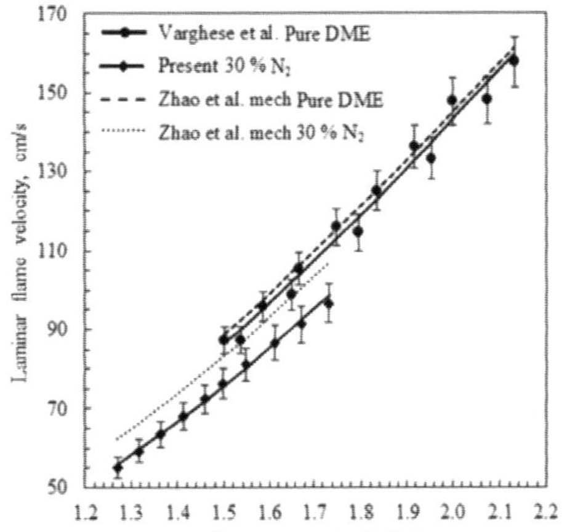

Figure 3. Laminar flame velocity of pure and 30 % N_2 diluted stoichiometric DME-air mixtures at elevated temperatures.

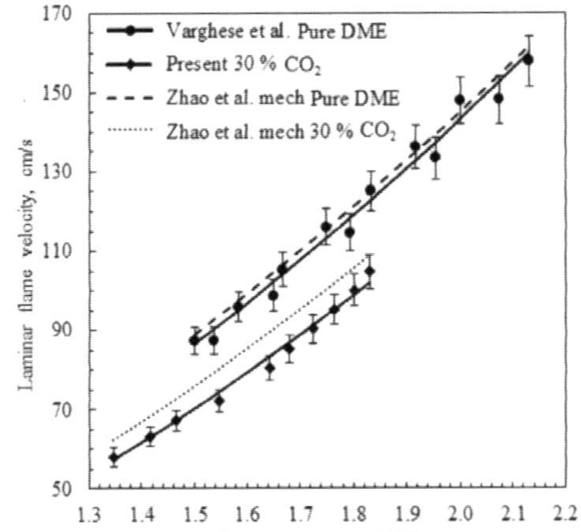

Figure 4. Laminar flame velocity of pure and 30 % CO_2 diluted stoichiometric DME-air mixtures at elevated temperatures.

mixtures. The computations predict higher magnitudes of laminar flame velocity for same initial mixture temperatures for diluted mixtures.

3.2. Effect of dilution on laminar flame velocity

The laminar flame velocity for DME-air mixture falls with an increase in dilution ratio Φ_d of an inert gas at all temperatures. Figure 5 shows the flame velocity of stoichiometric DME-air mixture at 300 K with the addition of different diluents. The addition of chemically inert diluents witnessed to reduce the flame velocity. This could be due to drop in the molar fraction of oxygen and fuel in the presence of inert diluents, lowering the likelihood of molecular collisions leading to the formation of final combustion products. Moreover, mixture dilution reduces the flame temperature as a result of increase in the specific heat of the mixture. These lead to the decrease the chemical reaction rate and hence the flame velocity. Thus diluent acts as better flame extinguishers.

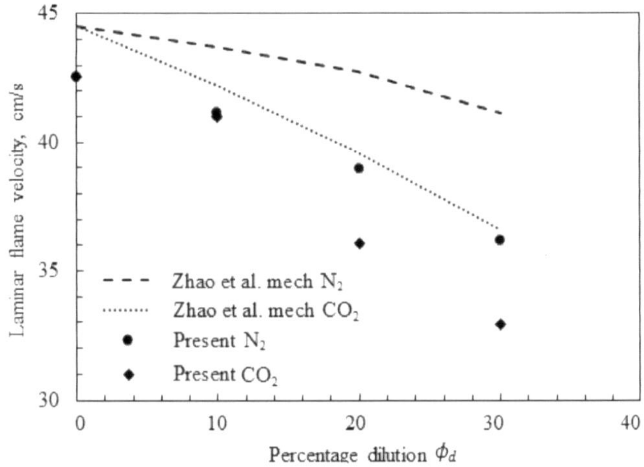

Figure 5. Effect of dilution on the flame velocity of the stoichiometric DME-air mixture at the mixture temperature of 300 K.

For a specific percentage of dilution, the laminar flame velocity with CO_2 dilution gives a smaller value than that with N_2 dilution, as can be seen in Fig. 5. This is due to the fact that specific heat ratio of CO_2 (triatomic molecular gas) is greater than that of N_2 (diatomic molecule gas) and hence, CO_2 will absorb the more heat released and consequently more drop in the flame temperature compared to N_2.

3.3. Effect of dilution on temperature exponent

Figure 6 shows the increase in the temperature exponent with the addition of dilution. Zhao et al. mechanism [10] slightly over-predicts the flame velocity at 300 K and under-predicts the temperature exponent. Under a specific percentage dilution, a higher increase in temperature exponent is observed for CO_2 addition as compared to N_2 addition case.

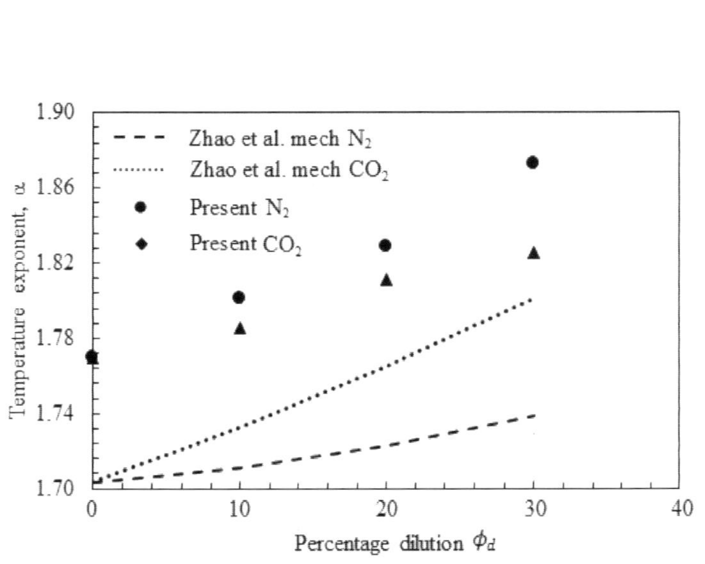

Figure 6. Variation of the temperature exponent of the stoichiometric DME-air with the addition of CO_2 and N_2 dilution at ambient conditions.

4. Conclusion

The effect of inert gas (N_2 and CO_2) dilutions on the laminar flame velocity of stoichiometric DME-air mixtures at elevated temperatures is computed and validated against the experimental results. Reaction mechanism slightly over-predicts the laminar flame velocity and over-predicts the temperature exponent for higher dilution cases. Dilution of these gases decreases the flame velocity of DME-air mixtures significantly. CO_2 dilution has a better extinguishing effect compared to N_2 dilution. A higher dilution of inert gases to the DME as in EGR should be considered cautiously.

5. References

[1] Verbeek R P, Doorn A van and Walwijk M van 1996 Global assessment of Dimethyl-ether as an automotive fuel (second edition)

[2] Daly C A, Simmie J M, Würmel J, Djebaïli N and Paillard C 2001 Burning velocities of dimethyl ether and air *Combust. Flame* **125** 1329–40

[3] Zhao Z, Kazakov A and Dryer F L 2004 Measurements of dimethyl ether/air mixture burning velocities by using particle image velocimetry *Combust. Flame* **139** 52–60

[4] Qin X and Ju Y 2005 Measurements of burning velocities of dimethyl ether and air premixed flames at elevated pressures *Proc. Combust. Inst.* **30** 233–40

[5] Huang Z, Wang Q, Yu J, Zhang Y, Zeng K, Miao H and Jiang D 2007 Measurement of laminar burning velocity of dimethyl ether-air premixed mixtures *Fuel* **86** 2360–6

[6] De Vries J, Lowry W B, Serinyel Z, Curran H J and Petersen E L 2011 Laminar flame speed measurements of dimethyl ether in air at pressures up to 10 atm *Fuel* **90** 331–8

[7] Akram M, Kishore V R and Kumar S 2012 Laminar Burning Velocity of Propane/CO2/N2 –Air Mixtures at Elevated Temperatures *Energy & Fuels* **26** 5509–18

[8] Kee R, Grcar J, Smooke M, Miller J and Meeks E 1985 PREMIX : A FORTRAN Program for Modeling Steady Laminar One-Dimensional *SANDIA Natl. Lab. ...* 1–87

[9] Robert J 1991 *Chemkin-II: A Fortran chemical kinetics package for the analysis of gas-phase chemical kinetics*

[10] Zhao Z, Chaos M, Kazakov A and Dryer F L 2008 Thermal decomposition reaction and a comprehensive kinetic model of dimethyl ether *Int. J. Chem. Kinet.* **40** 1–18

[11] Varghese R J, Kishore V R V, Akram M, Yoon Y and Kumar S 2017 Burning velocities of DME (dimethyl ether)-air premixed flames at elevated temperatures *Energy* **in review**

Experimental Study of an On-board Fuel Tank Inerting System

Fei Wu[1], Guiping Lin[1], Yu Zeng[1,*], Rui Pan[1] and Haoyang Sun[1]

[1] School of Aeronautic Science and Engineering, Beihang University, Beijing 100191, China

[*]zengyu@buaa.edu.cn

Abstract. A simulated aircraft fuel tank inerting system was established and experiments were conducted to investigate the performance of the system. The system uses hollow fiber membrane which is widely used in aircraft as the air separation device and a simplified 20% scale multi compartment fuel tank as the inerting object. Experiments were carried out to investigate the influences of different operating parameters on the inerting effectiveness of the system, including NEA (nitrogen-enriched air) flow rate, NEA oxygen concentration, NEA distribution, pressure of bleeding air and fuel load of the tank. Results showed that for the multi compartment fuel tank, concentrated flow washing inerting would cause great differences throughout the distribution of oxygen concentration in the fuel tank, and inerting dead zone would exist. The inerting effectiveness was greatly improved and the ullage oxygen concentration of the tank would reduce to 12% successfully when NEA entered three compartments evenly. The time span of a complete inerting process reduced obviously with increasing NEA flow rate and decreasing NEA concentration, but the trend became weaker gradually. However, the reduction of NEA concentration will decrease the utilization efficiency of the bleeding air. In addition, the time span can also be reduced by raising the pressure of bleeding air, which will improve the bleeding air utilization efficiency at the same time. The time span decreases linearly as the fuel load increases.

1. Introduction

Aircraft fuel system is an essential part of an aircraft. There is a large amount of flammable fuel vapor and air mixture in the gas space of the aircraft fuel tank. This mixture is easy to fire or explosion with an ignition source case which will cause serious crash. FAA (Federal Aviation Administration) research report published shows the commercial aircraft fuel tank will be safe when the ullage oxygen concentration is less than 12%, even if there is a fire in the fuel tank it will not be lit [1]. The main function of the aircraft fuel system is to reduce the risk of combustion or explosion of the fuel tank [2].

At present, the working mode of aircraft fuel inerting system widely used in the world is that the bleeding air is separated into nitrogen en-riched air (NEA) and oxygen enriched air through hollow fiber membrane, and then NEA is passed into the fuel tank to reduce the ullage (empty space above the fuel) oxygen concentration[3]. The effectiveness of the system is effected with the bleeding air pressure, NEA flow rate, NEA oxygen concentration and flow distribution. At the same time, if the fuel load is different, the inerting effect will be different [4]. According to the inertia of the aircraft fuel tank, the time required for the ullage oxygen concentration reducing to the safe range is very important in predicting the safety of the fuel tank [5-9].

Burns from FAA carried out theoretical and experimental research on rectangular fuel tank with single NEA vent and single air vent, considering the NEA flow, NEA concentration and the fuel tank temperature effect of the inerting effect. The study showed that VTE was 1.5 to decrease the ullage oxygen concentration to 8% with NEA of 95% concentration [10]. In order to study the influence of ventilation mode, William established an experimental research platform using a 24% scale CWT as the research object, whose top cover can move, thus changing the NEA distribution the ventilation mode. They obtained an optimal NEA distribution mode through a large number of experiments, which can reduce the amount of total NEA under the same inerting conditions [11]. Airbus and NASA installed this system in the warehouse of A320 and Boeing 747SP respectively, and the inerting effectiveness was studied [12]. The results showed that it's effective to inert the CWT by OBIGGS. However, the specific influence of bleeding air pressure, NEA flow rate and so on has not been studied deeply. This study attempts to figure out the influences of these parameters on the effectiveness of the aircraft fuel tank inerting system.

2. Equipment and procedures

2.1. Experimental setup

A simulated aircraft fuel tank inerting system has been established in this study, the schematic diagram of the corresponding experimental equipment is shown in Figure 1. High pressure air produced by the compressor passes through the filter, the pressure regulator and the heater to obtain the inlet gas in accordance with the experimental requirement, and then enters the hollow fiber membrane. A certain concentration of NEA from the hollow fiber membrane, is passed into the scale tank for inerting washing. At the same time, a small amount of gas from the tank is collected to measure the oxygen concentration, in order to detect the inerting effect of the fuel tank.

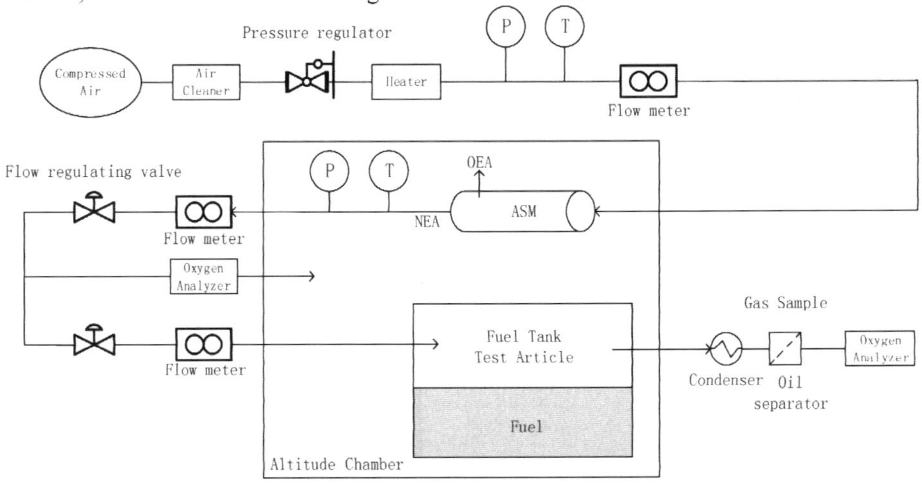

Figure 1. Schematic diagram of experimental setup.

The simulation fuel tank used in this experimental system is 25% length of a certain simplified aircraft fuel tank and the tank contains three compartments. The top of the fuel tank has a fuel inlet and a fuel outlet is arranged at the bottom, which is used for adding or reducing the amount of fuel. The left and right sides of the tank are arranged with one-way valves, which are designed for altitude experiments in the future. In ground tests, the two valve was used for gas emission in the tank. At the top of the tank, there are three inerting inlets, which are located in the center of each compartment. At the same time, there are 6 sample gas outlets at the top of the tank numbered 1~6, each compartment has two respectively. The size of the tank and the specific location of these inlets and outlets is shown in Figure 2. The sample gas outlets are distributed at the corner of each compartment, which can monitor the oxygen concentration of the area difficult to inert effectively. Large amount of fuel vapor in the sample gas will affect the measurement results of the oxygen concentration, so it is necessary to condense and separate the oil vapor from the sample gas. After that, the oxygen concentration is measured by the oxygen concentration measuring instrument.

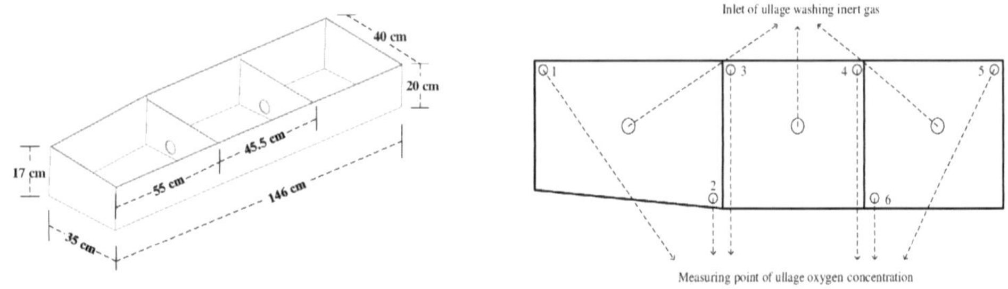

Figure 2. The size of the tank and location of inerting inlets and sample gas outlets.

2.2. Experimental procedures

Fuel tank inerting experiment requires that the oxygen concentration of the fuel tank before inerting is the same as the ambient oxygen concentration at first. So the ullage of the tank should be washed with air before inerting until the oxygen concentration of all the sample points are consistent with the surrounding air. Then start the air bleeding system, heat the compressed air and adjust the pressure, until the pressure and temperature at the entrance of the hollow fiber membrane reach the experimental requirements. Then regulate the flow control valve at the outlet of the hollow fiber membrane to make the oxygen concentration of NEA meets the requirement of the experiment. Regulate the flow regulating valve of the inerting branch so that the NEA flow rate reaches the required flow rate. At the same time, open the micro vacuum pump for sampling and monitoring the oxygen concentration of the sample points. With the inerting process, the oxygen concentration of the sample points continues to decline, until the oxygen concentration of all the sample points reaches 12%.

3. Results and analysis

3.1. Distribution of flow

To inert multi compartment tank using NEA of the same flow usually has two kinds of flow distribution mode, concentrated flow washing and uniform distribution of flow. When loaded 50%, inert the tank using NEA of 6% oxygen concentration and 3 kg/h of flow rate, and the bleeding air pressure is 0.3MPa. The variations of ullage oxygen concentration with time at different distribution mode of flow are presented in Figure 3.

As shown in the figure, the ullage oxygen concentration gradually decreased as time increases, but there is a great difference in the trend of the concentration of the sample points in two different ways. Figure 3(a) is the curve of washing inerting under uniform flow distribution, we can see the oxygen concentration change trend of every sample point is relatively consistent, and the final oxygen concentration reached 12% successfully. The oxygen concentration of point 1 decreased more slowly than the other sample points at the beginning, this is because the distance of sample point 1 from the inlet of inert gas is farther than other points. At the same time, the oxygen concentration of gas spread to point 1 is relatively high as the distance, so the oxygen concentration of point 1 reached 12% at last. This phenomenon appeared in later experiments too, so the oxygen concentration of point 1is regarded as the basis of the whole tank meeting the requirements of inerting.

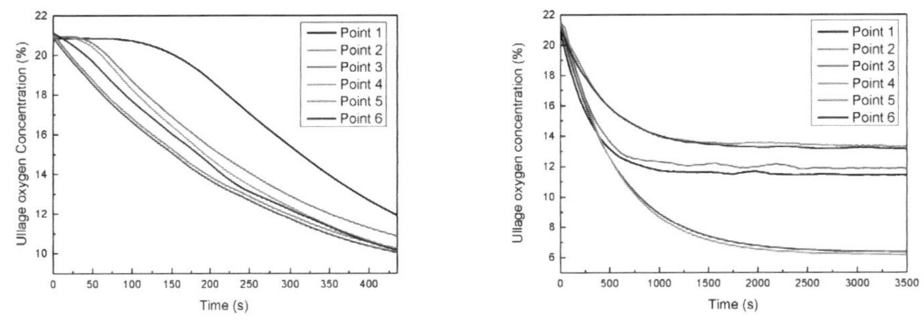

(a) (b)

Figure 3. Inerting effects under different conditions of NEA distribution.

Figure 3(b) is the curve of oxygen concentration when all the inert gas enter the tank from the middle compartment. As seen from the picture, the change of oxygen concentration of each sample point is quite different. The oxygen concentration of point 3 and point 4 located in the middle compartment decreased rapidly and reached 12% soon, and the final concentration was approximately stable at 6%. The concentration of point 1 and point 5 located in the compartments beside reduced slowly, and the final concentration can only drop to about 12%. The concentration of point 2 and point 6 decreased very slowly and finally stay at around 13%, which did not meet the inerting requirement. The phenomenon illustrates that concentrated flow washing will cause great differences throughout the distribution of oxygen concentration in the fuel tank and the emergence of inerting dead zone. Therefore, uniform distribution of flow should be chosen for the inerting of fuel tank with multi compartments. In the study of the influence of other factors, we have used this distribution mode of flow.

3.2. *Flow rate of NEA*

During the inerting experiments, the flow rate of the inert gas acts as an important factor and affects the time span of a complete inerting process directly. When loaded 50%, inert the fuel tank using NEA of 6% oxygen concentration, and the bleeding air pressure is 0.5MPa.

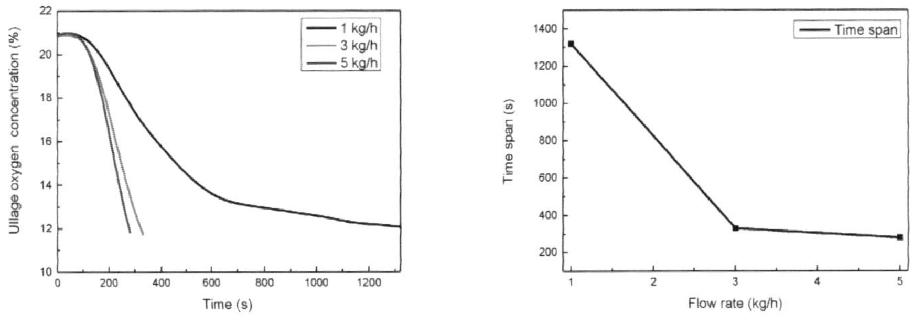

(a) (b)

Figure 4. Inerting effects under different conditions of NEA flow rate.

The variations of ullage oxygen concentration with time at different NEA flow rate are presented in Figure 4(a). It is seen from the picture that when the flow rate is higher, the time that needed to make the ullage oxygen concentration reaching 12% is obviously reduced. Figure 4(b) exhibits the relationship between the time span and NEA flow rate. The time span of a complete inerting process is 330s at the 3 kg/h flow rate, reducing by 75% of the time span at 1 kg/h flow rate which is 1320 s.

IOP Conf. Series: Materials Science and Engineering **187** (2017) 012010 doi:10.1088/1757-899X/187/1/012010

When the flow rate increased to 5kg/h, the time span is 280s, only reducing by 15% compared to the time span at the 3kg/h flow rate. It can be seen that with the increase in flow rate, the time span decreases but the trend becomes weaker gradually.

3.3. Concentration of NEA

During the inerting experiments, the oxygen concentration of NEA acts as another important factor. When loaded 50%, inert the fuel tank using NEA of 3kg/h flow rate, and the bleeding air pressure is 0.3MPa. The variations of ullage oxygen concentration with time at different NEA oxygen concentration are presented in Figure 5(a). It is seen from the picture that when the NEA oxygen concentration is lower, the time that needed to make the ullage oxygen concentration reaching 12% is obviously reduced. Figure 5(b) exhibits the relationship between the time span and NEA oxygen concentration. The time span of a complete inerting process is 320s at 6% concentration, reducing by 42% of the time span at 9% concentration which is 555 s. When the NEA concentration decreased to 3%, the time span is 260s, only reducing by 19% compared to the time span at 6% NEA concentration. It can be seen that with the decrease in NEA concentration, the time span decreases but the trend becomes weaker gradually.

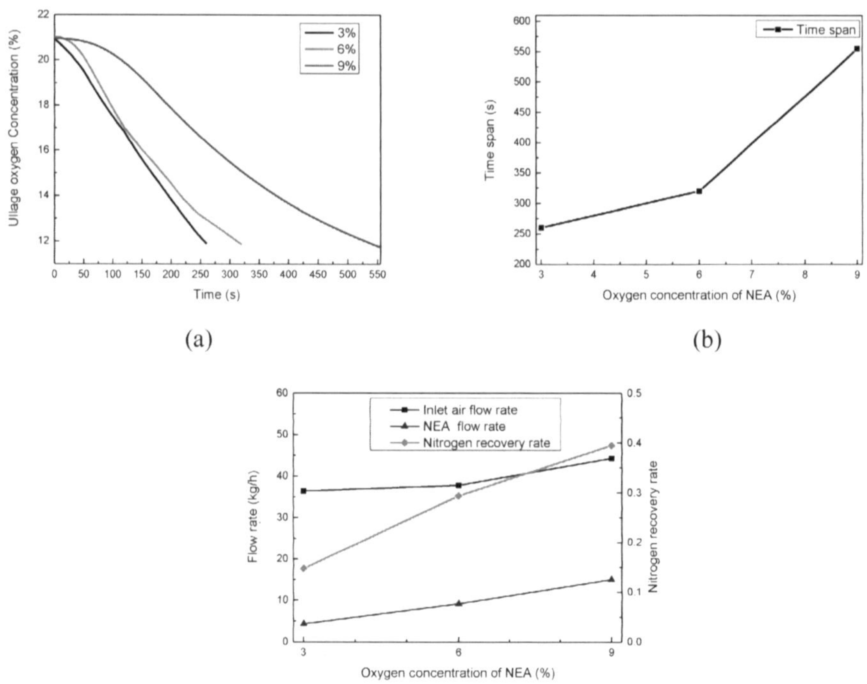

(a) (b) (c)

Figure 5. Inerting effects under different conditions of NEA concentration.

In addition, research showed that NEA oxygen concentration would have a marked impact on the performance of the hollow fiber membrane. So the efficiency of the bleeding air in the application of the fuel tank inerting system should also be taken into account. Therefore we define the nitrogen recovery rate of the hollow fiber membrane as θ and its calculation method is shown in equation(1). $N_2\%$ is the nitrogen concentration, Q is the flow rate. Higher nitrogen recovery rate means the higher utilization efficiency of the bleeding air.

$$\theta = \frac{(N_2\% \times Q)_{NEA}}{(N_2\% \times Q)_{Air}} \quad (1)$$

Figure 5(c) shows the relationship between the nitrogen recovery rate and NEA oxygen concentration. It can be known that the inlet air flow rate and the NEA flow rate of the hollow fiber membrane will decrease when the NEA oxygen concentration is decreased, and the nitrogen recovery rate will decrease. Moreover, it can be known that with the decrease of the NEA oxygen concentration, the decreasing trend of the nitrogen recovery rate is becoming stronger. This shows that when using NEA with lower oxygen concentration for inerting, the utilization efficiency of the engine bleeding air by the fuel tank inerting system is lower.

3.4. Pressure of bleeding air

The bleeding air can be supplied to the hollow fiber membrane at different pressure, and the air pressure has an important influence on the efficiency of the hollow fiber membrane. At the same time, the pressure of NEA at the outlet of the hollow fiber membrane is different, which will influence the inerting effect.

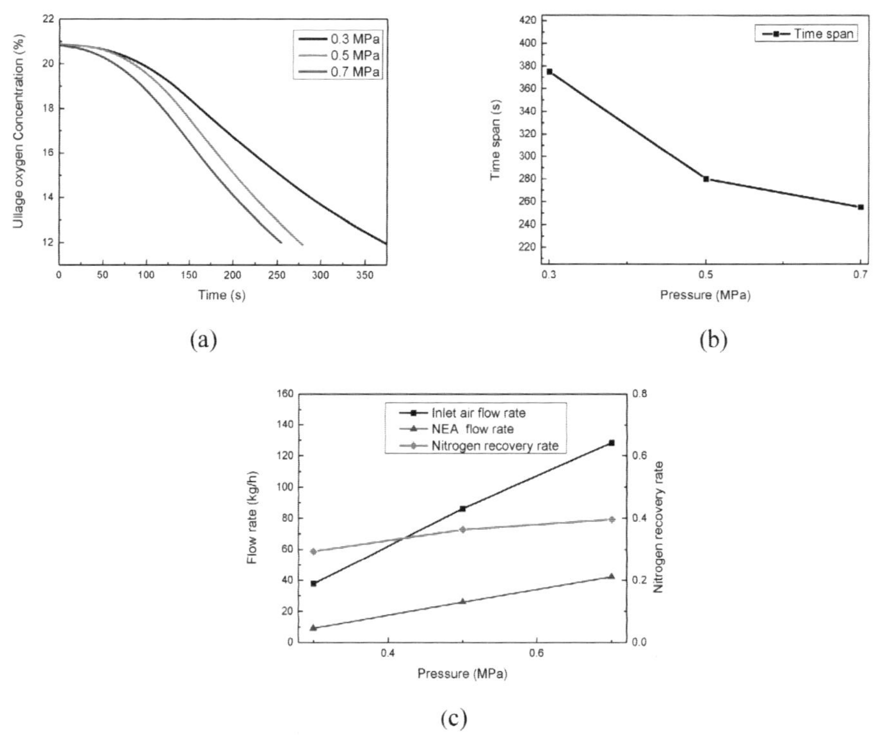

Figure 6. Inerting effects under different conditions of bleeding air pressure.

When loaded 50%, inert the tank using NEA of 6% oxygen concentration and 3 kg/h of flow rate. The variations of ullage oxygen concentration with time at different pressure of bleeding air are presented in Figure 6(a). It is seen from the picture that when the pressure is higher, the time that needed to make the ullage oxygen concentration reaching 12% is obviously reduced. Figure 6(b) exhibits the relationship between the time span and the pressure of bleeding air. The time span of a complete inerting process is 280s at the pressure of 0.5MPa, reducing by 25% of the time span at the pressure of 0.3MPa which is 375s. When the pressure increased to 0.7MPa, the time span is 255s, only reducing by 9% compared to the time span at the pressure of 0.5MPa. It can be seen that with the increase in pressure of bleeding air, the time span decreases but the trend becomes weaker gradually.

Figure 6(c) shows the relationship between the nitrogen recovery rate and the pressure of bleeding air. It can be known that the inlet air flow rate and the NEA flow rate of the hollow fiber membrane will increase when the bleeding air pressure is increased, and the nitrogen recovery rate will increase. This shows that when using NEA with higher bleeding air pressure for inerting, the utilization efficiency of the engine bleeding air by the fuel tank inerting system is higher.

3.5. Fuel load

The inerting effectiveness is investigated by experiments in this section with different fuel loads of the tank using NEA of 6% oxygen concentration and 3 kg/h of flow rate, and the pressure of bleeding air is 0.3MPa.

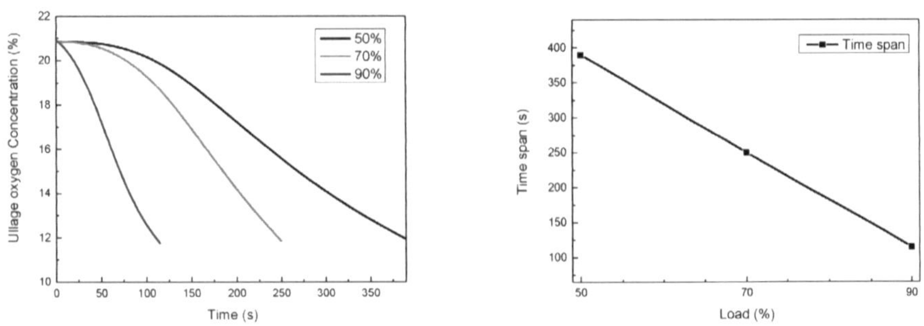

(a) (b)

Figure 7. Inerting effects under different conditions of fuel load.

Figure 7(a) presents the variation trends of ullage oxygen concentration as time increases when the fuel loads are 50%, 70% and 90% respectively. It is seen from the picture that when there is more fuel in the tank, the time that needed to make the ullage oxygen concentration reaching 12% is obviously reduced. The relationship between the time span and fuel load is shown in Figure 7(b). As we can see in the picture, the time span of a complete inerting process decreases linearly with the increase of fuel load.

4. Conclusions

In this research, experiments have been performed to inspect the influences of different operating parameters on inerting effectiveness of the established aircraft fuel tank inerting system. The following conclusions can be drawn.

For multi compartment fuel tank, concentrated flow washing inerting would cause great differences throughout the distribution of oxygen concentration in the fuel tank, and inerting dead zone would exist. When inerting under uniform flow distribution, the ullage oxygen concentration of the tank is consistent. The area far from the inlet of NEA is more difficult to inert, so the inlet of inert gas should be arranged in the middle of the compartment.

Results show that with the increase in NEA flow rate, the time span of a complete inerting process decreases but the trend becomes weaker gradually. When NEA oxygen concentration reduces, the time span of a complete inerting process decreases but the trend becomes weaker gradually. However, when using NEA with lower oxygen concentration for inerting, the utilization efficiency of the engine bleeding air by the fuel tank inerting system is lower. The time span can also be reduced by raising the pressure of bleeding air, which will also improve the bleeding air utilization efficiency. The time span decreased linearly with the increase of the fuel load.

5. References

[1] Summer S. M. Mass Loading Effects on Fuel Vapor Concentration in an Aircraft Fuel Tank Ullage[R]. DOT/FAA/AR-TN99/65, 1999
[2] Jiao Yan,Feng Lili. Review of American military jet fuels development[J].Journal of Rocket Propulsion，2008，34（1）：30-35
[3] Du Longxun，Xue Yan.Research Progress in Aviation Fuels Fire Protection[J].China Flame Retardant，2012（6）：174-180
[4] Cavage WM. The effect of fuel on an inert ullage in a commercial transport airplane fuel tank.Atlantic (NJ): Federal Aviation Administration; 2005. Report No.: DOT/FAA/AR-05/25.
[5] Zinn S. Jr. Inerted Fuel Tank Oxygen Concentration Requirements[R]. FAA-RD-71-42,1971
[6] Li Zhanguo.Aircraft Fuel System Design and Simulation Technology in the Future[J].Aeronautical Science and Technology，1996，3(25)：23-26
[7] Cavage W. M., Bowman T. Modeling In-Flight Inert Gas Distribution in a 747 Center Wing Fuel Tank[A]. 35th AIAA Fluid Dynamic Conference and Exhibit [C]. Toronto. Ontario, 2005: 1-13
[8] Stewart P. B., Starkman E. S. Inerting Conditions for Aircraft Fuel Tanks[M]. U.S.: University of California, Institute of Engineering Research, 1955
[9] Stewart P. B., Starkman E. S. Inerting Conditions for Aircraft Fuel Tanks[M]. U.S.: University of California, Institute of Engineering Research, 1955
[10] Li Yanjiang，Feng Zhenyu.Study on Simulation Analysis for Airplane Fuel System[J].Journal of Xi’an Technological University，2008，28（4）：401-405
[11] Burns M, Cavage W M. Ground and flight testing of a Boeing 737 center wing fuel tank inerted with nitrogen-enriched air[R]. WILLIAM J HUGHES TECHNICAL CENTER ATLANTIC CITY NJ, 2001.
[12] Cavage W M, Kils O. Inerting a Boeing 747SP Center Wing Tank Scale Model with Nitrogen-Enriched Air[R]. WILLIAM J HUGHES TECHNICAL CENTER ATLANTIC CITY NJ, 2002.

IOP Conf. Series: Materials Science and Engineering **187** (2017) 012011 doi:10.1088/1757-899X/187/1/012011

The Development Status and Key Technologies of Solar Powered Unmanned Air Vehicle

Li Sai*, Zhou Wei and Wang Xueren

601 Faculty, Xi'an Research Ins of Hi-Tech, Hong Qing Town, Xi'an. R.R. China

*946866025@qq.com

Abstract. By analyzing the development status of several typical solar powered unmanned aerial vehicles (UAV) at home and abroad, the key technologies involved in the design and manufacture of solar powered UAV and the technical difficulties need to be solved at present are obtained. It is pointed out that with the improvement of energy system efficiency, advanced aerodynamic configuration design, realization of high applicability flight stability and control system, breakthrough of efficient propulsion system, the application prospect of solar powered UAV will be more extensive.

1. Introduction

Solar powered UAV is a kind of aircraft based on solar radiation as a propulsion energy. During the solar powered UAV flight, the solar energy cell array converts the solar energy into DC output, then the Maximum Power Point Tracker(MPPT) controls the maximum output power of the solar cell and converges to the main power supply circuit. The power control device sends the power to the motor-propeller system. Solar radiation is the total energy source of UAV. During the daytime, solar radiation can be converted to electrical energy. The energy maintain power plants and airborne electronic equipment's work, and support the aircraft climbed to a certain height, also charge the energy storage system. At night, energy storage system release the stored energy, maintain the entire system of the normal operation. The aircraft gradually reduce to a certain height, by optimize the track to reduce the energy consumption of energy storage system[1]. If the energy absorbed during the daytime can meet the needs of the whole day flight, solar powered UAV can achieve uninterrupted flight day and night theoretically[2].

Solar powered UAV's prospects are very broad as solar radiation is a clean and renewable energy. Solar powered UAV can be used for communication relay, reconnaissance, surveillance, electronic countermeasures and so on as it can not consume energy, do not have to take off and land maintenance, can be sustained at high altitude flight. Google's Sky Bender project also plans to use solar unmanned aerial vehicle transmission 5G network, and in the future it may also achieve commercial manned flight. Many countries are in the relevant technology research as the solar aircraft has so many advantages.

2. Development status

2.1. Development status of foreign solar powered unmanned aerial vehicles

In November 4, 1974, Sunrise I successful maiden flight mark the first step in human solar flight unmanned aircraft [3]. Although Sunrise I is based on glider technology, advanced UAV design technology embodies less, but opened the door of solar flight. Solar powered UAV finally embarked on the rapid development of express Road with the solar cell conversion more efficiency , energy storage system energy density increasing, wide development and application in aircraft design technology, automatic control technology, microelectronics, new materials technology and so on [4].

2.1.1. "Helios" UAV. "Helios" UAV, developed by NASA and the Aeronautical Environment Corporation (UAS), use a large aspect ratio (31) wing layout to improve aerodynamic efficiency and is designed centre on the utilization of solar energy(as show in Figure 1). Wingspan 75.3m, area 183.6m2, with 62120 pieces of solar cells, can provide 40 kilowatts of power, driving 14 propellers flight, high-altitude model weight 720kg, long-time model weight 1053kg. In August 2001, the high-altitude model hit a record of 29,524 miter, it was the highest propeller-driven UAV in the world that time [5-7].

Figure 1 " *Helios* " UAV

Figure 2 " *Zephyr* 7"UAV

2.1.2. "Zephyr 7 "UAV. "Zephyr 7" UAV is designed by the United Kingdom Qinetiq Company. A large number of carbon fiber composite materials used in the solar powered UAV, and is the embodiment of the top technology of solar aircraft(as show in Figure 2). During the day, it use a thin film amorphous silicon thin-film batteries to absorb solar power to cruise at 18km high, at night it flight by the lithium-polymer battery witch energy is 2 times of the lithium-sulfur batteries [8]. In order to save energy, it glide down with dynamic. The wingspan is about 22.5m, total mass is 53kg, it has the lightest structure density in the unmanned aerial vehicles (0.8kg / m^2). In July 2010, 336 hours of continuous flight at 21.6km, creating a new UAV continuous flight time. Airbus said in a statement, the first "Zephyr" -8 UAV has been under construction, and will begin first flight in mid-2017 maiden voyage.

2.1.3. "Condor" unmanned aerial vehicle project. "Condor" UAV project proposed by the US Defense Advanced Research Projects Agency, the target aircraft flight height is 18 ~ 27km, sustained flight time 5 years, rated power 5kW, the payload can carry 450kg. If successful, will become the world's most advanced large-capacity high-altitude permanent flying solar unmanned aerial vehicles. The first phase of the project is the Aurora's Odysseus, Boeing's Solar Hawk and Loma's Norma, as shown in Figures 3, 4 and 5. September 2010 , "Condor" project using the "Solar Hawk" program, developed by Boeing using solar / power-driven unmanned aerial vehicle prototype. Boeing said the "Sun Hawk" program will continue flight of about 30 days of the first flight in 2014 to verification the power systems and structures and other key technologies . It is still in the validation phase, and there is no real load-carrying flight.

Figure 3 Aurora "Odysseus" UAV program

Figure 4 Loma's "Norma" UAV program

Figure 5 "Solar Hawk" UAV program

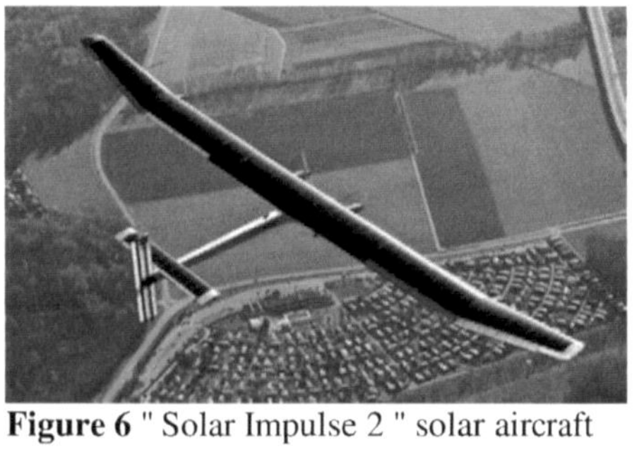

Figure 6 " Solar Impulse 2 " solar aircraft

2.1.4. "Solar Impulse 2" solar aircraft. "Solar Impulse 2" solar aircraft is designed by the sun team of high-altitude long-endurance manned aircraft, whoosh's flight required energy entirely from solar cells. A large number of new carbon fiber composite materials are used in the aircraft's manufacture(as show in Figure 6). The wingspan is 72 meters, weighing about 2300kg, the aircraft installed 17,248 thickness of only 135 micron monocrystalline silicon solar panels, can produce 340 kilowatt hours of electricity per day, driving four power 13.5 kilowatts of DC brushless, no Sense of engine work and charge 633 kg of lithium-ion battery. Beijing time March 9, 2015 to April 23, 2016 evening, "Solar Impulse 2" solar aircraft from the UAE capital Abu Dhabi departure, divided into eight sections of the Genesis manned world flight, and created the longest Distance single-person flight records. During the overheating of the battery problem, rest for 10 months. Local time at 4:00 on July 26,2016 " Solar Impulse 2" successfully landed in Abu Dhabi.

2.2. Development status of domestic solar powered unmanned aerial vehicles

After years of exploration and practice of China's solar energy application technology, mobile power storage, advanced aircraft design, modern control technology and advanced aviation composite materials, a certain foundation for China's independent research and development in solar unmanned aerial vehicles to provide a guarantee. However, we are still at the initial stage from the development of the field point of view.

2.2.1. "Soarer" UAV. "Soarer" UAV is China's first solar aircraft, design and manufacture by Dr. Li Xiaoyang and Professor Zhao Yong in 1992(as show in Figure 7). It is the first aircraft with original intellectual property rights of solar energy aircraft in the history of China. "Soarer" UAV's body and the wing using carbon fiber and Kevlar, balsa and other materials, the upper wing and horizontal tail are laid monocrystalline silicon solar cell sheet. The energy storage system is the nickel-metal hydride battery pack, propulsion system is a slow speed propeller. "Soarer" UAV's flight establish theoretical basis and accumulated experience in engineering practice of develop practical solar energy aircraft.

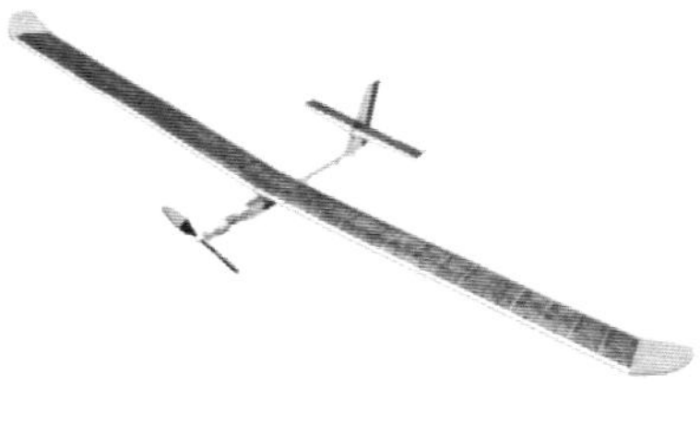

Figure 7 "*Soarer*" solar aircraft

Figure 8 "*Green Pioneer I*" verification machine

2.2.2. "Green Pioneer I" solar aircraft. "Green Pioneer I" solar aircraft is the green pioneer program's first generation of sample machine(as show in Figure 8). "Green Pioneer I" uses a layout and structure of new and reasonable composite flying wing aerodynamic layout. The whole machine, the lower wing surface set up a high conversion rate of flexible solar arrays as a power and control, task equipment, energy collectors. "Green Pioneer I" wind tunnel experiments and field flight experiments show that China's composite wing program than the current layout of other forms of advantage, is China's solar energy aircraft design a major initiative.

Figure 9 "*Rainbow*" solar powered UAV

2.2.3. "Rainbow" solar unmanned aerial vehicles. "Rainbow" solar powered UAV successfully test flight on May 22, 2015(as show in Figure 9). The solar unmanned aerial vehicle using large aspect ratio and lightweight structure, wingspan 40 meters, is the world's largest solar unmanned aerial vehicle following the United States NASA series. The success of the pilot opened a new chapter in China's solar unmanned aerial vehicle exploration.

At present, there has not a company can really realize the practical requirements of solar powered unmanned aerial vehicles, both solar power, long time alternating flight day and night, across the season and has a greater load capacity. Solar powered UAV research and development, there is still a long way to go. It is not only the focus of China's aviation industry in the new century, but also a new hot spot of national aviation industry research.

3. Key technologies and development trends

3.1. Energy system

From the solar panel to receive solar radiation, power system output power driving the aircraft flight, the efficiency of the entire system is only about 10% .

At present, most solar aircraft power generation system are silicon-based solar cells and compound solar cells. The most widely used monocrystalline silicon panels have high photoelectric conversion efficiency (15% -20%), but the structure quality is high. Amorphous silicon a-Si thin film solar cells as an alternative of monocrystalline silicon, the current conversion efficiency increased to 12% or more. Gallium arsenide (GaAs) solar cells have a higher conversion efficiency than silicon cells (43% theoretical efficiency, 28% efficiency), but expensive. Copper indium selenide (CIS) thin film solar cells also have great research value [8]. In addition, the solar aircraft on the laying of solar cell technology is also need to study. High altitude weather conditions are complex, the temperature difference between day and night changes, and the solar radiation intensity, require solar cells have good physical properties and high reliability. Solar cell is very thin, subject to greater external force will be severely damaged, requiring the wing can not have too much deformation. In addition, a large number of large solar UAV joints, will reduce the reliability of the circuit see the connection [4], its layout will also affect the efficiency of solar energy conversion [9][10].

Energy storage system is a very important part of solar powered unmanned aerial vehicles. At present, lithium batteries are widely used in solar airplanes, and some large-scale high-altitude solar airplanes use large-scale lithium ion batteries, lithium sulfur batteries ("Zephyr 7" UAV), fuel cells ("Helios" UAV). The weight of the energy storage system is generally 30% -50% of the total weight of the machine, which has a great impact in the ultimate aerodynamic layout and geometry of the UAV. The research of secondary battery with higher energy density, safety, reliability, cycle number and cheap price has become an important content of solar aircraft research. July 8, 2016, the world's first graphene lithium-ion battery was officially released, greatly enhancing the lithium-ion battery's applications.

During solar flight, the flight time, height, speed and attitude are constantly changing. The solar radiation intensity and temperature are also changing, which causes the open circuit voltage, short circuit current and maximum power of the solar cell to change [11][12]. The main function of the power control subsystem is to monitor the energy demand of each unit in real time and to allocate energy reasonably and efficiently. Today, MPPT controllers are widely used. It can match the input current of the solar cell and the current output to the load to achieve the maximum power output to achieve the best utilization of energy. According to the relevant units of the actual test, the use of MPPT controller power generation system than the traditional efficiency by 20% [13].

In summary, the future development trend of solar energy systems for unmanned aircraft is that: (1) improve the quality of silicon solar cell conversion efficiency and specific power; (2) new types of thin film solar cells, reduce quality, reduce costs, improve the conversion efficiency; (3) form of the solar cell module layout, layout, technology and mechanical characteristics, environmental characteristics, improve the applicability of the solar cell on the UAV; (4) improve the storage battery energy security and higher research and development security , better performance such as lithium polymer batteries, lithium - sulfur batteries, lithium-air batteries; (5) to carry out a more lightweight, efficient, safe, renewable fuel cell technology; (6) research and development of more powerful and efficient reliable intelligent power control system.

3.2. Overall integrated design technology

Solar powered UAV, which have a low Reynolds number are relatively small (about 10^5-10^6) in comparison with the Reynolds number, due to the high altitude flight of thin air. In order to obtain good lift characteristics, general solar aircraft have a large aspect ratio, in order to reduce the weight, carbon fiber, Kevlar and other advanced composites are widely used. At present, due to the unique parameters of solar aircraft, the conventional design method is difficult to meet the design requirements. Therefore, in the process of determining the overall parameters of solar aircraft, the energy balance should be integrated to optimize the efficiency, integration of subsystems, and system efficiency [14]. In addition, the integration of Multidisciplinary Design Optimization (MDO) is also a hot research [19].

At low Reynolds number, the separation and transition of solar UAV surface flow are sensitive to Reynolds number, pressure gradient and flow condition. Laminar flow separation occurs at the leading edge of the airfoil, and the boundary layer flows into turbulence to form a laminar separation bubble structure, which affects the aerodynamic characteristics of the airfoil. Due to the presence of laminar and turbulent flows in the wing surface, the state of the boundary layer must be improved to improve aircraft performance. Long flight UAV special flight conditions, the conventional airfoil is difficult to meet, so the airfoil must be designed for its doors. Good lateral and longitudinal stability is an important condition for maintaining high-altitude flight for a long time, and must be taken into consideration when carrying out integrated design. The traditional aerodynamic layout is relatively mature, the technical risk is low, but in the low-speed low Reynolds number flight conditions, there are obvious problems of aerodynamic efficiency, so there are aircraft flying wing and other new aerodynamic layout. Large aspect ratio wings (about30) enhance the aerodynamic efficiency, but also brings the problem of aeroelasticity, affecting the stability and safety of aircraft, need to be considered.

At present, light non-metallic materials because of its unique performance, in the application of solar UAV more carbon fiber composite materials because of its high strength characteristics of lightweight, multi-do for the load-bearing structure. Large aspect ratio and soft wings, will lead to unmanned aerial aircraft flying wing has a considerable deformation, and large deformation will cause changes in aerodynamic loads. This kind of aerodynamic / structural coupling makes the UAV have aero elastic problems, so that the flight quality of UAV will be affected, and even stability and security problems will be brought [22]. To improve the reliability of the system, the use of distributed propulsion system, propeller slip flow and aerodynamic interference of the wing is a difficult and difficult problem in the design of aircraft[25][26] The research of propeller wing Slip flow influence on the design of UAV aerodynamic layout has important significance.

3.3. Flight stability and control technology

The influence of the distributed propeller on the flight quality of the wing is more complicated. The flight stabilization and control module design is required for the solar powered unmanned aerial vehicle (UAV) to be subjected to aerodynamic forces during the flight, which results in great nonlinear deformation. Has brought a lot of restrictions, the conventional control law has not a very good solution to the problem. The effect of aeroelasticity on the flight control mode needs to be addressed. Power control subsystem and flight control system coordination, but also need to be addressed.

Due to the low flying speed of the UAV, low propeller speed, low power of the propulsion system, large aspect ratio of the wings and the distribution of the multiple propellers along the wings, the distributed propulsion system of the solar unmanned aerial vehicle Characteristics have a greater role, especially on the horizontal flight quality to bring a non-negligible impact [27]. Multiple-oars bring more options for heading control, and how to optimize the heading control without affecting the long life time is also a key problem worthy of study. There are two main problems in the design of thrust control and distribution system for solar unmanned aerial vehicle (UAV): (1) To improve the energy efficiency of the whole system on the basis of providing thrust and yaw moment required by the aircraft; (2) And the attenuation and hysteresis due to the slow dynamic characteristics of the propulsion system [28]. Stepanyan proposed a differential thrust adaptive control method in the case of a vertical tail fault[29]. Chen Huai-min, Wang Rui, and Zhu Xiaoping et al. Have studied propeller differential thrust to realize the solar-powered unmanned aerial vehicle heading control problem. Sun Chengxiao studied the pitch control method of multi-propeller solar unmanned aerial vehicle , and pointed out that the multi-propeller control realized by the change of motor speed could make the solar unmanned aerial vehicle satisfy the requirement of long flight time. Research and development to adapt to high-altitude long-endurance flight of unmanned aerial vehicles flying characteristics of the flight control system is the development of solar energy UAV key technology.

3.4. Propulsion system

For solar unmanned aerial vehicles, the propulsion system generally uses a motor-driven propeller approach. Direct-drive distributed propulsion systems are used in solar aircraft, and only in small solar-powered aircraft to reduce the drive to improve propeller efficiency. Solar unmanned aerial vehicle energy system output is DC, in order to reduce the conversion equipment and aircraft weight, should be preferred DC motor. The ordinary brush DC motor brush and commutation brush because of the existence of easy to produce sparks, and in high altitude conditions, heat dissipation performance is poor, poor reliability. Brushless DC motor can replace the brushless commutator drive circuit and commutator, fast response, large starting torque, and has the self-cooling effect, but the need to increase the DC - AC inverter, the circuit is more complex. Rare-earth permanent magnet brushless DC motor with high efficiency and less moving parts has been widely used in solar unmanned aerial vehicles (UAV), but its control is complex and expensive. Long-time solar unmanned aerial vehicles to reduce the moving parts, the use of more fixed-propeller. This also brings a certain problem: low-altitude characteristics of a good propeller, often at high altitude can not adapt to reduce the efficiency of different Reynolds number, the efficiency of change. The use of variable-propeller propeller to improve the adaptability of the UAV to maintain a high propulsion efficiency, but will increase the overall weight.

From the above analysis, the high-altitude long-term development of solar unmanned aerial vehicles can not be separated from its propulsion system of technological breakthroughs. One is to ensure the high-power torque characteristics of the drive motor, to improve the efficiency of the motor under the premise of reducing its size and weight; the second is to break the propeller and the high efficiency of the motor drive matching problem, research and development of efficient, Of the pitch-propeller, reducing costs and weight.

4. Conclusion

Since the first solar-powered aircraft first flight in November 1974, with the advancement of technology, mankind has realized the long flight of manned space flight. We believe that in the near future, with the solar cell conversion efficiency, energy storage system power density and propulsion system efficiency, and advanced aircraft design technology, automatic control technology, microelectronics technology, new materials technology development and application, the real The meaning of the permanent flight will be able to achieve, riding the sun will no longer be a dream flight.

5. References

[1] Ma D L, Bao W Z , Qiao Y H. Study of flight path for solar-powered aircraft based on gravity energy reservation [J]. Acta Aeronautica ET Astronautica Sinica ,2014,35 (2) :408-416.
[2] Romeo G, Frulla G. Aerodynamic and structural analysis of HAVE solar powered platform [R]. AIAA 2002-3504, 2002.
[3] Roberts C,Vaughan M,Bowman W J.Development of a solar powered micro air vehicle[R].AIAA 2002-0703,2002.
[4] Gao Guang lin, Song Bifeng, Li Zhanke, Ding Xiang. Key technologies of solar powered unmanned air vehicle [J]. Flight Dynamics,2010,28(1):1-4.
[5] Nicholas J, Colella S. Pathfinder and the development of solar rechargeable aircraft[R].Energy & Technology Review,1994:1-9.
[6] Ehemberger L J,Casey Donohue. A review of solar powered aircraft flight activity[R].NASA REPORT , 2004.
[7] Kirk Flittie,Bob Curtin. Pathfinder solar powered aircraft flight performance[R].AIAA 1998-4446,1998.
[8] Qu Peng, Wang Yin. Development status and prospect of solar power systems for UAVs[J]. Chinese Journal of Power Sources, 2015,39(04):864-870.
[9] Gao Jie , Li Qingdong, Zhao Yunhe. The Solar Cell Array Layout Optimization Based on Genetic Algorithme[C]. Yantai, China: Proceedings of 2014 IEEE Chinese Guidance,

Navigation and Control Conference,2014.
[10] Chang Min, Zhou Zhou, Cheng Ke. Wang Rui. Exploring the Characteristics of Power Density of Tracking PV Modules for High-altitude Stationary Solar-powered Airplanes[J]. Acta Aeronautica ET Astronautica Sinica , 2013,34(2):273-281.
[11] Chang Min, Zhou Zhou, Zheng Zhicheng. Flight Principles of Solar Powered Airplane and Sensitivity Analysis of Its Conceptual Parameters[J]. Journal of Northwestern Polytechnical University, 2010,28(5):792-796.
[12] Zhao Kai, Zhu Xiaoping, Zhou Zhou. Modeling and simulation of solar powered airplane for the conceptual design[J]. Flight Dynamics, 2011,29(1):13-16.
[13] Shi Ge, Hu Jingmin, Lou Hong, Yuan Shuhao. Design of an electronic speed controller for solar maximum power point tracking[J]. Journal of China University of Metrology, 2013,24(2):151-155.
[14] Cao Qing, Zhou Zhou, Chang Min. Design and research for conceptual parameters of solar-powered non-stop airplanes[J]. Flight Dynamics, 2014,32(02):132-136.
[15] David H W, Charles F D, Dimiceli E V, et al. A preliminary study of solar powered aircraft and associated power train[R].NASA-C R-3699,1983.
[16] Bailey D M, Bower V M. High altitude solar power platform[R].NASA-TM-103578, 1992.
[17] Noth A, Siegwart R, Engel W. Design of solar powered airplanes for continuous flight[D]. ETH ,2008.
[18] Zhang Qinling, Huang Jian, Liu Xiaoqian. General Design Method and Analysis of Long Endurance Solar Powered UAV[J]. Journal of Air Force Engineering University(Natural Science Edition), 2014,15(2):12-15.
[19] Deng Haiqiang. A Study on Integrated Aerodynamic/Structural/Propulsive Design of Solar Powered Aircraft[D].Nanjing, China: Nanjing University of Aeronautics and Astronautics,2006.
[20] Zheng Jun, Ping Lihao, Zhong Jianfeng, Zhan Dongdong. Discussion on Integration Thermal Design of Solar-powered Airplane[J]. Electro-Mechanical Engineering, 2014,30(2):5-8.
[21] Jiang Guangtai, Zhu Ming, Liang Haoquan, Wu Zhe. Reliability Based Multidisciplinary Design Optimization of Solar Powered Aircraft[J].Journal of Nanjing University of Aeronautics & Astronautics, 2012,44(4):464-471.
[22] Wang Hongwei, Yang Mao. Dynamic analysis on high aspect ratio composite wing[J].Computer Aided Engineering, 2008,17(2):21-23.
[23] Liu Xiangning, Xiang Jinwu. Study of Nonlinear Flutter of High-aspect-ratio Composite Wing[J]. Acta Aeronautica ET Astronautica Sinica , 2006,27(2):213-218.
[24] Wang Wei, Zhou Zhou, Zhu Xiaoping, Wang Rui. Static Aeroelastic Characteristics Analysis of a Very Flexible Solar Powered UAV with Geometrical Nonlinear Effect Considered[J]. Journal of Northwestern Polytechnical University, 2014, 32(4):499-504.
[25] Wang Kelei, Zhu Xiaoping, Zhou Zhou, Wang Hongbo. A study of distributed electric propulsion slipstream aerodynamic effects at low Reynolds number[J]. Acta Aeronautica ET Astronautica Sinica , 2015,36(x):1-10.
[26] Wang Hongbo, Zhu Xiaoping, Zhou Zhou, Xu Xiaoping. Aerodynamic Investigation on Propeller Slipstream Flows for Solar Powered Airplanes[J]. Journal of Northwestern Polytechnical University, 2015,33(6):913-920.
[27] Xiao Wei, Zhou Zhou, Wang Rui, Li Man. Effectively Determining Some Selected Effects of Distributed Propulsion System on Lateral Flight Quality of Solar UAV[J]. Journal of Northwestern Polytechnical University, 2012,30(6):868-873.
[28] Xu Mingxing, Zhu Xiaoping, Zhou Zhou, Xiao Wei, Feng Yin'an. Exploring an Effective Method of Thrust Allocation for Solar-Powered UAV with Multiple Propellers[J]. Journal of Northwestern Polytechnical University, 2013,31(4):505-510.
[29] V Stepanyan, K Krishnakumar, N Nguyen. Adaptive Control of a Transport Aircraft Using Differential Thrust[J].NASA Ames Research Center, Moffett Field, CA.NASA Publication,2009.

IOP Conf. Series: Materials Science and Engineering **187** (2017) 012012 doi:10.1088/1757-899X/187/1/012012

The Numerical Simulation of Jet Mixing With Steady and Antisymmetric Pulsed Jets in Different Frequency

R B Zhang[1], **L Hong**[1]

[1] School of Energy and Power Engineering, Beihang University, XueYuan Road No.37, HaiDian District, Beijing 100191, China

493719943@qq.com

Abstract. A parametric study of the effect of rectangular antisymmetric pulsed control jets on the jet mixing with different pulse frequency is conducted using RANS simulation. The mean centerline potential core length and the axial velocity contours of the main jet at different downstream positions (*x/D*) is investigated and discussed in detail. When the antisymmetric pulsed control jets are injected into the main flow, a starting vortex is observed in the axial velocity contours on x-z plane, which does not exist when the control jets are symmetric steady (*St*=0). The factors that have effect on the jet mixing are analyzed and described in detail, and the jet/ambient air interfacial area is found to play an important role in jet mixing. The best mixing effect can be reached at the same mass flow ratio of the main flow to the control jets if a proper pulse frequency of the control jets is used.

1. Introduction

The jet mixing enhancement is significantly useful for jet engines to reduce the jet potential core length [1,2] and the infrared emissions [3]. There are many ways to achieve it which could be characterized as the active control technology [4-7] and the passive control technology [8,9]. The active flow control technology [10] is introducing energy or auxiliary power into the flow [11,12]. The working parameters can be adjusted according to the state of the engine to achieve the desired mixing effect and the CJs can be switched off when not needed, which results in the reduction of the drag or thrust loss penalty [13] compared to the passive technology. Therefore, extensive investigations have been carried out in this area in the last few decades.

In 1994, R Howard [14] et al. studied the mixing effect when a pulsed control jet injected into a supersonic crossflow and found that the flow penetration is deeper than steady injection. D E Parekh [15] et al. used novel piezoelectric excitation devices for high frequency and high intensity injections into subsonic and supersonic shear layers and found that the mixing effect has a significant increased because of the flapping mode, and the optimum excitation Strouhal number is around 0.2 when the injection flow rates was as little as 1%. B F Jonathan [16] et al. found that the jet's development can be significantly altered when high-amplitude, low mass flux pulsed jets blowing normal to a jet's shear layer near the nozzle, and it appears to overwhelm the turbulence, resulting in nearly the same effect no matter how high and low the Reynolds numbers are. Besides, the jet mixing forcing at Sr = 0.2 is more effective, which is consistent with previous research conclusions. The research mentioned above are all studied using experiment, and B Parviz [17] et al. studied both experimentally and computationally. It is found that fluid tabs produce the same streamwise vortex formation process as solid tabs, which is confirmed in [18]. J K Sutariya [19] et al. experimentally investigated the supersonic mixing effect under different frequencies and pulse widths, and the results were compared with the results of continuous injection. Better mixing is found using pulsed injection and mixing is enhanced as pulse width decreases. In addition to the parametric research of the jet mixing effect which is conducted experimentally [20] and numerically (such as URANS [21] and Large Eddy [22]), the mechanics and the starting vortex are also studied [23-25].

The starting vortex discussed above is only studied in single pulsed jet. But in this paper, a starting vortex is observed when two antisymmetric pulsed control jets injected verticality into the main flow. And the current study parametrically investigates the jet mixing effect with different pulse frequencies using numerical simulations, providing a better understanding of the mechanisms that the influence on the jet mixing enhancement by using velocity contours and streamlines on different planes. In addition, the parameters affecting the mixing effect are summarized, and the optimum pulse frequency are found which can be used to achieve the best mixing effect at the same mass flow rate.

2. Computation scheme

2.1. Simulation methods

All the numerical simulations in the present work are conducted using ANSYS FLUENT software. And the Reynolds-Averaged Navier-Stokes (RANS) Renormalization Group (RNG) k-ε turbulence model is used in the current study.

2.2. Geometric model, Mesh generation and Boundary conditions

2.2.1. Geometric model. The cross-section of a 3D axisymmetric nozzle with CJs [26] is shown in figure 1. The flow direction of the main jet and CJs is indicated along X and Z direction respectively. The nozzle inlet diameter D_{in} is 203.2mm and the nozzle outlet diameter D_{out} equals to 101.6mm, a short parallel extension with the length of 127mm was connected to the nozzle outlet to smooth the main flow. The two actuators in green are distributed in the top and bottom of the nozzle outlet and attached to it. The value of width b of CJs is 43.18mm.

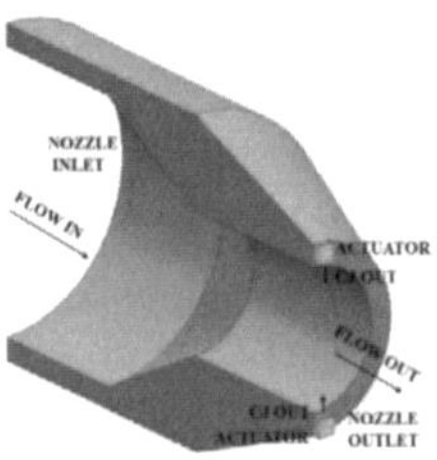

Figure 1. Schematic of 3D axisymmetric nozzle with CJs.

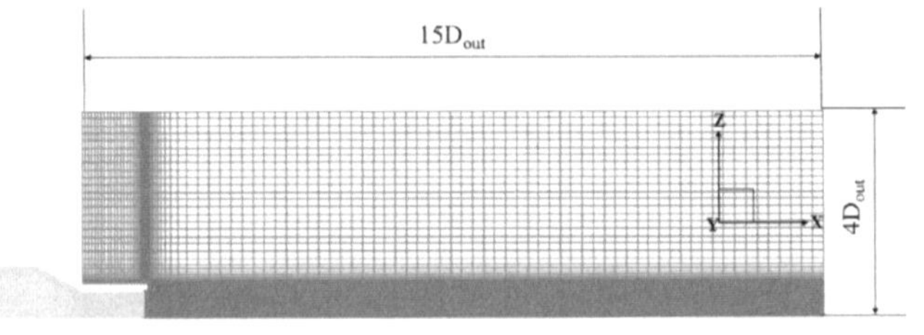

Figure 2. Cross-section of the solution domain and mesh structure: *x*-*z* plane.

2.2.2. Solution domain and mesh generation. The solution domain is shown in figure 2. It includes the nozzle and the entire jet flow area. In order to save the computation time, the simulation is conducted in a quarter of the entire domain since the nozzle is axisymmetric and the two actuators are symmetrically distributed on X-axis. In order to accurately simulate the interaction between the main jet and the CJ, the grids near the actuators are refined.

2.2.3. Boundary conditions. Incompressible ideal gas is used in the simulation because of the low flow velocity. The flow velocity at the nozzle inlet is 2.5m/s with the static temperature of 296.03K. The CJs are antisymmetrically injected into the main jet. Multiple CJ's pulse frequencies are used as shown in table 1 and the dimensionless pulse frequency of CJs *St* will be applied in the following passage. The pressure of the ambient air is 1 atm. The wall boundary condition is applied in the inner walls and the outer walls of the nozzle as well as the CJs.

Table 1. The geometry parameters and pulse frequencies of CJs.

a/mm	*b*/mm	*v*/m/s	*f*/Hz	*St* ($f*D_{out}/U_j$)
2.032	43.18	9.57	0	0
		19.15	5	0.05
			15	0.14
			20	0.19
			30	0.29
			40	0.38

3. Results and discussion

Numerical simulations of the jet mixing effect using the geometric model [26] have been performed for different pulse frequencies of antisymmetric CJs *St* at the same mass flow rate which equals to 1%. The predicted result with clean jet is also performed as the baseline to assess the jet mixing effect. The results and analysis of each case are described in detail in the following paragraphs.

3.1. Comparison between symmetric steady jets (St=0) and antisymmetric pulsed jets (St=0.14)

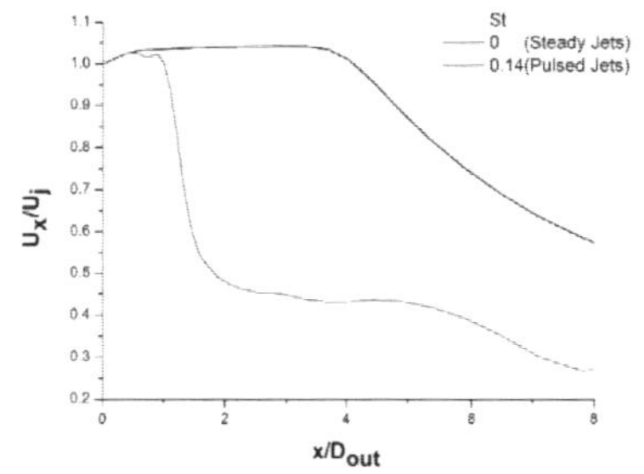

Figure 3. Normalized axial velocity of the primary jet centerline under steady jets (*St*=0) and pulsed jets (*St*=0.14).

The axial velocity distribution of the main jet centerline with nondimentional pulse frequency *St* of 0 and 0.14 are shown in figure 3 to characterize the jet mixing behavior. When *St* is 0, the CJs are essentially symmetric steady jets, which means that the excitation method is different from the pulsed jets. Besides, when St is 0.14, the axial velocity distribution of the main jet centerline is time-averaged. When *St* is 0, the centerline potential core length is $3.68D_{out}$, and when *St* increases to 0.14, the centerline potential core length decreases to $0.93D_{out}$. This indicates that the pulsed CJs have more pronounced effect on enhancing the jet/ambient mixing than the steady jets (*St*=0).

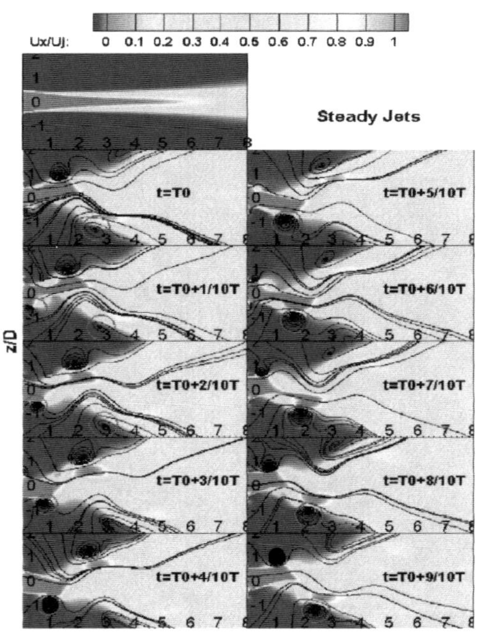

Figure 4. The axial velocity contour on x-z plane under steady jets (*St*=0) and pulsed jets (*St*=0.14) at different time instants.

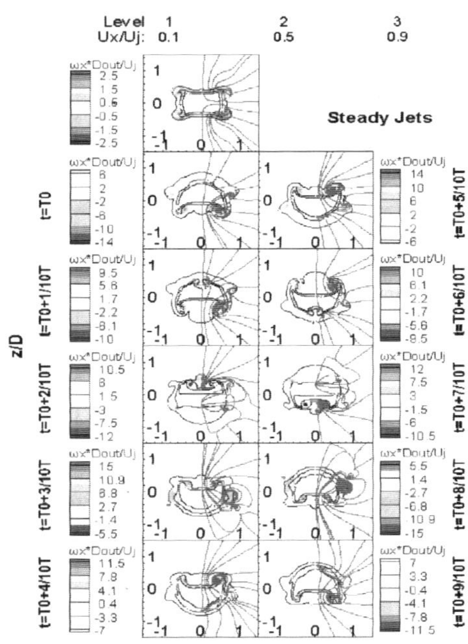

Figure 5. The axial velocity contour on y-z plane at x=Dout under steady jets (*St*=0) and pulsed jets (*St*=0.14) at different time instants.

In order to further analyze the effect of using symmetric steady jets (*St*=0) and antisymmetric pulsed jets on jet mixing, the contours of axial velocity on x-z plane are shown in figure 4. When the *St* is 0, there is a potential core contraction on x-z plane, but there is a flapping mode in the potential core area on x-z plane when the excitation method changes to antisymmetric pulsed jets. And the flapping frequency is consistent with the pulse frequency. Differ from symmetric steady jets (*St*=0), the expansion of the low-speed boundary of shear layer is more obvious along with a very large expansion area on x-z plane.

A vortex can be seen on x-z plane when the *St* is 0.14, which does not exist in symmetric steady jets (*St*=0). With the increase of time instant, the vortex is getting larger and moving to downstream on x-z plane. When the main jet is excited by the antisymmetric pulsed jets, downstream of the main jet breaks off with upstream of the main jet, which is equivalent to generating a new jet at downstream. The formation process of the vortex observed in figure 4 are similar to the vortex ring of starting jets [28], so it is called starting vortex. The ambient air will be entrained into the primary jet through the starting vortex, further improving the jet/ambient air mixing and further changing the shape of the primary jet.

The axial velocity contour lines, streamlines and the streamwise vorticity of symmetric steady jets (*St*=0) and antisymmetric pulsed jets at different time instants on y-z plane at $x=D_{out}$ are shown in figure 5. There are only two pair of counter rotating vortices when *St* is 0, but there are at least two pair of counter rotating vortices when the CJs changes to antisymmetric pulsed jets, the number of the streamwise vorticities is even up to four at some time instants. Although the streamwise vorticities of symmetric steady jets (*St*=0) reside on the low-speed boundary of the shear layer, the size and the intensity are smaller than the streamwise vorticities of antisymmetric pulsed jets. Besides, when *St* is 0.14, the shape of the primary jet deforms more significantly than using the steady jets (*St*=0) as CJs.

3.2. Comparison between antisymmetric control jets with different pulse frequency

The mixing metric U^* was defined as the average velocity for the antisymmetric pulsed control jets measured at $10D_{out}$ divided by the average velocity for the clean jets at $10D_{out}$. It can be seen in figure 6 that the optimum pulse frequency *St* is 0.14 in the present study and the mixing metric U^* decreases to 0.39. This indicates that the antisymmetric pulsed jets have a great effect on enhancing the jet/ambient mixing when they are injected into the main jet with *St* of 0.14.

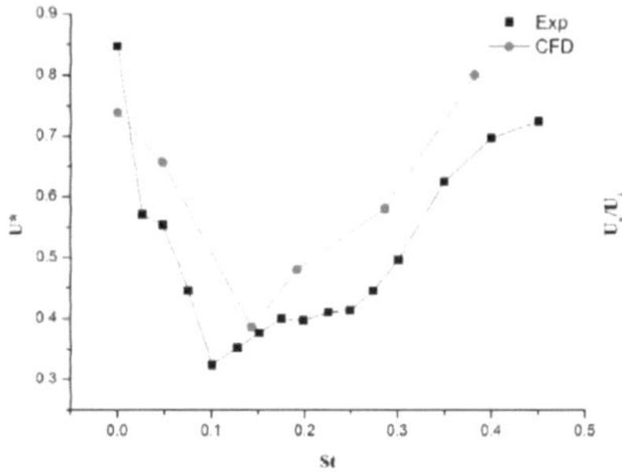

Figure 6. The mixing metric U^* with different *St*.

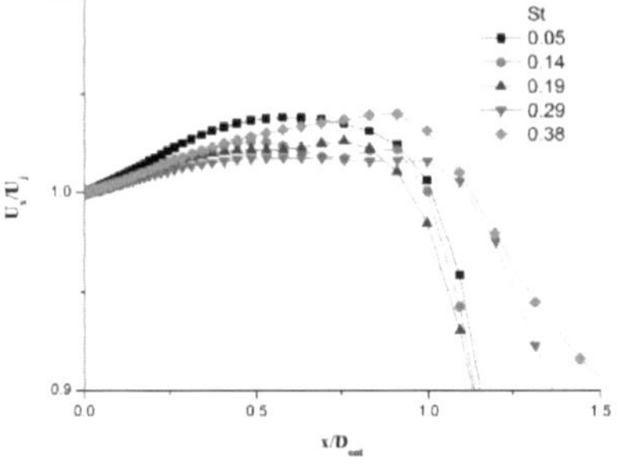

Figure 7. Normalized centerline axial velocity under different *St*.

As is shown in figure 7, the centerline potential core lengths with different pulse frequency St are all around D_{out}. The shortest centerline potential core length can be seen when St is 0.19, which is different from the mixing metric U^*'s optimum pulse frequency St.

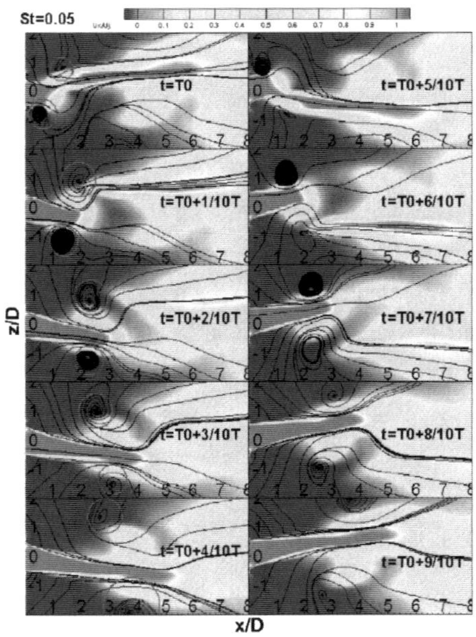

(a)

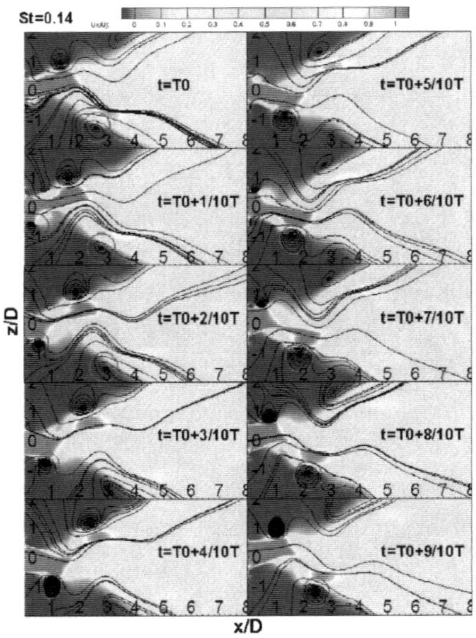

(b)

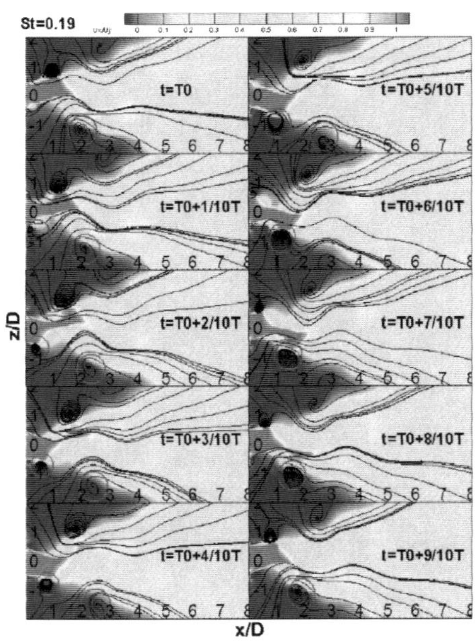

(c)

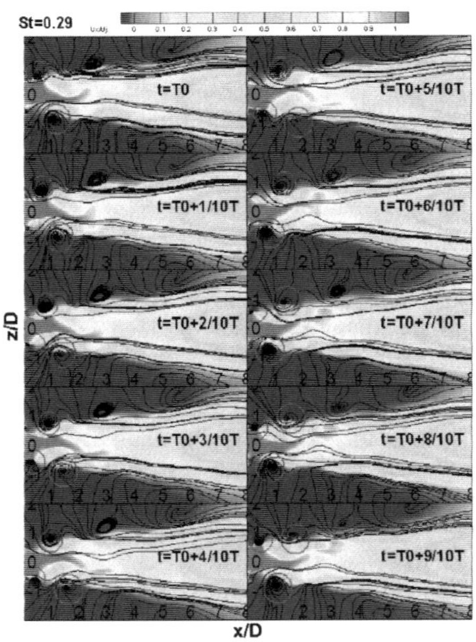

(d)

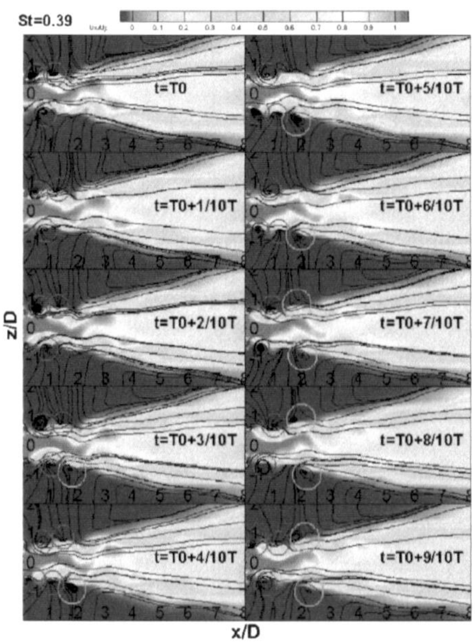

(e)

Figure 8. The axial velocity contour on x-z plane under different pulse frequency (*St*) at different time instants.

Figure 8 shows the axial velocity contour and streamlines on x-z plane under different pulse frequency (*St*) at different time instants. When antisymmetric pulsed jets are injected and the flapping frequency of the upstream potential core is consistent with *St*, under all the pulse frequencies. However, for downstream potential core which breaks off with the upstream because of the antisymmetric pulsed jets, the flapping frequency is consistent with the pulse frequency when *St* is 0.05, 0.14 and 0.19, and when St increases to 0.29 and 0.39, the flapping frequency is not consistent with the pulse frequency any longer. Besides, the flapping amplitude is obvious when *St* is 0.05, 0.14 and 0.19, and when *St* increases to 0.29 and 0.39, the flapping amplitude is very tiny.

When *St* is 0.05, the shape of the potential core changes a lot as the time instant increases, and the potential core length increases a lot, at the time instant of T0+4/10T, it even increases to more than $5D_{out}$. In addition, the potential core length of this case is the longest when compared with the other four pulse frequency cases at the same time instant. When *St* increases, both of the shape and the length of the potential core change a little as the time instant increases. The potential core length decreases at first, but then it begins to increase when *St* increased to 0.29.

In figure 8, an expansion can be observed in all the five cases on x-z plane. When *St* is 0.05, the expansion area is large which indicates that it is effective to spread the jet to downstream. As *St* increases to 0.14 and 0.19, the expansion area is getting larger, this means that spreading the jet to downstream is more effective. But when *St* continues to increase to 0.29, the expansion area decreases. When *St* increases to 0.39, the largest pulse frequency studied here, the expansion area increases compared with the expansion area when *St* is 0.29, but still smaller than the cases whose *St* is 0.05, 0.14 and 0.19, respectively. It is supposed that the expansion area is related to the flapping amplitude.

Starting vortex is observed in figure 8 using streamlines in all the five cases, but the details of the starting vortex are different. When *St* is 0.05, there are two starting vortex on both sides of the main jet, and both of the two starting vortex forms at the first time instant and close to the nozzle, however the starting vortex opposite to the working pulsed jet is closer to the nozzle than the other starting vortex, then the two starting vortex are getting bigger and moves to downstream as time increases, and at the fifth time instant, the starting vortex opposite to the working pulsed jets stops at x=4D_{out} and the other starting vortex stops at x=3D_{out}. When the other pulsed jet starts to work at the sixth time instant, both of the two starting vortex disappear and the phenomenon above appears once every half cycle. When *St* increases to 0.14 and 0.19, there is only one starting vortex forms at the second time instant, opposite to the working pulsed jet and the formation position is very close to the nozzle. Both of the two starting vortex stops at x=3D_{out} and the existence time is 1.2 cycles and 1.5 cycles, respectively. And when the other pulsed jet starts to work after half cycle, there will be the same developing process of the starting vortex, which is the same as the case when *St* is 0.05. When *St* increases to 0.29, the developing process of the starting vortex is more complicated. Because of the flapping frequency of the downstream is not consistent with the pulse frequency, and the starting vortex far from the nozzle has little effect on the jet mixing, so only the starting vortex marked with the red circle is concerned. The same as *St* of 0.14 and 0.19, there is only one starting vortex, but it forms at the third time instant, opposite to the working pulsed jet and the formation position is very close to the nozzle. Its stop position and existence time are not sure but the same developing process of the starting vortex once every half cycle. As for *St* of 0.39, the starting vortex is too complicated to find a developing law. The main starting vortex marked with red cycle forms at the fifth time instant, and its form frequency is consistent with the pulse frequency. In addition, it is worth to note that there is another starting vortex marked with green cycle, far from the nozzle, and it is supposed to be introduced when the main jet is spreading to downstream.

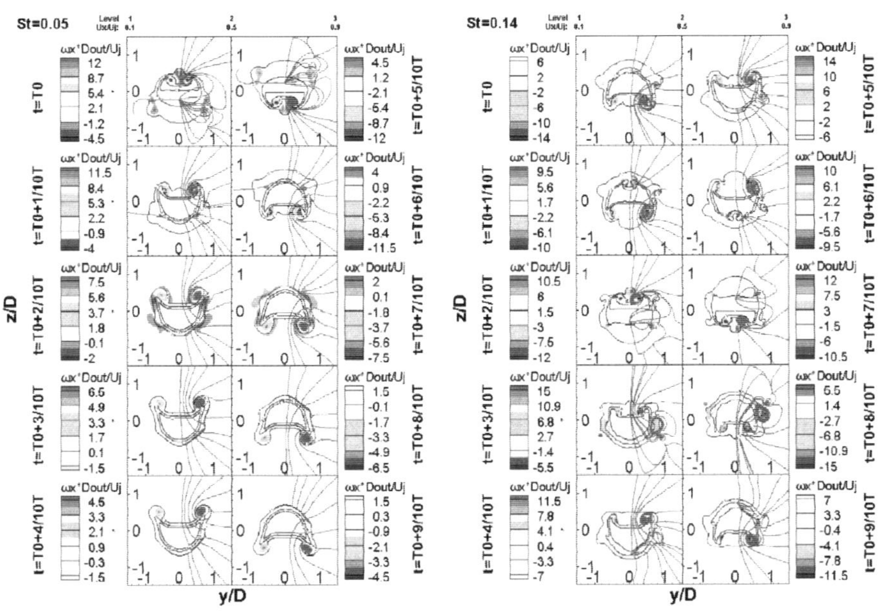

(a) (b)

IOP Conf. Series: Materials Science and Engineering **187** (2017) 012012 doi:10.1088/1757-899X/187/1/012012

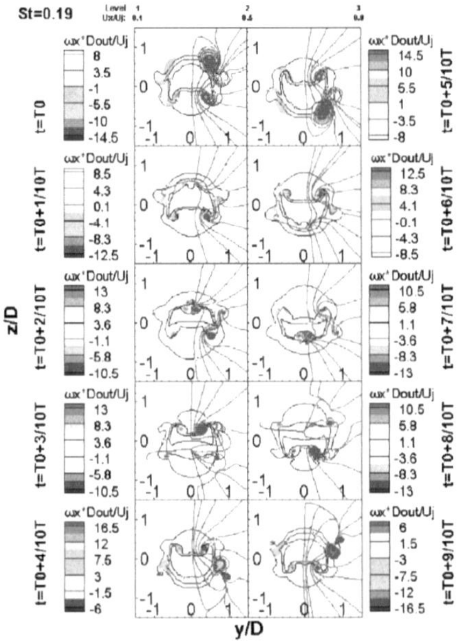

(c)

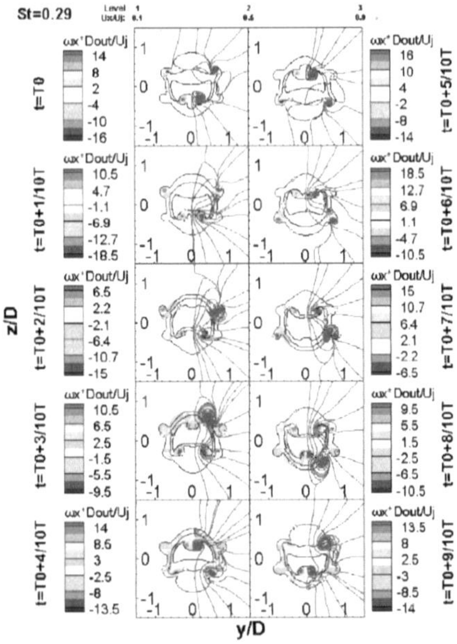

(d)

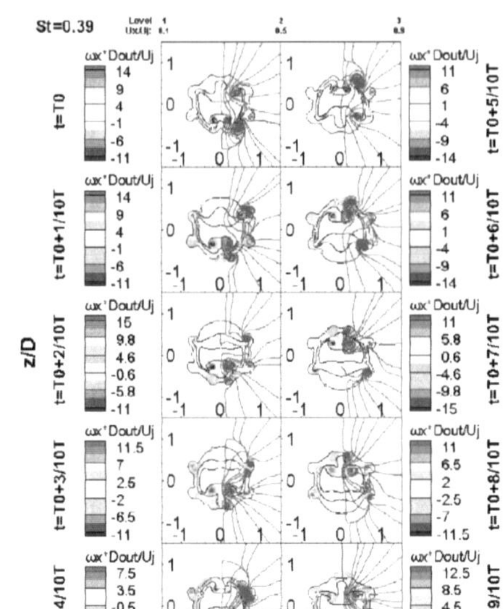

(e)

Figure 9. The axial velocity contour on y-z plane at $x=D_{out}$ under different pulse frequency (St) at different time instants.

Figure 9 shows the axial velocity contour on y-z plane at $x=D_{out}$ under different pulse frequency (St) at different time instants. It is found that all the main jet section shape of the five cases deformed dramatically when antisymmetric pulsed jets injected into the main jet. There is only one pair streamwise vortex in figure 9a, it resides on the low-speed side of the shear layer which can help to entrain large ambient air and result in a great mixing effect. There are at least two pair streamwise vortex when *St* is 0.14, 0.19, 0.29 and 0.39. However, only the vortex which is on the low-speed side of the shear layer can make more ambient air entrained which leads to a better mixing effect, so only the number and the size of these vortex have a great effect on jet mixing. When *St* is 0.14 and 0.19, the number and the size of the vortices which are close to the low-speed boundary of the shear layer is larger than other cases. This is an important reason that explains why the cases pulsed with *St* of 0.14 and 0.19 can have a better mixing effect when assessed with the mixing metric U^* and centerline potential core length.

4. Conclusions

A parametric study of the effects of antisymmetric pulsed CJs on the jet mixing of using different pulse frequency (*St*) have been conducted and the conclusions are listed as follows:

(1) When the main jet is excited with symmetric steady jets (*St*=0) at a mass flow rate of 1%, the mixing metric U^* is 0.74 and its centerline potential core length is $3.68D_{out}$, but when the excitation method changes to antisymmetric pulsed jets (*St*=0.14), the mixing metric U^* decreases to 0.39 and its centerline potential core length decreases to $0.93D_{out}$. Besides, a starting vortex is observed when antisymmetric pulsed jets (*St*=0.14) are injected into the main jet, and it does not exist when using the symmetric steady jets (*St*=0). So a better mixing effect is observed when using antisymmetric pulsed jets, no matter which evaluation method of mixing effect is used.

(2) When the whole jet/ambient mixing effect is concerned, then the minimum mixing metric U^* is 0.39 and the optimum pulse frequency *St* correspond to it is 0.14. When only infrared radiation of the main jet is concerned, the minimum centerline potential core length which reflects the mixing effect near the nozzle is observed when *St* is 0.19, which means that 0.19 is the optimum pulse frequency *St*.

(3) It is found that the jet/ambient interfacial area is an important factor to effect the jet mixing. And there are three factors that affect the jet/ambient interfacial area: 1) a flapping mode is generated by using antisymmetric pulsed jets, which results in a significant deformation of the main jet, and thus the jet/ambient interfacial area increases dramatically; 2) a starting vortex is found when the CJs are antisymmetric pulsed jets, and the longer it exists, the more ambient air will be entrained, as a result, the deformation of the primary jet is intensified, leading to a further increase in the jet/ambient air interfacial area; 3) the number and the size of the streamwise vorticities which are on the low-speed side of the shear layer are also important to the jet/ambient air interfacial area. The three poins above can explain why 0.14 and 0.19 are the optimum pulse frequency *St* when assessed with the mixing metric U^* and centerline potential core length, respectively.

5. References

[1] Valdis K, John D, David M S and Michael F M 1999 *30th AIAA Fluid Dynamics Conf.* 1999 Jun 28-Jul 1 Norfolk VA USA American Institute of Aeronautics and Astronautics 3507 pp 1-12
[2] Robert F, Ahuja K K, Brian C and Cacye B 2004 *10th AIAA/CEAS Aeroacoustics Conf.* 2004 May 10-12 Manchester England UK American Institute of Aeronautics and Astronautics 2874 pp 1-9
[3] Clarence F C, John D, David S and Valdis K 2000 *AIAA Fluid 2000 Conf .and Exhibit* 2000 Jun 19-22 Denver CO USA American Institute of Aeronautics and Astronautics 2471 pp 1-11
[4] Sun LY, Terrence W S, Susan C M, Mounir I, David G and Roy T 2008 *6th Int. Energy*

Conversion Engineering Conf. (IECEC) 2008 Jul 28-30 Cleveland Ohio USA American Institute of Aeronautics and Astronautics 5719 pp 1-17

[5] Samimy M, Kim J H, Kastner J, Adamovich I and Utkin Y 2007 Active control of a mach 0.9 jet for noise mitigation using plasma actuators *AIAA Journal* **45** 890-901

[6] Adnan E and Robert E B 2001 Structure, penetration, and mixing of pulsed jets in cross flow *AIAA Journal* **39** 417-23

[7] Ritchie B D, Mujumdar D R and Seitzman J M 2000 *38th Aerospace Sciences Meeting & Exhibit* 2000 Jan 10-13 Reno NV USA American Institute of Aeronautics and Astronautics 0404 pp 1-8

[8] Tong F and James J M 2006 *3rd AIAA Flow Control Conf.* 2006 Jun 5-8 San Francisco California USA American Institute of Aeronautics and Astronautics 3702 pp 1-12

[9] Callender B and Gutmark E 2003 *9th AIAA/CEAS Aeroacoustics Conf. and Exhibit* 2003 May 12-14 Hilton Head South Carolina USA American Institute of Aeronautics and Astronautics 3210 pp 1-16

[10] Linda D K Active flow control technology. *ASME Fluids Engineering Division Technical Brief* pp 1-28

[11] Jonathan B F and Parviz M 1998 *1998 ASME Fluids Engineering Division Summer Meeting* 1998 Jun 21-25 Washington DC USA The American Society of Mechanical Engineers 5235 pp 1-6

[12] Johari H, Pacheco T M and Hermanson J C 1998 Penetration and mixing of fully-pulsed turbulent jets in crossflow *AIAA Journal* http://arc.aiaa.org DOI: 10.2514/6.1998-2908

[13] Muhammad A K and James J M 2015 Unsteady predictions of mixing enhancement with steady and pulsed control jets *AIAA Journal* **53** 1262-76

[14] Howard R and Larry C 1994 Pulsed jets in supersonic crossflow *J. PROPULSION* **10** 746-8

[15] Parekh D E, Kibens V, Glezer A, Wiltse J M and Smith D M 1996 *AIAA 34th Aerospace Sciences Meeting and Exhibit* 1996 Jan 15-18 Reno NV USA American Institute of Aeronautics and Astronautics 0308 pp 1-24

[16] Jonathan B F and Parviz M 2000 Jet mixing enhancement by high-amplitude fluidic actuation *AIAA Journal* **38** 1863-70

[17] Parviz B and James J M 2004 *2nd AIAA Flow Control Conf.* 2004 Jun 28- Jul 1 Portland Oregon USA American Institute of Aeronautics and Astronautics 2401 pp 1-13

[18] Parviz B, Tong F and James J M 2008 Active flow control of jet mixing using steady and pulsed fluid tabs *IMechE* **222** 381-92

[19] Sutariya J K and Job K 2007 *43rd AIAA/ASME/SAE/ASEE Joint Propulsion Conf. & Exhibit* 2007 Jul 8-11 Cincinnati OH USA American Institute of Aeronautics and Astronautics 5032 pp 1-12

[20] Toshinori K, Noboru S, Muneo I and Sadatake T 2007 *43rd AIAA/ASME/SAE/ASEE Joint Propulsion Conf.& Exhibit* 2007 Jul 8-11 Cincinnati OH USA American Institute of Aeronautics and Astronautics 5405 pp 1-12

[21] Kouchi T, Sasaya K, Watanabe J, Sibayama H and Masuya G 2010 *46th AIAA/ASME/SAE/ASEE Joint Propulsion Conf. & Exhibit* 2010 Jul 25-28 Nashville TN USA American Institute of Aeronautics and Astronautics 6645 pp 1-10

[22] Ramya P, Joseph S, Jerry S, Jeff J and Suresh M 2010 *48th AIAA Aerospace Sciences Meeting Including the New Horizons Forum and Aerospace Exposition* 2010 Jan 4-7 Orlando Florida

USA American Institute of Aeronautics and Astronautics 1288 pp 1-12
[23] Geng T, Zheng F, Kuznetsov A V and Roberts W L 2007 *43rd AIAA/ASME/SAE/ASEE Joint Propulsion Conf. & Exhibit* 2007 Jul 8-11 Cincinnati OH USA American Institute of Aeronautics and Astronautics 5051 pp 1-17
[24] Yu S C M, Ai J J, Gao L and Law A W K 2008 Vortex formation process of a starting square jet *AIAA Journal* **46** 223-31
[25] Gao L and Yu S C M 2016 Formation of leading vortex pair in two-dimensional starting jets *AIAA Journal* **54** 1364-9
[26] Paul W 2007 Jet mixing enhancement by high amplitude pulse fluidic actuation [dissertation] Ann Arbor (48106-1346) Georgia Institute of Technology
[27] Muhammad A K and James J M 2011 Subsonic jet mixing via active control using steady and pulsed control jets *AIAA Journal* **49** 712-24
[28] Geno P, Carolina M C, Carlos M B and Pedro G H 2007 Experimental characterization of starting jet dynamics *Fluid Dynamics Research* **39** 711-30

Acknowledgments
The author gratefully acknowledge the support from the National Nature Science Foundation of China (Grants: 11572027).

Numerical Simulations of Jet Mixing Enhancement in a Round Jet Exhaust Using SJA

Liang Hong[1], Rubing Zhang[1], Zheng Liu[1]

[1] School of Energy and Power Engineering, Beihang University, Beijing, PRC

Abstract. Symmetric synthetic jet is used to stimulate a Mach number 0.9 and high temperature jet for the mixing enhancement. The movement of the piston-type synthetic jet is simulated by dynamic mesh based on URANS numerical methods. Jet flapping, streamwise vortex and geometric axis transformation are the jet mixing mechanism. However, the jet mixing enhancement effect depends on penetration area, which consists of penetration depth and spanwise. And the penetration area is determined by actuator nozzle diameter and blowing jet momentum. In addition, a 54% reduction in potential core length and maximum 45.8K decrease in potential temperature are achieved.

1. Introduction

Active flow control technology is a research hotspot in aerospace fields [1, 2]. Mixing enhancement is one of the methods of active flow control techniques. It has been widely used in aerospace. The propulsive jet plume is considered as a major source of inferred (IR) signature for military aircraft. In aero engine exhaust system, the use of the jet mixing can enhance the high temperature, high speed jet flow mixing with the surrounding ambient air, which can suppress the infrared signal [3-5] and plume noise [6-8]. Suppressing the infrared signal is the primary application of the current investigation.

The jet mixing enhancement methods can be identified as two major categories, passive flow control and active flow control technologies. The traditional passive flow control methods of the jet mixing enhancement include solid tabs [9-11], lobe nozzle [12, 13] and chevron nozzle [14-16]. These techniques can effectively alter the primary jet shear layers, but the performance of the aircraft exhaust system will decrease when this technique is in off-design conditions. The traditional passive flow control methods also have a thrust penalty on exhaust system. Nowadays more mature active flow control methods of the jet mixing enhancement include steady jet, pulsed jet and synthetic jet. These approaches are summarized in the review by Knowles and Saddington [17]. Active flow control methods are based on the use of the local energy inputting into the main flow, disturbing the jet flow shear layers, generating large-scale structure [18], which promotes jet mixing.

A periodic oscillation superimposed on a steady jet is the pulsed jet. The actuation frequency of the pulsation fall in a target range which is called optimal actuation frequency reflected by St. Raman et al.

[19] used two oscillating fluidic jets force rectangular primary jet, actuation at the first subharmonic of the preferred columnar instability mode founding that the CJs operating in-phase to generate a sinuous mode of forcing and out-of-phase to create a varicose mode. Parekh's et al. [20] experimental results showed that pulsed CJs actuating round jets were more effective in antisymmetric rather than symmetric actuation mode. Behrouzi's et al. [21] used 0.5% of primary jet flow CJs to excite subsonic jet shear layer finding that the performance of the pulsed CJs was strongly influenced by the CJs flow rate, pulse frequency, and phase.

Another time-periodic jet flow mixing enhancement active flow control method is the synthetic jet which is the main focus of this paper. Synthetic jet actuators are synthesized from the working fluid of the flow system, thus it can transfer momentum to the flow without a net mass injection while eliminating the need for an additional fluid source. Parekh et al. [22] used synthetic and pulsed jet from circular slit orifices injecting into low-speed shear layer, finding that tremendous increase in mixing by engaging the jet's flapping mode. Ritchie et al. [23] used nine coaxial synthetic jets to stimulate flow jet, finding that large-scale structures are more important in downstream and outside mixing layer and the small-scale mixing effects are strong in near field and downstream in inner mixing layer. David et al. [24] used a single synthetic jet to control axisymmetric free jet. The results showed that synthetic jet raise the primary jet's turbulent quantities to enhance mixing. From the mentioned above, the time-periodic jet has effects on jet mixing enhancement, although a thrust penalty. The main mechanism of this technology is flapping and streamwise vortex. Comparing with the steady jet, the time-periodic jet needs less or no bleeding air, but the thrust penalty for applying is still exist.

From the aforementioned, many investigations on time-periodic actuation of primary jet shear layers have been reported. However, much of the reported works have focused on low Re and low Mach number flow and the actuators they used also have a low-amplitude, others focused on jet vectoring and coaxial mixing. For aero engine application, it is important that the effectiveness of any enhancement technique will be demonstrated at high Re and Mach number values and the actuator must be "strong" enough. The investigation of using a high-amplitude synthetic jet forcing a high temperature primary jet to enhance jet mixing is seldom. So the piston-type SJA is used to actuate the high Re and high Mach number jet flow to study the mixing enhancement.

2 Numerical simulation model

2.1 *Domain Specification and Boundary Conditions and Computational Mesh*

The nozzle is a convergent nozzle with a convergent half angle of 11°, an outlet diameter d of 30 mm with a short parallel-walled section. The diameter of the nozzle inlet is 1.25d. The entire length of the axisymmetric convergent nozzle is 1.8d. There are four normal injection SJAs which placed 0.1d downstream away from the nozzle outlet. The symmetric actuation mode is used. Adjacent twin antiphase synthetic jet can also achieve actuation frequency double. 3D physical domain is modeled and the computational domain is shown in figure.1. Piston-type SJA a, d (b, c) has the same phase and a, c 180°out of phase from each other.

The inlet of nozzle with a total pressure of 1.7atm and a total temperature of 800K is defined as pressure inlet boundary condition, and $Re_d=4\times10^6$. The outlet of the computational domain is defined as pressure outlet at a static pressure as 1atm. The actuator nozzle diameter d_0 is 0.11d. The actuation frequency is set to 200 Hz.

The movement of the piston-type actuator is simulated by dynamic mesh. The growth of the piston wall is based on remeshes methods of mesh deforming. The unsteady time step size is based on the piston crank angle step size and actuation frequency.

Multi-block, structured meshes are created for the axisymmetric convergent nozzle's computational domain using the ICEM code. The grids as shown in Fig. 1 near the nozzle wall have been refined using the standard wall function. In the streamwise (x) direction, the initial spacing of the nodes is set at 0.5, with the interval count as 150 for the downstream flow field of the whole nozzle. An expanding grid with an expansion factor of 1.03 in the y and z directions is used.

The numerical computations were conducted by the finite volume method developed by Patankar via the commercial computation fluid dynamics code Fluent [25]. The conservation equations for momentum and density-based continuity are solved together by the coupled algorithm.

Figure 1. Computation domain boundary conditions and computational mesh

2.2 *Governing Equations and Turbulence Model*

The commercial CFD solver FLUENT with time step method is used to obtain the solution of unsteady compressible Reynolds averaged Navier-Stokes (URANS) equations in the present study. The inviscid flux vector is evaluated by second order upwind scheme with Roes flux difference splitting method. The ideal-gas model based air is set as the fluid material. The viscosity is calculated by Sutherland low, and the gravity effect can be neglected.

The turbulence model used is the RNG k-ε model. The RNG model is more responsive to the effects of rapid strain and streamlines curvature than the standard k-ε model, which explains the superior performance of the RNG model for certain classes of flows. According to the comparison between the PIV experiments of a dual synthetic jets actuator in quiescent air and numerical simulations, a RNG k-ε turbulent model [26] is most consistent with the experimental results. Therefore, the RNG k-ε turbulent model is selected for the current study.

3 Results and Discussion

3.1 *Characteristics of SJA*

Figure 2. shows the piston-type SJA which has been used in this investigation. A 19.5 mm piston within a matching cylinder with a stroke of 16.4 mm is used, such that the volume displacement is 5.5 cm^3. The cylinder cover plate with an axisymmetric orifice of diameter d_0=3.5mm. The thickness of the cylinder cover plate l is maintained l/d=2.0. The minimum clearance between the piston and the cylinder cover plate (at TDC) is 2.2 mm, yielding a maximum compression ratio of 8.45.

For the piston-type SJA, the Pr has shown a sharp increase in blowing phase, and it is even higher than the critical Pr (Pr_{cr} = 1.89, k = 1.4), thus a supersonic blowing jet exists in figure 3. That means the SJA can afford a high-momentum blowing jet.

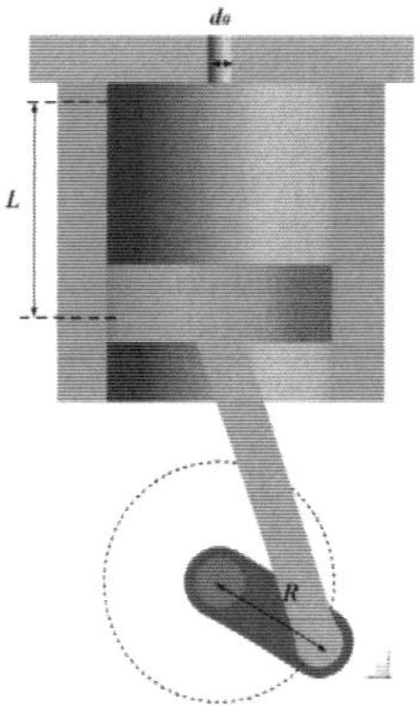

Figure 2. Schematic diagram of SJA

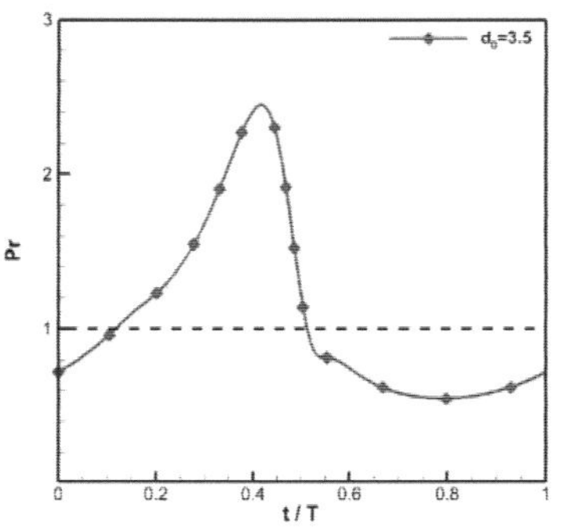

Figure 3. Time-periodic Pr distributions

3.2 *Mixing enhancement mechanism analysis*

3.2.1 *Primary jet flapping.* Figure 4. shows the instantaneous and time-averaged temperature field of the actuator plane. A jet flapping happens especially at t/T=0.25 and 0.75, which can reduce the potential core length and spanwise, so it plays an important role in jet mixing. Although the symmetric actuation mode is selected, but the position of the SJA is departure from the plume jet center, thus the jet flapping phenomenon appears. However, the primary jet flaps (even not violent enough) two times with the SJA actuation frequency. It can be conclude that the jet flapping is conducive to the enhancement of the jet flow expansion and mixing between the plume jet and ambient fluid. Compared with the unforced jet flow, the size of the time-average potential core is sharp decrease. Therefore, the synthetic jet has a great effect on the jet mixing enhancement, and sag area is also appearing in this figures.

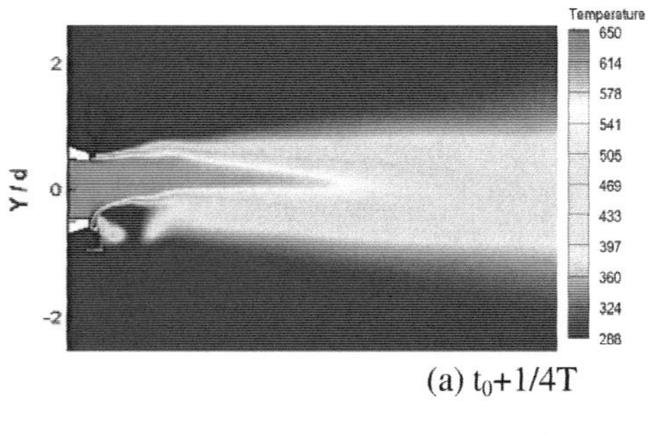

(a) t_0+1/4T

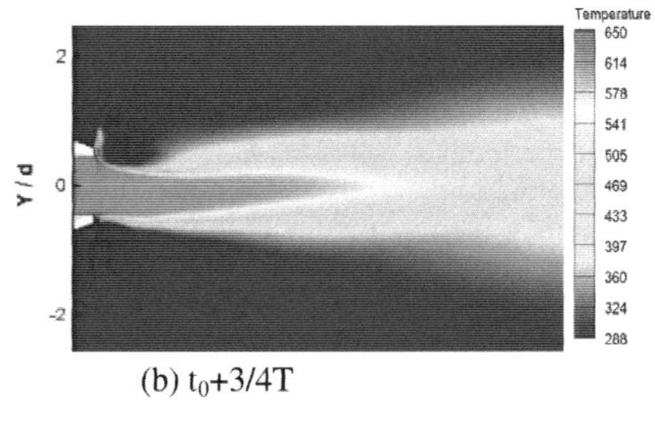

(b) t_0+3/4T

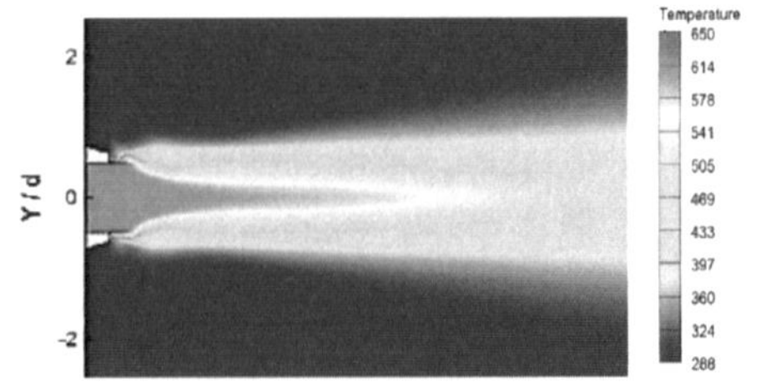

(c) time-average

Figure 4. Instantaneous temperature contours of centerline cross section of the actuation plane

3.2.2 *CVPs.* Streamwise vortices is defined as $\xi \cdot v / |v|$ [27]. And the magnitude of the streamwise vorticity has been normalized by v_e / d [28]. So, the dimensionless streamwise vorticity stv equation can be expressed as follows: $stv = \frac{d}{V_e}\left(\frac{\xi \cdot V}{|V|}\right)$. Figure 5 shows the evolution of the instantaneous stv distribution. The vorticity and position of the streamwise vortex are varying as the blowing jet momentum changes over cycle. When the synthetic jet begins to blow the primary jet, the formation of the CVPs is due to the interaction of the blowing jet and the primary jet. However, the vorticity of CVPs are not equal due to actuation of synthetic jet.

As a whole, the generation mechanism for the streamwise vortices produced by fluid actuation can be identified by two sources. The dominant source is the pressure hill formed upstream of the fluid actuation. The fluid actuation creates a pressure hill in the lateral direction, which causes the airflow to move around the fluid actuation. The result is generating a pair of streamwise vortices. The second source originates from the vortex lines shed from the sides of fluid actuation. As the jet propagates downstream, the vortex lines initially aligned parallel to the fluid actuation's side will be reoriented by the velocity gradients in the shear layer to become streamwise [28].

At 1/4T (figure 5. (c)), with blowing jet increases, the penetration depth further increases and synthetic jet blows into the primary jet, thus more ambient fluid is entrained into the primary jet, which changes the size of the potential core. The suction phase begins at 3/8T. However, the CVPs are still existing in the flow fields due to the duration of the blowing jet. At 1/2T, the CVPs disappears. And at 5/8T, the SJA repeat this process, and the evolution of the flow fields is similar to the mentioned above.

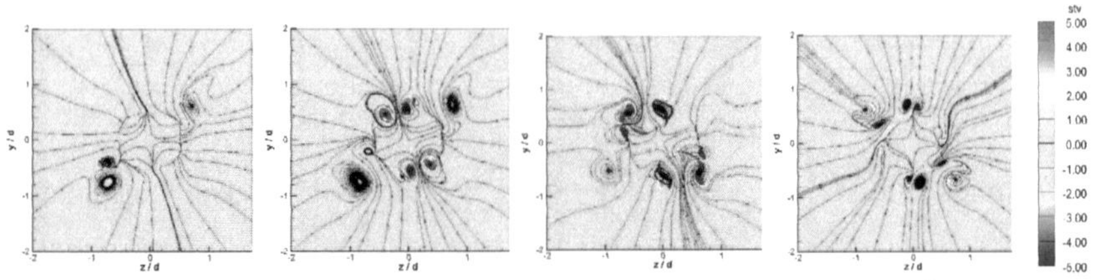

(a) t_0 (b) t_0+1/8T (c) t_0+1/4T (d) t_0+3/8T

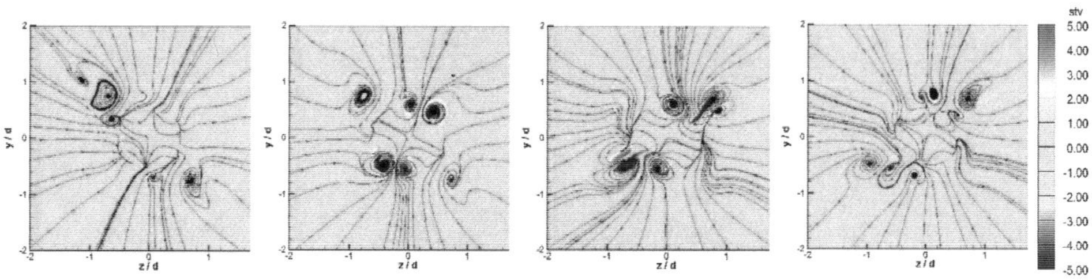

(e) t_0+1/2T (f) t_0+5/8T (g) t_0+3/4T (h) t_0+7/8T

Figure 5. Instantaneous Stv distribution at x=1.0 d.

3.2.3 *Geometric axis transformation.* The instantaneous temperature field of cross plane at x/d=1 is shown in figure 6. There is a low temperature "gap" in the potential core, due to the synthetic jet. The "gap" size increases and potential core size decreases obviously at t/T=1/4 and t/T=3/4. And the fluid of the "gap" is the ambient air, thus the temperature is low. However, the flow within the "gap" is complex, there are CVPs in this zone. As a result of this type actuation, at t=1/4T and t=3/4T, the "gap" is in the opposite direction, it can be named as geometric axis transformation. This phenomena may also promote the mixing of the ambient air and primary jet. However, the penetration area is determined by actuator nozzle diameter and blowing jet momentum.

The time-average temperature field distributions of the cross plane is show in figure 6(c). Compared with the unforced one, the size of the time-averaged potential core at cross plane is obvious decrease. And the CVPs lie in this zone, thus the CVPs play an important role in mixing enhancement.

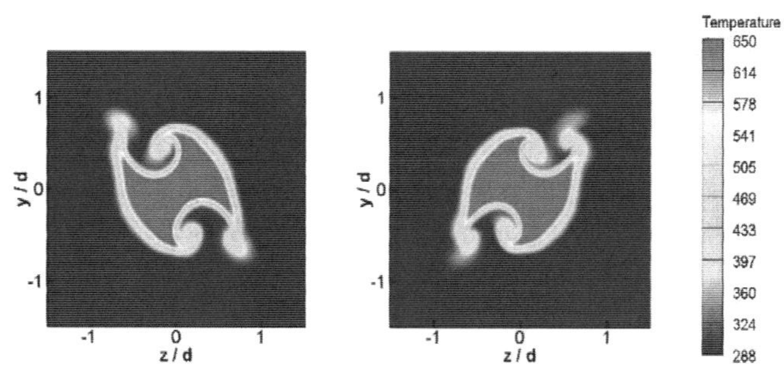

(a) t_0+1/4T (b) t_0+3/4T

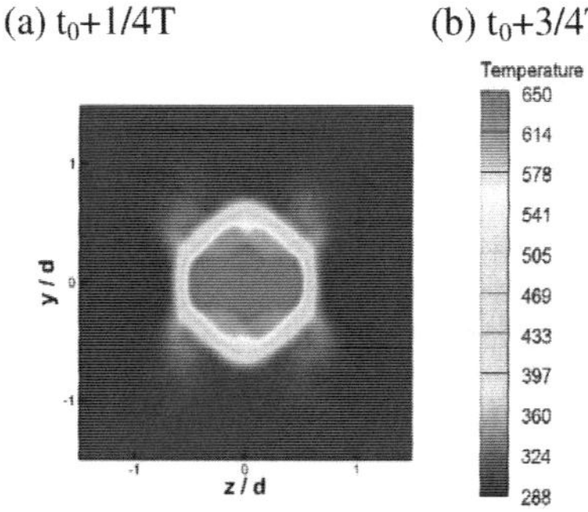

(c) time-average

Figure 6. Instantaneous temperature contours at x/d=1

3.2.4 *Evolution of 3D temperature field.* Figure 7 shows the evolution of 3D temperature field with the blowing jet at x/d=0, x/d=1, x/d=2 and x/d=3. The instantaneous temperature contour of the iso-surface is 650K. The blowing jet penetrates into primary jet at t=1/8T, changing the shape of the potential core. The moment of the blowing jet increases at t=1/4T, and the blowing jet has a strong impact on the primary jet. Compared with figure 7 (a), the sag area becomes larger and transports to the downstream, and the length of potential core becomes shorter apparently. The suction phase begins at t=3/8T. However, the preformed sag area transports alone the flow direction and the primary jet is also short. The influence of the synthetic jet on potential core is quite small at t=1/2T, and the length of potential core remains to the unforced status. After that, actuator a, d begins to blow the primary jet, and repeats this process. Compared with the unforced jet flow, the time-averaged temperature field shows that the SJA has a great effect on the jet mixing enhancement, and four sag area is seen obviously.

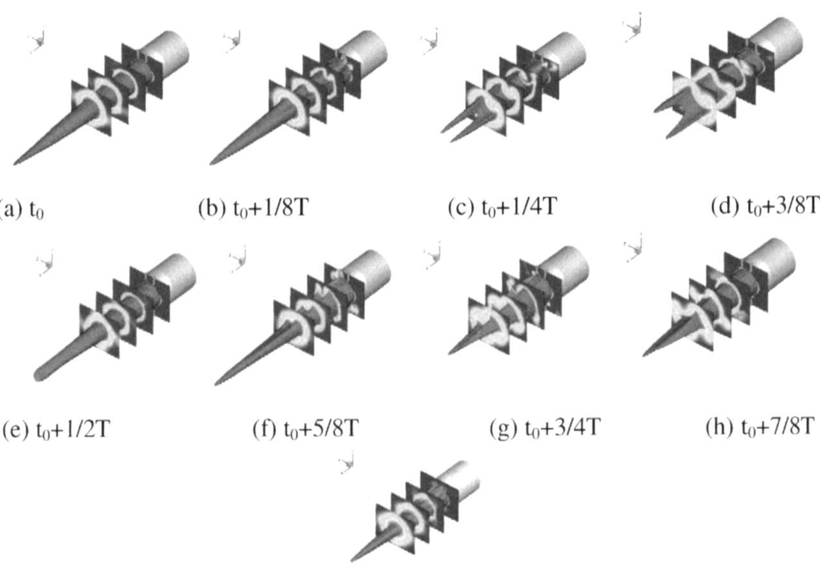

(a) t_0 (b) t_0+1/8T (c) t_0+1/4T (d) t_0+3/8T

(e) t_0+1/2T (f) t_0+5/8T (g) t_0+3/4T (h) t_0+7/8T

(i) time-averaged

Figure 7. Time sequence visualization of 3D iso-surfaces

3.2.5 *Comparison of the Jet's Centerline Temperature.* Figure 8 shows the time-averaged normalized temperature distribution of the primary flow centerline. The normalized temperature distribution of the primary jet centerline T' is used to access the mixing effectiveness. The definition of T' is as follows:

$$T' = \frac{T - T_a}{T_0 - T_a}.$$

With the SJA's actuation, the primary flow dimensionless centerline temperature decreases from x/d=2, and the length of the potential core reduces 54% and a maximum 45.8K decrease in potential core temperature. The vorticity of streamwise vortex is correlation to the penetration area, and a large penetration area can entrained more fluid into the primary flow. Thus a high-momentum and a larger nozzle diameter SJA is needed. From the analysis above, the jet flapping, CVPs and geometric axis transformation has the effects on jet mixing enhancement.

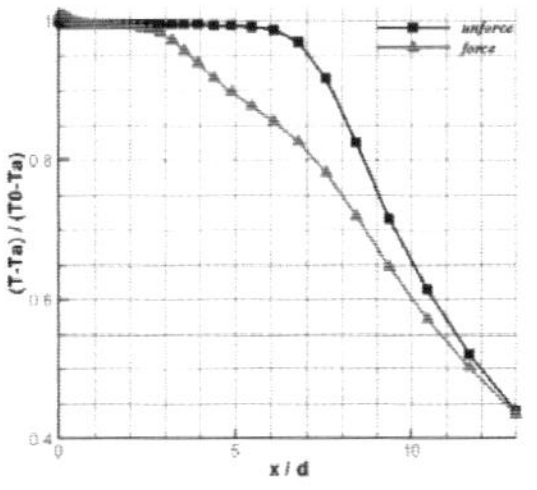

Figure 8. Comparison of the time-averaged primary jet centerline normalized temperature distributions.

4. Conclusion

Piston-type synthetic jet is used to enhance a subsonic, heat temperature jet mixing. Mixing enhancement effect is obvious. There are three main mixing mechanisms in this case, namely Jet flapping, streamwise vortex and geometric axis transformation.

Primary jet in actuator outlet plane flapping is caused by synthetic jet periodic blowing. This mechanism shortens the length of the potential core, accelerating the spread of potential core fluid to environment. CVPs plays an important role in mixing enhancement. This is due to the increase of the quality of the ambient fluid and primary jet potential core transportation, meanwhile enhancing the primary jet and environmental fluid mixing. Geometric axis transformation may promote the mixing of the ambient air and primary jet.

However, the jet mixing enhancement effect depends on penetration area. And the penetration area is determined by actuator nozzle diameter and blowing jet momentum. In addition, a 54% reduction in potential core length and maximum 45.8K decrease in potential temperature are achieved.

5. Reference

[1] Kamran M A, Manipulation of high Mach number shear layers using control jets. [D]. Loughborough Univ., U.K., 2009

[2] Edmnnd J. Field Kenneth F. Rossitto et al. Active control technology for enhanced performance operational capabilities of military aircraft, landing vehicles and sea vehicles [R].*ADA*395700. 2001. 1-1032

[3] Shripad P. Mahulikar, Hermant R. Sonawane, G. Arvind Rao. Infrared signature studies of aerospace vehicles [J]. *Progress in aerospace sciences* Vol.**43**, pp 218-245, 2007

[4] Timothy D. Smith, Alan B, Cain. Numerical simulation of Enhanced mixing in jet plumes using pulsed blowing [J]. *Journal of aircraft* Vol.**37**, No.3, 2001

[5] K Knowles, A J Saddington. A review of jet mixing enhancement for aircraft propulsion application [J]. *Journal of aerospace engineering* 2006 220:103

[6] Vinod G. Mengle. Relative clocking of enhanced mixing devices for jet noise benefit [R].*AIAA* 2005-0996

[7] Samimy M., Zaman K B M Q , and Reeder M F. Effect of tabs on the flow and noise field of an axisymmetric jets.[J] *AIAA Journal*, Vol.**31**, No.4, 1993, pp.609-619

[8] Brenton Greska, Anjaneyulu Krothapalli. High-speed jet noise reduction using microjets on a jet

engine [R]. *AIAA* 2004-2969

[9] Gretta W. J., Smith C. R., The flow structure and statics on a passing mixing tab [J]. *Fluid Eng.*, ASME. Vol.**115**, pp.255-263, 1993

[10] Zaman K. B. M. Q., Reeder M F, Samimy M. Supersonic jet mixing enhancement by delta-tabs. [R] *AIAA* 1992-3548

[11] Bradbury L J S., Khadem A H. Distortion of a jet by tabs. [J] *Journal of fluid mechanics*, Vol.4, 1975, pp.801-813

[12] Mengle V. G., Dalton W. N., Lobed mixer design for noise suppression [R]. NASA contract NAS3-27394, 1998

[13] R. Malecki, W. Lord Pratt. Navier-Stokes analysis of a lobe mixer and nozzle [C]. *AIAA* 1990-0453

[14] Samimy M., Kim J H. and Clancy P S. Passive control of supersonic rectangular jets via nozzle trailing-edge modifications.[J] *AIAA Journal*, Vol.**36**, No.7,1998, pp.1230-1239

[15] Callender B., Gutmark E., Martens S., A PIV flow field investigation of chevron nozzle mechanisms [C]. *AIAA* 2004-0191.

[16] Ali Uzun,M. Yousuff Hussaini., High-fidelity numerical simulation of a chevron nozzle jet flow [C]. *AIAA* 2009-3194.

[17] K Knowles, A J Saddington. A review of jet mixing enhancement for aircraft propulsion applications [J].*Journal of aerospace engineering*, 2006 220:113

[18] Ho C M., Huerre P. Perturbed free shear layer.[J] *Annual review of fluid mechanics*, Vol.**16**, 1987, pp. 365-424

[19] Raman G, Cornelius, D. Jet mixing control using actuation from miniature oscillating jets. [J]. *AIAA Journal*, Vol.**33**, No.2, 1995, pp 365-368

[20] Parekh D E, Kibens V. Innovative jet flow control mixing enhancement experiments [R]. *AIAA* 1996-0308

[21] Behrouzi P., Mcguirk J. Flow control of tabs mixing using a pulsed fluid tab nozzle.[R]. *AIAA*. 2006-3509

[22] Parekh, V.Kibens. Innovative jet flow control mixing enhancement experiments[R].*AIAA* 1996-0308

[23] B.D.Ritchie, D.B.Mujumdar, J.M.Seitzman. Mixing in coaxial jets using synthetic actuators [J]. *AIAA* 2000-0404

[24] David A. Tamburello, Michael Amitar. Active control of a free jet using a synthetic jet[J].*International journal of heat and fluid flow* Vol.**29**, pp 967-984, 2008

[25] Patankar, S. V., Numerical heat transfer and fluid flow [J]. McGraw Hill, New York, 1980, pp.89- 109.

[26] Zhen-bing Luo, Zhi-xun Xia, Wei Luo, et al. Principle of a novel dual synthetic jets actuator based continuous flow micro-pump. [C] *AIAA* 2008-4704

[27] Bluestein H B., "Synoptic dynamic meteorology in midlatitudes," New York, Oxford Univ press, 1992.

[28] Wan, C., and Yu, S. C. M., "Numerical Investigation of the Air Tabs Technique in Jet Flow," *Journal of Propulsion and Power*, Vol. **29**, No.1, 2013.

Extravehicular Space Suit Bearing Technology Development Research

Yan Pang[1], **Xiangyang Liu**[2, *], **Xie Guanghui** [3]

[123]China Astronauts Center, National Key Laboratory of Human Factors Engineering, Beijing 100094, China

[1]pangyan1992@126.com, [2, *]Lxyangacc@126.com (corresponding author)

Abstract. Pressure bearing has been acting an important role in the EVA (extravehicular activity) suit as a main mobility component. EVA suit bearing has its unique traits on the material, dustproof design, seal, interface, lubrication, load and performance. This paper states the peculiarity and development of the pressure bearing on the construction design element, load and failure mode, and performance and test from the point view of structure design. The status and effect of EVA suit pressure bearing is introduced in the paper. This analysis method can provide reference value for our country's EVA suit pressure bearing design and development.

1. Introduction

EVA is one of the critical technologies of manned space flight. EVA suit is the pivotal part to guarantee successful EVA accomplishment as individual protective equipment to improve EVA ergonomics meanwhile [1]. Protection ability and working efficiency are the two main functions of EVA suit [2]. Joint mobility of the suit is an important aspect of the EVA suit ergonomics and can reflect the technology level of space suit [3]. The joint mobility is achieved mainly through the soft fabrication and bearing structure. As space exploration reaching deep towards the moon, the mars and the asteroid, the space suit mobility requirements are higher [4].

The pressure bearing is the kind of bearing used in the EVA space suit sealed by the gas pressure. The pressure bearing is an important part making up and effecting the suit mobility along with the other join assembly. Currently used EVA suit, including American EMU (extravehicular mobility unit) suit [5], Russian Orlan space suit [6] served in the space station and our Feitian EVA suit, possesses pressure bearings assembly, usually applied to the shoulder, elbow, wrist and ankle joint to achieve good mobility [7-9]. The bearing structure is of good benefit to efficient suited walking and whole control of the suit.

Compared with the bearing of IVA (intravehicular activity) suit, the EVA suit bearing is faced up with more serious environment such as vacuum, low temperature, radiation and so on, and thus needs higher safety and reliability requirements. Therefore, the EVA suit bearing has more strict performance indexes on gas leakage, life of sealing element, or operation reliability in outer space [10]. So it's of great meaning to carry out investigation on EVA space suit pressure bearing.

2. Bearing construction structure design element

The space suit mobility is divided into two parts including ROM (range of motion) and degrees of freedom and mainly accomplished by soft fabric joint and pressure bearing. The soft joint structure can provide bending degrees while the single pressure bearing can provide rotary degrees. So the pressure bearings are applied to the joint like shoulder, elbow, wrist and ankle, characterized by rotation [11,12]. The usage of pressure bearing needs incorporation of human body kinematics and EVA suit joint requirements and reasonable design form of bearing and soft fabric joint according to these conditions [13]. The soft joint form includes the convolute joints, Wrinkling joints and hybrid joints. The bearing and the soft joint satisfy the requirements of extravehicular space suit joint ROM and degrees of freedom.

The followings are introductions from the aspects of structure format, material, dustproof techniques, seal, interface, lubrication, load and failure mode etc.

2.1 Pressure bearing structure format

The pressure bearing is customarily composed of inner ring, outer ring, the rolling part (usually rolling ball), the compression ring, the seal ring and other auxiliary component. The outer ring is of three purposes. First, it satisfies the bearing stiffness requirements as the primary structure together with the inner ring. Second, the outer ring is the connect part between the fabric layer and the bearing apparatus. Finally the outer ring provide a disconnect point for the all main parts and convenience for the astronauts donning the spacesuit. The bearing inner ring is mainly giving the connect point for the fabric layer and bearing contact surface and constitutes two rolling ball races along with the outer ring The inner ring aims at high strength and adaptive fit for the fabric layer. The compression ring imbedded to the bearing is the connect part of bearing and the seal ring. The compression ring is connected to the inner or outer ring by bolts or screw thread. The compression ring calls for highest compressive strength and fracture toughness. It also calls for low heat conductivity and coefficient of thermal expansion to fit the fabric, and low O_2 transmittance. The compression ring also has the function of sealing and friction [14]. Fig.1 shows the comparison of Russian and American shoulder bearing construction structure.

The pressure bearing, or called airtight bearing, is the kind of bearing sealed by gas pressure. The pressure bearing has the function of rotating and air sealing. The rotary function is achieved by the rolling part between the inner and outer ring. The sealing performance is accomplished by the contact force produced on the contact point between the sealing ring and inner ring. The sealing method of the pressure bearing is usually dynamic contact sealing.

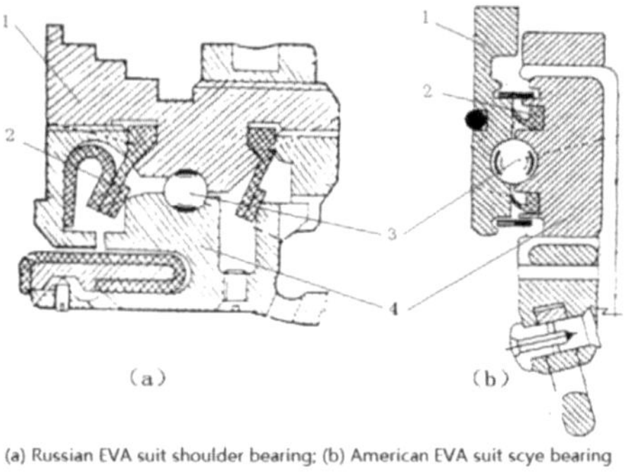

(a) Russian EVA suit shoulder bearing; (b) American EVA suit scye bearing

1 outer ring; 2 seal ring; 3 rolling ball; 4 inner ring

Fig. 1 typical pressure bearing structure of Russian and American suit

2.2 Pressure bearing material

The performance requests for the EVA suit bearing joint mainly consist of motion flexibility, air sealing and light weight. To meet the requirements, the materials are supposed to possess high abrasive resistance, low friction coefficient, high elasticity modulus and high size stability.

The stainless steel is applied as inner and outer ring structure in American EMU suit bearing to guarantee stiffness and wearing resistance [15]. The following focuses on material usage in American advanced extravehicular space suit.

In the 20th century 90s, ILC DOVER investigated soft space suit for multi gravity environment EVA named I-suit like micro gravity near-earth orbit and partial gravity environment (the moon and mars) [16]. This is an experiment suit designed for future EVA and it is developed by three generations [17]. The weight of I-suit is reduced through new material and design, while the number of pressure bearings does not increase. The stainless steel is replaced by titanium alloy as the wearing resistance material of inner and outer ring. And the aluminium alloy is replaced by the CF(carbon fiber)/epoxy composite material as the compression ring. The steps make the I-suit weigh only 29kg without the PLSS(portable life support system), much lighter than the NASA EMU suit served in the ISS(international space station) [18-20].

In 2009, NASA designed and developed new type PSS (planetary space suit) Z-1 for planetary exploration in the solar system. The design concept for Z-1 suit pressure bearing is light weight and low profile [21]. Current pressure bearing covers a big amount of weight of the space suit. Light pressure bearing is good for improving the astronauts EVA ability and reducing launching weight. And the low profile helps amending astronauts' suited comfort during long task. The pressure bearing of Z-1 suit in waist, shoulder, arm and wrist follows similar design construction. The section plot presenting the structure and material usage is as Fig.2. The compression ring is made from light CF/epoxy composite. The inner and outer ring is made from 440C stainless steel to ensure abrasion performance and rigidity. Many features of current space suit design are reserved to minimum technical risks. The total weight of the pressure bearing of waist, shoulder, arm and wrist is about 4kg, with 8.26kg reduced from the present EVA suit mobility bearing [22].

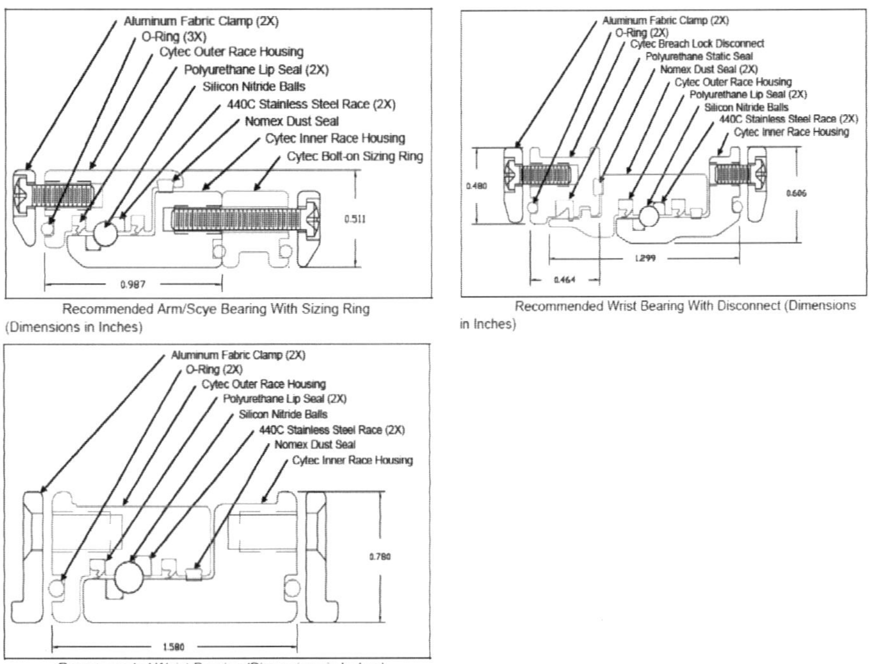

Fig. 2 recommended lightweight and low profile bearings

As for the mars exploration space suit, J.L. Marcy put forward a light bearing design project. The bearing joint is composed of three separate parts: outer bearing ring, inner bearing ring, and compression ring as Fig.3 shows [23].

The outer bearing ring is made from Ni-Ti SMA (shape memory alloy), just as its name implies, which possesses so called memory effect. The ring is fabricated into the desired size at room temperature in austenitic phase and recovers its original specific shape upon heating through transformation temperature range. The material will be used in the austenitic phase for the whole temperature range on mars. For the material of inner ring, Mg-CF shows integrated balance of performance indexes and light weight. Polyetheretherketone (PEEK)+60% carbon fiber (CF) exhibits the best overall property-based performance for the compression ring. A Teflon coating is used to reduce rotary friction to improve the pressure bearing mobility and could eliminate electro-static dust simulation. The rotary bearings are located at: the waist, to allow pelvis rotation and slight lean; the shoulder, to allow shoulder and arm rotation; the upper arm (a little higher than the elbow), to help arm rotation; the fore arm (a little higher than the wrist), to allow wrist rotation. The wrinkle joint are used at elbow, shoulder, knee, and ankle to allow flexibility of fundamental body joints. The whole fabric layer and bearing apparatus weigh only 20.88kg.

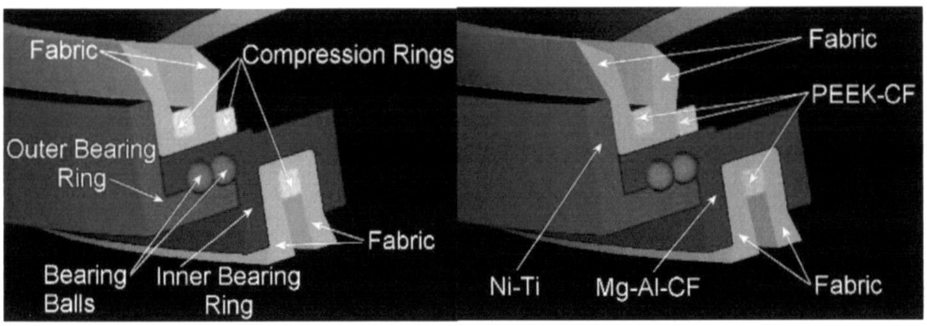

Fig. 3 the structure and materials of mars space suit bearing

2.3 Pressure bearing dustproof techniques

Dust is a serious challenge for the planetary EVA exploration [24]. Both the lunar and Martian surfaces contain deep covering layers of line dustlike soil composed of fine debris. Ranging from a medium sand (40 to 130 mm) to a medium silt (20mm), the regolith generally consists of an average particle size of 70 mm. The past Apollo lunar activity experience showed that the bearing performance rapidly decreased once intruded by the dust. Techniques for providing dust protection for bearings include separate environmental protective seals, labyrinth seals, and lubricant-impregnated felt seals or combinations of those mentioned above [25]. The techniques are illustrated in the following Fig.4. And certainly one of the main principle in the design process of the bearing dust protective system is not influencing the whole mobility torque.

In the dust seal concept shown, the outer environmental seal is loosely fitted into retaining grooves machined on both inner and outer races. The environmental seal serves as the primary outer barrier in providing protection against dust from entering the ball bearing seal and race areas. A secondary barrier to prevent any dust escaping past the outer environmental seal would be the labyrinth seal, which is a series of closely spaced machined grooves on both inner and outer races that serves as a maze in precluding dust from the seal and race area. Finally, a vacuum oil-impregnated felt strip, which is slightly compressed between inner and outer races， serves as the final dust barrier by capturing any dust particles on the oil surface of the felt strip. Because the contact pressure between the felt and the race is low and the felt is impregnated with a lubricant, friction will be low.

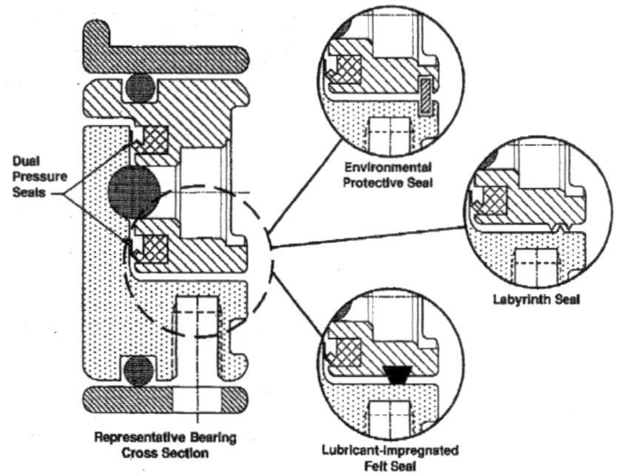

Fig. 4 pressure bearing dustproof techniques

2.4 Pressure bearing seal

The design of pressure bearing involves the construction design and material choice. Y lip seal was applied in the Shuttle EMU space suit. During the tests leakage and failure of the seal was discovered. The test results showed that bearing ball speed changes could cause the seal to drape. This is because the seal is acting as a part that can both separate air and rolling balls. The bearing speed change is effected by the contact angel and ball load. Thus the bearing structure and seal is developed by Joseph V. Welch using FEM (finite element method). Joseph V. Welch improved the load distribution by redesigning the construction to make the contact angel in line with the load direction while did not eliminate the leakage of the Y lip seal thoroughly [26]. The shuttle EMU Y lip seal waist bearing and improved prototype bearing are shown in Fig.5 [27,28].

Polyurethane lip seal is chosen as the lightweight and low profile bearing subassembly by J. Fricker. The seal lip is the same as EMU bearing seal and therefore the risk are technically low. Two seal rings are applied in the bearing and located on both sides of the rolling ball. The advantages of the seal rings are that their size is small so the room can be saved. It is a technical challenge to prevent leakage and guarantee seal and maintain minimum friction meantime on operational pressure that need balance and optimum design.

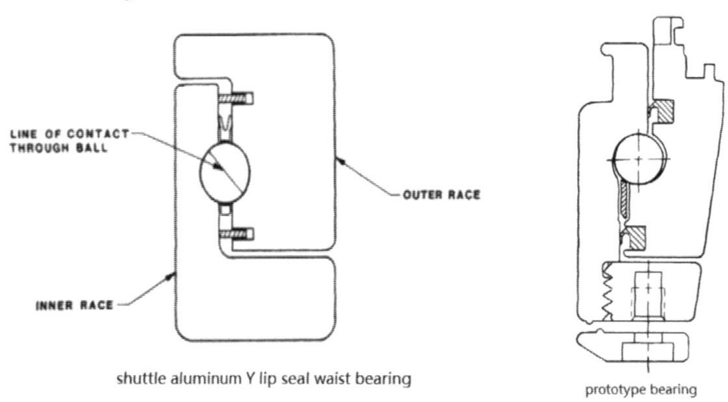

Fig. 5 shuttle aluminium Y lip seal waist bearing and prototype bearing

2.5 Pressure bearing interface

Common interfaces between bearing apparatus and SSA (space suit assembly) are axial webbing bracket and restrict flange. The axial load is transmitted to the bearing through the webbing fastened to the bracket. The webbing is usually mounted at the opposite 180° position around the cylinder joint as shown in Fig.6. The interface of bearing to the hard torso closure is snap ring and connect ring with thread [29]. The hardness characteristic of the interface tends towards distributing the axial load more evenly. The bending moments and out of plane displacement are smaller than webbing bracket under evener pressure load. The snap ring and connect ring help distribute even bending torque and deformation and hence the requirements of bearing interface with soft joint are higher than with hard apparatus representing the most challenge of the bearing interface design purpose.

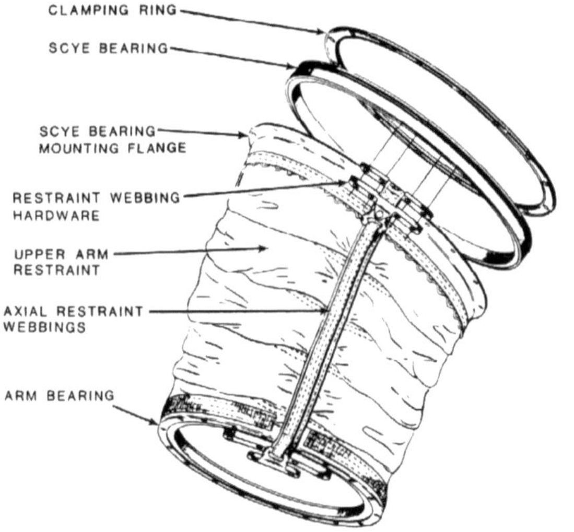

Fig. 6 shuttle SSA upper arm

Aluminium fabric clamp ring interface for the lightweight and low profile bearing is selected as Fig.2 shows. In the connections, a molded flange is connected to the suit pressure bladder and compressed between the fabric clamp and pressure bearing with blots and then sealed by the O-ring. This interfaced is proved to be highly reliable and easy to assemble and disassembly. Aluminium alloy is chosen instead of Cytec composite is due to its low density and ductility after all there are many connection points on the compression ring while the Cytec composite is hard and crisp existing fracture failure. Cytec composite could be an option of light material for the future space suit before performance modification and strict tests.

2.6 Pressure bearing lubrication

The twirling torque of the pressure bearing comes from the friction among the parts inner ring, outer ring, ball, seal ring and compression ring. Smooth operation of the bearing needs lubrication among the rolling races to decrease mobility torque and improve working efficiency which is of benefit to the wetting of the dust seal and hereby the dust protection is better. The Brayco 814 oil (a perfluoropolyether oil) is selected for the lubrication of lightweight and low profile bearing. Tests show the Brayco 814 oil is compatible with all the materials of the bearing recommended choice. Krytox 142 oil is used in the dustproof seal tests, the same kind of oil as Brayco 814, just another brand. For the underwater conditions, the Brayco 814 oil should be replaced by Mobil-28 oil which is less effected by water and harder to be washed out.

Speaking of our country's EVA space suit bearing, the design experience and results could not be imported blindly due to different construction structure. Self-lubricating material has been hot spot in the tribology at present and many colleges and research institutes has been doing much research [30-

32]. S.F Tian in HIT (Harbin Institute of Technology) researched the self-lubricating mechanism of TiB_2/Al material and derived the effects of load, velocity and roughness on the self-lubricating performance. However there is still a distance for the material wide application in the industry and space career. Self-lubricating material is a probable trend of future bearing design due to its ability of reducing rotary torque [33,34].

3. Pressure bearing failure and load

The hertz stress means the contact stress of the balls to the races [35]. The possible failure could occur on static conditions, including brinelling, or on dynamic conditions, such as fatigue. Brinelling causes the pits of races or balls due to high static load. Brinelling is not catastrophic failure that bearings are able to continue operation but brinelling leads to reduced life circle and increased torque. As the new materials including aluminium alloy and composites are widely applied in the pressure bearing instead of stainless steel, the possibility of brinelling is much lowered. The load is much smaller when fatigue failures happen whereas the running time is relatively long for thousands of circles which can also reduce pressure bearing life. The static load is the primary determinant because the desired life circles are relatively low which means it will not be possible to happen when the pressure bearing operates under small load for very long time.

The load that the bearing endures comes from pressure load and man load. The pressure load can be divided into hoop load and axial load according to the direction of acting force which is induced by the gas pressure of pressing specific space suit joint. The determination of pressure load needs the information of cross section area of the joint and pressure level. Nominal operation pressure level and maximum failure operation pressure level are two typical pressure used to verify the bearing intensity in case of failures. Besides the pressure load brings a radial force when the area of one bearing joint is bigger than the adjacent one (for example the scye bearing and the arm bearing). The man load occurs when the man interacts with the suit and is usually greater than pressure load. The man load is suit specific and depends on the suit construction and man-suit fit. Besides the man load is also influenced by the relative strength of crew man. The load analysis is an important part to validate bearing structure design [36].

4. Pressure bearing performance and experiment

The main function performance of pressure bearing includes sealing, mobility and life [37,38]. Aimed at the three indexes, S X.HE carried out series of tests on the sealing, mobility and life of Feitian EVA suit pressure bearing and derived the leakage and torque under different operation pressure and after required life circles. However the performance is yet to be researched from the construction design aspect [39]. The test principle is as the following Fig.7 [40].

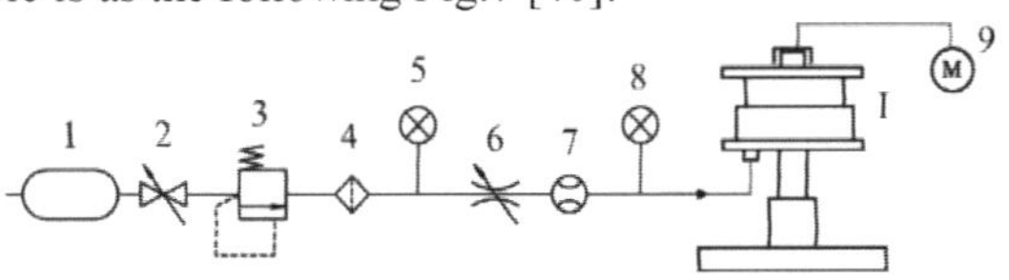

1. gas bottle; 2. on-off valve; 3. reductor; 4. filter; 5. Manometer; 6. Modulatory valve; 7. Flowmeter; 8. Manometer; 9. torque instrument; I. assembly part of airtight bearings and instruments

Fig. 7 the principle of pressure bearing sealing and mobility tests

Compared with soft fabric joint, the bearing joint possesses almost unlimited ROM, covering 360° rotation, which is hardly possible for the soft joint. And in the whole ROM, the bearing is supposed to exhibit the equal torque when it rotates towards one direction or the other [41,42]. J.Y Wang derived the mobility characteristics of the shoulder bearing, elbow bearing, and wrist bearing in their individual ROM of joint using the EVA suit arm joint mechanic characteristic test platform as Fig.8

shows [43]. There are two measurements numbered 1 and 2 each for shoulder bearing flexion and extension. While the torque curves are asymmetrical about x axis due to gravity effect.

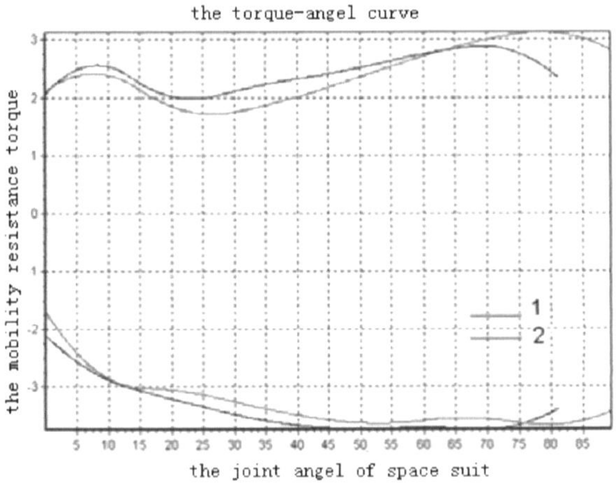

Fig. 8 the shoulder bearing joint angle-torque curve of spacesuit

In order to determine the design values of the joint for the advanced Constellation space suit [44], Matty J performed the mobility tests of 16 joints each suit including EMU, ACES (advanced crew escape suit), I-suit, D-suit, EM-ACES (enhanced mobility ACES) and Mark III [45,46]. The following Fig.9 is the test rests of scye bearing flexion/extension mobility torque under 30kPa of EM-ACES [47, 48].

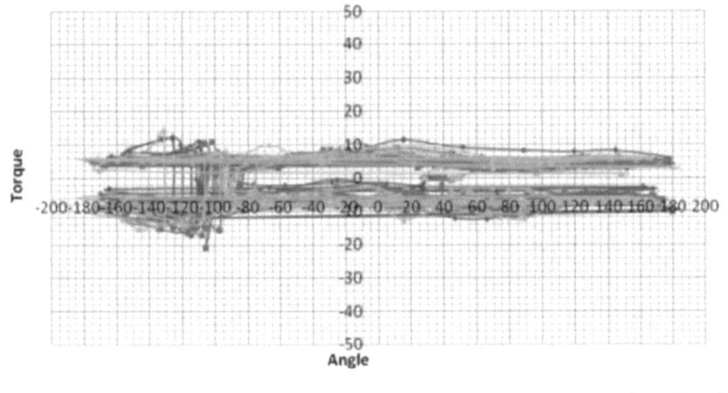

Fig. 9 torque profile for EM-ACES shoulder bearing at 4.3 psi during flexion/extension

Besides the mobility tests[49], after the Apollo program, NASA carried out experiments on the lunar dust invasion influence on the suit pressure bearing seal ring abrasion. The wrist bearing is installed in the test equipment and the lunar dust pours in as the wrist bearing rotates. The test equipment can simulate whether the lunar dust would lower the pressure bearing performance as shown in Fig. 10.

Fig. 10 the test equipment of lunar dust effects on pressure bearing

5. Conclusion

This paper introduces the pressure bearing construction structure design element, load and failure modes and performance and test related researches home and abroad from the aspect of EVA suit structure design. The separate analysis will provide good reference value for the pressure bearing and EVA suit design.

The pressure bearing and the soft fabric joint exhibit different characteristic for providing mobility. The soft joints can help accomplish the flexion/extension of the joints under pressure through fabric and the mobility angel is much smaller than 360°. The torque is different at the same angel in the process of flexion and extension, existing so-called hysteresis. The pressure bearing, made from alloy or composites, provide 360° rotation mobility. The torque hardly changes with the angel. Our country's EVA suit bearing inherits and develops Russian design concept the advantages of which are relatively easy fabrication and reliability and exhibits the problem of low structure stiffness, unstable size, bad abrasion resistance and slippage resistance at the same time. The pressure bearing is the main source of EVA suit leakage and weight-reduce aim part and needs to satisfy the function of bearing, sealing and rotation.

The design level of pressure bearing reflects and influencing the development of the future EVA suit. EVA suit bearing design involves many subjects such as material, seal, tribology and so on and possesses strict requirements of process, test equipment and so on. The development of pressure bearing needs the development of theory and manufacture. The design of pressure needs coordination with the requirements of the entire EVA suit.

6. Reference

[1] JIA S G, CHEN J S. The Status Quo and Development of Eva Suit System[C]// China Aviation Association Meeting.2003.
[2] CHEN J S, LI T Q, BIAN J M. space suit engineering[M], Beijing, National Defence Industry Press, 2002, 1-23, 205-206.
[3] ZHOU Q X. The Work Efficiency of Spacesuit for EVA and its Research Way[J]. Aerospace Shanghai, 2005, 22(3):47-51.
[4] Jacobs S, Tufts D. Space Suit Development for the Multi-Purpose Crew Vehicle[M]. 2012.
[5] Schmidt, Barrett P. An investigation of space suit mobility with applications to EVA operations[J]. Massachusetts Institute of Technology, 2001.
[6] Abramov I, Skoog I. Russian Spacesuits[M]. Springer Verlag, 2003.
[7] Abramov. I,Moiseyev. N and Stoklitsky. A. Concept of Space Suit Enclosure for Planetary Exploration[R].SAE 2002-01-2168.
[8] Kenneth S. Thomas, Harold J. McMann. U. S. Spacesuits[M].America:Springer Science 2012, 310-314.
[9] Harris G L. The origins and technology of the advanced extravehicular space suit[J]. Isis, 2002, 93(4):764-764.

[10] GUAN C L, LI M, CHEN J S. Advances in the American/Russian Lunar Extravehicular Space Suit [J]. Manned Space Flight, 2010, 16(02):45-52.
[11] Bradshaw H. Analysis of Scye Bearing Motion as Applicable to the Design of a Morphing Spacesuit[J]. Dissertations & Theses Gradworks, 2011.
[12] Hill T R, Johnson B J. EVA space suit architecture: Low earth orbit Vs. Moon Vs. Mars[C]// IEEE Aerospace Conference. 2010:1-28.
[13] Bradshaw H, Akin D, Humbert J S. A Range of Motion Study and Human-Suit Correlation Analysis Supporting the Morphing Spacesuit Concept[C]// International Conference on Environmental Systems. 2013.
[14] Some Thoughts on the Architecture of a Planetary Extravehicular Spacesuit.
[15] Dover I, Ferl J, Hewes L, et al. System Considerations for an Exploration Spacesuit Upper Torso Architecture[J]. System, 2006.
[16] Klaus D, Bamsey M, Schuller M, et al. Defining Space Suit Operational Requirements for Lunar and Mars Missions and Assessing Alternative Architectures[C]// International Conference On Environmental Systems. 2006.
[17] Dover I, Ross A J, Graziosi D, et al. Evaluation of the Rear Entry I-Suit during Desert RATS Testing[J]. Evaluation, 2006.
[18] D.Graziosi, R. Lee. I-Suit Advanced Spacesuit Design Improvements and Performance Testing[J].SAE 2003-01-2443.
[19] Hoffman S J, Hoffman S J. Advanced EVA Capabilities: A Study for NASA's Revolutionary Aerospace Systems Concept Program[J]. 2004.
[20] Graziosi D, Ferl J, Splawn K, et al. Evaluation of a Rear Entry System for an Advanced Spacesuit[J]. Evaluation, 2005.
[21] Ross A J. Z-1 Prototype Space Suit Testing Summary[C]// International Conference on Environmental Systems. 2013.
[22] Fricker J, Mancuso D, Hedgecock J, et al. Lightweight/Low-Profile Spacesuit Bearings[C]// International Conference On Environmental Systems. 2007.
[23] Marcy J L, Shalanski A C, Yarmuch M A R, et al. Material choices for mars[J]. Journal of Materials Engineering & Performance, 2004, 13(2):208-217.
[24] AIAA. NDX-2: Development of an Advanced Planetary Space Suit Demonstrator System for the Lunar Environment[J]. Civil Engineering & Environmental Systems, 2011, 30:263-267.
[25] Kosmo J J. Design considerations for future planetary space suits[J]. Environmental Control Systems, 1990.
[26] Welch J V. Analysis of Space Suit Mobility Bearings Using the Finite Element Method[J]. 1991.
[27] FU P, CHANG D G. Seal Design Manual[M].Chemical Industry Press,2011.
[28] WU X L, YUAN L J. Introduction to Mechanical Seal Design. Chemical Industry Press,2013
[29] Abramov I, Moiseyev N, Stoklitsky A. Concept of Space Suit Enclosure for Planetary Exploration[C]// 2001.
[30] YAO S J, QIU M, ZHAGN Y Z. Research Progress on Friction and Wear Properties of Self-Lubrication Spherical Bearings [J]. Bearing, 2008(11):38-42.
[31] WANG M. Study on Tribological Properties of NiAl Matrix Self-Lubricating Composites[D]. Wuhan University of Technology, 2013.
[32] ZHEN C, HE X M, YANG M S. The Research Development of Solid Self-Lubrication Composite Coatings' Manufacturing Technology [J].Journal of Soochow University,2011, 31(5):19-23.
[33] TIAN S F. Study on Self-Lubrication Action Mechanism for TiB2/Al Composite[D].Harbin Institute of Technology,2013
[34] Ross A, Rhodes R, Graziosi D, et al. Z-2 Prototype Space Suit Development[C]// 2014.
[35] Fan L.FEA on Stress Field of Rolling Element Bearing Based on Explicit Dynamics [J].Journal of Beijing Jiaotong University, 2006-08,30(04):110-112.
[36] SI S J. Study on Bearing Preload of Space Joint and its Dynamic Characteristic [D]. Harbin

Institute of Technology,2010

[37] Valish D, Eversley K. Space Suit Joint Torque Measurement Method Validation[C]// International Conference on Environmental Systems. 2013.

[38] Kosmo J J, Ross A. Space Suit Mobility Evaluations in Lunar/Mars Gravity Environments[C]// International Conference On Environmental Systems. 1998.

[39] Graziosi D, Ferl J G. Performance Evaluations of an Advanced Space Suit Design for International Space Station and Planetary Applications[J]. 1999.

[40] HE S X, LIU X Y, LI T Q. Design of Airtight Bearings in Feitian EVA Space Suit. Space Medicine and Medical Engineering, 2013, 26-5, 388-390.

[41] Jacobs S E, Benson E A, Akin D L. Kinematic Analysis of a Robotically Augmented Pressure Suit for Planetary Exploration[J]. Hand, 2007.

[42] Philosophies Applied in the Selection of Spacesuit Joint.

[43] WANG J Y. research on robotic system for eva spacesuit joint's damping parameters measuring[D]. Harbin Institute of Technology,2008

[44] Tufts D, Jacobs S, Barry D. The Demonstrator Suit: Incorporating Waist, Hip and Shoulder Mobility in a Full Pressure Suit[C]// International Conference on Environmental Systems. 2010.

[45] Matty J. Results and Analysis from Space Suit Joint Torque Testing[C]// International Conference on Environmental Systems. 2010.

[46] Menendez V, Labourdette X, Baez J M. Performance of EVA Suit Mobility Joints Influence of Driving Parameters[C]// International Conference On Environmental Systems. 1993.

[47] Massimiliano Di Capua, David L. Akin, y Justin C. Brannanz.MX-3 Design and Technology Development of an Advanced All-Soft High Mobility Planetary Space Suit[J]. 40th International Conference on Environmental Systems. AIAA 2010-6176.

[48] Pablo de Leónand Gary L. Harris.NDX-2 Development of an Advanced Planetary Space Suit Demonstrator System for the Lunar Environment[J]. 41st International Conference on Environmental Systems. AIAA 2011-5013.

[49] Kosmo J J, Ross A. Remote Field Site Space Suit Mobility Study Results[C]// International Conference On Environmental Systems. 1999.

The Application of Leap Motion in Astronaut Virtual Training

Xie Qingchao[1], **Chao Jiangang**[2]

China Astronaut Research and Training Centre, Beijing 100094, China

[1]xqch@mail.ustc.edu.cn, [2] xjtucjg@163.com

Abstract. With the development of computer vision, virtual reality has been applied in astronaut virtual training. As an advanced optic equipment to track hand, Leap Motion can provide precise and fluid tracking of hands. Leap Motion is suitable to be used as gesture input device in astronaut virtual training. This paper built an astronaut virtual training based Leap Motion, and established the mathematics model of hands occlusion. At last the ability of Leap Motion to handle occlusion was analysed. A virtual assembly simulation platform was developed for astronaut training, and occlusion gesture would influence the recognition process. The experimental result can guide astronaut virtual training.

1. Introduction

With a boom of computer vision, virtual reality has been applied in many fields including games, entertainments and simulation training. Additionally, it is appropriate that using VR in astronaut training compared with expensive real simulators. There are two main parts in astronaut virtual training [1]. First significant part is to structure simulation virtual environments, and second important part is human-computer interaction. Most training task are completed by hand manipulation, so gesture interaction is one of important way of human-computer interaction in astronaut virtual training.

In order to meet the increasingly growing demands of training, gesture recognition based virtual reality training system has been used in astronaut training, such as virtual glove-box(GVX) system of NASA [2-4] and EVA astronaut virtual system of China astronaut center [5]. Both of these systems use data gloves to recognize gestures. However, data gloves are uncomfortable, complex and expensive. So advanced tools are essential for more convenient, non-contact, unencumbered interaction.

Recently many consumer-grade optic equipment for hand gesture recognition arose, which can gain hand gesture information easily without touching operators. Those advanced equipment, such as Kinect and Leap Motion, can help to realize natural human-computer interaction.

Kinect is consumer-grade depth sensor, which was designed for natural human-computer interaction. The Kinect sensor can capture depth and color images simultaneously. Kinect has been used in many fields including mapping, human body pose estimation, 3D modeling and hand gesture recognition. However, Kinect was designed for far human body recognition only with low-resolution of depth. So it is difficult using Kinect to achieve full 3D gestures.

Leap Motion controller allows the acquisition of full 3D gestures. With Leap Motion as a tool for gesture input, operator could communicate freely with virtual environments. There has been attempt combining Leap Motion with astronaut virtual training [6]. Leap Motion can provide precise and fluid tracking of hands, fingers, and small objects with sub-millimeter accuracy [7][8]. But Leap Motion could not offer steady and accuracy tracking of hands, when occlusions happen.

The main goal of this paper is to analyze the ability of Leap Motion to handle hands occlusions. This paper also presents some interesting features of Leap Motion. The main contributions of this paper are the following:

- Astronaut virtual training system based Leap Motion was structured.
- Mathematics model of hands occlusion was established.
- Ability of Leap Motion to handle occlusion was analyzed.

2. Astronaut virtual training system

Traditional astronaut ground training employs the construction of the simulation of the real scene to simulate the space environment. Generally the physical simulator is expensive, and it is difficult to simulate the distortion of the scene and the space scene. Additionally we cannot repeat the destructive test by physical simulator. Excitingly, virtual reality technology can break through the limitations above. The virtual reality technology is appropriate to be introduced into the astronaut's ground training. The astronaut virtual training system can be constructed. Depending different tasks various virtual space scenes are created using virtual reality simulator. In astronaut virtual training system stereoscopic vision is displayed to astronaut through helmet or cave. And astronaut could interact with virtual scenes using gesture. During the training process, feedback including visual, auditory and force, is provided to astronaut.

The astronaut virtual training system framework based on natural gesture is showed in fig.1. The system mainly includes virtual reality simulator and human-computer interface. The mission of virtual reality simulator is structure virtual scenes, which are presented to astronaut through helmet-mounted displays. Meanwhile head motion is tracked by sensor. Another important part is human-computer interface. Leap Motion is suitable to be used as gesture input device because Leap Motion controller is small and precise.

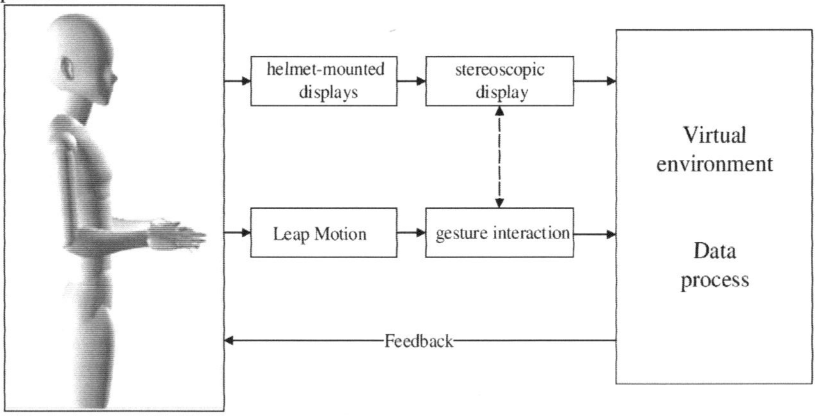

Fig 1. The astronaut virtual training system based leap motion

3. Experiment

The majority of applications for Leap Motion are gesture-based user interfaces. There are some researches about accuracy of Leap Motion, which showed high accuracy of Leap Motion for hand tracking. However, as optic equipment Leap Motion cannot resume full 3D gestures when occlusion happens. Experiment was designed to explore the features of Leap Motion.

3.1. Leap Motion controller

The Leap Motion Controller uses infrared (IR) imaging to track and recognize objects in the recognition space in real time. Leap motion sensor is located at one side of the surface. The detection region is a conical space, and the visual range is an inverted pyramid whose spire is in the center of the equipment. Detection region ranges from 25mm to 600 mm in front of the device. Leap Motion can provide precise and fluid tracking of hands. However, there hardly are details about technology because of patent and trade secret restrictions.

The coordinate system (left) and hand model (right) in Leap Motion are showed in fig.2.

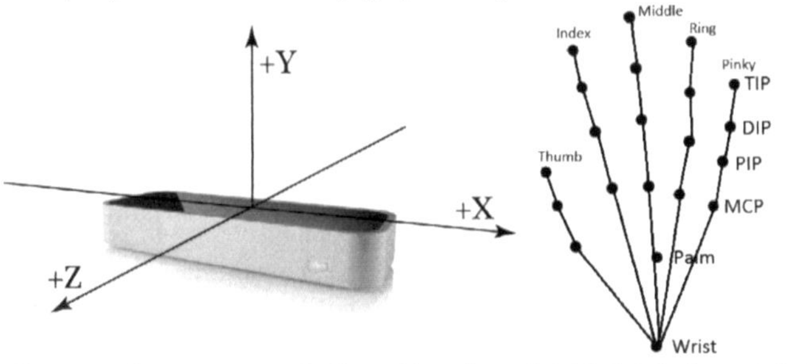

Fig2. The coordinate system (left) and hand model (right) in Leap Motion

3.2. Mathematics Model of Hands Occlusion

Human hand is composed of more than 50 bones. Complex structure and high degree of freedom are significant features of hand. When operator manipulates virtual objects by gesture, occlusion easily takes place. The occlusion of one hand includes three cases. The first case is that one finger bends, and one part of the finger is occluded by remain part of the finger. Second case is that one finger is occluded by another finger. The last case is that the finger is occluded by palm. Sometimes two hands will help finish tasks better. However, cooperative hands holding on the top of Leap Motion would interfere each other.

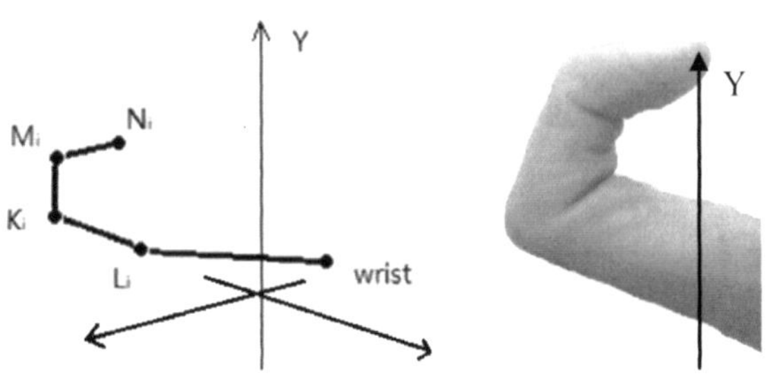

Fig3. One finger bends and self-occlusion

As fig.3 shows, finger i bends. And the bone M_iN_i is occluded by $\overrightarrow{K_iL_i}$. Mathematical expressions are as follow:

$$\begin{cases} \overrightarrow{M_iN_i} \cdot \overrightarrow{K_iL_i} > 0 \text{ and } \overrightarrow{K_iL_i} \cdot \vec{y} > 0 \text{, finger keeps bending.} \\ \text{Projection of bone } M_iN_i \text{ and } K_iL_i \text{ are same place in surface XOZ. } \vec{y} \text{ is the positive unit vector} \end{cases}$$

of Y axis.

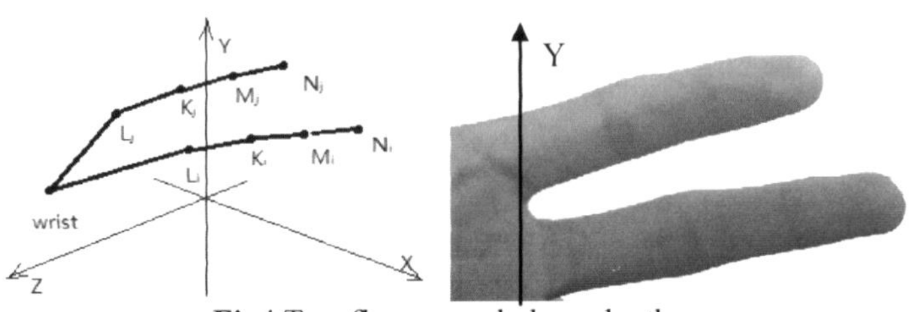

Fig4. Two fingers occlude each other

As fig.4 shows, finger i is below finger j. As an example, the surface which constructed by M_iN_i and M_jN_j, is perpendicular to the surface XOZ. And M_jN_j is occluded by M_iN_i. Mathematical expressions are as follow:

$$\begin{cases} \overrightarrow{M_iN_j} \cdot \vec{n} = 0, \overrightarrow{M_jN_i} \cdot \vec{n} = 0, \vec{n} \text{ is the normal vector to surface } M_iN_iN_jM_j. \\ \vec{y} \cdot \vec{n} = 0, \vec{y} \text{ is the positive unit vector of Y axis.} \end{cases}$$

When the finger is occluded by palm, its mathematical expressions are as follow:

$$\begin{cases} \vec{y} \cdot \vec{p} > 0.\ \vec{p} \text{ is the normal vector of palm.} \\ (N_{ix}, 0, N_{iz}) \in S_{xz}.\ S_{xz} \text{ is projection of palm in the surface XOZ.} \end{cases}$$

The mathematics model of hands occlusion will help us to understand occlusion better. Obviously, occlusion will decrease accuracy of Leap Motion, and even hand tracking is unavailable. So we want to know tracking ability of Leap Motion when occlusion happens. The following experiment presented the performance of Leap Motion under occlusion.

3.3. Experiment design

Leap Motion Visualizer is an application SDK provides, which can display tracking results of hands in real time, as fig.5 shows. Also the SDK provides coordinate of hand joints. Table 1 showed position of fingertips given by Leap Motion We experimented with Leap Motion controller under various occlusion cases. First, operator put one hand in the recognition region of Leap Motion, and spread fingers to initialize. Then operator made designed gestures which include all case in mathematics model of occlusion. With the help of Leap Motion Visualizer, the tacking results were observed. Position of fingertips were recorded to analyse Leap Motion accuracy. The experiment process is showed in Fig 6 and Fig 7.

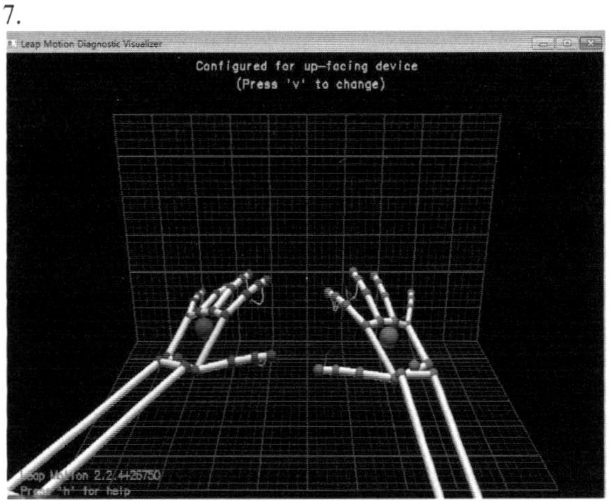

Fig5. Leap Motion Visualizer

Table 1. Fingertipsposition given by Leap Motion (mm)

Frames	195123	195138	195146
Thumb	(-12.2372, 234.425, -21.8778)	(-12.3027, 234.463, -21.8936)	(-12.4073, 234.463, -21.8259)
Index	(-3.32502,220.292, -93.3202)	(-3.56228,220.323, -93.3642)	(-3.73435, 220.222, -93.3804)
Middle	(-3.49694, 188.152, -114.144)	(-3.74823,188.109, -114.197)	(-3.96643, 188.029, -114.186)
Ring	(1.35593, 159.87, -108.419)	(1.13183, 159.93, -108.439)	(0.905643, 159.965, -108.407)
Little	(-10.9444, 117.279, -93.5915)	(-11.1747, 117.291, -93.6007)	(-11.3961, 117.382, -93.5931)

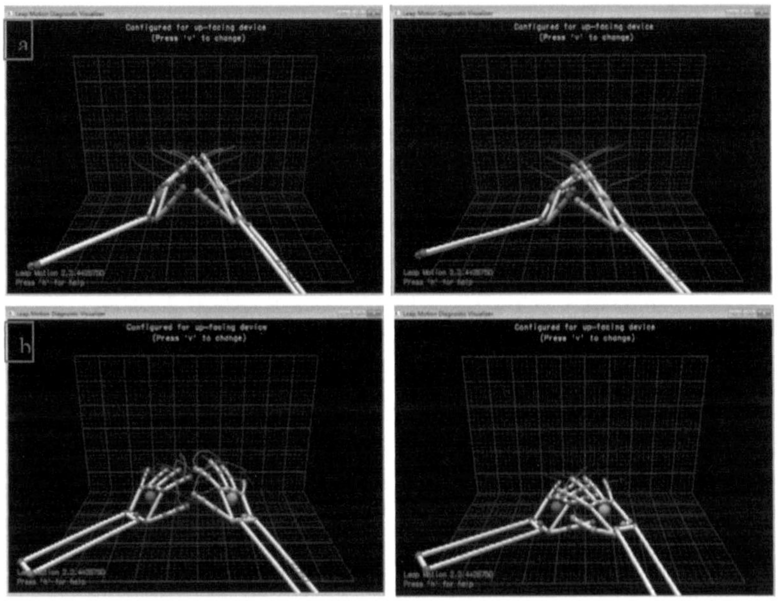

Fig 6. Two hands form non-occlusion to occlusion

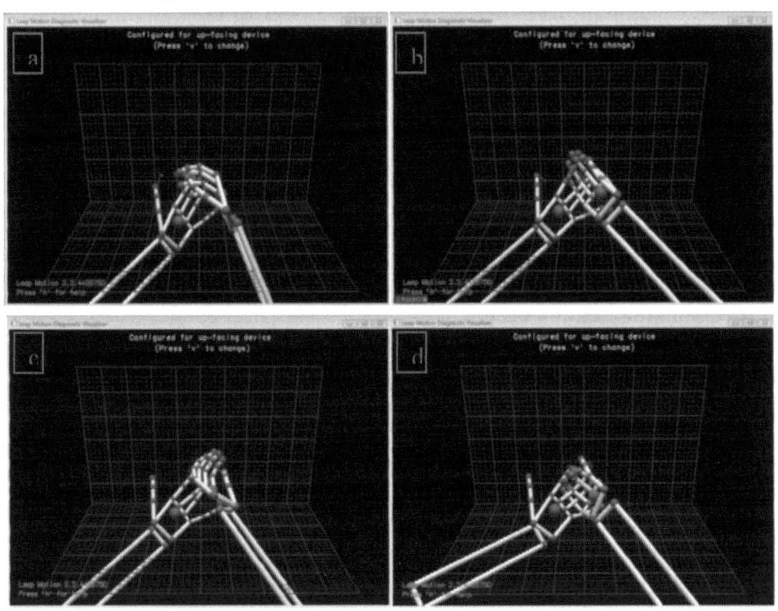

Fig 7. Two hands keep continuous occlusion

3.4. Application

We combined Leap Motion with astronaut training program, as input equipment. And a virtual hand was constructed in virtual environment, which was driven by hand pose data from Leap Motion. Rules were designed to catching and moving virtual components. So a virtual assembly simulation platform was developed to carry on test. Virtual environment was constructed by Unity 3D, which concluded virtual some work pieces and a virtual hand. The virtual hand was driven by data form Leap Motion.

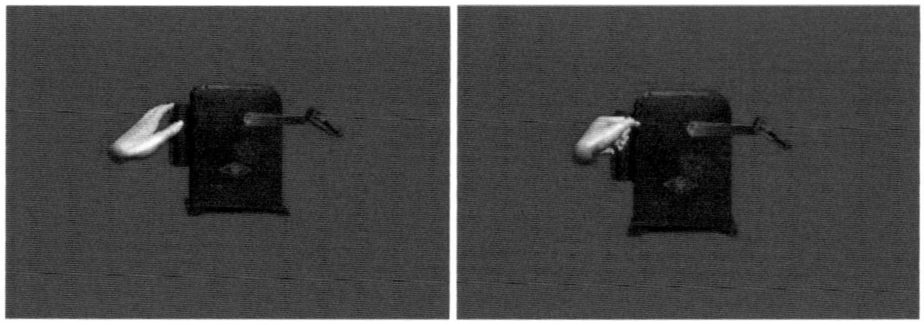

Fig 8. Virtual hand moves to work piece and grab

We designed grab gesture to catch and move work piece like in real world, as Fig 8 shows. And as Fig 9 shows, bounding box was used to judge if hand has capture the virtual work piece, which materially is collision detection. When the bounding box of virtual gets in touch with bounding box of component, they collide each other.

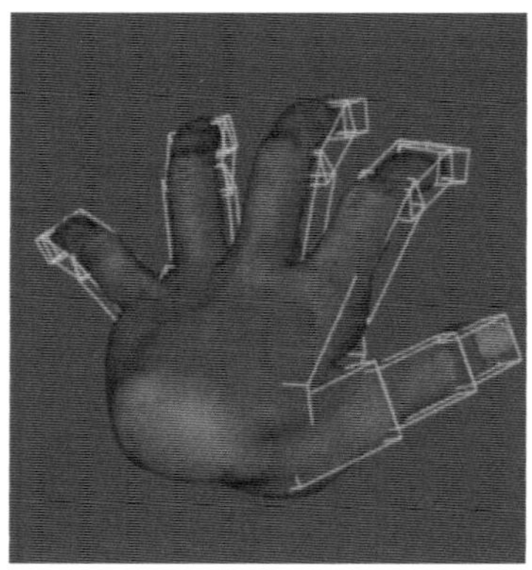

Fig 9. Bounding box of virtual hand

As Fig 10 shows, an integrated process of grab task concluded: a. click and choose a correct part; b. grab the part and move it; c. move the part to appointed position.

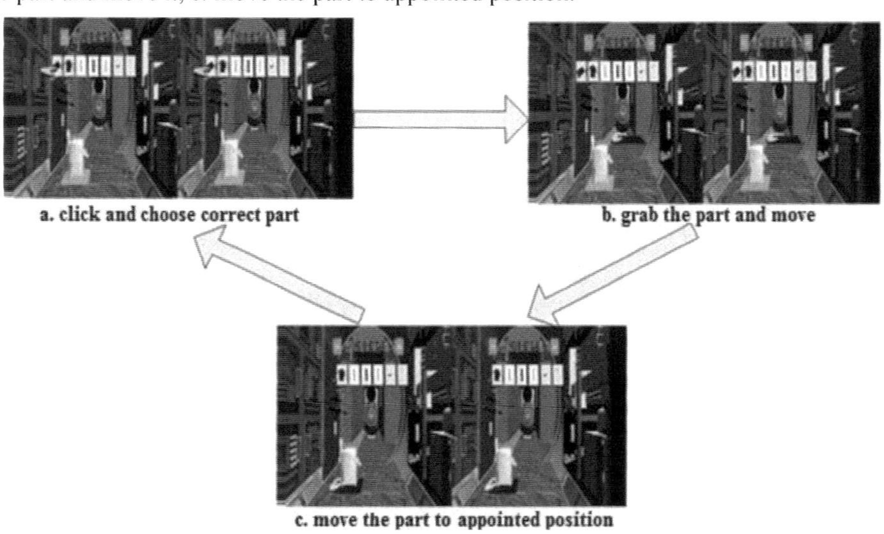

Fig 10. An integrated process of grab task

4. Results and analysis

Through the experiment in 3.3 we found the following direct results:

- The current frame of hand image Leap Motion drew was relevant to the state of the previous frame. The state of virtual hand can be kept a state before occlusion happened.
- The motion of the fingers is restricted by physical movement rules.
- When the finger is blocked, Motion Leap cannot identify the occluded part. At this point the virtual hand cannot truly reflect the real hand pose.
- When hand continued to move and occlusion disappeared, Motion Leap could immediately recovery exact tacking.
- When two hands are separate, the tracking process is similar to the process of one hand. When two hands occlude each other, Motion Leap would lose tracking to hands.

At the same time, we performed grab task using two different gesture: a. index finger and thumb knead, and the rest of fingers were straight; b. index finger and thumb knead, the other fingers were curled, in which self-occlusion took place, as Fig 11 shows.

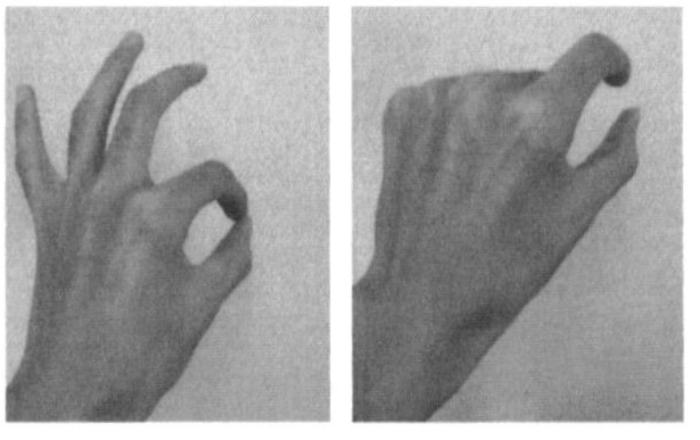

Fig 11. No occlusion (left) and self-occlusion gesture (right)

We recorded the time required to grab and move one work piece to the particular location. The result was showed in Table 2. Averagely 4.41s was cost to complete an integrated grab task using self-occlusion gesture, which spend more time compared with using no occlusion gesture.

Table 2 Time of succeed catching and moving the work piece to appointed position

Time	Average Time	Standard Deviation
No occlusion gesture	4.41	3.36
Self-occlusion gesture	3.32	2.91

The experimental results indicated that self-occlusion would influence astronaut's performance. So when carries out a training task, astronaut should try to avoid perform self-occlusion gesture. In other words, appropriate operating gestures should be designed in astronaut virtual training.

5. Summary

The experiment results showed robustness of Leap Motion, no matter hand self-occlusion or two hands occlusion. We guessed that Kalman filtering is used to predict hand motion and smooth tracking process. But it was powerless when long time continuous occlusion happened. Therefore, to use Leap Motion in astronaut virtual training we can simplify the operation of tasks.

We also developed a virtual assembly simulation platform combining Leap Motion and VR. And in future, appropriate operating gestures should be designed in astronaut virtual training.

Additional equipment or method can be used to assist in solving hand occlusion problem. For example, we can use multiple Leap Motion controllers which are placed in different viewpoints. We believe that there must be more advanced tools in future.

IOP Conf. Series: Materials Science and Engineering **187** (2017) 012015 doi:10.1088/1757-899X/187/1/012015

6. References

[1] Hu H, Chao J G, Zhao Z Q. Study of vision-based hand gesture recognition system for astronaut virtual training[J]. Advanced Materials Research, 2014, 998-999:1062-1065.

[2] Twombly I A, Smith J, Bruyns C, et al. NASA virtual glovebox: An immersive virtual desktop environment for training astronauts in life science experiments[J]. 2003.

[3] Smith J D, Gore B F, Dalal K M, et al. Optimizing biology research tasks in space using human performance modeling and virtual reality simulation systems here on earth[J]. Sae Technical Papers, 2002.

[4] Smith J. The Virtual Glovebox: Emerging Simulation Technology for Space Station Experiment Design, Development, Training and Troubleshooting[J]. Aiaa Journal, 2003.

[5] AIAA. VR simulation system for EVA astronaut training[C]// AIAA SPACE 2010 Conference & Exposition. 2010.

[6] Hu H, Chao J G, Lin W H, et al. Construction method of virtual hand based on Leap Motion and its application in spaceflight training [J]. Manned Spaceflight, 2015, 21(03):257-262.

[7] Weichert F, Bachmann D, Rudak B, et al. Analysis of the accuracy and robustness of the leap motion controller.[J]. Sensors, 2013, 13(5):6380-6393.

[8] Guna J, Jakus G, Pogačnik M, et al. An analysis of the precision and reliability of the leap motion sensor and its suitability for static and dynamic tracking[J]. Sensors, 2014, 14(2): 3702-3720.

IOP Conf. Series: Materials Science and Engineering **187** (2017) 012016 doi:10.1088/1757-899X/187/1/012016

Leader-Follower Formation Control for Quadrotors

Falin Wu[a], Jiemin Chen[b] and Yuan Liang[c]

School of Instrumentation Science and Opto-electronics Engineering, Beihang University, No.37 XueYuan Road, HaiDian District, Beijing, China

[a] falin.wu@buaa.edu.cn, [b] chenjiemin0417@buaa.edu.cn, [c] liangyuan@buaa.edu.cn

Abstract. Quadrotors are gaining an increasing interest in public and extensively explored in recent years. In many situations, a team of quadrotors is desired to operate in a certain shape, which is also called formation. In this paper, a linear PID controller is used to control each single quadrotor and a slide mode controller is adopted to solve the formation flying problem which employs the leader-follower structure. The formation simulations are run in the Matlab/Simulink environment to evaluate the performance of control laws.

1. Introduction

Quadrotor, or commonly known as drone and quadcopter, is a simple aerial vehicle which can operate in the narrow space with four propellers. With the pace of electronic device industry accelerating, quadrotors are being extensively explored and studied by an increasing number of scholars over the past few years. Carrying with sensors and other necessary devices, they make a difference to numerous civilian and military applications, including disaster mitigation, environment preservation, reconnaissance and academic research [1].

However, in many application scenarios, a team of unmanned aerial vehicles (UAVs) needs to follow the preset trajectory while maintaining a specific geometric shape. Compared with the conventional system, formation flight can increase the anti-interference performance and efficiency, improve the probability of success in search tasks, expand the region of surveillance and reduce the expense of military missions. As a result, the cooperation of multiple quadrotors is an active and popular research topic as formation has broad applications in accomplishing complicated tasks in real-world domains [2-4].

There are various methods to realize formation cooperation, such as leader-follower strategy [3; 5], virtual structure approach [6] and behavior-based method [7].

The leader-follower method is the most common hierarchical structure used in the formation flight. Each follower aircraft is controlled to maintain its position and velocity to a designated leader aircraft which flies along some predefined trajectory. This structure is characterized by its easy reconfiguration and simple expandability to new objects, for its function can be quickly taken over by other UAVs [8]. However, this strategy does not use direct feedback from the followers in general.

Virtual architecture approach enables quadrotors to fly as if they were embedded in a rigid structure. The formation has a so-called geometry center, which is a virtual point determined by shape, flying speed, orientation of the formation or some other factors. If one of the UAVs in the team loses its position, the overall flock can change the flight trajectory to help the lost one. The literature [9] describes this approach applied in formation of space crafts. However, this strategy is relatively difficult to be implemented in practical applications.

The behavior-based approach prescribes the desired behaviors and generates a vector based on a weighted average of sensory inputs. Possible behaviors include formation keeping [7], trajectory tracking and obstacle avoidance. This strategy is suitable for applications with multiple agents and has an advantage in communication among agents. However, it is difficult to assure some of the characteristics of the formation.

In this paper we adopt a slide mode controller [10] to manipulate formation flight of three UAVs, which takes the leader-follower structure. PID control laws are used and preformed for each of the three UAVs. To verify the control effect, simulations with six-degrees-of-freedom state space UAV model are presented, which are run in Matlab/Simulink software.

The outline for this paper is as follows: Section 2 introduces the dynamic model of the quadrotor. Section 3 lays out the controller strategy, including PID controller to manipulate a single quadrotor and a formation controller to guide the whole team of UAVs. The simulation results are summarized in Section 4 which evaluates the performance of the strategy. Section 5 is the conclusions.

2. Dynamic model of single quadrotor

The kinematic of quadrotor is formalized in earth frame E and body-fixed frame B. The position and Euler angles (roll, pitch and yaw) of the vehicle are defined as $\boldsymbol{r} = \begin{bmatrix} x & y & z \end{bmatrix}^T$ and $\eta = \begin{bmatrix} \phi & \theta & \psi \end{bmatrix}^T$ respectively, with mass m and inertia matrix $\boldsymbol{I}$. The angular velocity $\omega = \begin{bmatrix} p & q & r \end{bmatrix}$ is defined in the body-fixed frame B.

Some reasonable assumptions need to be considered to simplify the design which can be referred in literature [11]. Firstly, the quadrotor is a symmetric rigid body with the center of mass and body-fixed frame origin coinciding. So the inertia matrix $\boldsymbol{I}$ can be simplified as $\boldsymbol{I} = diag\begin{bmatrix} I_{xx} & I_{yy} & I_{zz} \end{bmatrix}$. Secondly, the quadrotor operates in the hover state where roll and pitch are small ($c\phi \approx 1$, $c\theta \approx 1$, $s\phi \approx \phi$, $s\theta \approx \theta$), so the angular velocity ω can be seen equal to $\dot{\eta}$ and the gyroscopic torque can be neglected.

The translational and rotational dynamics equations are as follows:

$$m\ddot{\boldsymbol{r}} = -mg\boldsymbol{e}_z + T_B \boldsymbol{R}_B^E \boldsymbol{e}_z \tag{1}$$

$$\boldsymbol{I}\dot{\boldsymbol{\omega}} = -\boldsymbol{\omega} \times \boldsymbol{I}\boldsymbol{\omega} + \tau \tag{2}$$

where g is the gravity acceleration; $\boldsymbol{e}_z = \begin{bmatrix} 0 & 0 & 1 \end{bmatrix}^T$ is the unit vector w.r.t. earth frame E; T_B is the total force generated by the four rotors w.r.t. body-fixed frame B; $\boldsymbol{R}_B^E \in SO(3)$ is the rotation matrix from body-fixed frame to earth frame; τ is the control torque in three dimensions obtained by changing the rotor speeds. The speeds of four rotors are manipulated by adjusting the total force T_B and torque τ, both of which actually are the components of inputs $\boldsymbol{u}$ of the system. The relationship between inputs and the speed of rotors is as follows:

$$\boldsymbol{u} = \begin{bmatrix} u_1 \\ u_2 \\ u_3 \\ u_4 \end{bmatrix} = \begin{bmatrix} T_B \\ \tau_\phi \\ \tau_\theta \\ \tau_\psi \end{bmatrix} = \begin{bmatrix} k & k & k & k \\ 0 & -lk & 0 & lk \\ lk & 0 & -lk & 0 \\ -d & d & -d & d \end{bmatrix} \begin{bmatrix} \Omega_1^2 \\ \Omega_2^2 \\ \Omega_3^2 \\ \Omega_4^2 \end{bmatrix} \tag{3}$$

where k is the thrust factor; l is the distance from the center of mass to any one of the rotors; d is the drag coefficient; Ω_i is the speed of the rotor i.

3. Control strategy

In this section, a PID law is utilized to make the single UAV move along the predefined trajectory. Then we focus on leader-follower formation control according to the geometry of formation and the relative dynamics.

3.1. PID control

The controller is designed to calculate the attitude and velocity needed for UAVs to follow the desired trajectory. Each quadrotor is controlled independently by the nested feedback loops shown in figure 1. The outer and inner loops control position and attitude respectively. The reference signals for velocity control are the sums of the reference values given by reference trajectory generator and correction signals calculated by rigid body dynamics. The superscript d represents the corresponding value is desired.

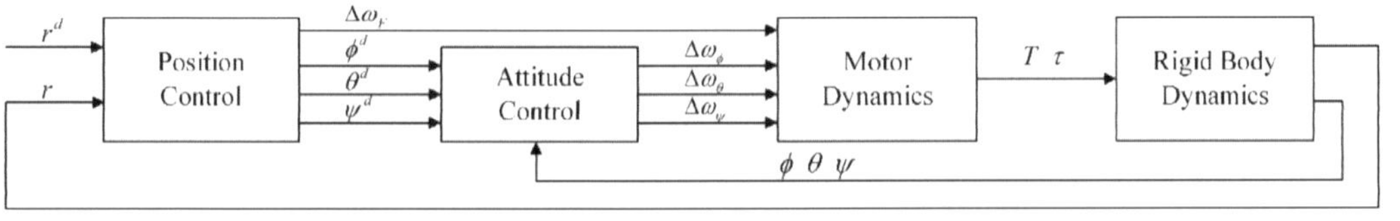

Figure 1. The nested control strategy block diagram

3.1.1. Attitude control. The attitude controller is designed at a point where roll and pitch are small. According to the assumptions, the products of inertia are zero and $I_{xx} \approx I_{yy}$ because of the symmetry then:

$$I_{xx}\dot{p} = u_2 - qr\left(I_{zz} - I_{yy}\right) \tag{4}$$

$$I_{yy}\dot{q} = u_3 - pr\left(I_{xx} - I_{zz}\right) \tag{5}$$

$$I_{zz}\dot{r} = u_4 \tag{6}$$

The component of the angular velocity r is very small. So the rightmost terms in (4) and (5) can be ignored compared with other terms. The system is presented in the hover state, so $\omega \approx \dot{\eta}$. For those reasons, simple proportional derivative control laws can be employed as follows:

$$u_2^d = k_{p,\phi}(\phi^d - \phi) + k_{d,\phi}(\dot{\phi}^d - \dot{\phi}) \tag{7}$$

$$u_3^d = k_{p,\theta}(\theta^d - \theta) + k_{d,\theta}(\dot{\theta}^d - \dot{\theta}) \tag{8}$$

$$u_4^d = k_{p,\psi}(\psi^d - \psi) + k_{d,\psi}(\dot{\psi}^d - \dot{\psi}) \tag{9}$$

3.1.2. Position control. We use pitch and roll to control position in XY plane, u_4 to control yaw angle, and u_1 to control height. The desired position is denoted as r_i^d, which can be calculated from PID controller

$$(\ddot{r}_{i,T} - \ddot{r}_i^d) + k_{d,i}(\dot{r}_{i,T} - \dot{r}_i) + k_{p,i}(r_{i,T} - r_i) + k_{i,i}\int(r_{i,T} - r_i) = 0 \tag{10}$$

where $\dot{r}_{i,T} = \ddot{r}_{i,T} = 0$ for hover.

By linearizing equation (1), the relationship between the desired acceleration and angles (roll and pitch) is obtained according to following assumptions:

a. The roll and pitch are small, so $c\phi \approx 1$, $c\theta \approx 1$, $s\phi \approx \phi$, $s\theta \approx \theta$.

b. The yaw keeps unchanged, which means $\psi = \psi_T = \psi_0$, where ψ_0 is the initial yaw and ψ_T is the desired yaw.

c. In hover state $\sum T_B \approx mg$.

After linearization, we can get

$$\phi^d = \frac{1}{g}(\ddot{r}_1^d \sin\psi_T - \ddot{r}_2^d \cos\psi_T) \tag{11}$$

$$\theta^d = \frac{1}{g}(\ddot{r}_1^d \cos\psi_T + \ddot{r}_2^d \sin\psi_T) \tag{12}$$

$$u_1^d = m\ddot{r}_3^d \tag{13}$$

3.2. Leader-follower formation control

The process of formation flight is as follows: the leader flies along a predefined trajectory which is stipulated by the trajectory tracking controller. Then the formation controller is designed to keep a specific shape in XY plane (e.g. the shape shown in figure 3), according to the relative kinematics between the leader and the follower. After the formation controller generating the desired velocity for the follower, the follower can track this velocity so as to keep the relative distance λ and orientation angle φ constant at the same height. The leader tracks the predefined trajectory of the formation while the follower tracks the desired velocity to keep the shape, but the same tracking control strategy can be used for both UAVs. The formation flying strategy is shown in figure 2.

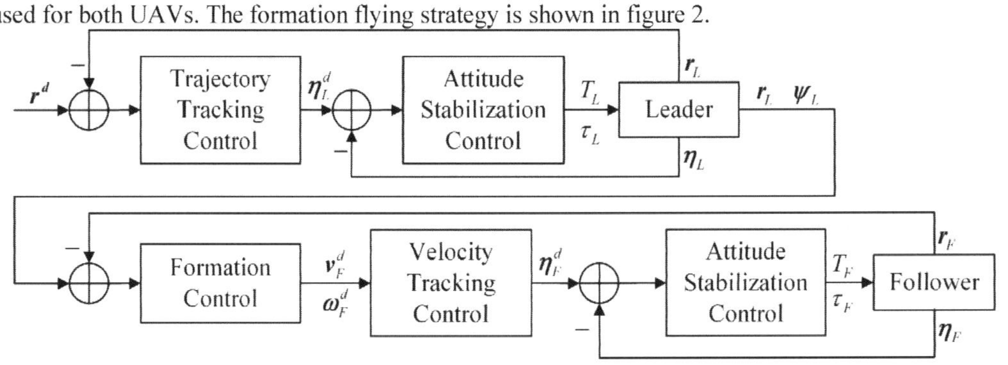

Figure 2. Control strategy block diagram

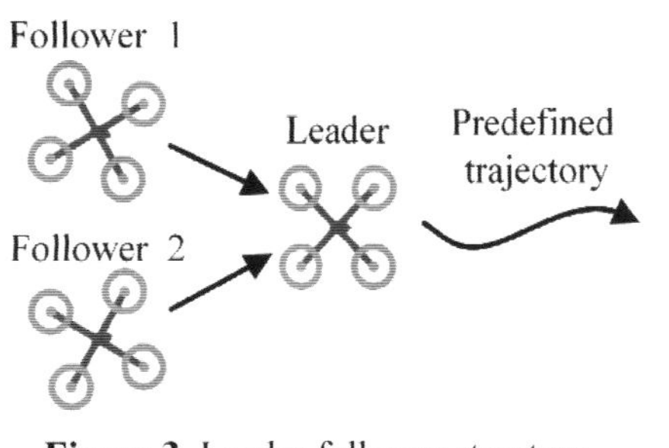

Figure 3. Leader-follower structure

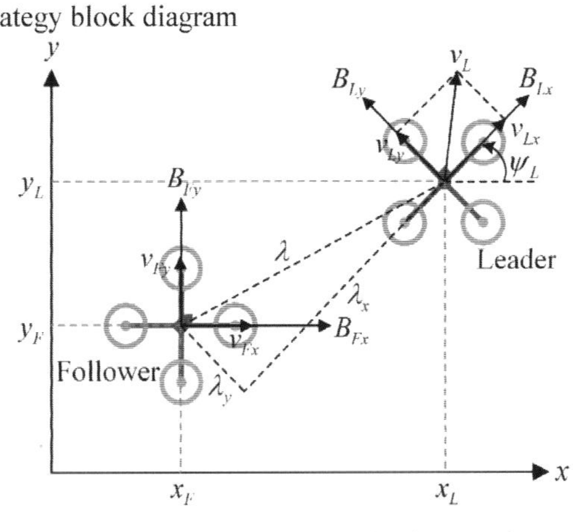

Figure 4. Quadrotors formation in XY plane

The leader-follower structure for three UAVs is shown in figure 3, with a geometry called 'V' shape. The position of UAV can be determined by velocity v_i and angle velocity ω_i for the yaw angle, which are the reference inputs to the trajectory tracking controller of the follower. More specifically, v_{ix} and v_{iy} are the velocity components in X and Y direction respectively w.r.t. body-fixed frame coordinate B, the subscript i denotes the leader ($i = L$) or the follower ($i = F$). ψ_i is the angle between the heading direction and the x axis of earth frame E.

Considering the dynamic characteristics of the quadrotor, it can be found that its vertical subsystem and horizontal subsystem are completely decoupling. Therefore, it is possible to design these two subsystems separately. For vertical subsystem,

$$\dot{z}_i = v_{iz}, i = \{F, L\} \tag{14}$$

where v_{iz} is the velocity component in z axis w.r.t. body-fixed frame B and $\dot{z}_i$ is the velocity component w.r.t. earth frame E.

Then we mainly focus on the relative translational kinematics in XY plane at the same altitude, which is demonstrated in figure 4. $\dot{x}_i$ and $\dot{y}_i$ are the velocity components w.r.t. earth frame E, while v_{ix}, v_{iy} are the velocity components w.r.t. body-fixed frame B. The translational dynamics in the XY plane can be described as

$$\dot{x}_i = v_{ix}\cos\psi_i - v_{iy}\sin\psi_i \tag{15}$$

$$\dot{y}_i = v_{ix}\sin\psi_i + v_{iy}\cos\psi_i \tag{16}$$

$$\dot{\psi}_L = \omega_L \tag{17}$$

Similarly, $\dot{x}_i$, $\dot{y}_i$ w.r.t. E can be transformed to v_{ix}, v_{iy} w.r.t. B.

$$v_{ix} = \dot{x}_i\cos\psi_i + \dot{y}_i\sin\psi_i \tag{18}$$

$$v_{iy} = -\dot{x}_i\sin\psi_i + \dot{y}_i\cos\psi_i \tag{19}$$

As shown in figure 4, λ is the distance from the mass center of the leader to one of the followers, λ_x, λ_y are the x, y coordinates in the leader's body-fixed frame(B_L). Then,

$$\lambda_x = -(x_L - x_F)\cos\psi_L - (y_L - y_F)\sin\psi_L \tag{20}$$

$$\lambda_y = (x_L - x_F)\sin\psi_L - (y_L - y_F)\cos\psi_L \tag{21}$$

Differentiating (20) w.r.t. time and using (15), (16), (17), (18) and (19), we can obtain

$$\dot{\lambda}_x = \lambda_y\omega_L + \dot{x}_F\cos\psi_L + \dot{y}_F\sin\psi_L - v_{Lx} \tag{22}$$

By defining $e_\psi = \psi_F - \psi_L$, using the equation (15), (16) with the trigonometric identities, one gets

$$\dot{\lambda}_x = \lambda_y\omega_L + v_{Fx}\cos e_\psi - v_{Fy}\sin e_\psi - v_{Lx} \tag{23}$$

Similarly, $\dot{\lambda}_y$ can be written as

$$\dot{\lambda}_y = -\lambda_y\omega_L + v_{Fx}\sin e_\psi + v_{Fy}\cos e_\psi - v_{Ly} \tag{24}$$

The desired distance λ^d and angle φ^d are constant, so $\dot{\lambda}_x^d = \dot{\lambda}_y^d = 0$. Define the formation error

$$e_x = \lambda_x^d - \lambda_x \tag{25}$$

$$e_y = \lambda_y^d - \lambda_y \tag{26}$$

$$e_\psi = \psi_F - \psi_L \tag{27}$$

Differentiating (25), (26), (27) w.r.t. time, one gets

$$\dot{e}_x = -\left(\lambda_y^d - e_y\right)\omega_L - v_{Fx}\cos e_\psi + v_{Fy}\sin e_\psi + v_{Lx} \tag{28}$$

$$\dot{e}_y = \left(\lambda_x^d - e_x\right)\omega_L - v_{Fx}\sin e_\psi - v_{Fy}\cos e_\psi + v_{Ly} \tag{29}$$

$$\dot{e}_\psi = \omega_F - \omega_L \tag{30}$$

From the formation dynamics (28), (29) and (30), a control law is designed to make the errors converge to zero. For doing this, v_{Fx}, v_{Fy} and ω_F are considered as the formation control inputs and the dynamics of the formation error is written as

$$\dot{\chi} = \boldsymbol{F}(\chi) + \boldsymbol{G}(\chi)\upsilon \tag{31}$$

where

$$\chi = \begin{bmatrix} e_x & e_y & e_\psi \end{bmatrix}^T \tag{32}$$

$$\upsilon = \begin{bmatrix} v_{Fx} & v_{Fy} & \omega_F \end{bmatrix}^T \tag{33}$$

$$F(\chi) = \begin{bmatrix} e_y \omega_L + \gamma_1 \\ -e_x \omega_L + \gamma_2 \\ -\omega_L \end{bmatrix} \tag{34}$$

$$G(\chi) = \begin{bmatrix} -ce_\psi & se_\psi & 0 \\ -se_\psi & -ce_\psi & 0 \\ 0 & 0 & 1 \end{bmatrix} \tag{35}$$

with

$$\gamma_1 = v_{Lx} - \omega_L \lambda_y^d \tag{36}$$

$$\gamma_2 = v_{Ly} + \omega_L \lambda_x^d \tag{37}$$

The slide mode technique is used to design a robust controller. The switching function is

$$\sigma_2 = \chi + \boldsymbol{k}_f \int \chi dt \tag{38}$$

where $\boldsymbol{k}_f$ is a constant matrix. The dynamics of the formation error are expected to keep on a surface defined by σ_2=0 through choosing $\boldsymbol{k}_f$ appropriately, on which the errors converge to zero. On this surface $\dot{\sigma}_2$=0, so we can get

$$\dot{\sigma}_2 = \dot{\chi} + \boldsymbol{k}_f \chi = \boldsymbol{F}(\chi) + \boldsymbol{G}(\chi)\upsilon_{eq} + \boldsymbol{k}_f \chi = 0 \tag{39}$$

υ_{eq} can be obtained by (39) as

$$\upsilon_{eq} = \boldsymbol{G}^{-1}(\chi)\left(-\boldsymbol{F}(\chi) - \boldsymbol{k}_f \chi\right) \tag{40}$$

To make sure the system keep on the surface σ_2=0 considering perturbations, it can be obtained

$$\dot{\sigma}_2 = \dot{\chi} + \boldsymbol{k}_f \chi = \boldsymbol{F}(\chi) + \boldsymbol{G}(\chi)\upsilon_{eq} + \boldsymbol{k}_f \chi = -Lsgn(\sigma_2) \tag{41}$$

where L is a positive constant. Then the control input can be obtained by

$$\upsilon_{eq} = \boldsymbol{G}^{-1}(\chi)\left(-\boldsymbol{F}(\chi) - \boldsymbol{k}_f \chi - Lsgn(\sigma_2)\right) \tag{42}$$

The $sgn(x)$ is defined as

$$sgn(x) = \begin{cases} 1 & x > 0 \\ -1 & x < 0 \end{cases} \tag{43}$$

and the vector

$$sgn(\sigma_1) = \begin{bmatrix} sgn(\sigma_{11}) & sgn(\sigma_{12}) & sgn(\sigma_{13}) \end{bmatrix}^T \tag{44}$$

4. Simulation and test results

The main aim of the paper is a numerical simulation which is run in Matlab/Simulink software. The controller system is highly sensitive to parameters. Improper parameters result in large overshoot, as a consequence, the expected distance among UAVs cannot be maintained. So the gains for PID controller are tuned experimentally to provide as short as possible convergence time and non-oscillatory character of the system response.

The quadrotors used in the simulation are all the same, with mass m = 0.58 kg, and moments of inertia $I_{xx} = I_{yy}$ = 0.1147 $kg{\cdot}m^2$ and I_{zz} = 0.0522 $kg{\cdot}m^2$. The products of inertia are assumed to be zero. These are typical values for small UAVs. In the simulation, the leader and the followers are set to follow the similar trajectory which is an s-shape curve actually. The velocity of the leader is provided to the followers. The initial position of quadrotor formation is $\begin{bmatrix} 0 & 0 & 0 \end{bmatrix}^T$ and the selected distance $\lambda_x = \lambda_y = 5\,m$. The simulation time is 250 seconds. The red line is the leader's trajectory while the black and blue line describe the trajectory of Follower 1 and Follower 2 respectively.

Simulation results of the quadrotor formation flight trajectory tracking in three dimensions and XY plane are presented in figure 5 and figure 6 respectively, indicating that the agents try to follow the leader's trajectory fairly well.

The position errors of Follower 1 in X and Y direction in the first 10 seconds are shown in figure 7 and figure 8. Due to the fact that the three quadrotors start from the same position, the initial position errors in X and Y direction are both $5\,m$ according to the predefined λ_x and λ_y. Then the Follower 1 moves to (5, 5) while the Follower 2 moves to (-5, -5) in the XY plane w.r.t. body-fixed frame B of the leader. So e_x and e_y converge to zero during the first 5 seconds. As can be seen from the figures, the errors converge to zero at a fast rate and fluctuate around zero after 5 seconds. The position errors of Follower 2 in X and Y direction are similar to Follower 1. In addition, the error of yaw also oscillates between -0.04 *rad* and 0.08 *rad* in the first 5 seconds and then fluctuates around zero after 5 seconds, which is not illustrated in the figure. The error is relatively small without considering the communication delay and disturbance. The analysis of error demonstrates that the performance of the formation control is quite satisfactory under the circumstance where the roll and pitch are small.

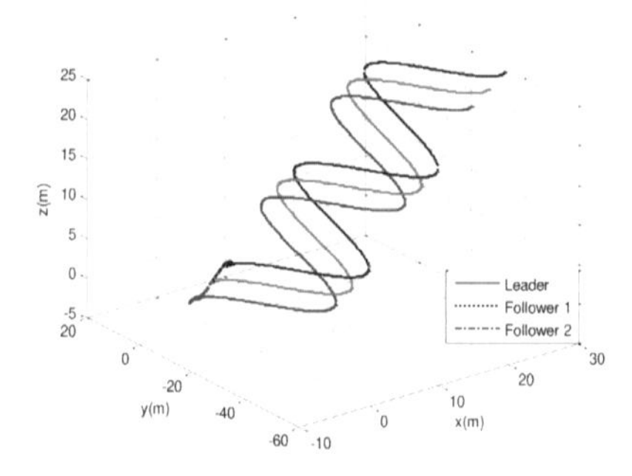

Figure 5. Quadrotors formation flight trajectory in three dimensions

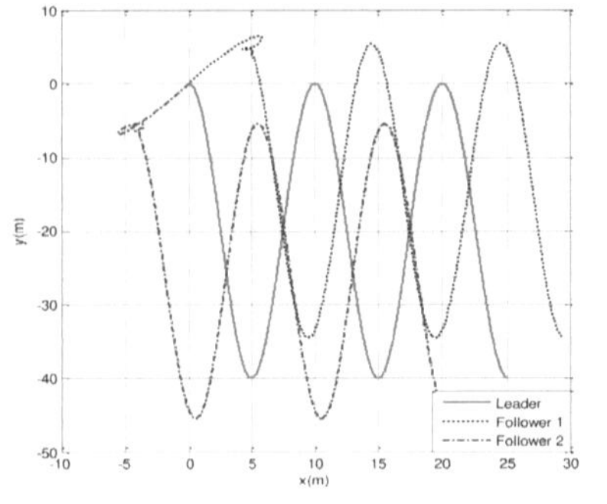

Figure 6. Quadrotors trajectory tracking in XY plane

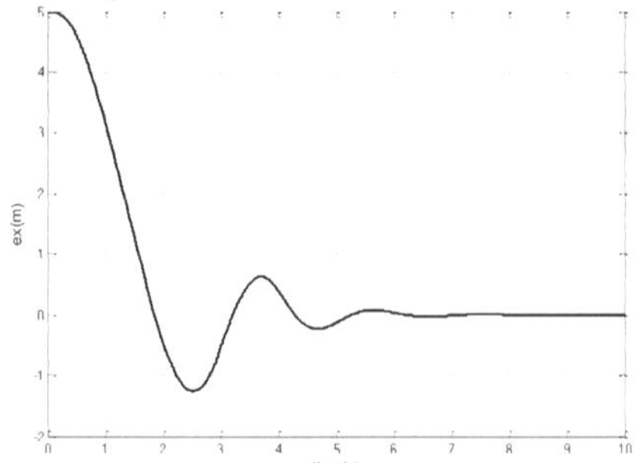

Figure 7. Position error in X direction

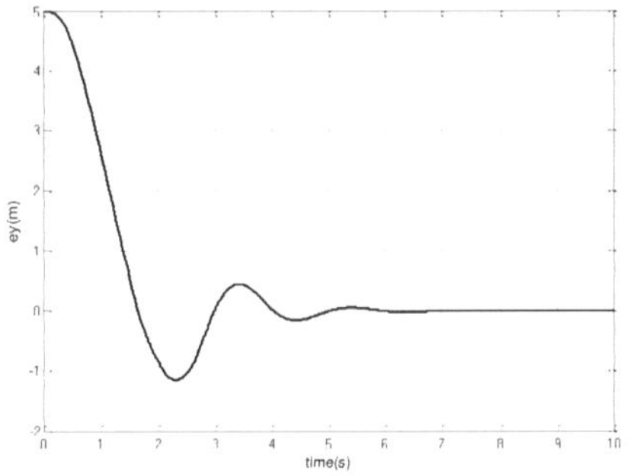

Figure 8. Position error in Y direction

5. Conclusions

The paper focuses on formation control with a leader-follower structure designed for three quadrotors. PID control law is presented to manipulate a single vehicle and a slide mode controller is adopted for formation control. The experiments illustrate the effective performance of control laws. However, this research is restricted to simulations without considering the practical noise and communication delay. Future research will focus on the hardware platform for formation flight experiments.

6. References

[1] Floreano D and Wood RJ 2015 Science, technology and the future of small autonomous drones *Nature* **521** 460-6

[2] Kendoul F 2012 Survey of advances in guidance, navigation, and control of unmanned rotorcraft systems *J. Field Robot* **29** 315-78

[3] Yun B, Chen BM, Kai YL and Tong HL 2010 Design and implementation of a leader-follower cooperative control system for unmanned helicopters *J. Control Theory Appl.* **8** 61-8

[4] Nathan PT, Almurib HA *et al* 2011 A review of autonomous multi-agent quadrotor control techniques and applications. *4th Int. Conf. on Mechatronics* (Kuala Lumpur) pp 1-7

[5] Roldão V, Cunha R, Cabecinhas D, Silvestre C and Oliveira P 2014 A leader-following trajectory generator with application to quadrotor formation flight *Robot. Auton. Syst.* **62** 1597-609

[6] Lewis MA and Tan K-H 1997 High precision formation control of mobile robots using virtual structure *Autonomous Robot* **4** 387-403

[7] Balch T and Arkin RC 1998 Behavior-based formation control for multirobot teams *IEEE Trans. Robot. Automa.* **14** 926-39

[8] Weitz LA, Hurtado JE and Sinclair AJ 2008 Decentralized cooperative-control design for multivehicle formations *J. Guid., Control, Dyna.* **31** 970-9

[9] Ren W and Beard R 2004 Decentralized scheme for spacecraft formation flying via the virtual structure approach *J. Guid., Control, Dyna.* **27** 73-82

[10] Mercado DA, Castro R *et al* 2013 Quadrotors flight formation control using a leader-follower approach *Proc. European control conf.* (Zurich) pp 3858-63

[11] Basri MAM, Husain AR and Danapalasingam KA 2015 Stabilization and trajectory tracking control for underactuated quadrotor helicopter subject to wind-gust disturbance *Sadhana* **40** 1531-53

Acknowledgements

This work was supported by Postgraduate Innovation Research Fund of Beihang University under Grant YCSJ-01-2016-10.

IOP Conf. Series: Materials Science and Engineering 187 (2017) 012017 doi:10.1088/1757-899X/187/1/012017

Temperature and Runback Ice Prediction Method for Three-Dimensional Hot Air Anti-Icing System

Ying Zhou[1], Guiping Lin[1], Xueqin Bu[1, 3, *], Zuodong Mu[1], Rui Pan[1], Qimo Ge[2] and Xudong Qiao[2]

[1] School of Aeronautic Science and Engineering, Beihang University, Beijing 100191, China

[2] Jiangxi Hongdu Aviation Industry Group Corporation Limited, AVIC, Nanchang, Jiangxi, 330024, China

[*]buxueqin@buaa.edu.cn

Abstract. A prediction method of surface temperature and runback ice for a three-dimensional hot air anti-icing system was proposed. Computational approach to realize this method was introduced. Both the external and internal flows were separately calculated, results of which were set as boundary conditions of heat conduction computation in airfoil skin. The results of external and internal flow calculations show that the effect of surface temperature on convective heat transfer coefficients and local droplet collection efficiency is negligible and the calculations can be decoupled. The prediction method based on heat flux was used to calculate surface temperature and runback ice results. The results show that, the effects of LWC and Mach number are much more significant than the effect of external flow temperature. The surface temperature at impinging interaction point is more sensitive to the change of external conditions than that at stagnation point. The surface temperature changes significantly with changing Mach number because both the mass rate of droplet and the impact limit are changed.

1. Introduction

Ice accretion on the aircraft surface often leads to both weight and drag increasing, lift reduction and a degradation of the aerodynamic and handling performance, therefore, is a safety hazard for flight. To protect aircraft from such a hazard, anti-icing systems are employed. One of the most widely used anti-icing systems on commercial aircrafts is the hot air anti-icing system, which uses the high-pressured hot air, bleeding from the compressor of the engine, to heat the wing leading edge. The hot air is transported through small round holes in a piccolo tube installed inside the anti-icing cavity.

In order to comprehensively understand the hot air anti-icing process, a lot of valuable researches has been done for the last decades. Those studies were mainly concentrated in the following aspects: 1, droplet collection characteristics and water film models; 2, parametric investigation of the piccolo tube and anti-icing cavity; 3 simulation method to predict the surface temperature and runback ice.

Both Lagrange [1] and Euler [2] method were utilized to calculate the water droplet trajectory. Water film models were investigated by Al-Khalil [3] and Myers [4] to obtain more accurate and physical descriptions of the water film motion process in the latter case.

Saeed [5, 6] simulated the thermal performance of 3-D hot-air array jetting on concave surfaces by using normal semi-circular concave surfaces instead of anti-icing cavity. Extensive experiments were conducted by Papadakis and See-Ho [7] to investigate the system performance and efficiency.

Three-dimensional hot air anti-icing system was studied computationally under dry air flow condition by Hugh H. T. Liu [8, 9] and See-Ho Wong [10]. The external and internal flows in their researches were coupled by heat conduction through the solid skin. Two-dimensional system with wet surface was computed by Francois Morency [11] and Rodrigo H. Domingos [12] using decoupled computations of external and internal flows. Lee [13] developed a procedure to integrate the internal and external solutions in Three-dimensional system.

A prediction method of the surface temperature and runback ice for three-dimensional hot air anti-icing system was proposed in this paper. Following the method applied in previous researches, both the external and internal flows were separately calculated, and the calculation results were used as boundary conditions of heat conduction computation of airfoil skin. By calculating the heat conduction iteratively, the results of temperature and runback ice were obtained.

2. Geometric model

Figure 1 shows a geometry model of the present CFD simulation. The three-dimensional wingspan segment is composed from a 90-mm span, 1800-mm chord NACA 23012 airfoil. The hot air anti-icing system is inserted into the first 160-mm chord of the wing. The thickness of the leading edge skin is 1.6 mm.

Figure 2 shows details of the piccolo tube in the anti-icing cavity, which has two horizontal rows of 1.32mm diameter round holes. The adjacent spacing between two holes along spanwise is 30 mm. The rows are placed at 45 degrees circumferentially above and below the piccolo tube axis. Tow thin channels (1.27mm) at both upper and lower sides of the leading edge skin allow hot air to exit the cavity.

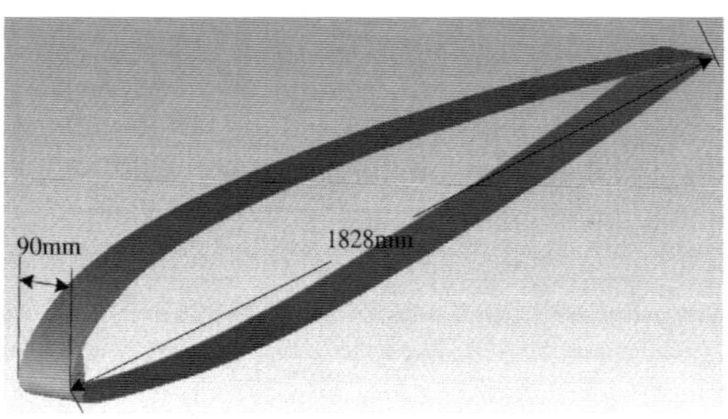

Figure 1 Geometric model

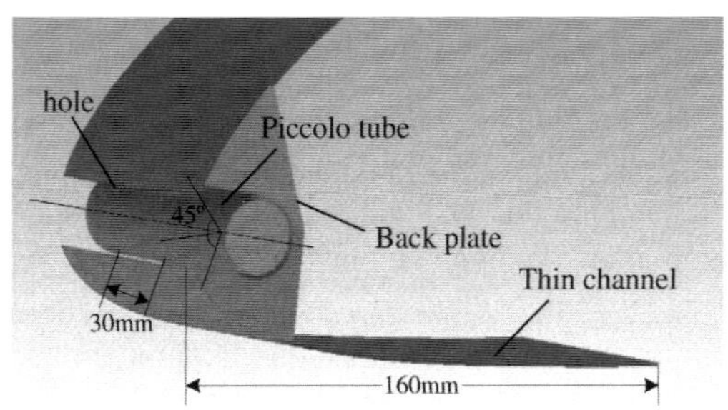

Figure 2 Anti-icing system details

3. Methodology

The surface temperature prediction process was divided into three regions: the external flow region including dry air convection heat transfer and water droplets impingement, the internal flow region of piccolo tube heat transfer with the inner surface, and the airfoil skin region conducting heat from internal flow to external.

The external flow and internal flow were computed separately by a commercial CFD flow solver. The results of the convective heat transfer coefficient of both external and internal skin surface and the local collection efficiency of droplet impingement were stored respectively to provide boundary conditions to airfoil skin thermodynamic equilibrium. Detailed introduction of the external and internal flow computations were given in [14, 15]. The present work was mainly focused on the airfoil skin thermodynamic equilibrium consisting of the heat conduction in the skin and the heat transfer of the phase-changing water film over it.

The conservation equations for thermodynamic equilibrium of phase-changing water film considering runback ice are as follows.

3.1. Mass conservation equation

For a water film control volume (CV) on the leading edge skin shown in **Figure 3**, the mass flow of water flowing into the CV is equal to the quality of the water flowing out. The mass conservation equation can be established as equation(1):

$$\dot{m}_{\text{in}} + \dot{m}''_{\text{imp}} \cdot \Delta s = \dot{m}''_{\text{evap}} \cdot \Delta s + \dot{m}_{\text{out}} + \dot{m}''_{\text{ice}} \cdot \Delta s \quad (1)$$

$\dot{m}_{\text{in}}$, $\dot{m}_{\text{out}}$ represent the water mass flow into and out of the control volume respectively, *kg/s*. $\dot{m}''_{\text{evap}}$ is the rate of water evaporating, *kg/s•m²*. $\dot{m}''_{\text{imp}}$ is the rate of droplets impinging on the leading edge surface, *kg/s•m²*. $\dot{m}''_{\text{ice}}$ represents the mass rate of water freezing, *kg/s•m²*.

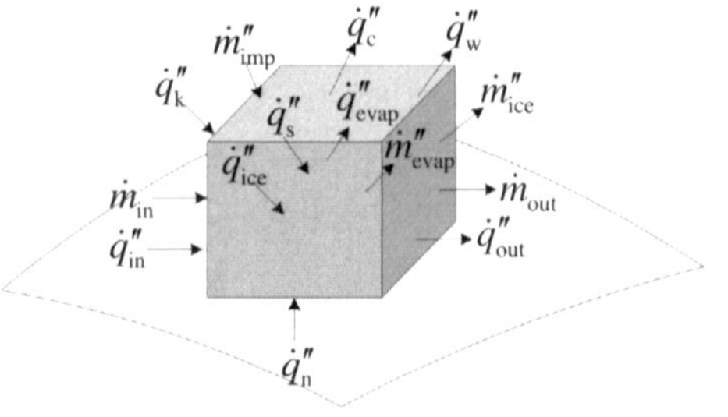

Figure 3 Mass and energy fluxes

3.2. Energy conservation equation

According to the steady state energy balance for the water film control volume on the external surface, the energy gets in and energy gets out of the control volume are equal. So the conductive heat flux $\dot{q}''$ conducted from the leading edge skin keeping the control volume thermodynamic equilibrium can be calculated by equation(2):

$$\dot{q}''_{\text{n}} + \dot{q}''_{\text{in}} + \dot{q}''_{\text{k}} + \dot{q}''_{\text{ice}} + \dot{q}''_{\text{s}} = \dot{q}''_{\text{out}} + \dot{q}''_{\text{c}} + \dot{q}''_{\text{evap}} + \dot{q}''_{\text{w}} \quad (2)$$

$\dot{q}''_{\text{in}}$ and $\dot{q}''_{\text{out}}$ are the energy flow into and out of the CV. $\dot{q}''_{\text{k}}$ is the kinetic energy released by the droplets at impingement. $\dot{q}''_{\text{ice}}$ is the latent solidification heat flux of the water. $\dot{q}''$ is the heat exchange released when the ice temperature is below freezing point. $\dot{q}''_{\text{c}}$ is the convective heat flux to the external flow. $\dot{q}''_{\text{evap}}$ is the latent evaporation heat flux of the water. $\dot{q}''_{\text{w}}$ is the heat exchange absorbed when the water temperature is above freezing point. The calculation formulae of each heat flux term were elaborately recommended in [15].

After impinging on the leading edge, water droplet is partially vaporized in the protected area, while the rest droplets flow along the airflow on the surface and may freeze downstream. Therefore, the calculation area can be divided into three cases: temperature above freezing point, freezing point and temperature below freezing point:

(1) $T_s > 273.15$K

When the leading edge external surface temperature is higher than freezing point, there is no frozen water. $\dot{m}''_{\text{ice}} = 0, \dot{q}''_{\text{ice}} = 0$.

$$\dot{q}''_{\text{n}} + \dot{q}''_{\text{k}} + \dot{q}''_{\text{in}} = \dot{q}''_{\text{c}} + \dot{q}''_{\text{evap}} + \dot{q}''_{\text{w}} + \dot{q}''_{\text{out}} \quad (3)$$

(2) $T_s = 273.15$K

When the leading edge external surface temperature is equal to freezing point, both water and ice exist on the surface. $\dot{q}''_{\text{s}} = \dot{q}''_{\text{out}} = 0$.

$$\dot{q}''_{\text{n}} + \dot{q}''_{\text{k}} + \dot{q}''_{\text{in}} + \dot{q}''_{\text{ice}} = \dot{q}''_{\text{c}} + \dot{q}''_{\text{evap}} + \dot{q}''_{\text{w}} \quad (4)$$

(3) $T_s < 273.15$K

When the leading edge external surface temperature below freezing point, the heat release by water condensation and ice cooling must be considered. Due to completely freezing on the surface, there is no outflow of water and energy. $\dot{m}_{\text{out}} = 0$, $\dot{q}''_{\text{out}} = 0$.

$$\dot{q}''_{\text{n}} + \dot{q}''_{\text{k}} + \dot{q}''_{\text{in}} + \dot{q}''_{\text{ice}} + \dot{q}''_{\text{s}} = \dot{q}''_{\text{c}} + \dot{q}''_{\text{evap}} + \dot{q}''_{\text{w}} \tag{5}$$

4. Numerical solution method

According to Section 3, the water state on the anti-icing surface can be determined by the conductive heat flux $\dot{q}''_{\text{n}}$ required for thermodynamic equilibrium of the water film independently. A numerical method based on heat flux $\dot{q}''_{\text{n}}$ was proposed in the present study. The method is that 1) use the conductive heat flux $\dot{q}''_{\text{n}}$ obtained from the heat conduction of the leading edge skin to calculate the surface temperature via the mass and energy conservation equations of the water film, 2) set the temperature as the Dirichlet boundary condition for the heat conduction calculation to update $\dot{q}''_{\text{n}}$, 3) calculate iteratively until the result converges.

It is necessary to evaluate the water state on the anti-icing surface before calculating. Assuming T_s = 273.15K, the required $\dot{q}''_{\text{n}}$ for f=1 and f=0 are determined respectively.

For f = 1: $\dot{q}''_{\text{n1}} = \dot{q}''_{\text{c}} + \dot{q}''_{\text{evap}} + \dot{q}''_{\text{w}} - \dot{q}''_{\text{k}} - \dot{q}''_{\text{in}} - \dot{q}''_{\text{ice}}$

For f = 0: $\dot{q}''_{\text{n2}} = \dot{q}''_{\text{c}} + \dot{q}''_{\text{evap}} + \dot{q}''_{\text{w}} - \dot{q}''_{\text{k}} - \dot{q}''_{\text{in}}$

By comparing the known conductive heat flux $\dot{q}''_{\text{n}}$ with $\dot{q}''_{\text{n1}}$ and $\dot{q}''_{\text{n2}}$, the surface temperature on a control volume can be divided into three cases:

(1) $\dot{q}''_{\text{n}} < \dot{q}''_{\text{n1}}$, the conductive heat flux is so small that the water is completely frozen. f = 1, T_s < 273.15K;

(2) $\dot{q}''_{\text{n}} > \dot{q}''_{\text{n2}}$, the conductive heat flux is large enough to dry up the surface. f = 0, T_s > 273.15K;

(3) $\dot{q}''_{\text{n1}} < \dot{q}''_{\text{n}} < \dot{q}''_{\text{n2}}$, T_s = 273.15K.

5. Results

5.1. External flow region

The Reynolds-averaged Navier-Stokes equations and the Transition SST turbulence model were applied to compute the external air flow to gain the convection heat transfer on the airfoil skin. Structured meshes involving hexahedral elements were used to model the external flow by Pointwise as shown in **Figure 4**. Pointwise is a commercial software to generate high quality orthogonal grids. The wall-adjacent cell distance (cell center-to-wall distance) in terms of the non-dimensional distance $y+$ was kept below 1. Pressure-far-field boundary condition was used for the air flow with an air pressure of 101,325Pa, a temperature of 263K and a Mach number of 0.2. The airfoil wall was modeled by using the isothermal wall boundary condition with $T_{\text{wall,ex}}$ = 270K, 280K, 290K, 300K, 310K, 320K, 330K, 340K. The convective heat transfer coefficient h_{ex} can be obtained by equation(6):

$$h_{\text{ex}} = q_{ex} / (T_{\infty} - T_{\text{ad}}) \tag{6}$$

T_{ad} is the local temperature when the airfoil is under adiabatic condition, K. q_{ex} is the surface heat flux to the external flow. T_{∞} is the external flow temperature, K.

The distributions of h_{ex} for different $T_{\text{wall,ex}}$ are presented in **Figure 6**. The results indicate that the effect of external surface temperature on the convective heat transfer coefficient under the same inlet air condition (air speed, temperature and pressure) is negligible. Thus, the convection heat transfer computation can be decoupled, and the convective heat flux to the external flow in equation (2) can be calculated by:

$$q_c = (T_{\infty} - T_{\text{ad}}) \cdot h_{\text{ex}} \tag{7}$$

The droplet collection characteristics were calculated by Euler method based on the previous external air flow field. The distributions of local droplet collection efficiency β for different T_{wall} with Ma = 0.2, 0.3 and MVD = 20μm were shown in **Figure** 7. The effect of T_{wall} on β is also negligible, and the droplet collection characteristics computation can be decoupled as well seen from **Figure** 7.

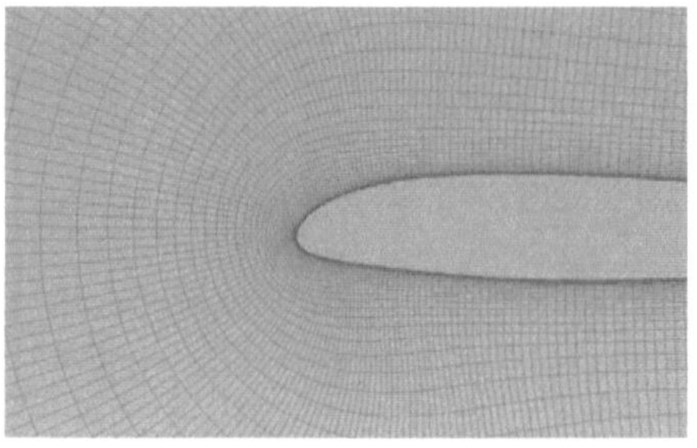

Figure 4 Grids of external flow

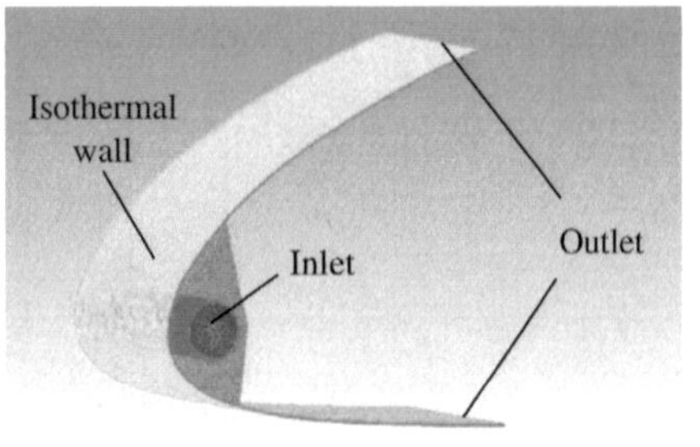

Figure 5 Grids of internal flow

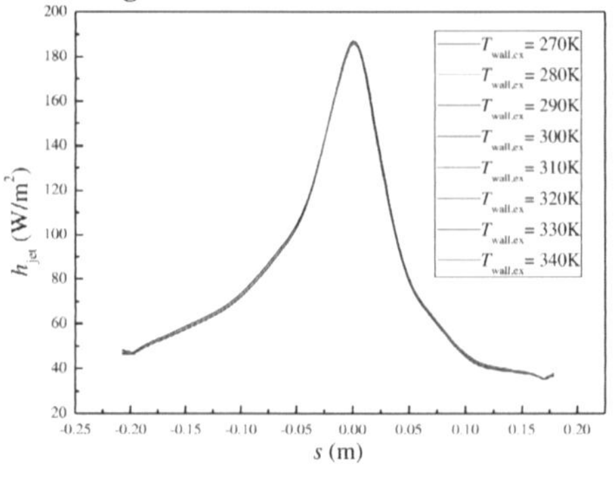

Figure 6 External convective heat transfer coefficient

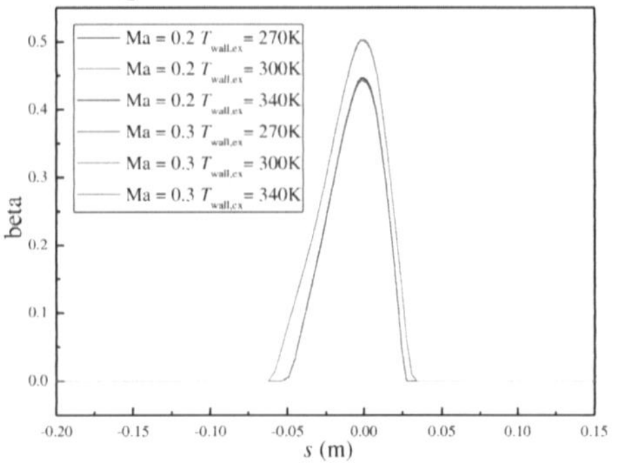

Figure 7 Local droplet collection efficiency

5.2. *Internal flow region*

The Realizable k-ε turbulence model with near-wall treatment method was conducted to calculate the convection heat transfer of jets impinging on the anti-icing cavity inner surface. Unstructured meshes were generated by ICEM software (in Figure 5) and the non-dimensional distance was $30 < y+ < 300$. The hot air inlet used the mass-flow-inlet boundary with a 0.003 kg/s mass flow-rate and a 453K total temperature. The exit flow was defined as a pressure-outlet boundary with the pressure of 101,325 Pa. The impingement wall was modeled isothermal with $T_{wall,in}$ = 270K, 280K, 290K, 300K, 310K, 320K, 330K, 340K, respectively. The convective heat transfer coefficient h_{jet} can be calculated as follows:

$$h_{jet} = q_{jet} / (T_{jet,tol} - T_{wall,in}) \tag{8}$$

q_{jet} is the heat flux provided by the impinging jets, W/m^2. $T_{jet,tol}$ is the total temperature in the piccolo tube, K. T_{wall} is the temperature of the inner surface, K.

The distributions of h_{jet} for different $T_{wall,in}$ were shown in Figure 8. It is indicated that the effect of inner surface temperature on h_{jet} is negligible. So h_{jet} and $T_{jet,tol}$ can be used as the Robin boundary condition for a given hot air property in the piccolo tube (including the total temperature and mass flow rate). As a consequence, the internal flow computation can be decoupled.

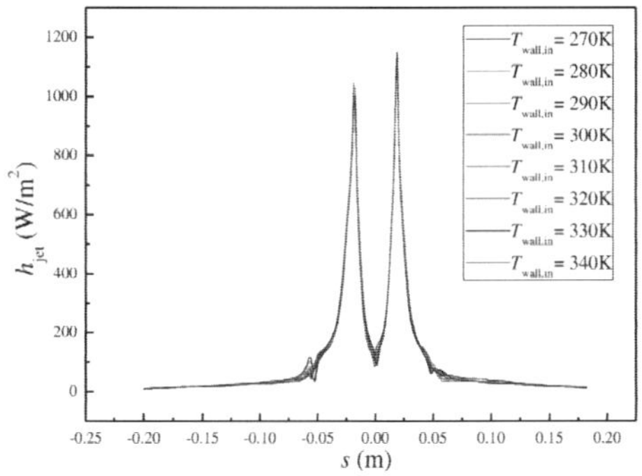

Figure 8 Internal convective heat transfer coefficient

5.3. Prediction of temperature and runback ice

The performance of the simulation method based on heat flux was evaluated by analyzing the temperature and runback ice results under different conditions. The external and internal conditions used in the validation cases are summarized in Table 1.

Table 1. The external and internal conditions used in the validation cases

	External flow				Internal flow	
case	Mach number	Temperature (K)	MVD (μm)	LWD (g/kg)	Total temperature (K)	Mass flow rate (kg/s)
1	0.2	263	20	1	453	0.003
2	0.2	263	20	2	453	0.003
3	0.3	263	20	1	453	0.003
4	0.3	263	20	2	453	0.003
5	0.3	263	20	3	453	0.003
6	0.3	258	20	3	453	0.003
7	0.3	268	20	3	453	0.003

The surface temperature distribution is presented in **Figure 9**. The maximum temperature is observed at the stagnation point of the internal impingement contributed by its intense heat transfer performance. The runback water freezes on the surface out of the protected area as shown in **Figure 10**. H_{ice} is the ice accretion rate calculated as follows:

$$H_{ice} = \dot{m}''_{ice} / \rho_{ice} \quad (9)$$

where ρ_{ice} is the density of ice, kg/m^3.

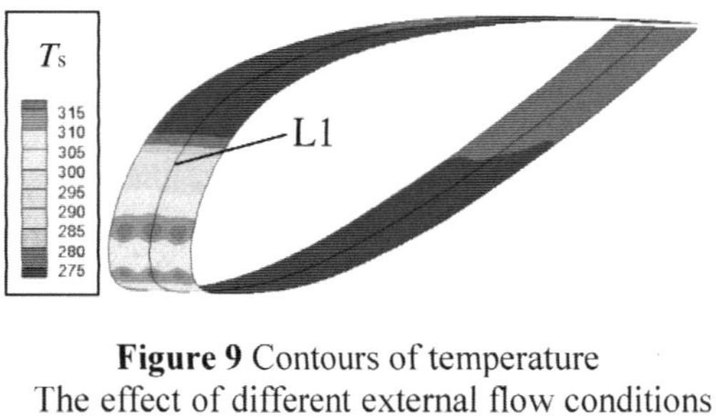

Figure 9 Contours of temperature

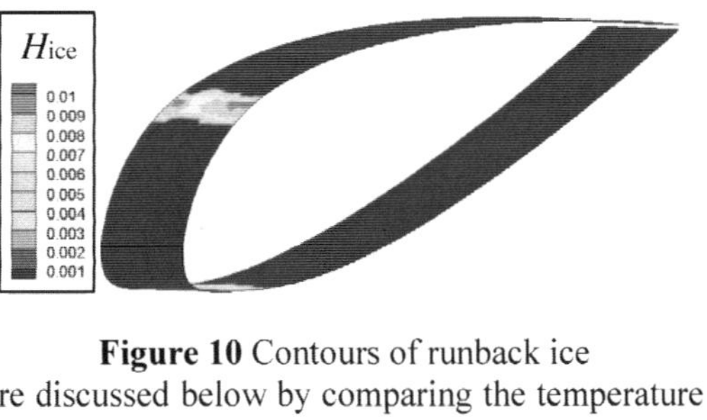

Figure 10 Contours of runback ice

The effect of different external flow conditions are discussed below by comparing the temperature distributions on L1 as shown in **Figure 9**.

5.3.1. Effect of LWC. The temperature distributions for different LWC are shown in Figure 11. When LWC increases from 1.0 g/kg to 2.0 g/kg, temperature changes at the stagnation points are ΔT_1 = 3.04K and ΔT_2 = 4.04K, at the impinging interaction point (s = 0) is ΔT_3 = 5.36K. When LWC increases to 3.0 g/kg, temperature changes are ΔT_1 = 6.05K, ΔT_2 = 8.06K and ΔT_3 = 10.08K. It is indicated that the effect of LWC on temperature at the interaction point is greater than that at the stagnation point.

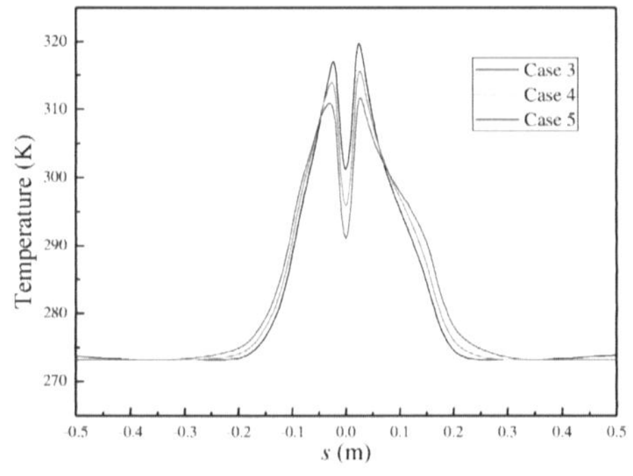

Figure 11 T_s distributions for varying LWC

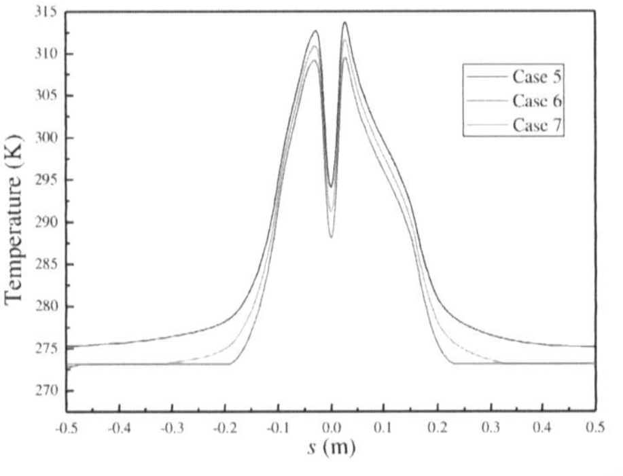

Figure 12 T_s distributions for varying external flow temperature

5.3.2. Effect of external flow temperature. Figure 12 shows the temperature distributions under different external flow temperatures. Compared with LWC, the effect of external flow temperature is much smaller. The temperature at impinging interaction point increases by ΔT_3 = 2.98K and 6K respectively with the external flow temperature rising from 258K to 263K and 268K.

5.3.3. Effect of Mach number. Figure 13 presents the temperature distributions for different Mach number. The effect of Mach number on surface temperature is significant seen from the figure. The surface temperature drops by ΔT_1 = 4.7K, ΔT_2 = 4.75K and ΔT_3 = 5.2K when Mach number increases from 0.2 to 0.3 for LWC = 1.0 g/kg. And the temperature drops by ΔT_1 = 5.89K, ΔT_2 = 6.18K and ΔT_3 = 6.8K for LWC = 2.0 g/kg. The maximum change of temperature occurs at s = -0.14mm with a change of 18.01K and 18.12K for LWC = 1.0 g/kg and 2.0 g/kg. It is because both the mass rate of droplet and impact limit are changed with Mach number as shown in Figure 7. So does the heat transfer coefficient of the external surface.

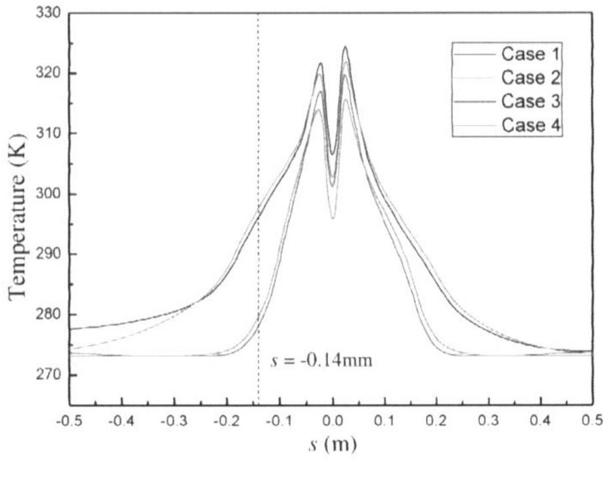

Figure 13 Distributions of temperature for different Mach number

6. Conclusion

A prediction method of surface temperature and runback ice for a three-dimensional hot air anti-icing system was proposed and was implemented based on the thermodynamic equilibrium of the water film and airfoil skin heat conduction.

To evaluate the performance of the simulation method presented in this paper, external and internal flow computations were conducted under different skin temperature. The results show that the effect of skin temperature is negligible so that the calculations can be decoupled. Temperature and runback ice results under different external conditions were obtained and analyzed to verify the effectiveness and rationality of the method.

The results indicate that the effects of LWC and Mach number are much more significant than external flow temperature. The surface temperature at impinging interaction point is more sensitive to the change of external conditions than that at stagnation point. The surface temperature changes a lot with changing Mach number because both the mass rate of droplets and impact limit are changed.

7. References

[1] Wright W. B. Users Manual for the Improved NASA Lewis Ice Accretion Code LEWICE 1.6[C]. June 01. 1995. NASA-Contractor Report-198355

[2] Bourgault Y., Boutanios Z., Habashi W.G. Three-Dimensional Eulerian Approach to Droplet Impingement Simulation Using FENSAP-ICE, Part 1: Model, Algorithm, and Validation[J]. Journal of Aircraft. 2000. 37(1): 95-103

[3] K. M. Al-Khalil, T. G. Keith Jr., K.J. De Witt. Development of an Anti-Icing Runback Model[C].28th Aerospace Sciences Meeting. 1990. AIAA 90-0759

[4] Tim G. Myers, Chris P. Thompson. Modeling the Flow of Water on Aircraft in Icing Conditions[J]. AIAA Journal, Vol. 36, No. 6 (1998), pp. 1010-1013

[5] Mathieu Fregeau. F. Saeed. I. Paraschivoiu. Surface Heat Transfer Study for Ice Accretion and Anti-Icing Prediction in Three Dimension[C]. 42nd AIAA Aerospace Sciences Meeting and Exhibit. 2004. AIAA 2004-0063

[6] Mathieu Fregeau. F. Saeed. I. Paraschivoiu. Numerical Heat Transfer Correlation for Array of Hot-Air Jets Impinging on 3-Dimensional Concave Surface[J]. Journal of Aircraft. 2005. Vol. 42 No. 3: 665-670

[7] M. Papadakis, S.J. Wong, H.W. Yeong, S.C. Wong, Icing tunnel experiments with a hot air anti-icing system[J], AIAA paper 2008-444, 2008

[8] Hugh H. T. Liu and Jun Hua. Three-Dimensional Integrated Thermodynamic Simulation for Wing Anti-Icing System[J]. Journal of Aircraft. Vol. 41, No. 6, November–December, 2004

[9] Jun Hua and Hugh H. T. Fluid Flow and Thermodynamic Analysis of a Wing Anti-Icing System[J]. Canadian Aeronautics and Space Journal. Vol. 51, No. 1, March 2005

[10] See-Ho Wong, Michael Papadakis, Alonso Zamora. Computational Investigation of a Bleed Air Ice Protection System[C]. 1st AIAA Atmospheric and Space Environments Conference. 22 - 25 June 2009, San Antonio, Texas. AIAA 2009-3966

[11] Francois Morency, Fatih Tezok, Ion Paraschivoiu. Anti-Icing System Simulation Using CANICE[J]. Vol. 36, No. 6, November–December 1999

[12] Rodrigo H. Domingos, Michael Papadakis, Alonso O. Zamora. Computational Methodology for Bleed Air Ice Protection System Parametric Analysis[J]. AIAA Atmospheric and Space Environments Conference. 2 - 5 August 2010, Toronto, Ontario Canada. AIAA 2010-7834

[13] Lee, J., Rigby, D., Wright W. and Choo, Y. Analysis of Thermal Ice Protection System (TIPS) with Piccolo Tube using State-of-the-Art Software[C]. 44th AIAA Aerospace Sciences Meeting and Exhibit. 9 - 12 January 2006, Reno, Nevada. AIAA 2006-1011

[14] Bu Xue-qin, Lin Qui-ping, Peng You-xin, Yu Jia. New Method for Calculation of Anti-icing Heat Loads[J]. Chinese Journal of Aeronautics. Vol. 127 No. 12. March 2006

[15] Bu Xue-qin, Lin Qui-ping, Yu Jia. Three-dimensional conjugate heat transfer simulation for the surface temperature of wing hot-air anti-icing system[J]. Journal of Aerospace Power. Vol .24 No .11. Nov .2009

IOP Conf. Series: Materials Science and Engineering **187** (2017) 012018 doi:10.1088/1757-899X/187/1/012018

Integrated GPS/DR Vehicle Navigation System Based on Sequential and Square Root Kalman Filters

MOSTAFA Elzoghby[1, *], USMAN Arif [1], FU Li [1] and XI Zhi Yu [1]

[1] School of Automation Science and Electrical Engineering, Beihang University, Beijing 100191, China

*E-mail: mostafa@buaa.edu.cn

Abstract. Global Positioning System (GPS) has become part of many applications in life. In mountainous terrains and around buildings, GPS reception is compromised. In dense urban canyons, signals bounce off the buildings creating multipath reception and provide erroneous measurements. To overcome GPS bandwidth and signal fading problems, Navigation solutions are built on GPS measurements fused with inertial sensors to provide dead reckoning (DR) based position solution. Solution for land vehicle Navigation System using GPS, inertial sensor and odometer is presented. The sensors fusion is performed based on conventional, sequential (SKF) and square root Kalman (SRKF) filters. SRKF based on Cheolesky factorization for covariance matrix P. Simulations are performed on real data, with precisely known covariance's to simulate mathematical stability, performance and processing time required by each method on a high end microprocessor. The results demonstrate integrated system using SRKF has better performance in stability and estimation accuracy than conventional and sequential filter.

1. Introduction

Land vehicle navigation systems are one the fastest growing area in-vehicle electronics market [1]. Generally, the land Vehicle navigation system gets the positioning information via GPS. Hence the GPS can provide sufficient positioning information but suffers seriously especially in tunnels or urban canyons where GPS signal gets faded or corrupted; therefore degrading the availability and accuracy for the GPS based positioning information. Many high end GPS antennas and amplifiers have been designed to solve this problem, but these systems increase the cost of the Navigation Solution. There are four main parameters that describe the performance of a navigation system: availability, continuity, integrity and accuracy [2].In order to fulfill these requirements, the most widely accepted approach is to integration of DR with GPS which increased the integrity and availability of Navigation Solution [3].

Other The principle of DR is to provide the current position of the vehicle based on previous position and velocity, and provides continuous real time navigation information. Using only DR has a drawback: the navigation error accumulates over time and cannot keep up the accuracy requirements. Therefore for a complete navigation solution, DR can provide the vehicle positioning information when GPS is not available, while GPS can correct the accumulated error of DR [4]. Most commonly used DR sensors for land vehicle navigation include odometer, gyro and magnetic compass. In our work, we have utilized a single axis gyroscope which provides the angular rate of change in the heading direction while Odometer which provides absolute velocity information as Dead Reckoning sensors. Another approach like map matching algorithms with DGPS used for Land Vehicle position determination in[5].The low cost IMU Integrated with GPS System for Land Vehicle Navigation by using kalman filter is proposed in[6,7]. Cascade filters implemented for DR navigation solution in [8] via using a wheel encoders to provide distance and a magnetic compass to provide direction. Also, differential wheel speed sensor integrated with an inertial measurement and GPS are used in [9] to reduce the errors from sideslip and eliminate the large errors in vehicle motions by using two nonlinear kalman filter forms. And some of another Kalman filter forms used to estimate vehicle position by using heading sensors combined with speed sensors during GPS unavailability in [10-13].

The key problem of GPS/DR integrated system is how to achieve the optimal estimation for the position, moving and states of interest parameters. Our work involves designing land based navigational solution with our prescribed model and investigating still provides any superiority over conventional or batch Kalman filtering as well as to compare the computational time required by each method. We have chosen SKF and SRKF and compared the results with conventional KF in our problem of designing navigation solution for a land vehicle on a high end processor. In the following sections, we begin with describe the mathematical model for GPS/DR integration then a brief summary about the algorithms of the sequential and square root Kalman Filter. The part of the simulation and discussion to show the effect of (KF), (SKF) and (SRKF) on the integration, Simulation results and a comparison of these methods are presented in section 2. Finally, in section 3, these methods are evaluated in theory and simulation.

2. Formulations and equations

2.1 GPS/DR Mathematical Model

2.1.1 The state equations. *The east position* E_x, *north position* N_y, *heading* ψ, *speed* v, *gyro rate* $\dot{\psi}$, *bias* , B *and odometer scale* S *can be expressed in state space form as:*

$$E_x(k+1) = E_x(k) + V(k)T\sin(\psi\ (k)) + w_1 \tag{1}$$

$$N_y(k+1) = N_y(k) + V(k)T\cos(\psi\ (k)) + w_2 \tag{2}$$

$$\psi\ (k+1) = \psi\ (k) + T\dot{\psi}\ (k) + w_3 \tag{3}$$

$$V\ (k+1) = V\ (k) + w_4 \tag{4}$$

$$\dot{\psi}(k+1) = \dot{\psi}\ (k) + w_5 \tag{5}$$

$$B\ (k+1) = B\ (k) + w_6 \tag{6}$$

$$S\ (k+1) = S\ (k) + \frac{V(k) - N(k)S(k)}{N(k)} + w_7 \tag{7}$$

A processes noise $\boldsymbol{w_k}$ is white, zero-mean, uncorrelated, and has known covariance matrix $\mathbf{Q}_k$.

2.1.2 The observation equations. *The measurements equations are expressed as:*

$$E_{gps} = E_x(k) + v_1 \tag{8}$$

$$N_{gps} = N_y(k) + v_2 \tag{9}$$

$$\psi_{\ gps} = \psi\ (k) + v_3 \tag{10}$$

$$V_{speed}(k) = n(k)\ S(k) + v_4 \tag{11}$$

$$\dot{\psi}_{gyro}(k) = \dot{\psi}\ (k) + B + v_5 \tag{12}$$

The measurement noise $\mathbf{v_k}$ is a zero mean white noise and the measurement noise covariance matrix $\mathbf{R_k}$.

2.2 Sequential Kalman Filter

This technique is also called "sequential-sensor method" [14]. The approach is to implement the kalman filter while avoiding the matrix inversion during the computation of the kalman gain and considering every sensor observation as an independent and separate realization. At each time-step the kalman gain and the new prediction are computed at every observation from every sensor.

2.2.1 *The system and measurement equations.*

$$\boldsymbol{X}(k) = \boldsymbol{F}(k, k-1)\boldsymbol{X}(k-1) + \boldsymbol{w}(k-1) \tag{13}$$
$$\boldsymbol{Z}(K) = \boldsymbol{H}(k)\boldsymbol{X}(k) + \boldsymbol{v}(k) \tag{14}$$

Where w_k and v_k uncorrelated and white noise. The measurement covariance matrix R_k is a diagonal and given by $R_k = \mathrm{diag}(R_{1k}, \ldots, R_{rk})$, Uncorrelated measurements in Z(K) can be taken as accomplished step by step.

$$\begin{bmatrix} \boldsymbol{Z}_k^1 \\ \boldsymbol{Z}_k^2 \\ \vdots \\ \boldsymbol{Z}_k^r \end{bmatrix} = \begin{bmatrix} \boldsymbol{H}_k^1 \\ \boldsymbol{H}_k^2 \\ \vdots \\ \boldsymbol{H}_k^r \end{bmatrix} \boldsymbol{X}_k + \begin{bmatrix} \boldsymbol{v}_k^1 \\ \boldsymbol{v}_k^2 \\ \vdots \\ \boldsymbol{v}_k^r \end{bmatrix} \tag{15}$$

2.2.2 Realization of SKF. *The state prediction and the corresponding covariance at any one time step for the first observation Z_k^1 are given by*

$$\widehat{\boldsymbol{X}}_k^0 = \widehat{\boldsymbol{X}}(\boldsymbol{k} \mid \boldsymbol{k} - \boldsymbol{1}) \tag{16}$$
$$\boldsymbol{P}_k^0 = \boldsymbol{P}\,(k \mid k - 1) \tag{17}$$

The next new state estimation made by fusing Z_k^1 which achieved at time k by the first sensor is given as

$$\widehat{\boldsymbol{X}}_k^1 = \widehat{\boldsymbol{X}}_k^0 + K_k^1(\boldsymbol{Z}_k^1 - \boldsymbol{H}_k^1\widehat{\boldsymbol{X}}_k^0) \tag{18}$$
$$\boldsymbol{P}_k^1 = (\boldsymbol{I} - K_k^1\boldsymbol{H}_k^1)\boldsymbol{P}_k^0 \tag{19}$$
$$K_k^1 = \boldsymbol{P}_k^0{\boldsymbol{H}_k^1}^T[\boldsymbol{H}_k^1\boldsymbol{P}_k^0{\boldsymbol{H}_k^1}^T + R_k^1]^{-1} \tag{20}$$

To fuse the observation from the second sensor so, it requires $\widehat{\mathbf{X}}_k^1$ to produce a prediction. Generally, the estimation of $\widehat{\mathbf{X}}_k^i$ at time $\mathbf{k}$ on the basis of the observations from all the sensors until time $(\mathbf{k}-\mathbf{1})$ and through first $\mathbf{i}$ sensors until time $\mathbf{k}$ can to calculate from the estimation of $\widehat{\mathbf{X}}_k^{i-1}$. Update sequentially until the last observation $\mathbf{Z}_k^r$ then the estimation of $\widehat{\mathbf{X}}_k^r$ at time $\mathbf{k}$ on the basis of the observations from all the sensors until time $(\mathbf{k}-\mathbf{1})$ and through first $\mathbf{r}$ sensors until time $\mathbf{k}$ can to calculate from the estimation $\widehat{\mathbf{X}}_k^{r-1}$ as kalman throughout

$$\widehat{\boldsymbol{X}}_k^r = \widehat{\boldsymbol{X}}_k^{r-1} + K_k^r(\boldsymbol{Z}_k^r - \boldsymbol{H}_k^r\widehat{\boldsymbol{X}}_k^{r-1}) \tag{21}$$

$$\widehat{\boldsymbol{X}}_k^r = \widehat{\boldsymbol{X}}(k \mid k) \tag{22}$$
$$\boldsymbol{P}_k^r = (\boldsymbol{I} - K_k^r\boldsymbol{H}_k^r)\boldsymbol{P}_k^{r-1} \tag{23}$$
$$\boldsymbol{P}_k^r = \boldsymbol{P}\,(k \mid k) \tag{24}$$
$$K_k^r = \boldsymbol{P}_k^{r-1}{\boldsymbol{H}_k^r}^T[\boldsymbol{H}_k^r\boldsymbol{P}_k^{r-1}{\boldsymbol{H}_k^r}^T + R_k^r]^{-1} \tag{25}$$

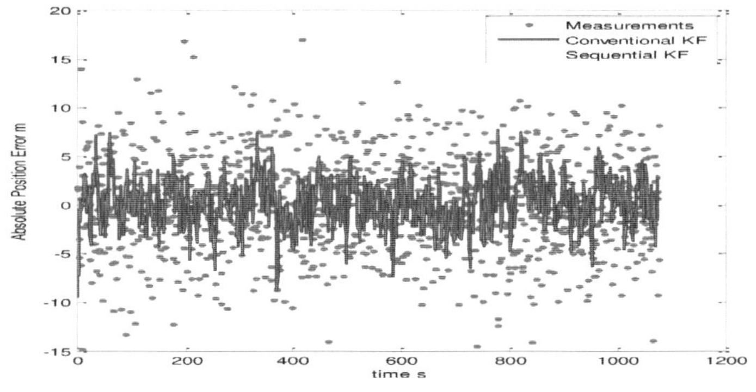

Figure 1. Conventional and Sequential KF Results for absolute position error

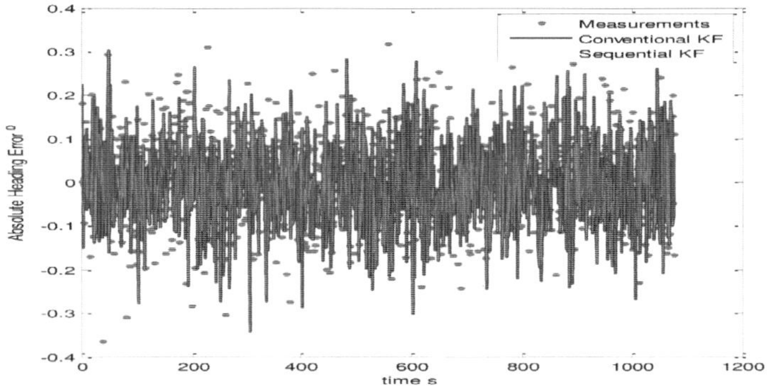

Figure 2. Conventional and Sequential KF Results for absolute heading error

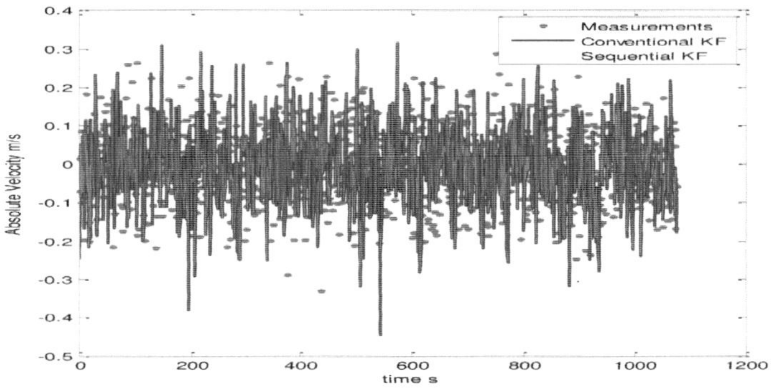

Figure 3. Conventional and Sequential KF Results for absolute velocity error

2.3 Square Root Kalman Filter

SRKF effectively increases the accuracy of the Kalman filter by reducing the divergence and instability, but on the other side increase the computational cost [15]. Generally, square root filter is stable numerically more than the conventional filter; also, it is less affected by numerical problems [16]. This technique is also known as the covariance square root filter. It works by calculating square root of covariance matrix P using Cholesky decomposition, given by:

$$\boldsymbol{P} = \boldsymbol{S}\boldsymbol{S}^T = \begin{bmatrix} S_{11} & \cdots & 0 \\ \vdots & \ddots & \vdots \\ S_{n1} & \cdots & S_{nn} \end{bmatrix} \begin{bmatrix} S_{11} & \cdots & S_{n1} \\ \vdots & \ddots & \vdots \\ 0 & \cdots & S_{nn} \end{bmatrix} \tag{26}$$

Where

$$S_{ii} = \sqrt{P_{ii} - \sum_{j=1}^{i-1} S_{ij}^2} \qquad (i < j) \tag{27}$$

$$S_{ij} = \sqrt{\frac{1}{S_{jj}}(P_{ij} - \sum_{k=1}^{j-1} S_{ik}S_{jk})} \quad (i > j) \tag{28}$$

$$S_{ij} = 0 \quad (i < j) \tag{29}$$

Mathematically, there is no difference between the square root and conventional filter time update equation for P but the difference is to update and propagate S instead of P [15]. In ill-conditioned cases this approach can yield double effective precision as compared to the standard filter [17]. And successfully keeps the error covariance as a positive semi definite which yields the numerical stability as well. The difficulties in the numerical calculations maybe cause P_k to be indefinite or non symmetric but cannot cause the same for SS^T[15]. Numerical problems may arise and The Error covariance matrix can become ill conditioned if one state variable has a much higher uncertainty than another. The square root Kalman filter works effectively especially if the estimation problem is ill conditioned or computer word length is limited.

2.3.1 The square root time update equations. The filter of Square root covariance propagate Cholesky factor of a priori error covariance matrix P (k | k − 1).

$$\boldsymbol{P}(k \mid k-1) = \boldsymbol{S}(k \mid k-1)\boldsymbol{S}^T(k \mid k-1) \tag{30}$$

Where $S(k \mid k-1)$ is square root of a priori error covariance matrix $P(k \mid k-1)$ and chosen to be lower triangular. As well, $S(k \mid k)$ is square root of a posteriori error covariance matrix $P(k \mid k)$ and initially $S(0 \mid 0)S^T(0 \mid 0) = P(0|0)$. However, there are 2 cases related with the system model if it has a dynamic noise for all time $(Q_k \neq 0)$ or zero process noise$(Q_k = 0)$. The time propagation with neglecting control inputs in case of $(Q_k = 0)$ will be as follows:

$$\hat{\boldsymbol{X}}(k \mid k-1) = \boldsymbol{F}(k, k-1)\hat{\boldsymbol{X}}(k-1 \mid k-1) \tag{31}$$
$$\boldsymbol{P}(k \mid k-1) = \boldsymbol{F}(k)\boldsymbol{P}(k-1 \mid k-1)\boldsymbol{F}^T(k) \tag{32}$$

By letting $\boldsymbol{P}(k \mid k-1) = \boldsymbol{S}(k \mid k-1)\boldsymbol{S}^T(k \mid k-1)$ and $\boldsymbol{P}(k-1 \mid k-1) = \boldsymbol{S}(k-1 \mid k-1)\,\boldsymbol{S}^T(k-1 \mid k-1)$, equation (32) can be rewritten as

$$[\boldsymbol{S}(k \mid k-1)]\,[\boldsymbol{S}^T(k \mid k-1)] = [\boldsymbol{F}(k)\,\boldsymbol{S}(k-1 \mid k-1)]\;[\boldsymbol{S}^T(k-1 \mid k-1)\,\boldsymbol{F}^T(k)] \tag{33}$$

From this equation it is clear that the appropriate time propagation for the square root filter would be

$$\hat{\boldsymbol{X}}(k \mid k-1) = \boldsymbol{F}(k, k-1)\hat{\boldsymbol{X}}(k-1 \mid k-1) \tag{34}$$
$$\boldsymbol{S}(k \mid k-1) = \boldsymbol{F}(k)\,\boldsymbol{S}(k-1 \mid k-1) \tag{35}$$

The time propagation with neglecting control inputs in case of $(Q_k \neq 0)$ will be as follows:

$$\hat{\boldsymbol{X}}(k \mid k-1) = \boldsymbol{F}(k, k-1)\hat{\boldsymbol{X}}(k-1 \mid k-1) \tag{36}$$
$$\boldsymbol{P}(k \mid k-1) = [\boldsymbol{F}(k)\,\boldsymbol{S}(k-1 \mid k-1)]\;[\boldsymbol{S}^T(k-1 \mid k-1)\,\boldsymbol{F}^T(k)] + \boldsymbol{G}(k)\boldsymbol{Q}(k)\boldsymbol{G}^T(k) \tag{37}$$

IOP Conf. Series: Materials Science and Engineering **187** (2017) 012018 doi:10.1088/1757-899X/187/1/012018

In this case will compute P (k | k − 1) firstly and then generate S(k | k − 1) as its Cholesky lower triangular square root and this method called root sum square (RSS) and this case is simulated in the paper.

2.3.2 The square root measurements update equations.

$$\widehat{\boldsymbol{X}}(k \mid k) = \widehat{\boldsymbol{X}}(k \mid k-1) + \boldsymbol{K}(k)(\boldsymbol{Z}(k) - \boldsymbol{H}(k)\widehat{\boldsymbol{X}}(k \mid k-1)) \quad (38)$$

$$\boldsymbol{K}(k) = b_k \boldsymbol{S}(k \mid k-1) a_k \quad (39)$$

$$\boldsymbol{S}(k \mid k) = \boldsymbol{S}(k \mid k-1)[\boldsymbol{I} - b_k r_k a_k a_k^{T}] \quad (40)$$

$$a_k = \boldsymbol{S}^T(k \mid k-1)\boldsymbol{H}^T(k) \quad (41)$$

$$b_k = [a_k^{T} a_k + R_k]^{-1} \quad (42)$$

$$r_k = [1 + [b_k R_k]^{\frac{1}{2}}]^{-1} \quad (43)$$

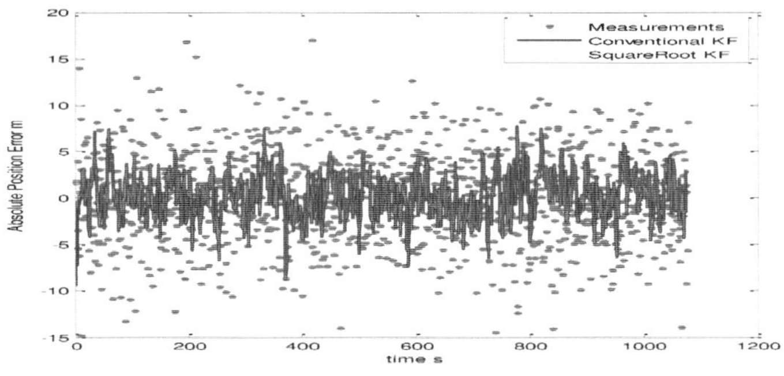

Figure 4. Conventional and Square Root KF Results for absolute position error

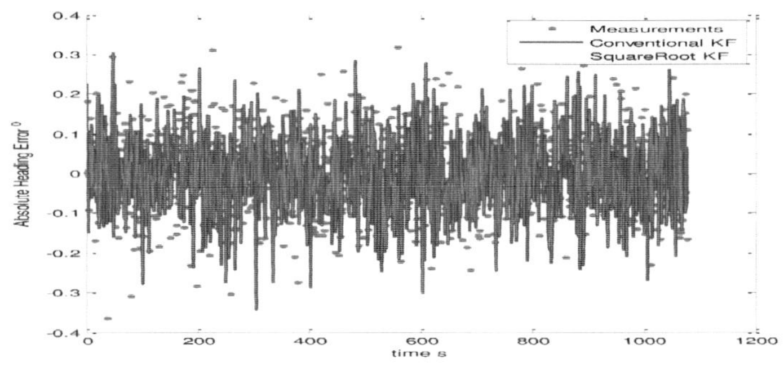

Figure 5. Conventional and Square Root KF Results for absolute heading error

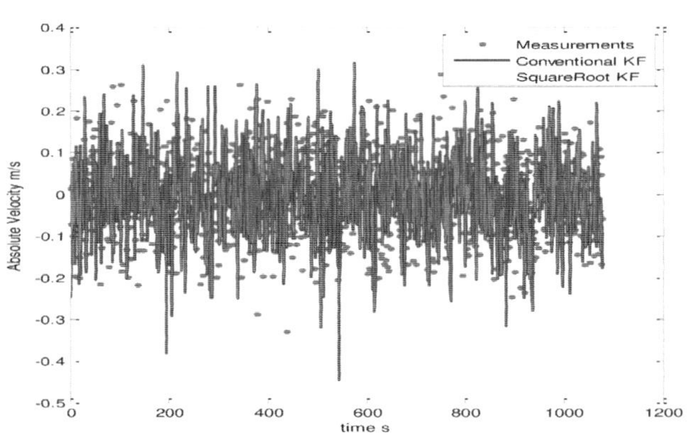

Figure 6. Conventional and Square Root KF Results for absolute velocity error

3. Experimentation and Results

A GPS receiver is mounted on a land vehicle which provides east and north position, velocity, and heading angle in the navigation frame. MEMS gyroscope sensor is used for yaw rate measurement along with odometer for speed measurement. The measurement noise covariance is already known precisely. The data is recorded and is processed offline on a mini processing unit. Three separate Kalman filters namely conventional, SKF and SRKF are implemented separately with same process and measurement covariance matrices.

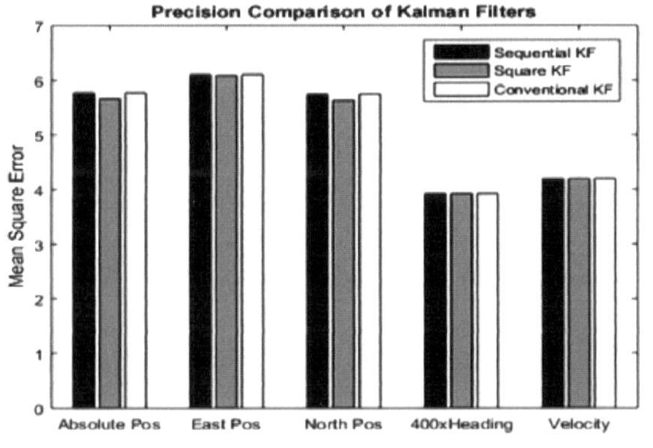

Figure 7. Mean Square error Comparison between Three Filters

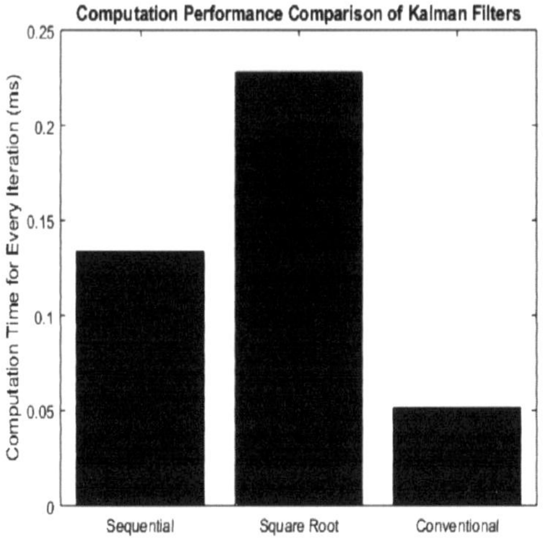

Figure 8. Computational Time Comparison of Three Filters

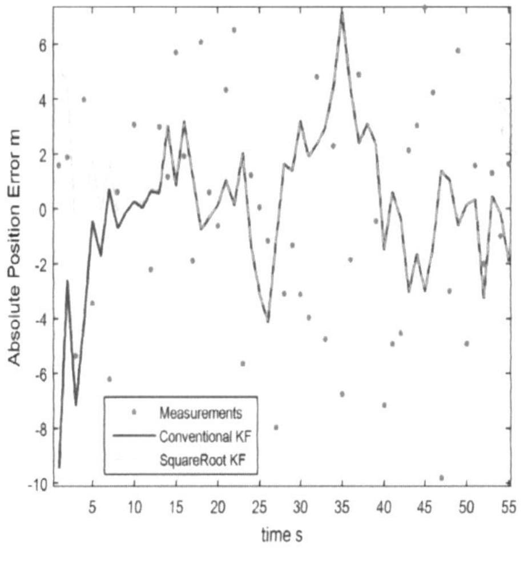

Figure 9. Convergence Vs Time for Conventional and Square Root KF

The absolute position, velocity and heading are shown in the figures (1, 2 and 3) for conventional and SKF. In figures (4, 5 and 6), a comparison of conventional with SRKF is shown. A complete comparison for all the states for the three sorts of Kalman filter methods is summarized in the bar graph shown in figure 7. It is clear that the SRKF has slightly better performance than the other two Kalman Filters. Figure8 summarizes the computation time utilized by each method for single iteration.

Computation time of conventional is smallest while SRKF is largest. In figure 9 the convergence speed comparison of SRKF and conventional is shown. The superiority of SRKF over conventional is due to its lower convergence time. The convergence time of SKF and conventional is same.

4. Conclusion

The results clearly show that the SRKF method based on the Cholesky factorization performs slightly better than the other two methods which perform equally. This is because the convergence rate of SRKF is better than conventional and SKF. If we increase the number of states, the SRKF method may provide an increased efficiency. The computation time for SRKF is four times that required for conventional Kalman filter. Therefore, if computational time requirements are not stringent and numbers of states are more, then we can use SRKF while for less number of states, ordinary Kalman filter performs quite well. These results are only valid for high end processor based applications. For low computational power embedded systems and SRKF will obviously work far better, even for reduced number of states.

5. References

[1] Wang J H 2006 Intelligent MEMS INS/GPS Integration for Land Vehicle Navigation, Department of Geomatics Engineering, Calgary, Alberta.
[2] Ashkenazi V, Moore T, Hill C J, Ochieng W Y and Chen W 1995 *Proceedings of the 8th International Technical Meeting of the Satellite Division of The Institute of Navigation* 463-472.
[3] Zhao L and Yuan Z 2012 *Journal of Information & Computational Science* **9** 2771–2779.
[4] Yan G, Yan W and Xu D 2008 *Journal of Chinese Inertial Technology* **16** 253-264.
[5] Saab S S and Kassas Z M 2002 *Intelligent Vehicle Symposium IEEE* **1** 209 - 214.
[6] Maklouf O, Ghila A, Abdulla A and Yousef A 2013 *Engineering and Technology International Journal of Electrical, Computer, Energetic, Electronic and Communication Engineering* **7** (2).
[7] Zhao Y 2011 GPS/IMU Integrated System for Land Vehicle Navigation based on MEMS, Royal Institute of Technology, Stockholm Sweden.
[8] Rogers R M 1997 Land Vehicle Navigation Filtering for a GPS/Dead-Reckoning System *Proceedings of the 1997 National Technical Meeting of The Institute of Navigation* 14 – 16.
[9] Hazlett A C, Crassidis J L and Fuglewicz D B 2011 Differential Wheel Speed Sensor Integration with GPS/INS for Land Vehicle Navigation *AIAA Guidance, Navigation, and Control Conferenc.*
[10] Abbott E and Powell J D 1999 *Proc. IEEE* **87** 145–162.
[11] Rogers R 1999 Improved Heading Using Dual Speed Sensors for Angular Rate and Odometry in Land Navigation *Proceedings of the Institute of Navigation National Technical Meeting* 353–361.
[12] Stephen J and Lachapelle G 2001 Development and Testing of a GPS-Augmented Multi-Sensor Vehicle Navigation System *J. Nav.* 268-278.
[13] Stephen J 2014 Development of a Multi-Sensor GNSS Based Vehicle Navigation System, Department of Geomatics Engineering, the University of Calgary.
[14] Hoseini S A and Ashraf M R 2013 *International Journal of Chaos, Control, Modeling and Simulation* **2** (2).
[15] Simon D J 2006 Optimal State Estimation (John Wiley & Sons, Inc).
[16] Schutz, Tapley, and Born 2004 Statistical Orbit Determination (Academic Press, Inc.).
[17] Maybeck P S 1982 Stochastic models, estimation, and control (Academic Press, Inc.).

Acknowledgements

This study is supported by a grant from the National Natural Fund of China, Grant No. 61375082. Special Thanks for the reviewers for their valuable comments.

IOP Conf. Series: Materials Science and Engineering **187** (2017) 012019 doi:10.1088/1757-899X/187/1/012019

Investigation of Adaptive Robust Kalman Filtering Algorithms for GPS/DR Navigation System Filters

MOSTAFA Elzoghby[1,*], USMAN Arif [1], FU Li [1] and XI Zhi Yu [1]

[1] School of Automation Science and Electrical Engineering, Beihang University, Beijing 100191, China

[*]E-mail: mostafa@buaa.edu.cn

Abstract: The conventional Kalman filter (KF) algorithm is suitable if the characteristic noise covariance for states as well as measurements is readily known but in most cases these are unknown. Similarly robustness is required instead of smoothing if states are changing abruptly. Such an adaptive as well as robust Kalman filter is vital for many real time applications, like target tracking and navigating aerial vehicles. A number of adaptive as well as robust Kalman filtering methods are available in the literature. In order to investigate the performance of some of these methods, we have selected three different Kalman filters, namely Sage Husa KF, Modified Adaptive Robust KF and Adaptively Robust KF, which are easily simulate able as well as implementable for real time applications. These methods are simulated for land based vehicle and the results are compared with conventional Kalman filter. Results show that the Modified Adaptive Robust KF is best amongst the selected methods and can be used for Navigation applications.

1. Introduction

Since Kalman filter was established in 1960; it has been utilized in almost every prediction, estimation and filtering problem. It has proven its grounds in the field of inertial navigation, trajectory estimation, radar tracking, global navigation satellite systems (GNSS), Magnetic detection and Location as well as controllers and observer models [1].

After the initial work on Kalman filter, a number of variants of Kalman filters have been developed. The main categories include linear Kalman filters, adaptive Kalman filters, and nonlinear Kalman filters. But basic idea is same, estimate the state by fusing the information of measurements with linear/nonlinear state prediction equations. The fusion is performed based on the information provided by process noise matrix Q and measurement noise matrix R [2]. Therefore the Q and R matrices play a vital role in Kalman filtering state estimation. If Q and R are known, Kalman filter gives the optimal estimate. R can be obtained using data sheets of the measurement sensors which provide general statistical properties for the class of sensors, while Q is gathered analytically, by the experience of the designer. But if these are not chosen properly, the Kalman filter could lead to a non optimal solution. Therefore many modified Kalman filter algorithms have been proposed which can estimate the R and Q along with the states.

Usually the matrices Q and R are considered as constants, but in order to get optimal state, Q and R must adjust themselves according to the changing states as well as sensor accuracies, as sensors have different accuracies under different static and dynamic conditions. Such kind of Kalman filters are called adaptive Kalman filters. Similarly if state is changing abruptly, e.g. in target tracking and motion estimation, robustness becomes more essential as than smoothing. The robustness can be obtained by changing Kalman gain K, Process covariance P, adjusting measurement noise matrix R, or combination of these. Such filters are called as robust Kalman filters.

For real time applications, adaptiveness as well as robustness is required to obtain optimal state. A number of adaptive and robust Kalman filters have been proposed. [3] Have devised a robust Kalman filter for outlier detection. It removes the outliers from robotic sensors data by introducing weights for each data sample. Similarly [4] has opted an adaptively robust Kalman filter for GPS/INS integration. This filter performs GPS/INS integration based on outlier detection. Another commonly used adaptive Kalman filter is Sage Husa which estimates optimal Q and R on alternate iterations. [5] Has utilized Sage Husa KF for INS initial alignment. Similarly in [6] have used Sage Husa KF to de-noise gyroscopic signal. Most applications of Sage Husa include static data, where robustness requirement is very less. When robustness is required, Sage Husa can diverge, instead of converging. This is because Sage Husa Kalman Filter is an adaptive filter, not a robust filter. Similarly, in [7] strong tracking algorithm is proposed which is robust but not adaptive. An adaptive robust algorithm is also proposed in [8]. But the problem with this method is that it is not real time implementable, [9] has used it and it took 96 hours to complete the simulation.

In our work, we have selected three Kalman filters namely Adaptive Sage Husa KF, Modified Adaptive Robust Kalman filter, which is based on Sage Husa KF for adaptivity and Strong Tracking KF for robustness, and another adaptively robust method given in [10] .They are implemented, and compared with linear/ conventional version of Kalman Filter. Data for navigation is collected via land based vehicle which has GPS, heading rate gyro and an odometer. The initial values for Q and R matrices are based on analytical assumptions only. These three filters are implemented and simulated. The results are compared to investigate the performance of each method for real time applications.

2. Formulations Material and Methods

2.1. GPS/DR Problem Description

Our problem is a ground vehicle that is travelling on surface of the earth with linear velocity V. The position of ground vehicle is described in NED reference frame, with N_y denoting position of vehicle along true North and E_x denoting position of vehicle along true East respectively.

The yaw angle is denoted by ψ, which gives the angle between the North axis of NED reference frame and the vertical line along the length of the vehicle. The vehicle speed is measured by the odometer. The relationship between odometer and vehicle speed is:

$$V = SN \tag{1}$$

Where N is odometer readings in counts per second and S is scale of odometer.

2.1.1 The state equations. The east position E_x, north position N_y, heading ψ, speed v, gyro rate $\dot{\psi}$, bias , B and odometer scale Scan be expressed in state space form as:

$$E_x(k+1) = E_x(k) + V(k)T\sin(\psi\ (k)) + w_1 \tag{2}$$
$$N_y(k+1) = N_y(k) + V(k)T\cos(\psi\ (k)) + w_2 \tag{3}$$
$$\psi\ (k+1) = \psi\ (k) + T\dot{\psi}\ (k) + w_3 \tag{4}$$
$$V\ (k+1) = V\ (k) + w_4 \tag{5}$$
$$\dot{\psi}(k+1) = \dot{\psi}\ (k) + w_5 \tag{6}$$
$$B\ (k+1) = B\ (k) + w_6 \tag{7}$$
$$S\ (k+1) = S\ (k) + \frac{V(k) - N(k)S(k)}{N(k)} + w_7 \tag{8}$$

A processes noise $\boldsymbol{w_k}$ is white, zero-mean, uncorrelated, and has known covariance matrix $\mathbf{Q_k}$.

2.1.2 The observation equations. The sensors which attached on the body of the vehicle are GPS, Single axis gyro for yaw rotation measurement and odometer. Then, the measurements equations are expressed as:

$$E_{gps} = E_x(k) + v_1 \tag{9}$$

$$N_{gps} = N_y(k) + v_2 \tag{10}$$

$$\psi_{\ gps} = \psi\ (k) + v_3 \tag{11}$$

$$V_{speed}(k) = n(k)\, S(k) + v_4 \tag{12}$$

$$\dot{\psi}_{gyro}(k) = \dot{\psi}\ (k) + B + v_5 \tag{13}$$

The measurement noise $\mathbf{v_k}$ is a zero mean white noise and the measurement noise covariance matrix $\mathbf{R_k}$.

2.2 Adaptive Sage Husa Kalman Filter

Sage Husa is one of the most commonly used methods. [5, 11] used it for initial INS alignment. [6] used it for de noising FOG measurements. In all these examples, the data and states are static and Sage Husa proves to be very efficient in estimating correct Q and R for static data. In case of dynamic data, where states are changing e.g in INS where vehicle is moving or in case of target tracking, where the target is changing its states quite fast, Sage Husa fails to deliver optimal values of Q and R, and can lead to diverged states. We have also conducted a series of simulations to observe this. Here we are including Sage Husa for reference and it would be compared with the other adaptive filters later.

The Sage Husa consists of Sate prediction, State correction and Statistical properties Q and R update stages. The state prediction and correction are same as for linear Kalman filter.

2.2.1 Q and R Update on Every Iteration

$$\hat{R}(k+1) = [1 - d(k)\,]\hat{R}(k) + d(k)[\tilde{Z}(k+1)\tilde{Z}(k+1)^T - H(k+1)P(k+1|k)H(k+1)^T] \tag{14}$$

$$\hat{Q}(k+1) = [1 - d(k)\,]\hat{Q}(k) + d(k)[K(k+1)\tilde{Z}(k+1)\tilde{Z}(k+1)^T K(k+1)^T + P(k+1) - \Phi(k+1|k)\,P(k|k)\Phi^T(k+1|k)] \tag{15}$$

$$\hat{r}(k+1) = [1 - d(k)\,]\hat{r}(k) + d(k)[Z(k+1) - H(k+1)\hat{X}(k+1|k)] \tag{16}$$

$$\hat{q}(k+1) = [1 - d(k)\,]\hat{q}(k) + d(k)[\hat{X}(k+1) - \Phi(k+1|k)\hat{X}(k)] \tag{17}$$

Where d is attenuation factor and b is the forgetting factor, the value of d is given as

$$\boldsymbol{d}(\mathrm{k}) = \frac{1-\mathrm{b}}{1-b^{k+1}} \quad 0 < b < 1 \tag{18}$$

The selection of forgetting factor b is of prime importance. Whether suitable for selecting of forgetting factor is the key of the filter divergence and precision [12]. In general, trial and error analysis is used mostly; while some scholars use the optimal estimation theory in filter and iteratively find the forgetting factor in real time,. The value of b is usually chosen between 0.95 to 0.99 [11]. But also another method has been proposed in [11] to estimate b.

In case if data is static and also if we readily know Q, we can quantify R. Similarly if we know R, we can Quantify Q, using Sage Husa adaptive algorithm. If both are unknown, we can calculate Q and R using periodic switching between Q and R calculations. In our case, we have wrongly initialized Q and R initially, and let Sage Husa converge to optimal R and Q respectively.

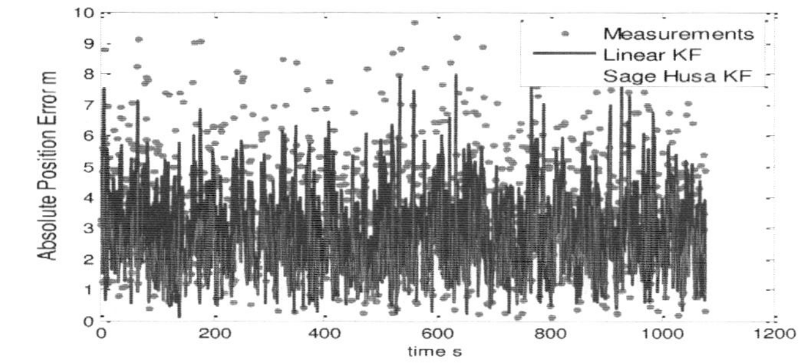

Figure 1. Linear KF and Sage Husa KF Results for absolute position error

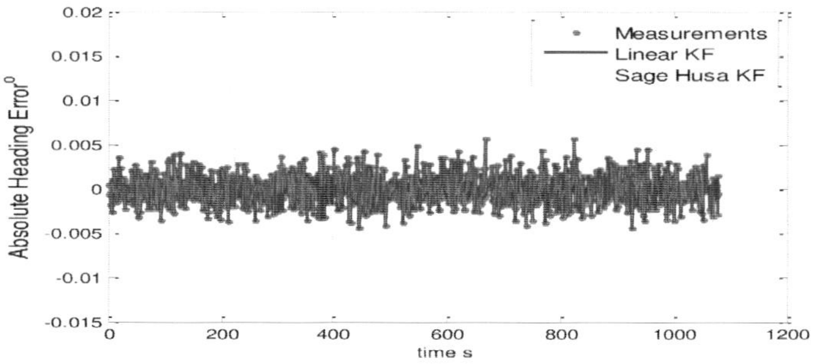

Figure 2. Linear KF and Sage Husa KF Results for absolute heading error

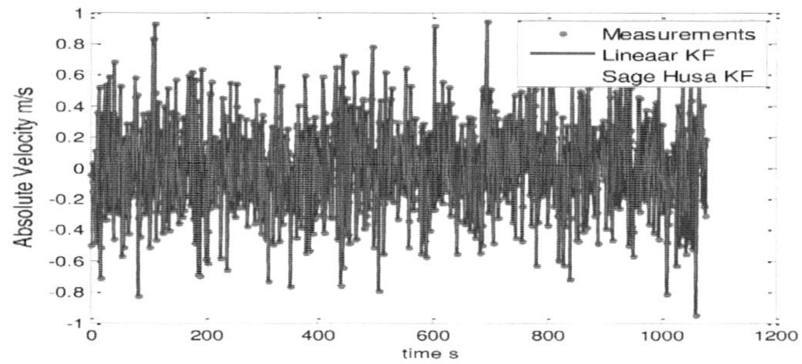

Figure 3. Linear KF and Sage Husa KF Results for absolute velocity error

2.3 Strong Tracking Kalman Filter

As previously described, the Sage Husa Adaptive Kaman filter can estimate optimal Q and R if new measurements have less information and system is less dynamic. Kalman filter works by calculating the optimal gain K which depends on R and P. In case if information/measurement is changing very fast, R and Q or may be P must be updated in such a way so that our Kalman filter could utilize the new measurement to give a real time effect. This can be done through following.

- Choose optimal *R* - Choose optimal *P* - Choose optimal *R* and *P*

Strong tracking is an algorithm that can be used to update *P* by depressing the accuracy so that the states are optimal as well as real time.

The generic algorithm for strong tracking algorithm is given as follows:

2.3.1 Prediction Stage.

$$\hat{\boldsymbol{X}}(\boldsymbol{k}+\mathbf{1}|\boldsymbol{k}) = \boldsymbol{\Phi}(\boldsymbol{k}+\mathbf{1}|\boldsymbol{k})\,\hat{\boldsymbol{X}}(\boldsymbol{k}|\boldsymbol{k}) + \boldsymbol{B}(\boldsymbol{k})\boldsymbol{U}(\boldsymbol{k}) \tag{19}$$

When the measurements arrive, calculate the following:

$$\lambda(k+1) = \mathrm{diag}[\lambda_1(k+1), \lambda_2(k+1), \lambda_3(k+1), \ldots \lambda_n(k+1),] \tag{20}$$

Where $\boldsymbol{\lambda_i(k+1)}$ can be obtained as

$$\lambda_i(k+1) = \begin{cases} a_i C_{k+1} & if\ (a_i C_{k+1} > 1) \\ 1 & if\ (a_i C_{k+1} \leq 1) \end{cases} \quad (21)$$

Where C_{k+1} can be found by

$$C_{k+1} = \frac{Tr[V_{o(k+1)} - R_{(k+1)} - H_{(k+1)} Q_k H^T_{(k+1)}]}{Tr\{\sum_{i=1}^{n} a_i [\Phi_{k+1,k}\ P_k\ \Phi^T_{k+1,k}\ H^T_{k+1,k}\ H_{k+1}]\}} \quad (22)$$

And

$$V_{o(k+1)} = \begin{cases} \widetilde{Z_1}\,\widetilde{Z_1}^T & if (k = 0) \\ \dfrac{pV_{o(k)} + \widetilde{Z_{k+1}}\,\widetilde{Z_{k+1}}^T}{1+p} & if (k \geq 1\ and\ 0 \leq p < 1) \end{cases} \quad (23)$$

From [7] the value of p is chosen to be 0.95. And by Using the above variables, calculate $P_{k+1,k}$

$$P_{k+1,k} = \lambda_{k+1,k} \Phi_{k+1,k} P_k \Phi_{k+1,k}{}^T + Q_k \quad (24)$$

2.4 Hybrid between SAGE Husa and Strong Tracking (Modified Adaptive Robust KF)

Strong tracking is necessary when state is changing quickly and we need more information, while Sage Husa works under less dynamic conditions. Therefore we must check whether we must update R and Q using Strong tracking or Sage Husa. This check is provided by $\tilde{Z}(k/k-1)$ and $\tilde{Z}(k/k-1)\tilde{Z}(k/k-1)$ provides the mean square error (MSE). When $\tilde{Z}(k/k-1)$ is large, therefore the MSE is large, and we need strong tracking to make $\tilde{Z}(k/k-1)$ small, otherwise use Sage Husa.

$$\tilde{Z}(k,k-1) = Z(k,k-1) - HX(k,k-1) \quad (25)$$

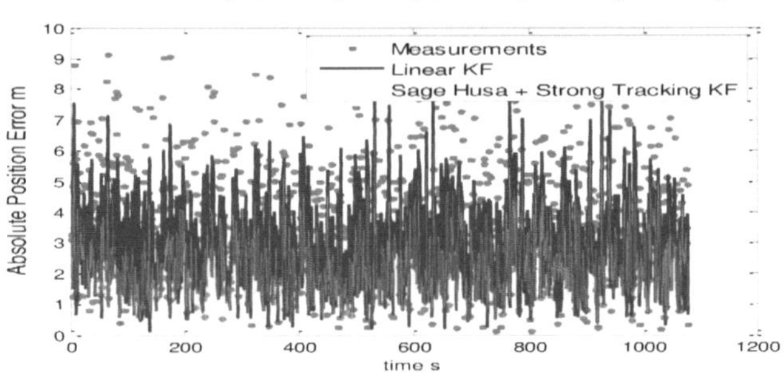

Figure 4. Linear KF and Modified Adaptive Robust KF Results for absolute position error

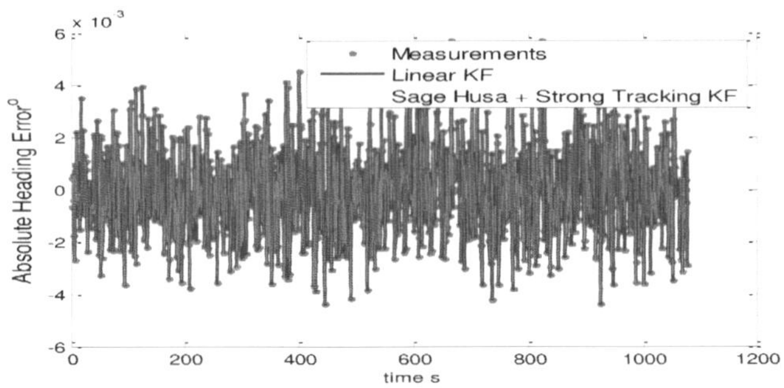

Figure 5. Linear KF and Modified Adaptive Robust KF Results for absolute heading error

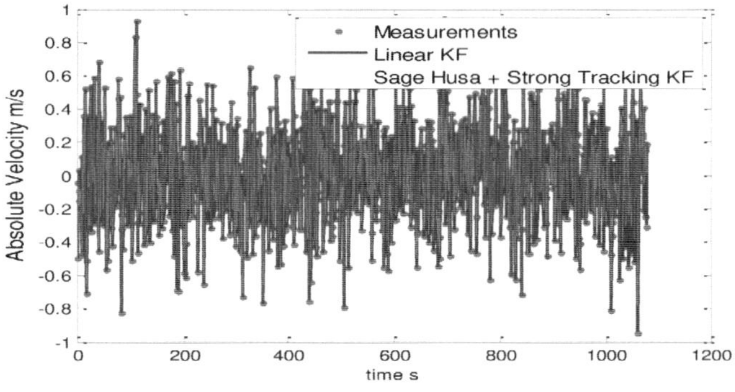

Figure 6. Linear KF and Modified Adaptive Robust KF Results for absolute velocity error

2.5 Adaptively Robust Kalman Filter

The adaptively robust Kalman Filter is described in [10] and is based on switching algorithm depending on adaptive factor α which decides either adaptivness or robustness is required. Robustness is provided by least square method while adaptiveness is provided by modified equations of P and K where as Q and R are updated using Sage Husa Kalman Filtering equations. The complete equations for implementation are describe in [10] and are discussed below.After prediction, estimate residual given by:

$$\tilde{Z}(\mathrm{k}\,,\mathrm{k}-1) = \mathrm{Z}(\mathrm{k},\mathrm{k}-1) - \mathrm{HX}(\mathrm{k},\mathrm{k}-1) \tag{26}$$

2.5.1 Calculate Learning Statistics S. Different learning statistic are given in the literature [10], based on different conditions. Here we took $\tilde{\mathrm{Z}}(\mathrm{k}+1)$ and $\hat{\mathrm{X}}(\mathrm{k}+1|\mathrm{k})$ as two different observations, and calculate learning statistic S based on Ratio of Variances method.

$$S = \frac{\mathrm{V_x}'\,\mathrm{P_x V_x}}{\mathrm{m}\,\mathrm{V_k}'\,\mathrm{P_k V_k}} \tag{27}$$

2.5.2 Calculate α. Adaptive factor α is calculated as an exponential function [10]. We have taken c=1.

$$\alpha = \begin{Bmatrix} 1 & S \le c \\ e^{-(S-c)^2} & S > c \end{Bmatrix} \tag{28}$$

- If $\alpha = 0$, Use Least Square Estimation given by:

$$X = \frac{H^T\ Z}{H\,(\mathrm{P_k}H)^T} \tag{29}$$

- If $\alpha = 1$, Use standard Kalman Filter for state correction

- If $0 < \alpha < 1$, Use adaptive filter, whose correction equations are given as:

$$P_k = \tfrac{1}{R} \text{ and } P_{Xk} = \tfrac{1}{P} \tag{30}$$

$$\hat{X}(k+1|k+1) = [H^T(k+1)P_k H(k+1) + \alpha P_{Xk}]^{-1}[H^T(k+1)P_k Z(k+1) + \alpha P_{Xk}\hat{X}(k+1|k)] \quad (31)$$

$$K(k+1) = \frac{1}{\alpha}\left\{P(k+1|k)H^T(k+1)[\frac{1}{\alpha}H(k+1)P(k+1|k)H^T(k+1) + R(k+1)]^{-1}\right\} \quad (32)$$

$$P(k+1|k+1) = P(k+1|k) - K(k+1)H(k+1)P(k+1|k) \quad (33)$$

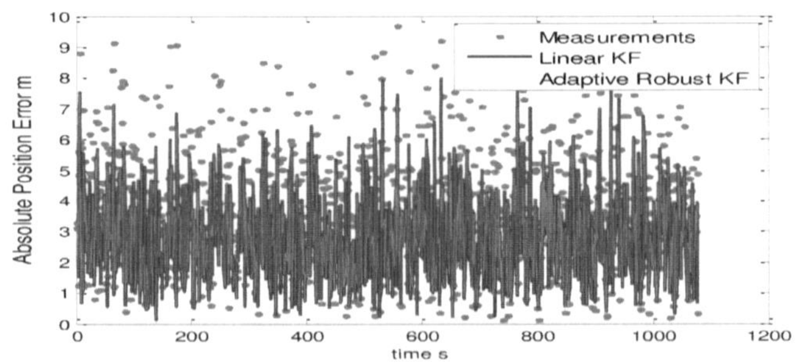

Figure 7. Linear KF and Adaptive Robust KF Results for absolute position error

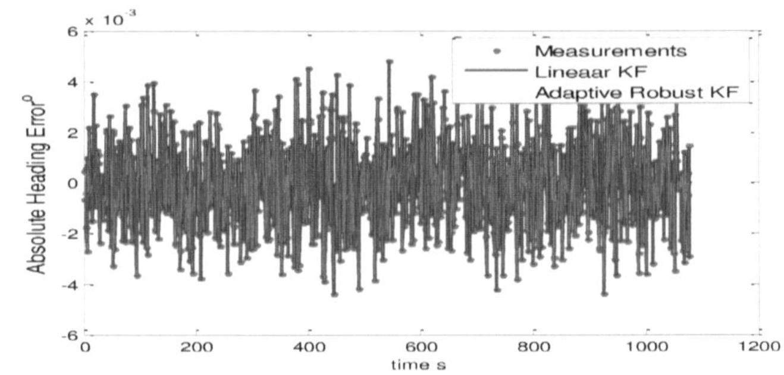

Figure 8. Linear KF and Adaptive Robust KF Results for absolute heading error

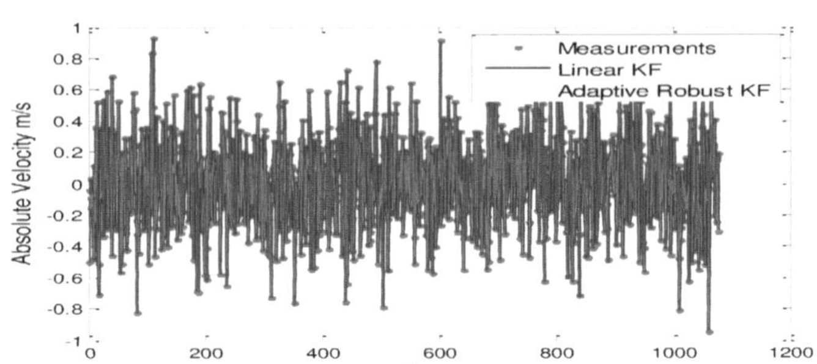

Figure 9. Linear KF and Adaptive Robust KF Results for absolute velocity error

3. Results and observations

The absolute position, velocity and heading are shown in the figures (1, 2 and 3) for Linear KF and Sage Husa. In figures (4, 5 and 6), a comparison of Linear KF and Modified Adaptive Robust KF is shown. Also in figures (7, 8 and 9) the comparison between Linear KF and Adaptive Robust KF is shown.The summarized results are shown in the Figure 10. The graph gives the complete overview of the results. As can be seen, the best position results are obtained by Sage Husa KF while it diverges in velocity calculation. The overall best results are obtained by the Modified Adaptive Robust KF which converges for all states and give the best results. Adaptively Robust Kalman filter has same accuracy as linear Kalman filter. This is because the simulations are made for land based vehicle whose states don't change abruptly, except for velocity. If states are changing fast like in the application of target tracking or aerial vehicle navigation, adaptively robust KF can prove efficient over linear Kalman filter.

Results clearly show that the Modified Adaptive Robust method gives the best convergence. Even if state changes fast like velocity or state is changing smoothly, like east and north positions, Modified Adaptive Robust Kalman filter converged to the optimal values of state. During our simulation work, we made following observations:

- While implementing Sage Husa Q and R must not be updated simultaneously but alternately on each iteration.
- Sage Husa works very well if either one of Q or R is known, and the other is updated.
- The best approach is to take static data with known covariance R. Use Sage Husa to update Q. Then use this Q and Sage Husa to update R for dynamic data.

If data is static, Sage Husa gives best convergence while if robustness is also required, modified adaptive must be used.

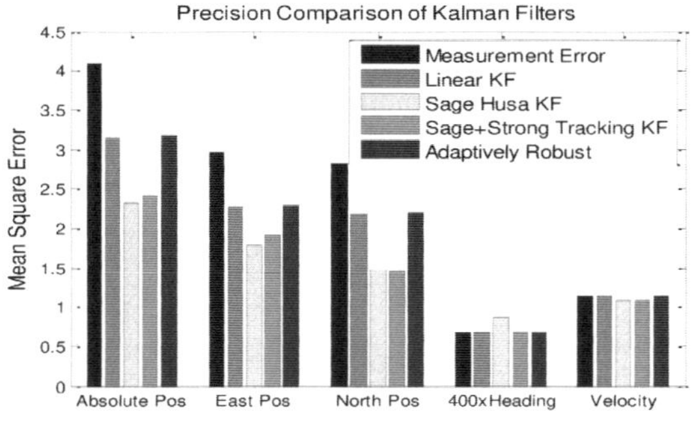

Figure 10. Mean Square error Comparison between Filters

4. Conclusion and Error Analysis

The results; the noise added to the Kalman filter is truly Gaussian, whose mean is zero. The gain K settles to a constant value, thus the system of conventional Kalman Filter in our case becomes a complementary filter. Therefore if we decrease Q and increase R, the system acts like an averaging low pass filter, thus smoothing out all the white Gaussian noise. This is due to additive nature of added noise that the conventional Kalman filter is giving us good results.

Analyzing Sage Husa Kalman Filter, the values for the variance of noise added to the North position, East Position, Heading and velocity are 8.5, 8.5, $1\text{x}10^{-5}$ and 0.1. As seen in the following figures, for initial wrong estimates of these parameters, the covariance values of these parameters converge approximately to the real values except heading. Heading converges to the correct value up till 200th iteration. At 200th iteration, there is some abnormality in the heading measurement. Therefore convergence of heading is lost here and covariance values in R for heading takes some abnormal value. As it is clear from the results of the Figure 10, heading error is more in case of Sage Husa KF as compared to other Kalman filters. The problem can be solved using some check for these abnormal values.

In case of Fusion between strong tracking and Sage Husa, the problem defined above is solved by using $\tilde{Z}(k/k-1)\tilde{Z}(k/k-1)$ as a check condition. When abnormality arrives, this term gets larger, the abnormality is taken as information and the Kalman Filter moves to strong tracking case. But shown in Figure 10, the absolute position error of this filter is a little larger as compared to the "only Sage Husa" case. This is because the values of R and Q had not been converged to the true values and filter switched to strong tracking algorithm. This is caused by limited data. Therefore it is strongly recommended that while implementing such a filter, only Sage Husa must be used until convergence of R and Q is achieved and then move to Sage Husa + Strong Tracking Filtering.

Adaptively Robust Kalman filter depends on Learning Statistics S and Adaptive Factor α. We have treated $\tilde{Z}(k/k-1)$ and $X(k/k-1)$ as two separate measurements and calculated s using ratio of variance components. The value of α is calculated as an Exponential Function of s. α=1 implies that $\tilde{Z}(k/k-1)$ is not high and we should continue with conventional Kalman Filter. 0<α<1 brings the filter into robust state. Kalman gain is increased by the factor of 1/α and system relies more on measurements. The filter acts as a high pass filter, utilizing latest information. Thus in our case its accuracy is lower than conventional Kalman Filter. The Adaptively Robust Filter would perform better in the scenarios where much robustness is involved and states and measurements are varying abruptly.

Table.1 Converged Values of Covariance in Three Selected Adaptive Filters

Parameters	Original Variances	Initialized Variance	Variance by Sage Husa	Variance by Strong Tracking+ Sage Husa	Variance by Adaptive Robust
North Position	9.4	50	8.87	2.24	10.56
East Position	8.9	50	8.73	1.82	10.40
Velocity	3.25e-06	0.01	0.0082	0.0013	0.0088
Heading	0.097	1	0.27	0.21	0.275

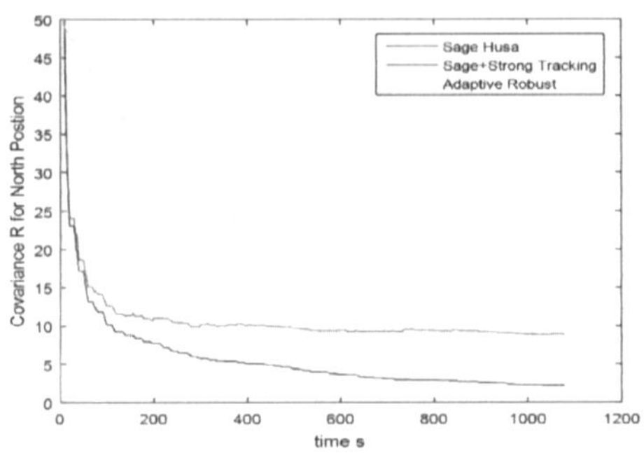

Figure 11. Covariance **R** for North Position

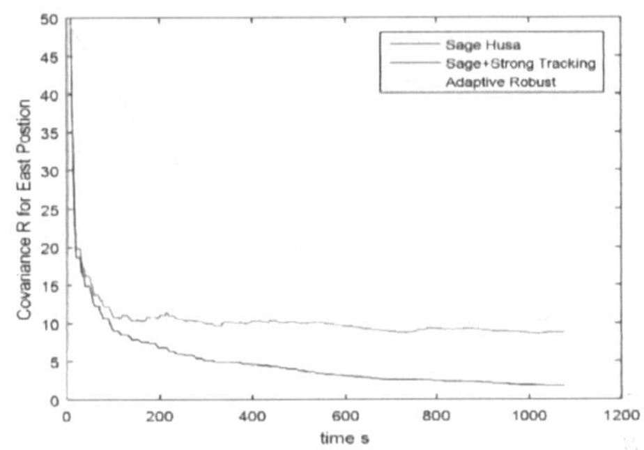

Figure 12. Covariance **R** for East Position

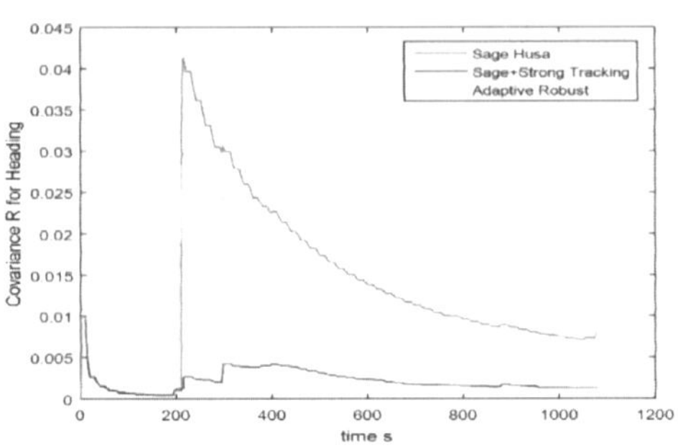

Figure 13. Covariance **R** for Heading

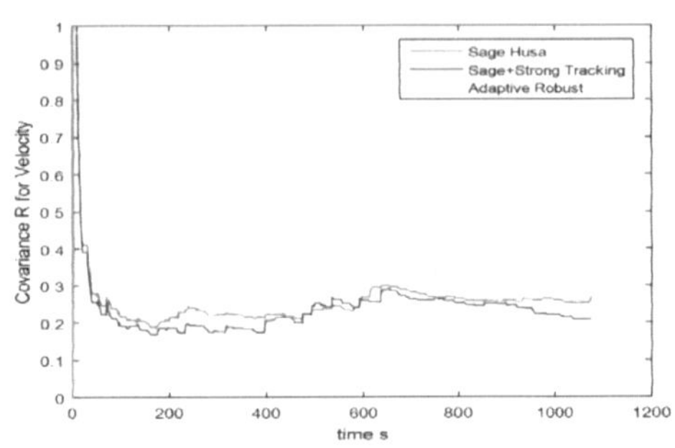

Figure 14. Covariance **R** for Velocity

Table.1 clarifies the Converged Values of Covariance in Three Selected Adaptive Filters. North and East position indicate low dynamic data while Velocity and Heading constitutes fast varying data. Therefore Combination of Fused Strong Tracking and Sage Husa can better estimates the covariance in fast varying data case while adaptive robust and Sage Husa have better convergence for slow

IOP Conf. Series: Materials Science and Engineering **187** (2017) 012019 doi:10.1088/1757-899X/187/1/012019

varying data sets. In figures (11, 12, 13 and 14) we can notice the varying in the Covariance **R** for position, velocity and heading with respect to the three filters.

5. References

[1] Grewal M S and Andrews A P 2010 Applications of Kalman Filtering in Aerospace 1960 to the Present *IEEE Control Systems Magazine* **30** 69–78.

[2] Welch G and Bishop G 2006 An Introduction to the Kalman Filter *UNC-Chapel Hill, TR 95-041.*

[3] Ting J A, Theodorou E and Schaal S 2007 A Kalman Filter for Robust Outlier Detection *IEEE Conference on Intelligent Robots and Systems.*

[4] He W, Lian B and Tang C 2014 GNSS/INS Integrated Navigation System Based on Adaptive Robust Kalman Filter Restraining Outliers *International Conference on Communications in China.*

[5] Xin S W 2014 Application of Sage-Husa Adaptive Filtering Algorithm for High Precision SINS Initial Alignment *11th International Computer Conference on Wavelet Active Media Technology and Information Processing.*

[6] Narasimhappa M, Rangababu P, Sabat S L, Nayak J 2012 A modified Sage-Husa adaptive Kalman filter for denoising Fiber Optic Gyroscope signal *Annual IEEE India Conference* 1266 - 1271.

[7] Gupta M and Kumar R 2014 *International Journal of Advance Research In Science And Engineering* **3**.

[8] Simon D J 2006 Optimal State Estimation (John Wiley & Sons, Inc).

[9] Kosanam S 2004 Kalman Filtering For Uncertain Noise Covariances, Cleveland State Univerisity.

[10] Yang Y 2010 Adaptively Robust Kalman Filters with Applications in Navigation (Springer-Verlag Berlin Heidelberg).

[11] Zhang H 2011 Application of a New Adaptive Kalman Filitering Algorithm in Initial Alignment of INS *International Conference on Mechatronics and Automation, Beijing.*

[12] D.Huang 1990 Inertial Navigation System (Publishing Company of National Industry Press).

Acknowledgements

This study is supported by a grant from the National Natural Fund of China, Grant No. 61375082.

System Identification and Control Design of an Unmanned Helicopter Using a PI-MPC Controller

Quang Le Tri [a] and Ying-Chih Lai [b]

Department of Mechanical and Aeronautical Engineering, No. 100 Wenhwa Rd., Seatwen Dist., Taichung, Taiwan 40724, R.O.C.

[a]quangletri@gmail.com; [b]yingclai@fcu.edu.tw

Abstract. This paper presents the study of the system identification and controller design for an unmanned helicopter using the integration of Proportional Integral (PI) and Model Predictive Control (MPC). Since the dynamic model of a helicopter is highly nonlinear and contains many uncertainties, the system identification and control are challenging and complicated. To accelerate the development, the autonomous flight and trajectory tracking of an unmanned helicopter, this study first setup a software simulation environment of the helicopter using the X-Plane flight simulator. The prediction-error minimization (PEM) and subspace methods were applied in this study to identify the dynamic model of the interested flight trim conditions. The lateral, longitudinal, heave, and yaw dynamic models were predicted by using the System Identification Toolbox of MATLAB. To enhance the stability and eliminate the uncertainty of the control system, the Integration of Proportional Integral (PI) and MPC were introduced. The developed control system was then applied to perform the trajectory tracking of a helicopter. The simulation results show that the performance of the proposed approach can track the desired trajectory.

1. Introduction

Comparing with fixed-wing aircrafts, helicopters have many advantages since they are a type of rotorcraft in which lift and thrust are supplied by rotors. Therefore, helicopters have the capability to take off and land vertically, to hover, and to fly forward, backward, and laterally. With the capability of high mobility, unmanned helicopters are applied to many applications such as rescue, transportation, and agriculture. However, it is well-known that the dynamic model of a helicopter is complex and highly nonlinear. The modeling and control of a helicopter are very complicated and need many efforts.

Before designing a model-based controller, it is required to study the system dynamics. The first step is to determine the system model. Many system identification methods have been investigated in [1]. Several identification methods and software packages are used for helicopter model estimation such as CIFER® (Comprehensive Identification from Frequency Responses) [2], System Identification Programs for Aircraft (SIDPAC) [3], extended and unscented Kalman filter, neural networks, and so on. These techniques often are applied to large systems. Moreover, there are also some popular system identification methods used to determine the helicopter model such as PEM method and subspace method [4]. These methods are available in the system identification toolbox of MATLAB. As mentioned above, the aerodynamic characteristics of helicopters are highly nonlinear. There are forty unknown parameters of the helicopter dynamic model used in [5]. It is difficult to determine all the parameters at the same time. The helicopter model can be divided into sub-models, which are lateral, longitudinal, heave, and yaw dynamic models [5]. System identification of these sub-models can be done by using system identification toolbox. After getting the dynamic model of the helicopter, the model-based controller can be designed and implemented. Many control methods have been adopted to perform flight control of helicopters, such as MPC, PID, Linear-Quadratic Regulator (LQR), Linear-Quadratic-Gaussian (LQG), fuzzy logic, and neural networks. In this paper, a PI-MPC controller was proposed. MPC has been applied to many systems with highly complex, multivariable processes and real-time applications [6,7]. In this study, MPC was integrated with PI to enhance the performance of the flight control system for the unmanned helicopter. The used control scheme consists of the inner and outer loops. PI was used in the outer-loop to improve the response time, and MPC was used in the inner loop to increase the robustness of flight control. In this study, four PI-MPC controllers were applied to the sub-models.

The rest of this paper is organized as follows. Dynamic model, system identification, flight control design for the helicopter presented introduced in Section 2, 3, and 4, respectively. Results and discussions are described in section 5. Finally, conclusions are presented in Section 6.

2. Dynamic model of the small-scale helicopter

The linearized model of a helicopter has the following form:

$$\dot{x} = Ax + Bu \quad (1)$$

$$Y = Cx \quad (2)$$

where x is the state vector, which contains $[u\ v\ p\ q\ \phi\ \theta\ a\ b\ w\ r\ r_{fb}]^T$, and u, v, and w are the longitudinal, lateral, and vertical velocities, respectively. p, q, and r are roll, pitch and yaw angular rates, respectively. a and b are longitudinal and lateral flapping angles, respectively. ϕ and θ are roll and pitch angles, respectively. u is the control input, $[\sigma_{lat}\ \sigma_{lon}\ \sigma_{col}\ \sigma_{ped}]^T$, where σ_{lat} , σ_{lon} , σ_{ped} are roll, pitch, and yaw control inputs, respectively. σ_{col} is height control. Y is the output vector, which contains $[u\ v\ w\ p\ q\ r\ \phi\ \theta]^T$. The adopted A and B matrices are presented in Equation (3) and (4). Referring to the study [5], the entire model can be divided into sub-models as lateral, longitudinal, heave, and yaw dynamic models as shown in Equation (6), (7), (8), (9), respectively.

$$A = \begin{bmatrix} Xu & 0 & 0 & 0 & 0 & -g & -g & 0 & 0 & 0 & 0 \\ 0 & Yv & 0 & 0 & g & 0 & 0 & g & 0 & 0 & 0 \\ Lu & Lv & 0 & 0 & 0 & 0 & La & Lb & 0 & 0 & 0 \\ Mu & Mv & 0 & 0 & 0 & 0 & Ma & Mb & 0 & 0 & 0 \\ 0 & 0 & 1 & 0 & 0 & 0 & 0 & 0 & 0 & 0 & 0 \\ 0 & 0 & 0 & 1 & 0 & 0 & 0 & 0 & 0 & 0 & 0 \\ 0 & 0 & 0 & -1 & 0 & 0 & -1/\tau_a & Ab & 0 & 0 & 0 \\ 0 & 0 & -1 & 0 & 0 & 0 & Ba & -1/\tau_a & 0 & 0 & 0 \\ 0 & 0 & 0 & 0 & 0 & 0 & Za & Zb & Zw & Zr & 0 \\ 0 & 0 & Np & 0 & 0 & 0 & 0 & 0 & Nw & Nr & Nrf \\ 0 & 0 & 0 & 0 & 0 & 0 & 0 & 0 & 0 & Kr & Krf \end{bmatrix} \quad (3) \qquad B = \begin{bmatrix} 0 & 0 & 0 & 0 \\ 0 & 0 & 0 & 0 \\ 0 & 0 & 0 & 0 \\ 0 & 0 & 0 & 0 \\ 0 & 0 & 0 & 0 \\ 0 & 0 & 0 & 0 \\ Alat & Alon & 0 & 0 \\ Blat & Blon & 0 & 0 \\ 0 & 0 & Zcol & 0 \\ 0 & 0 & Ncol & Nped \\ 0 & 0 & 0 & 0 \end{bmatrix} \quad (4)$$

$$C = \begin{bmatrix} 1&0&0&0&0&0&0&0&0&0&0 \\ 0&1&0&0&0&0&0&0&0&0&0 \\ 0&0&0&0&0&0&0&0&1&0&0 \\ 0&0&1&0&0&0&0&0&0&0&0 \\ 0&0&0&1&0&0&0&0&0&0&0 \\ 0&0&0&0&0&0&0&0&0&1&0 \\ 0&0&0&0&1&0&0&0&0&0&0 \\ 0&0&0&0&0&1&0&0&0&0&0 \end{bmatrix} \quad (5)$$

$$\begin{bmatrix} \dot{v} \\ \dot{p} \\ \dot{\phi} \\ \dot{b} \end{bmatrix} = \begin{bmatrix} Yv & 0 & g & g \\ Lv & 0 & 0 & Lb \\ 0 & 1 & 0 & 0 \\ 0 & -1 & 0 & -1/\tau_a \end{bmatrix} \begin{bmatrix} v \\ p \\ \phi \\ b \end{bmatrix} + \begin{bmatrix} 0 \\ 0 \\ 0 \\ Blat \end{bmatrix} \delta_{lat} \quad (6)$$

$$\dot{w} = Z_u u + Z_w w + Z_{col}\delta_{col} \quad (7)$$

$$\begin{bmatrix} \dot{u} \\ \dot{q} \\ \dot{\theta} \\ \dot{a} \end{bmatrix} = \begin{bmatrix} Xu & 0 & -g & -g \\ Mu & 0 & 0 & Ma \\ 0 & 1 & 0 & 0 \\ 0 & -1 & 0 & -1/\tau_a \end{bmatrix} \begin{bmatrix} u \\ q \\ \theta \\ a \end{bmatrix} + \begin{bmatrix} 0 \\ 0 \\ 0 \\ Alon \end{bmatrix} \delta_{lon} \quad (8)$$

$$\begin{bmatrix} \dot{r} \\ \dot{r}_{fb} \end{bmatrix} = \begin{bmatrix} N_r & N_{rf} \\ K_r & K_{rf} \end{bmatrix} \begin{bmatrix} r \\ r_{fb} \end{bmatrix} + \begin{bmatrix} N_{ped} \\ 0 \end{bmatrix} \delta_{ped} \quad (9)$$

3. System identification

3.1. System identification procedure

The adopted system identification procedure is illustrated in Figure 1. Logging data includes two parts, estimation data and validation data. Estimation data are used to identify the dynamic model. Validation data are utilized in the last step to perform model validation. The identified model was validated by using different data. The validation data were used to compare the response of the real system and estimated model. If the outputs of the real system and estimated model are similar, the identified model is precise. In contrast, an alternative procedure needs to be performed. In this paper, subspace and PEM methods were used to determine the helicopter model. Subspace method was employed to determine the initial model, and PEM method was used to improve the accuracy of the identified model. The details of subspace method can refer to the study [8].

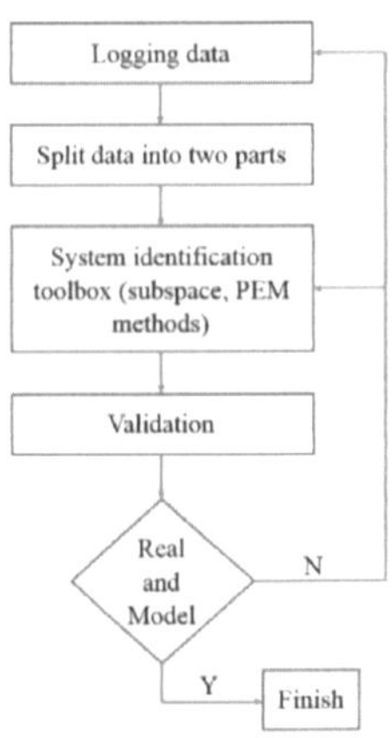

Figure 1. The used system identification procedure

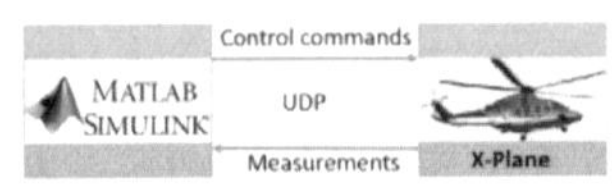

Figure 2. Communication between Simulink and X-Plane

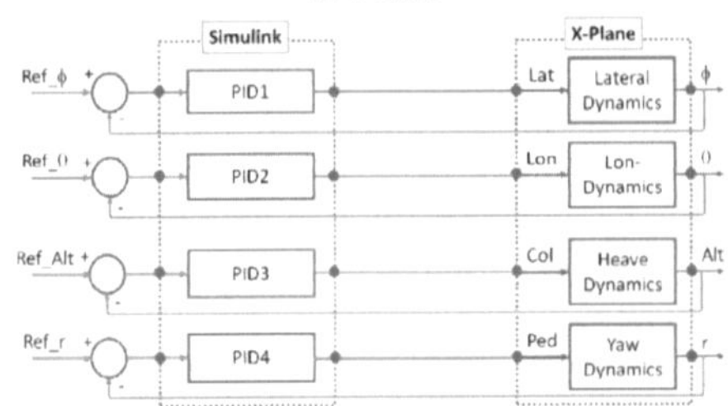

Figure 3. Structure of the used software-in-the-loop

3.2. Experimental setup and logging data

The simulation was performed by using a virtual helicopter on X-Plane. The adopted control system was developed by Simulink. Figure 2 presents the communication between Simulink and X-Plane via UDP protocol. Before collecting data, the helicopter must operate in trim condition. Referring to [9], four PID controllers were designed. The structure of the adopted controllers is shown in Figure 3. PID1, PID2, PID3, and PID4 controllers correspond to the lateral, longitudinal, heave, and yaw dynamic model, respectively. The control responses of the helicopter in hovering are showed in Figure 4. The designed inputs of the system identification procedure are shown in Figure 5. The collected flight data will be used to identify the used four sub-models.

3.3. PEM method

PEM is based on numerical optimization to minimize the *cost function*, a weighted norm of the prediction errors, defined as follows for scalar outputs [10]:

$$V_N(G,H) = \sum_{t=1}^{N} e^2(t) \quad (10)$$

where *e(t)* is the difference between the measured output and the predicted output of the model. For a linear model, the error is defined as:

$$e(t) = H^{-1}(q)[y(t) - G(q)u(t)] \quad (11)$$

where the cost function $V_N(G,H)$ is a scalar value. The subscript N denotes that the cost function is a function of the number of data samples and becomes more accurate for larger values of N.

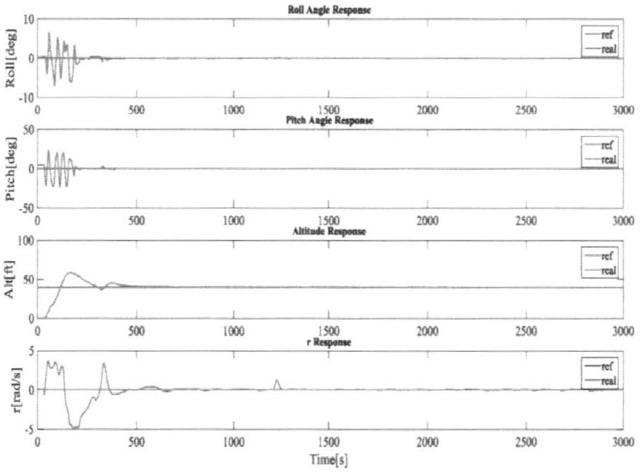

Figure 4. The responses of the helicopter in simulation

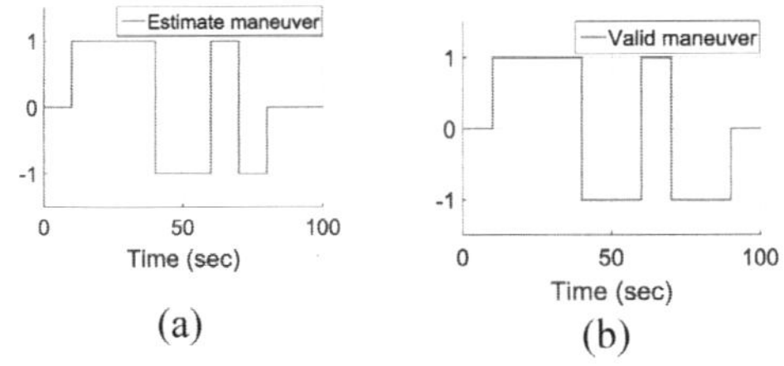

Figure 5. Designed Inputs for (a) estimation and (b) validation

4. Flight control design

The trajectory tracking control was applied to verify the developed system identification procedure and controller design. The structure of the developed control system is illustrated in Figure 6. The system includes six blocks. Reference block creates a reference trajectory. Inertial to Body block transfer from inertial coordinates to body coordinates. PI- MPC controller block processes signals and creates control commands. Plant block includes four sub-models, which was determined in section 3. Body to Inertial block transfers vectors from body coordinates to inertial coordinates. In this paper, the control commands of the developed PI-MPC are the velocities of the helicopter in body coordinate frame. Figure 7 presents the structure of the developed PI-MPC controller. V_{ref} is the target velocity; e is the difference between actual and target velocity; u_{ref} is the control input of MPC controller; u is the control inputs of lateral, longitudinal, heave, and yaw dynamic model; V is the output velocity of the Plant block. There are two feedback signals. One is the measured velocity for PI controller, and the other is the measured state vectors (attitude, velocities, and angular rates) for MPC controller. The designed control parameters are presented in Table 1.

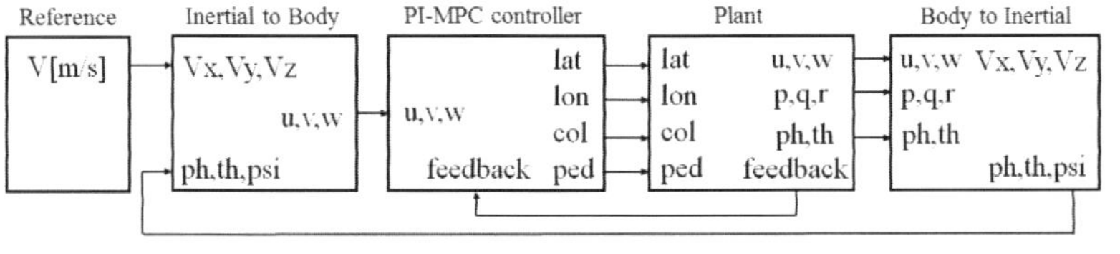

Figure 6. The structure of the system

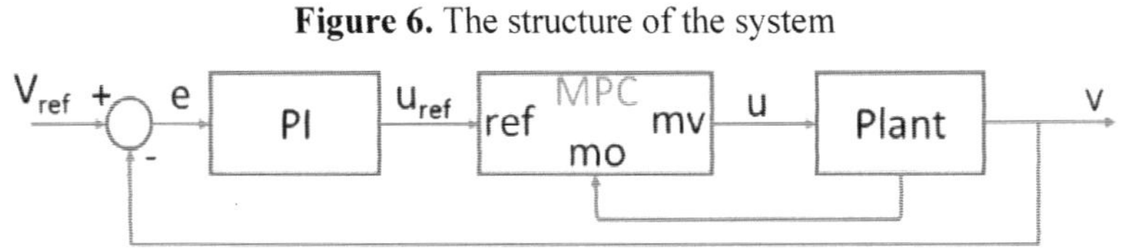

Figure 7. The structure of the controller

Table 1. Values of Kp, Ki

	Lateral	Longitudinal	Heave	Yaw
Kp	0.2	0.4	0.3	0.3
Ki	4.5	3.5	0.96	2.0

5. Results and discussions

The identified system matrices of lateral, longitudinal, and heave dynamic models are shown as follows:

$$Alat = \begin{bmatrix} -0.4214 & 0.7099 & 9.441 & -4.812 \\ 0.7506 & 0.0843 & -4.11 & 4.128 \\ 1.594 & -6.022 & -62.16 & 27.8 \\ -1.615 & 2.353 & 31.24 & -16.15 \end{bmatrix} \quad (12) \qquad Blat = \begin{bmatrix} 0.6425 \\ -0.3183 \\ -6.052 \\ 3.007 \end{bmatrix} \quad (13)$$

$$Alon = \begin{bmatrix} -0.02961 & 0.1246 & 0.6048 & -0.2666 \\ -0.02099 & 0.008143 & -0.02384 & -0.01824 \\ -2.783 & 1.179 & -5.878 & -5.204 \\ -1.007 & 0.458 & -0.6565 & -3.06 \end{bmatrix} \quad (14) \qquad Blon = \begin{bmatrix} -0.1317 \\ 0.03987 \\ 8.562 \\ 2.559 \end{bmatrix} \quad (15)$$

$$Aheave = [9.632e-05] \quad (16) \qquad Bheave = [-1.583e-05] \quad (17)$$

$$Ayaw = \begin{bmatrix} 0.1321 & -1.089 \\ 0.4828 & -0.5013 \end{bmatrix} \quad (18) \qquad Byaw = \begin{bmatrix} 0.1109 \\ 0.04702 \end{bmatrix} \quad (19)$$

Comparing these system matrices with those presented in (6, 7, 8, 9), it reveals that they are different since the discrete models were used. Although the discrete models are usable for controller design, it needs to improve the precision of the models for dynamics analysis. It needs to find out the good initial values of A and B matrices and then fix values of some elements to estimate others. Identification results of the lateral and longitudinal dynamic models are shown in Figure 8, and 9, respectively. The performance of the developed PI-MPC controllers is shown in Figure 10, 11, 12, and 13.

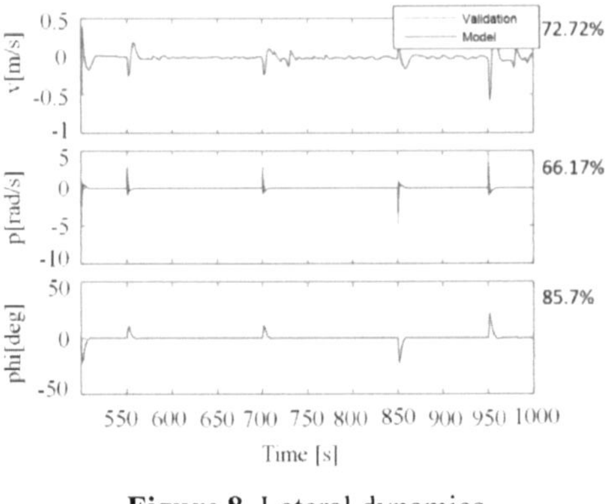

Figure 8. Lateral dynamics

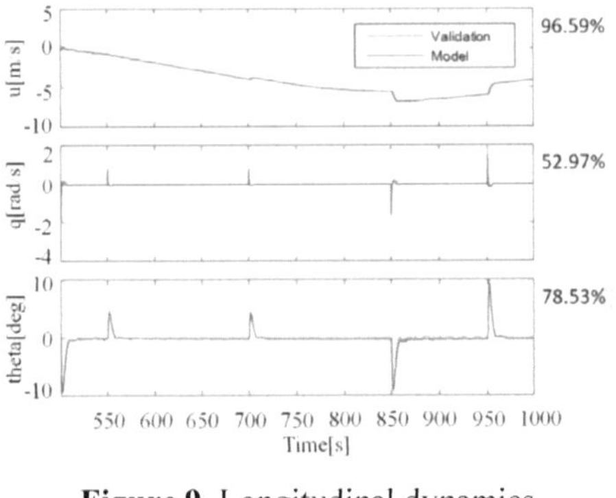

Figure 9. Longitudinal dynamics

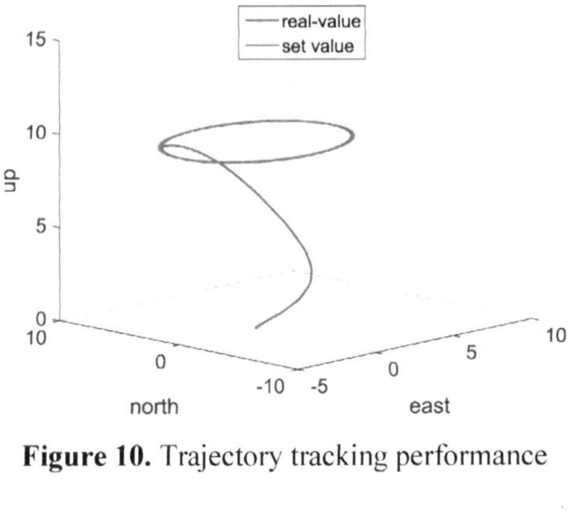

Figure 10. Trajectory tracking performance

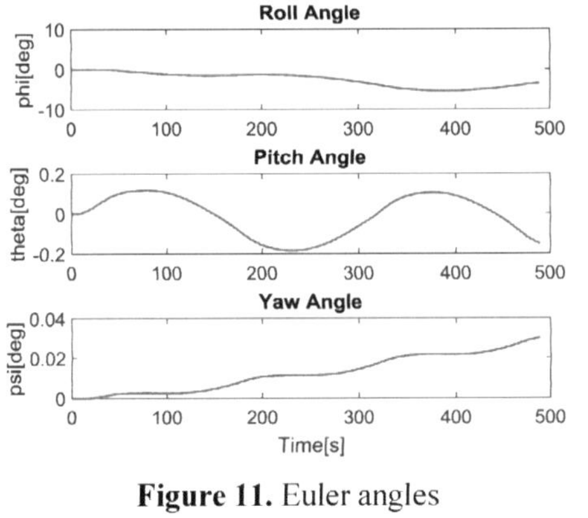

Figure 11. Euler angles

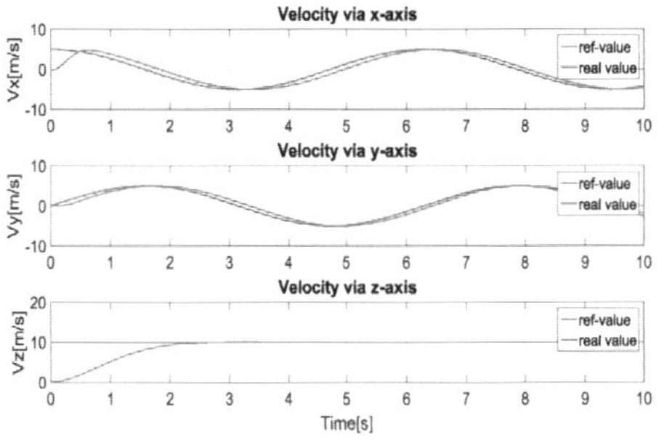

Figure 12. Velocity tracking performance

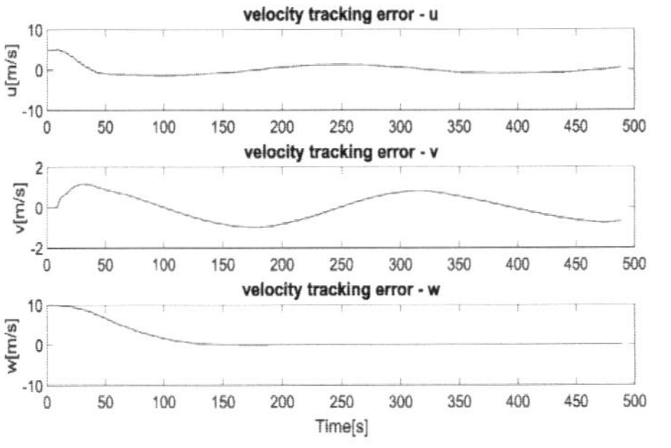

Figure 13. Velocity tracking error

6. Conclusions

A system identification and control design approach for unmanned helicopters has been developed in this paper. A proposed PI-MPC control scheme was proposed for the control design of the used sub-models, which are lateral, longitudinal, heave, and yaw dynamic models. The performance of the designed PI-MPC flight controller is viable to perform trajectory tracking control of the helicopter. The results of software-in-the-loop simulation show that the develop system identification procedure and developed control scheme can achieve the goal of flight control design of the helicopter. Although, the identification results are acceptable, the precision of the identified model need to be improved in further study for dynamics analysis.

7. References

[1] Li Fu and Pengfei Li 2013 *Int. Conf. on The Research Survey of System Identification Method* vol 2 (IEEE) p 397

[2] Tischler, Mark B, and Robert K Remple 2006 *Aircraft and rotorcraft system identification: Engineering Methods with Flight Test Examples* (AIAA education series)

[3] Wei Yuan and Jay Katupitiya 2011 *Proc. Conf. on A Time-Domain Grey-Box System Identification Procedure for Scale Model Helicopters (Australia)* p 1-8

[4] Wenbing Chen, Ming Liu, Yunjian Ge 2007 *Int. Conf. on Subspace Model Identification Methods for Identifying the Small-Scale Unmanned Helicopter Dynamics* (IEEE) p 31

[5] Mettler B 2003 *Identification Modeling and Characteristics of Miniature Rotorcraft* (Kluwer Academic Publishers)

[6] Cunjia Liu, Wen-Hua Chen, John Andrews 2012 *Tracking control of small-scale helicopters using explicit nonlinear MPC augmented with disturbance observers,* **20** 258

[7] Liuping Wang 2009 *Model Predictive Control System Design and Implementation Using MATLAB* (Springer Science & Business Media)

[8] S Joe Qin 2006 *An overview of subspace identification* **30** 1502

[9] KimonP. Valavanis 2007 *Advances in Unmanned Aerial Vehicles: state of the art and the road to the autonomy* (Springer Science & Business Media)

[10] Ljung, Lennart 1999 *System Identification: Theory for the User (2nd Edition)* (Prentice Hall PTR)

Acknowledgments

This study was supported in part by the Ministry of Science and Technology of Taiwan under grant number MOST 103 - 2632 - E - 035 - 001 - MY3.

IOP Conf. Series: Materials Science and Engineering **187** (2017) 012021 doi:10.1088/1757-899X/187/1/012021

An Optimal Moving Horizon Estimation for Aerial Vehicular Navigation Application

Haris Ubaid Gul[1] and Yang Dong Kai[2]

School of Information & Electronic Engineering, Beihang University Beijing China

[1]harisubaidgul@gmail.com; [2]yangdongkai@sina.com

Abstract. In this article, an optimal state is estimated using the moving horizon estimation technique (MHE), based on the minimizing the deterministic cost function defined for moving window with a finite number of samples at specific time interval. The optimal moving horizon observer was designed and implemented for the non-linear dynamic problem of aerial vehicle integrated navigation. The low grade commercial inertial measuring instrument (IMU) equipped with accelerometers and gyros sensors instrumented on-board in the strapdown configuration, is employed for collection of the real time experimental data. The data fusion algorithm of moving horizon estimation is realized and the results are collected from the offline algorithm testing on the Matlab software platform. Essential data processing and cleaning of data processing was conducted before algorithm application i.e. solving the multi rate sensors data synching and removing high frequency unwanted contents. Finally, the aerial vehicle dead reckoning integrated navigation was performed with recursive observer using IMU/GPS avionics. Contrary to the widely practiced extended Kalman filter results, recursive observer of MHE exhibited performance enhancement in the response and precision aspect, regardless of environmental noise and failure scenarios.

1. Introduction

Global positioning and satellite system has versatile applications in civil and military fields, with certain deficiencies of the environmental signal attenuations, signals unlock in presence of hurdles i.e. roof, trees for land vehicle navigations, and the low refresh rates, limits its use for certain high dynamic systems. Despite the above mentioned drawbacks, is its precision makes it attractive as the aiding source for high rate IMU systems. In literature the optimal state for data mixing/fusion is computed using the unscented filtering [1], particle filtering and widely practiced extended type filtering [2]. Kalman Filter serves as the optimal state estimator and various approaches are based on the Kalman filtering technique, a data fusion algorithm central difference Kalman filter was designed by Zhao [3]. Errors are introduced due to the linearization of the system model in Kalman filter based techniques and distribution function selection is difficult in the particle type of filtering. Currently the moving horizon estimation serves an optimal tool for handling the state estimation of dynamic systems despite of non-linaerities and disturbances. Recursive moving horizon technique minimizes the cost function in the defined window of fixed number of measurements samples, composed of prediction error calculated on recent batch of samples and 2nd is the fresh measurements response at current time [4]. In [5], MHE observer was designed and implemented for a simple case of land vehicle 2D navigation. MHE is also termed as the sliding window estimation, as at each time interval a fixed number of sample is processed for optimal state calculation by minimizing the cost function equation and on the subsequent interval new measurements are sampled for recursive continuous operation. Here the actual aerial vehicle flight data is used for the optimal state estimation extending MHE based linear observer [6] for nonlinear complex navigation problem, further this optimal state is processed for the navigation parameter computations for onwards use in the flight computer. In literature MHE is being studied for different applications, like the model predictive controllers [7], spacecraft attitude estimation and sensors calibration [8] and integrated navigation filtering [9].

In this article the optimal state is estimated by designing the nonlinear estimator, the mentioned estimator is validated offline in the Matlab environment. For algorithm validation, the real time raw data was collected from an aerial vehicle flight, containing the onboard instrumented inertial measurement unit (IMU) accelerometers and gyros sensors along 3-axis, and GPS receiver installed with smart antenna set. Various supporting avionics for data collection and processing were also installed i.e. analog/digital modules, power source, conditioners and data logger module. Data processing and cleaning was conducted with the raw data from sensors, including the data matching and synching processing with the low rated GPS data, unwanted high frequency noise rejection with the high order Butter worth filtering [10] and highly precise integration scheme of RK-4 [11] replacing the simple integration for avoiding the unwanted computation errors. The algorithm stage comprises of three different stages, one is the dead reckoning stage using the cleaned/processed sensors data and computing the free inertial solution with the singularity avoiding quaternions technique, MHE optimal state estimation based on the nonlinear system model with the measurement matrix, incorporating the optimal state for computing the final navigations parameter and updating the latest states. Finally, for data evaluations two references were available; one is the real time GPS recorded data and the other is the extended Kalman filter implemented side by side with the MHE observer with same starting and data inputs ensured. The complete mutirate package of data cleaning, navigation algorithm implementation, optimal estimation, parameters correction with optimal states, reference extended Kalman filtering and data evaluation was performed on the Matlab platform with homogenous environmental conditions, the data comparison stage witnesses the performance enhancement with MHE.

Consequently, this manuscript is presented comprising the following sections; sections 2 discusses the strap down inertial navigation system model formation, section 3 discusses the moving horizon observer, section 4 is reserved for the algorithm response with the help of the experimental data in comparison to the reference EKF and conventionally article ends with paper conclusion.

2. Non-Linear Strapdown INS Error Model

In this article the navigation reference coordinate system followed is the north east down or in other words NED geographic reference coordinate system. The sensors installed on the vehicle are considered in the body reference coordinate system and the transformation from body to navigation frame is realized with the quaternion rotation matrix formation [12].

$$\begin{pmatrix}\dot q_0\\ \dot q_1\\ \dot q_2\\ \dot q_3\end{pmatrix}=\frac{1}{2}\begin{pmatrix}0 & -\omega_x & -\omega_y & -\omega_z\\ \omega_x & 0 & \omega_z & -\omega_y\\ \omega_y & -\omega_z & 0 & \omega_x\\ \omega_z & \omega_y & -\omega_x & 0\end{pmatrix}\begin{pmatrix}q_0\\ q_1\\ q_2\\ q_3\end{pmatrix}, C_b^n(q_n^b)=\begin{pmatrix}q_1^2+q_0^2-q_2^2-q_3^2 & 2(q_1q_2+q_3q_0) & 2(q_1q_3-q_2q_0)\\ 2(q_1q_2-q_3q_0) & q_2^2+q_0^2-q_1^2-q_3^2 & 2(q_2q_3+q_1q_0)\\ 2(q_1q_3+q_2q_0) & 2(q_2q_3-q_1q_0) & q_3^2+q_0^2-q_1^2-q_2^2\end{pmatrix}^T \quad (1)$$

The strap down differential equations [13] are given as;

$$\begin{aligned}&\delta\dot V^n = f^n\times\Phi^n+\nabla^p-(2\delta\omega_{ie}{}^n+\delta\omega_{en}{}^n)\times V^n-(2\omega_{ie}{}^n+\omega_{en}{}^n)\times\delta V^n\\ &\delta\dot L=\frac{\delta V_n}{R_m+h},\delta\dot\lambda=\frac{\delta V_e}{(R_n+h)}\sec L+\frac{V_e}{(R_n+h)}\tan L\sec L\delta L,\delta\dot h=\delta V_u \quad (2)\\ &\dot\Phi^n=\delta\omega_{ie}{}^n+\delta\omega_{en}{}^n-(\omega_{ie}{}^n+\omega_{en}{}^n)\times\Phi^n+\varepsilon^n\end{aligned}$$

Where the state vector is $x_{state} = (\varphi_n, \varphi_e, \varphi_d, \delta v_n, \delta v_e, \delta v_d, \delta Lat, \delta Lon, \delta Alt)$ with $\varphi_n, \varphi_e, \varphi_d$ are the attitude, $\delta v_n, \delta v_e, \delta v_d$ are the velocity error states and $\delta Lat, \delta Lon, \delta Alt$ are the position states.

$$.Z_p(t) = [(L_I - L_G)R_m; (\lambda_I - \lambda_G)R_n \cos L; h_I - h_G; V_{In} - V_{Gn}; V_{Ie} - V_{Ge}; V_{Id} - V_{Gd}]^T \quad (3)$$

The observation/measurement matrix of system depends on the number of available measurement sources. Here the three position and three velocities parameters are received from the GPS receiver forming the system observation matrix. The state space compact representation takes the form;

$$X(k+1) = f(x(k), u(k)) + w(k), Z(k) = h(x(k), u(k)) + v(k) \quad (4)$$

3. Optimal Moving Horizon State Estimator

This section explains the structure of MHE state estimator, which is based on minimizing the estimation cost function over a moving fixed sized window of samples. MHE is the convex optimization framework, incorporating the current measurement information along with the preceding measurements set to estimate the upcoming optimal state. MHE can incorporate the system constraints directly and can be formulated for nonlinear problems, the system cost function comprised of prediction error based on current data and arrival cost summarizing the past samples, given as [8];

$$J_t = \sum_{k=0}^{t} \hat{v}^T{}_k R^{-1} \hat{v}_k + \hat{w}^T{}_k Q^{-1} \hat{w}_k + (\hat{x}_0 - \bar{x}_0) Pc_0^{-1} (\hat{x}_0 - \bar{x}_0)^T \quad (5)$$

The equation Eq. (5) can further be expressed in two times interval to formulate the recursive state estimator valid for continuous recursive state estimation.

$$J_t = \sum_{k=t-N+1}^{t} (\hat{v}^T{}_k R^{-1} \hat{v}_k + \hat{w}^T{}_k Q^{-1} \hat{w}_k) + \sum_{k=0}^{T-N} (\hat{v}^T{}_k R^{-1} \hat{v}_k + \hat{w}^T{}_k Q^{-1} \hat{w}_k) + (\hat{x}_0 - \bar{x}_0) Pc_0^{-1} (\hat{x}_0 - \bar{x}_0)^T \quad (6)$$

Subject to, $\hat{x}_k \in \Omega_x, \hat{v}_k \in \Omega_v, \hat{w}_k \in \Omega_w$, N is the sample size considered at a time, depending on the computation capacity of the resources available onboard, P is the positive definite covariance weighting matrix, Q and R are the system matrices. The pictorial framework of MHE estimator can be explained in Fig.1. By minimizing the Eq. (6), the least squares optimal state solution is given as [6],

$$x(k) = F^N x(k-1) + F^N (P_0 + P_N)^{-1} M_N^T * Wt * [(Z(k-N) - M_N x(k-N))]^T \quad (7)$$

Putting $M_N = [(HF^N)^T, (HF^{N-1})^T, ..., (H)^T]^T$, and solving, optimal state equation forms as;

$$\hat{x}(k+1) = (F - L_N HF)\hat{x}(k) + L_N y(k+1) - L^{-}_N (y(k-N) - HF^N \hat{x}(k)) \quad (8)$$

Where; $L_N = F^N Pc_{N-1} (HF^N)^T * (I + (HF^N) Pc_{N-1} (HF^N)^T)^{-1} + R_v^{-1}$ and $L^{-}_N = a * F^N Pc_N (HF^{-1})^T$. F is the system transition matrix and H is the output observation matrix, a is the tuning parameter with value variation from 0~1. L_N is the system gain matrix and the $(y(k-N) - HF^N \hat{x}(k))$ can be viewed as the appropriate weighting matrix.The covariance propagation matrix is derived as;

$$Pc_N = (Pc_{N-1} - Pc_{N-1} (HF^N)^T * (I + (HF^N) Pc_{N-1} (HF^N)^T)^{-1} * (HF^N) Pc_{N-1}) + Q_w \quad (9)$$

For comparison purpose the extended Kalman filter (EKF) was implemented, in brief the implemented Kalman filter [3] can be expressed as;

TimeUpdate

$$\chi_{k+1} = \Phi_{k+1} \chi_k,$$
$$P_{k+1}^- = \Phi_k P_k \Phi_k^T + Q_k$$

Measurments

$$K_k = P_k^- H_k^T \left[H_k P_k^- H_k^T + R_k \right]^{-1}$$
$$\hat{\chi}_k = \hat{\chi}_k^- + K_k \left[z_k - H_k \hat{\chi}_k^- \right]$$
$$P_k = \left[I - K_k H_k \right] P_k^-$$

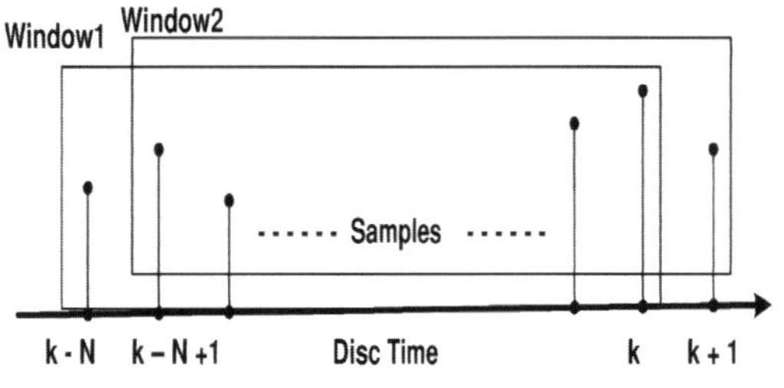

Figure. 1. MHE Framework

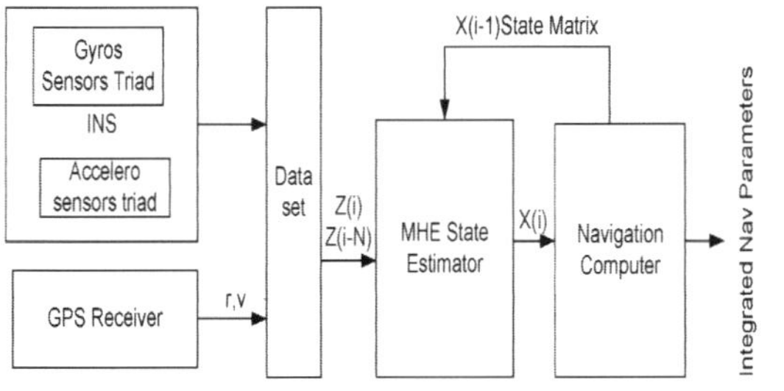

Figure. 2. Instrumentation & Algorithm

In the first part of the algorithm, the data is acquired from the sensors modules and after solving the unit's dichotomy, synchronization issues, initialization, the free inertial navigations parameters are computed using the quaternions. In the 2nd part the raw necessary inputs for the estimator is prepared i.e. observation matrix computation, Jacobian system transition, and output matrices and then data is fed to the MHE observer for optimal state estimate, subsequently the optimal state is incorporated into the navigation parameters resulting in the integrated navigation parameters ready to be used for guidance and control of the aerial vehicle.

4. Experimental Setup and Analysis

The algorithm was validated with the actual flight test data of a small unmanned vehicle. This experiment was conducted with the low-grade commercial IMU system, installed with the commercial-grade GPS receiver with the data rate of 1hz, for continuous visibility multiple smart antennas was fabricated on the vehicle body. The continuous clean power supply was ensured with help of onboard source and dc-dc power conditioner.

4.1 Performance Analysis

In order to assess the performance of the MHE state estimator, a prototype of the GPS/IMU integrated navigation was implemented for aerial vehicle dynamic navigation. Figure.2 shows the system configuration, in parallel an extended Kalman filter was implemented with equal inputs, initializations and environmental disturbances for meaningful comparison. Contrary to the spherical model of the earth, ellipsoidal earth model was implemented, also the latitude dependent gravity model was exploited to avoid the errors. SI unit's data processing and representation are used throughout the article. The navigation parameters velocities and positions were initialized from the available GPS values and for the attitude initialization, the coarse alignment routine was followed. MHE and Kalman filter were both aided with the available GPS 6 measurements, for more insight following results are presented. In figures.3~5, the velocity error state and velocity parameter response was presented in the normal flight, under the similar data input, initializations and disturbance condition, it is shown that the MHE velocity states error with respect to GPS was optimal as compared to the extended Kalman filter response because MHE has more data information by considering a batch of samples for next state estimate.

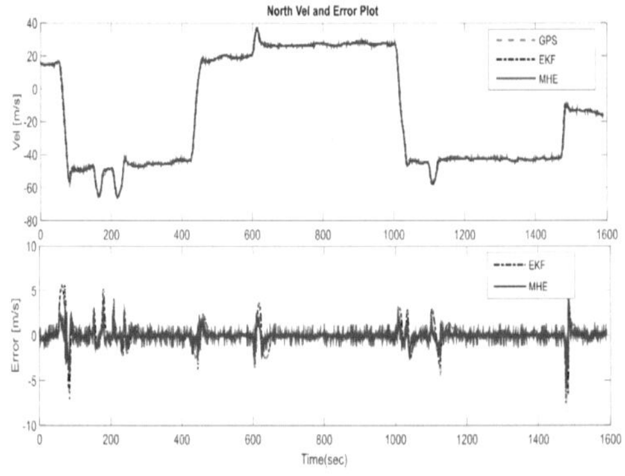

Figure. 3. North Velocity and Velocity Error Plot w.r.t. GPS

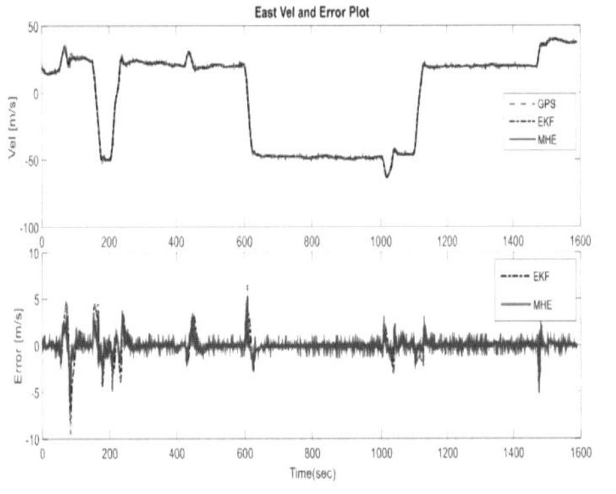

Figure. 4. East Velocity and Velocity Error Plot w.r.t. GPS

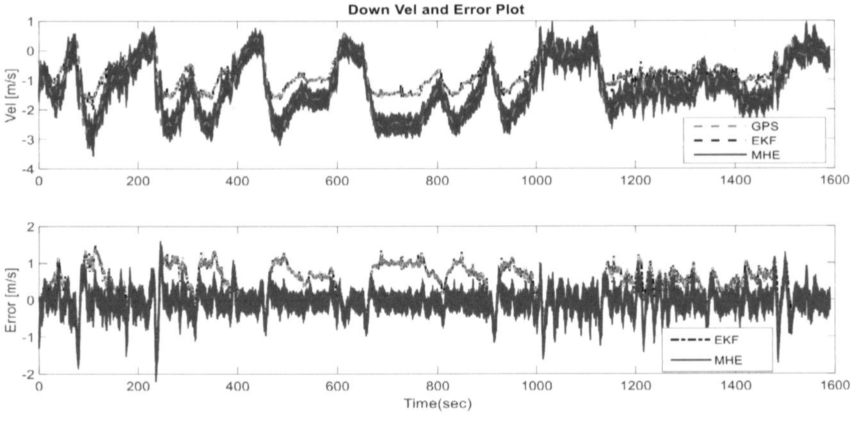

Figure. 5. Down Velocity and Velocity Error Plot w.r.t. GPS

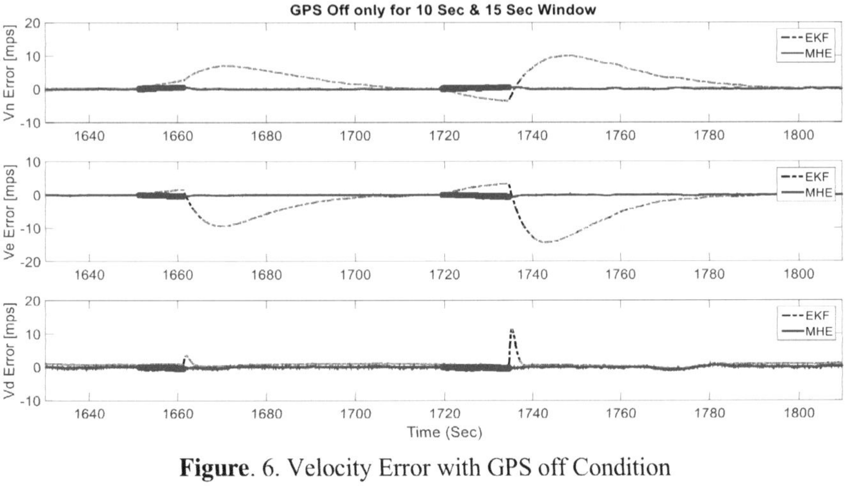

Figure. 6. Velocity Error with GPS off Condition

In the failure scenario simulation as shown in Fig.6, GPS data to the algorithm was stopped and the measurement vector was updated from the previously calculated states. The GPS off interval is highlighted for easier comparative analysis, also in the GPS off failure scenario simulation MHE proved to be more robust as compared EKF. EKF response was comparable to the MHE during a normal profile run, but in the failure mode, the course divergence & slower convergence was observed.

5. Conclusion

In this manuscript, moving horizon estimation technique is implemented for the small aerial vehicle navigation. The quaternions based nonlinear state and observation models, and focal moving horizon observer were developed for GPS/IMU integrated navigation system. In the navigation, the three set of integrations i.e. quaternions coefficient integration, velocity integration and position integration were replaced with the high order Runge-Kutta integration. Experiments and simulation analysis demonstrate that the proposed moving horizon based estimator enhance the navigation accuracy for the GPS/IMU integrated navigation as indicated by results, for normal and failure scenarios. Future research will focus on the GPS/IMU coupling schemes i.e. tightly coupled systems, resources optimization and implementation for a multisensory-based fusion schemes, and real-time on-board algorithm validation.

6. References

[1] Wang, Q., Jiang, W., Yang, C. and Lei, Y., 2010, August. A novel particle filter for tracking fast target. In *Advanced Computational Intelligence (IWACI), 2010 Third International Workshop on* (pp. 288-293). IEEE.

[2] Becerra, V.M., Roberts, P.D. and Griffiths, G.W., 2001. Applying the extended Kalman filter to systems described by nonlinear differential-algebraic equations. *Control Engineering Practice*, *9*(3), pp.267-281.

[3] Zhao, L. and Yuan, Z., 2012. GPS/DR Vehicle Integrated Navigation System Based on Central Difference Kalman Filter⋆. *Journal of Information and Computational Science*, *9*(10).

[4] Alessandri, A., Baglietto, M., Battistelli, G. and Zavala, V., 2010, December. Advances in moving horizon estimation for nonlinear systems. In *49th IEEE Conference on Decision and Control (CDC)* (pp. 5681-5688). IEEE.

[5] Omar, H.M., Yanzhong, Z., Bo, Z. and Gul, H.U., 2016, May. Integration of GPS and dead reckoning navigation system using moving horizon estimation. In *Information Technology, Networking, Electronic and Automation Control Conference, IEEE* (pp. 553-556). IEEE.

[6] Ling, K.V. and Lim, K.W., 1999. Receding horizon recursive state estimation. *IEEE Transactions on Automatic Control*, *44*(9), pp.1750-1753.

[7] Frick, D., Domahidi, A., Vukov, M., Mariéthoz, S., Diehl, M. and Morari, M., 2012, September. Moving horizon estimation for induction motors. In *Sensorless Control for Electrical Drives (SLED), 2012 IEEE Symposium on* (pp. 1-6). IEEE.

[8] Vandersteen, J., Diehl, M., Aerts, C. and Swevers, J., 2013. Spacecraft attitude estimation and sensor calibration using moving horizon estimation. *Journal of Guidance, Control, and Dynamics*, *36*(3), pp.734-742.

[9] Polóni, T., Rohal-Ilkiv, B. and Johansen, T.A., 2015. Moving Horizon Estimation for Integrated Navigation Filtering. *IFAC-PapersOnLine*, *48*(23), pp.519-526.

[10] Release, M.O.T., 2012. The MathWorks, Inc., Ndtick, Mdssdchusetts, Unitea Stdtes."lsqcurvefit" solver.

[11] Kaps Kaps, P. and Rentrop, P., 1979. Generalized Runge-Kutta methods of order four with stepsize control for stiff ordinary differential equations. *Numerische Mathematik*, *33*(1), pp.55-68.

[12] Zhong, Y., Gao, S. and Li, W., 2012. A quaternion-based method for SINS/SAR integrated navigation system. *IEEE Transactions on Aerospace and Electronic Systems*, *48*(1), pp.514-524.

[13] Titterton, D. and Weston, J.L., 2004. *Strapdown inertial navigation technology* (Vol. 17). IET.

IOP Conf. Series: Materials Science and Engineering **187** (2017) 012022 doi:10.1088/1757-899X/187/1/012022

A Study on Rotordynamic Characteristics of Swirl Brakes for Three Types of Seals

Wanjun Xu[a] and Jiangang Yang[b]

National Engineering Research Center of Turbo-Generator Vibration, Southeast University Nanjing, 210096, China

E-mail: [a]230149022@seu.edu.cn, [b]jgyang@seu.edu.cn

Abstract. In order to understand swirl brakes mechanisms and their influence on rotordynamic characteristics for different types of seals, a three-dimensional flow numerical simulation was presented. Three typical seals including labyrinth seal, fully partitioned damper seal and hole-pattern seal were compared under three inlet conditions of no preswirl, preswirl and preswirl with swirl brakes. FAN boundary condition was used to provide inlet preswirl. A modified identification method of effective damping was proposed. Feasibility of the swirl brakes on improving performance of damper seals was discussed. The results show that the swirl brakes influence the seal stability characteristics with whirl frequency. For the labyrinth seal the swirl brakes reverse the sign of effective damping at low frequency and improve the seal stability performance in a wide frequency range. The swirl brakes also improve the damper seals' stability performance by increasing the low frequency effective damping and reducing their crossover frequency. Further results indicate the swirl brakes affect the rotational direction of the maximum (minimum) pressure positions and enhance the stability of the seals by reducing tangential force in each cavity.

1. Introduction

Annular seals in turbomachinery are necessary components to restrict leakage flow through rotor-stator clearances between different pressure regions [1]. These components, while crucial for minimizing leakage of process fluid or lubricant, also drive instability in a rotor system [2]. The unfavorable excited force produced in seals has been recognized as a major destabilizing factor [3]. Stable seals are critically required with equipment's development towards high power. Many new types of seals were proposed by researchers. The honeycomb seal was one of the most representative inventions. The excited force in honeycomb is effectively reduced due to the special hexagon structures. Subsequently hole-pattern seal and pocket damper seal were invented. The three seals are commonly called damper seals which can increase the damping of the inside fluid [4].

Besides damper seals, swirl brakes were the first proposed method for improving the stability of labyrinth seal [5]. Swirl brakes are a series of vanes provided at seal entrance to impede or direct the entrance flow of circumferentially rotating fluid, as shown in figure 1. Practical applications of swirl brakes have been reported, such as high-pressure fuel turbopump seal [6], high pressure oxygen turbopump turbine seal [7] and centrifugal-compressor impeller eye seal [4]. Benckert and Wachter [5] firstly demonstrated that introducing swirl brakes at the seal entrance can greatly reduce the destabilizing force and destabilizing cross-coupled stiffness. Sivo [8] investigated the effects of swirl brakes on shrouded centrifugal pumps and found installing swirl brakes at the leakage entrance

IOP Conf. Series: Materials Science and Engineering **187** (2017) 012022 doi:10.1088/1757-899X/187/1/012022

significantly reduced the destabilizing tangential force. Kwanka [9] experimentally showed that a small amount of swirl brakes (8 swirl brakes) effectively reduced the induced force to a quarter of the initial value. Kimura [10] carried out a numerical simulation of leakage flow in a centrifugal impeller. The swirl brakes suppressed the swirl of the leakage flow which caused rotor instability. The axial thrust on the impeller can be controlled by adjusting the shape of swirl brakes. Nielsen [11] compared aerodynamic swirl brakes and nonaerodynamic swirl brakes by numerical simulation. The nonaerodynamic swirl brakes had better performance than the aerodynamic one because of strong separation vortices within the swirl brakes. Brown [12] tested a hole-pattern seal with negative preswirl. High negative inlet preswirl produced negative cross-coupled stiffness over a frequency range brand and improved the stability of the seal. Childs [13] investigated negative-swirl brakes for labyrinth seal. The negative-swirl brake changed the sign of cross stiffness and produced a marked increase in effective damping.

Although damper seals are considered as the alternative to traditional labyrinth seal, researchers have found damper seals have the shortcoming of potential instability at low frequency [14]. Studies on optimizing damper seals performance have been carried out for ten years. Smalley [15] compared the stability between a straight honeycomb seal and a diverging taper honeycomb seal. The diverging taper improved seal effective damping over a wide frequency range. Childs [16] investigated rotordynamic characteristics of hole-pattern seals with varying hole depth axially. The various depth patterns significantly improved seal performance in terms of both increasing the seal's effective damping and reducing its crossover frequency. Vannarsdall [17] tested a hole-pattern seal with large diameter holes and found the seal had high and positive effective damping without crossover frequency for all tested frequencies.

Figure 1. Swirl brakes in a centrifugal compr essor [4].

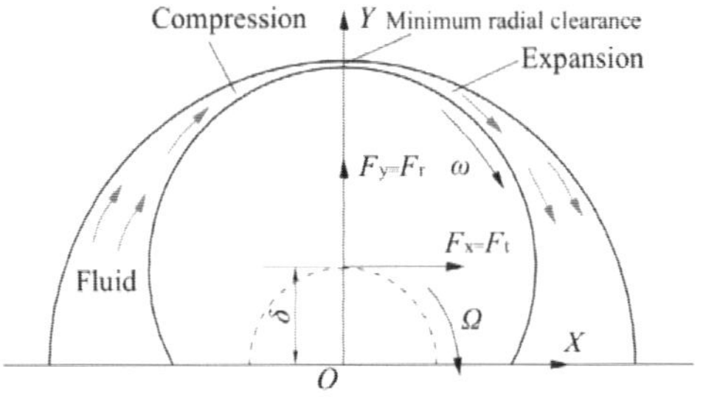

Figure 2. Rotor eccentric position and coordinate direction definition.

Swirl brakes were originally proposed for labyrinth seals. To author's knowledge, applying swirl brakes on damper seals has not been reported before. In order to further improve the stability of damper seals, the paper studied the stability of damper seals by considering swirl breaks installed at seal entrance. A three-dimensional flow numerical analysis was presented to compare the performance of three typical seals including labyrinth seal, fully partitioned damper seal and hole-pattern seal. The three seals were compared under three inlet conditions of no preswirl (NPS), preswirl (PS) and preswirl with swirl brakes (PSSW). The purpose is to understand the swirl brakes rotordynamic mechanisms and to explore the feasibility of the swirl brakes on improving performance of damper seals.

2. Methods

2.1. Numerical Calculation Methods

Many successful applications of fluid software for solving seal rotordynamic coefficients have been reported [18-21]. In the paper, fluid software FLUENT 14.0 is used to solve the full eccentric three dimensional seal model. Rotating reference frame method [18] is employed to simulate the whirl movement of rotor. The entire fluid zone rotates with the reference frame with speed Ω. The rotor walls rotate with speed ω-Ω and the stator walls rotate with speed -Ω. Standard k-ε turbulence model and standard wall function are selected. The working fluid is treated as ideal gas. Convergence residuals are set to 10^{-5}.

The rotor is placed at the positive Y axis direction for eccentric and rotated in a clockwise direction, as shown in figure 2. In this case, the sign of the tangential force F_t primarily determines the seal stability. If the sign is positive, the tangential force points to positive X direction and accelerates the whirl movement of the rotor. The seal system is unstable under the situation.

Compared with another fluid software CFX, FLUENT can directly specify inlet preswirl with FAN boundary condition. In CFX, specifying inlet preswirl usually needs a special approach. Pugachev [21] proposed that a velocity boundary condition set at seal inlet and a pressure boundary condition set at a bypass. This approach sometimes could lead to improper flow condition when the area or velocity of the boundary condition is set to an unreasonable parameter. In FLUENT the FAN boundary condition is an interior face representing a virtual fan with known characteristics. The FAN boundary condition is placed at 1/3 length of seal entrance away from inlet for current analysis.

2.2. *Identification methods of effective damping*

Rotor of rotating machinery usually works in synchronous mode or asynchronous mode. Rotor self-excited vibration often occurs in asynchronous mode [4]. In order to comprehensively evaluate the stability of seals, the effective damping C_{eff} is used as evaluation criterion. It is defined as [17]

$$C_{\mathrm{eff}} = C - k/\Omega \tag{1}$$

The effective damping is a function of rotor whirl frequency Ω and thus it can evaluate the seal stability with different Ω.

The relation between tangential force and rotordynamic coefficients, C and k, is expressed as [18]

$$F_t = F_x = k\delta - \Omega C\delta \tag{2}$$

To obtain the effective damping, the rotordynamic coefficients, C and k, usually need to be obtained firstly. In simulation, the identification method of the rotordynamic coefficients is to solve Eq. (1) with different Ω. It implies the simulation should be carried out at least two times. However, the rotordynamic coefficients are generally frequency dependent, especially for damper seals [22]. This method may introduce error. For avoiding solving rotordynamic coefficients, the effective damping is directly obtained by substituting C and k from Eq. (2) into Eq. (1), as follows:

$$C_{\mathrm{eff}} = C - k/\Omega = -F_t/(\Omega\cdot\delta) \tag{3}$$

All the seals are evaluated with the effective damping for current analysis.

2.3. *Seals and swirl brakes parameters*

Labyrinth seal and fully partitioned damper seal are both chosen from literature [14]. Hole-pattern seal (2668 holes) is chosen from literature [23]. Figure 3 and 4 show parameters of swirl brakes and meshes of the three seals. 72 swirl brakes are uniformly distributed along the circumference at seal entrance. Seal body mesh and seal entrance mesh are separately generated with hexahedral grids. Then the two independent meshes are connected with the help of the interface (see figure 4). The radial clearance at tooth tip is divided into 10 layer nodes for all seals. The circumferential direction is divided into 350 layer nodes for the labyrinth seal and the fully partitioned damper seal. For the hole-pattern seal, the mesh is generated with 1 mm node distance and its mesh number is approximate 7 million which is 8 times of the other two seals. Compared with Hirano's mesh density [18], the present mesh density increases circumferential number and reduces elements' aspect ratio.

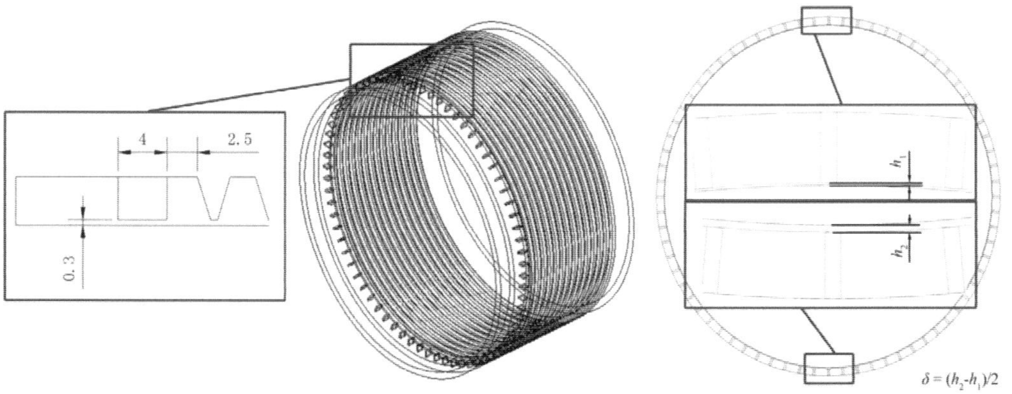

Figure 3. Model and parameters of swirl brakes for labyrinth seal.

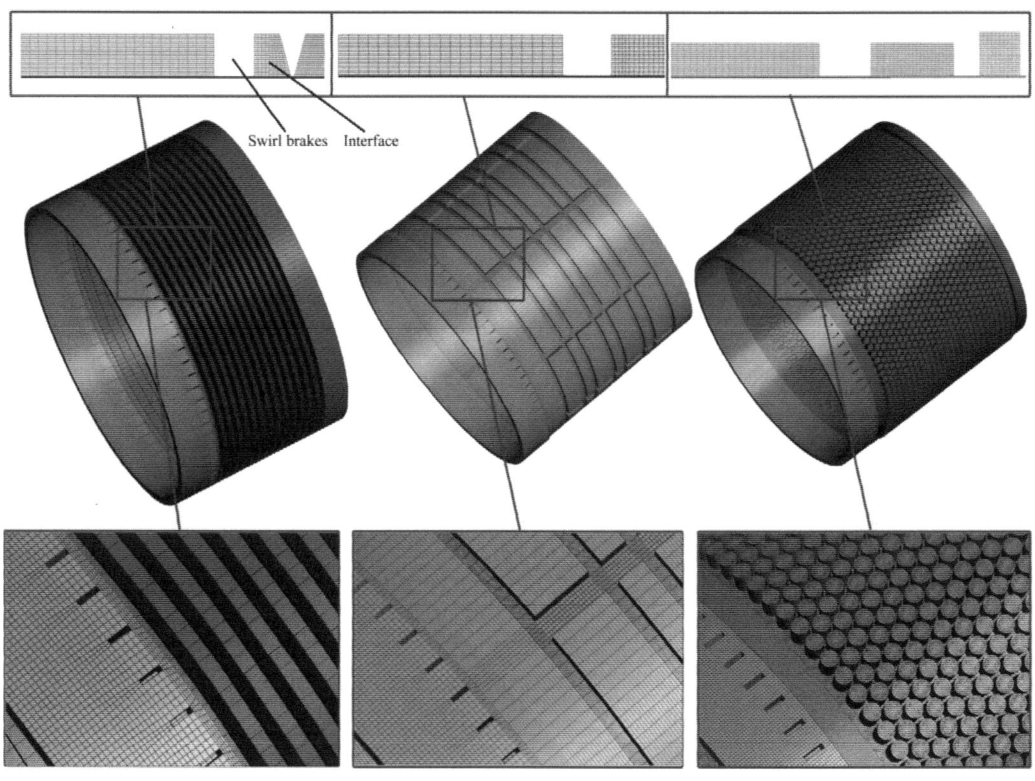

Figure 4. Meshes of the three seals.

The labyrinth seal and fully partitioned damper seal are calculated with 6.9 bar total pressure inlet, 7000 rpm rotational speed and 60 m/s preswirl. The hole-pattern seal is calculated with 7 MPa total pressure inlet, 3.15 MPa pressure outlet, 20200 rpm rotational speed and 72.5 m/s preswirl.

3. Numerical method validation

Figure 5 shows the comparison of the effective damping between present results and literature data for the three seals. It can be seen that the effective damping is generally well predicted by the present numerical method for all three seals. The error mainly comes from the different eccentric ratio of seals. The eccentric ratio in experiment is not a constant but a changing value with time, while the eccentric ratio in simulation always keeps a constant. The eccentric ratio affects the amplitude of the tangential force and hence introduces error into the results.

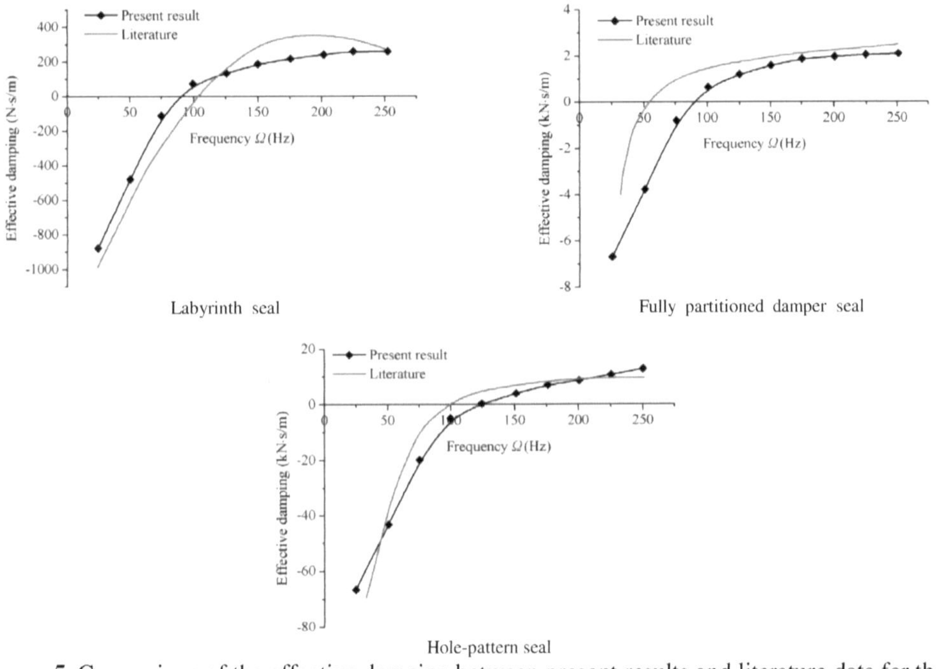

Figure 5. Comparison of the effective damping between present results and literature data for the three seals.

4. Results and discussion

Figure 6 shows the effective damping of the three seals. Each graph displays results with NPS, PS and PSSW. Obviously, the highest effective damping of the three seals all appears at NPS condition. That is to say, NPS is relatively stable compared to PS and PSSW. However, NPS doesn't exists in reality because partial admission or rotating blades can generate inlet preswirl velocity at seal entrance in rotating machineries.

In PS condition the effective damping becomes high as whirl frequency increases for all seals. It implies the rotor tangential force becomes low and seal system becomes stable. The working fluid undergoes compression and expansion when entering the minimum radial clearance (see figure 2), which causes non-uniform pressure distribution in circumferential direction (see figure 10). If considering rotor whirl movement, the minimum clearance position changes with time in circumferential direction, and the compression and expansion of fluid are delayed. The non-uniform pressure distribution is reduced by the delayed effect and thus the effective damping is increased.

The labyrinth seal possesses negative effective damping at low frequency under PS condition and positive effective damping throughout the entire calculated frequency range under NPS and PSSW condition. The negative effective damping is unstable for seal system. The preswirl velocity is the main reason of causing flow excited force. The swirl brakes reverse the sign of effective damping at low frequency and an improved stability in a wide frequency range is obtained. Installing swirl brakes at labyrinth seal entrance is encouraged. Unlike the labyrinth seal, the damper seals yield negative effective damping at low frequency and transform to positive effective damping as frequency increases under all three inlet conditions. The swirl brakes are also helpful for improving the damper seals' stability performance by increasing the low frequency effective damping and reducing their crossover frequency. Installing swirl brakes at damper seals entrance is also encouraged.

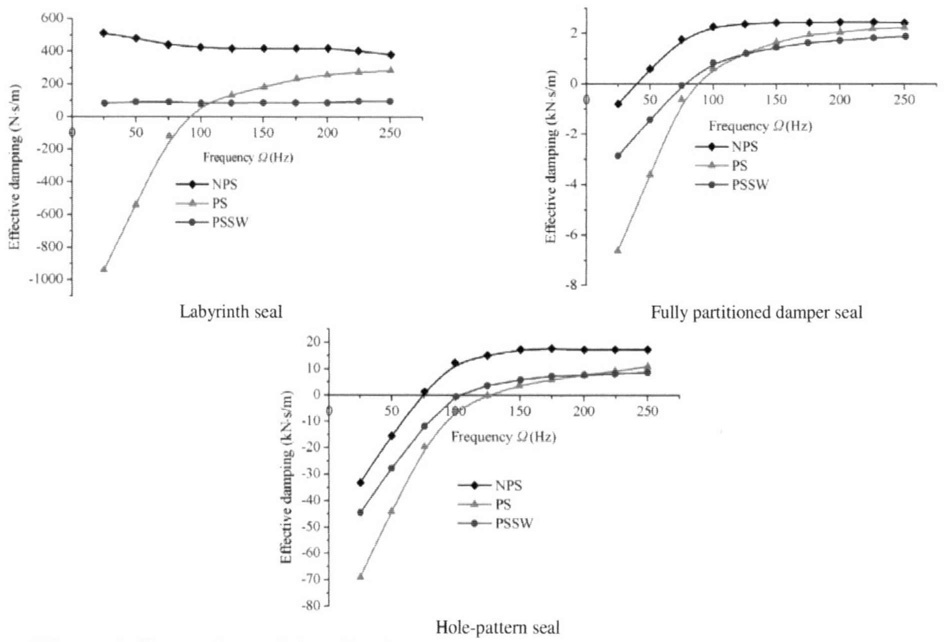

Figure 6. Comparison of the effective damping for the three seals with NPS, PS and PSSW.

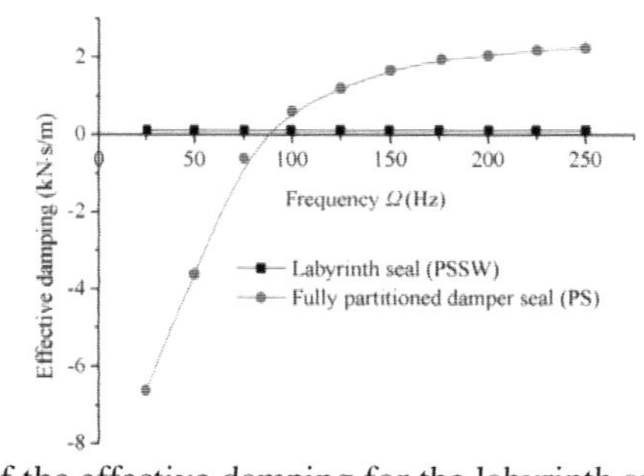

Figure 7. Comparison of the effective damping for the labyrinth seal with PSSW and the fully partitioned damper seal with PS.

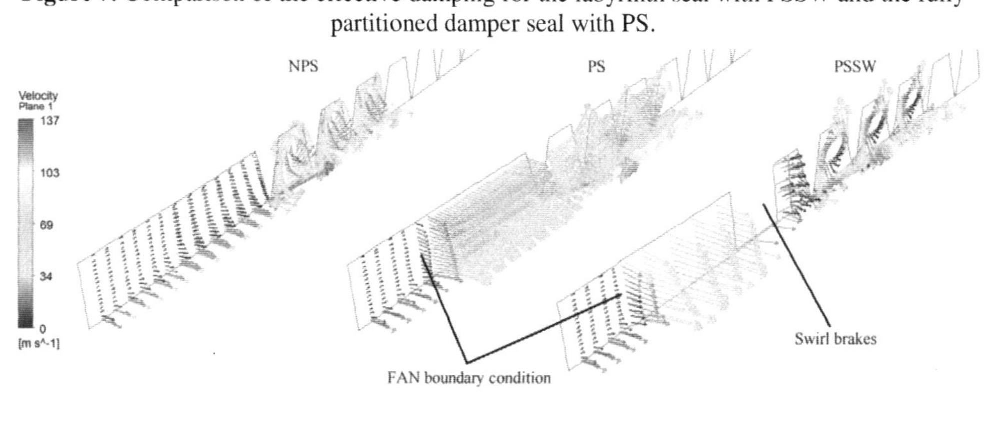

Figure 8. Velocity vectors in the center plane of the labyrinth seal with NPS, PS and PSSW.

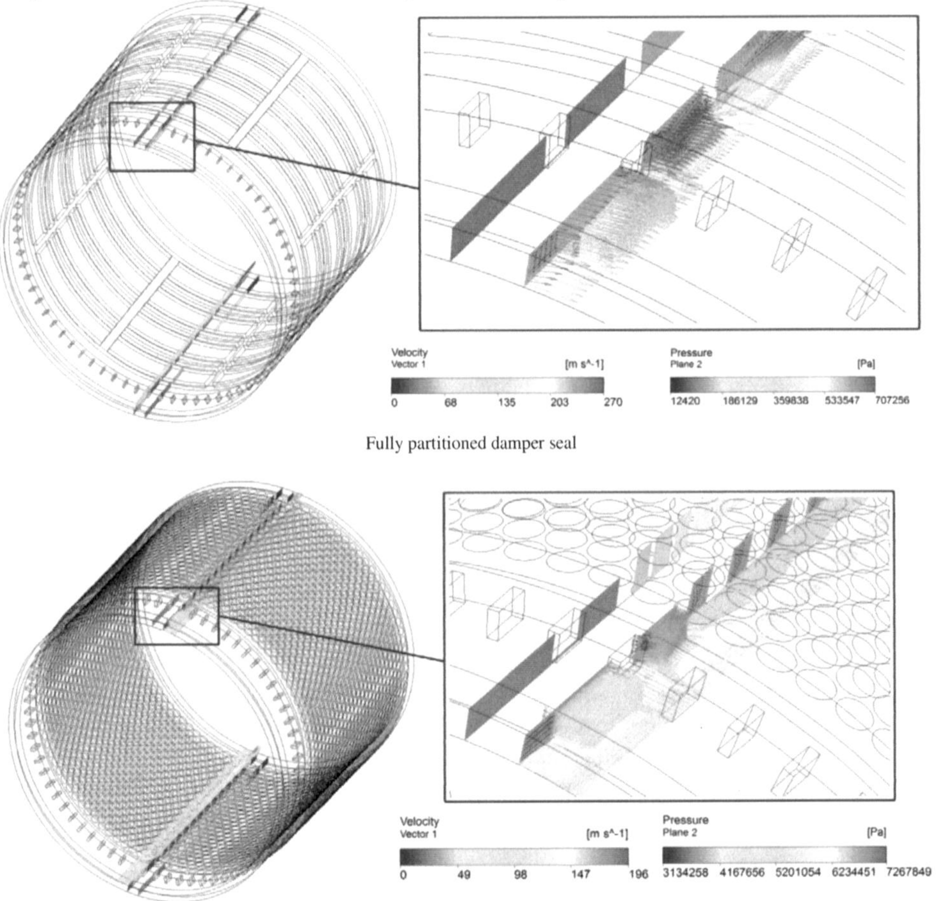

Hole-pattern seal

Figure 9. Pressure distribution and velocity vectors in the center plane of damper seals with PSSW for whirl frequency 50 Hz.

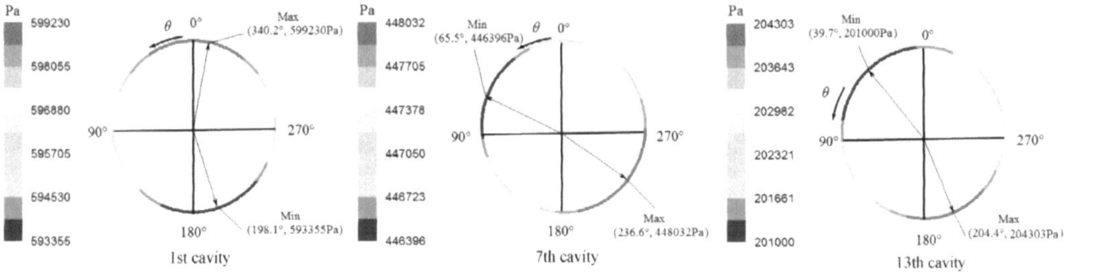

NPS condition

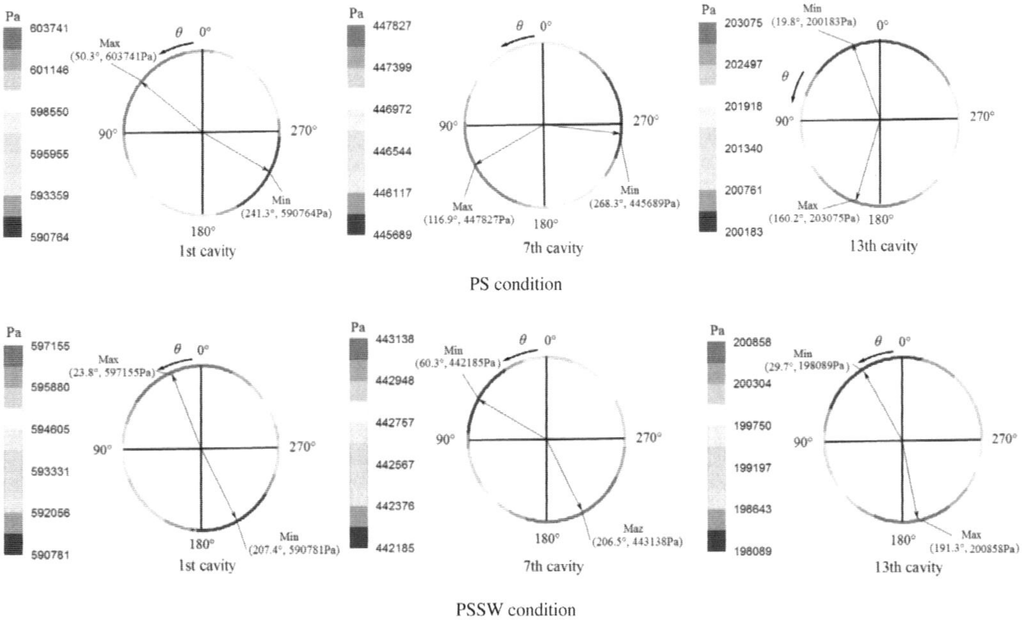

Figure 10. Pressure distributions on the rotor surface at 1st, 7th and 13th cavity center planes of the labyrinth seal for whirl frequency 50 Hz.

It is noted that the effective damping of the labyrinth seal is positive over the whole frequency, while that of the damper seals is negative at low frequency under NPS and PSSW condition. Different from labyrinth seal, damper seals contain baffles in seal cavity. The fluid cannot rotate through the entire circular seal cavity smoothly. At low frequency their pressure distribution inside seal cavity may be more sensitive to rotor rotational speed which causes the negative effective damping (see results of literature [14]).

The effects of swirl brakes on the three seals are not strictly consistent, but the benefits can be regarded as improving the seal stability between NPS condition and PS condition. The function of swirl brakes is to decrease the flow circumferential velocity at seal entrance. Therefore, swirl brakes cannot bring more stability improvement than seal with NPS.

Figure 7 shows the comparison of the effective damping for the labyrinth seal with PSSW and the fully partitioned damper seal with PS. The labyrinth seal with swirl brakes is more stable than the damper seal at low frequency; however, its mid-high frequency stability performance is not desirable, which is at least 10 times lower than that of the damper seal. Optimizing the stability of damper seal is significant. Similar results were described by literature [22].

Figure 8 shows the velocity vectors in the center plane of the labyrinth seal with NPS, PS and PSSW. In NPS condition, the circumferential velocity is produced by the rotor rotational movement. The velocity value is weak and decreases with radial direction rapidly. The FAN boundary condition successfully provides stronger circumferential velocity at seal entrance for PS and PSSW. The circumferential velocity is approximately linearly distributed in radial direction. Compared to the PS, the swirl brakes partly reduce the inlet circumferential velocity by their flow guiding function. Similarly, the swirl brakes have the same effects on the fully partitioned damper seal and hole-pattern seal, as shown in figure 9.

To explain how the swirl brakes suppress the spiral flow of fluid at low frequency, figure 10 shows the pressure distributions on the rotor surface at 1st, 7th and 13th cavity center planes of the labyrinth seal for whirl frequency 50 Hz. It is noted that the displayed pressure is acting on rotor surface and the displayed surface thickness is amplified for facilitate observation. Maximum and minimum pressure always exists in center plane of each cavities. Extreme pressure positions are marked with symbol θ which denotes circumferential coordinate. The main feature of figure 10 is the different rotational direction of maximum (minimum) pressure positions with the main flow direction for the three inlet conditions. The rotational directions are clockwise and the same as the rotational direction of the rotor for the seal with NPS and PSSW. They are from 340.2° to 236.6° and to 204.4°, and from 23.8° to 206.5° and to 191.3°, respectively. Contrary to this, the seal with PS shows the counter clockwise rotational direction as from 50.3° to 116.9° and to 160.2°. The difference of the rotational direction is consistent with the sign of effective damping. The effective damping of the seal with NPS and PSSW at 50 Hz is positive (see figure 6(a)). It means that the tangential force is negative and points to the negative X direction (see figure 2). Therefore, most of the maximum pressure positions should be distributed between 180° and 360°, which are shown in figure 10(a) and (c). Correspondingly, most of maximum pressure positions of the seal with PS are distributed between 0 and 180° as shown in figure 10(b). It can be seen that the swirl brakes suppress the spiral flow of fluid by reversing the rotational direction of extreme pressure positions at low frequency.

Table 1 lists the tangential force of the labyrinth seal for whirl frequency 50 Hz. The tangential force decreases along the main flow direction for the seal with PS. The maximum tangential force appears at the first cavity and accounts for 20.2% of the total tangential force. This is because the first seal cavity possesses the strongest circumferential velocity. The tangential force of the seal with NPS and PSSW is negative and uniformly distributed in each cavity. The swirl brakes drop the tangential force in each cavity to enhance the stability of the seals.

Table 1. Tangential force of the labyrinth seal for whirl frequency 50 Hz.

		Cavity 1	Cavity 2	Cavity 3	Cavity 4	Cavity 5	Cavity 6	Cavity 7
Seal with NPS	F_t (N)	-0.70	-0.62	-0.72	-0.77	-0.77	-0.72	-0.66
	Percent	8.8%	7.9%	9.1%	9.8%	9.8%	9.1%	8.3%
Seal with PS	F_t (N)	6.41	4.25	3.52	3.01	2.63	2.30	2.03
	Percent	20.2%	13.4%	11.1%	9.5%	8.3%	7.3%	6.7%
Seal with PSSW	F_t (N)	-0.15	-0.14	-0.16	-0.16	-0.15	-0.16	-0.15
	Percent	9.6%	9.0%	10.3%	10.3%	9.6%	10.3%	9.6%
		Cavity 8	**Cavity 9**	**Cavity 10**	**Cavity 11**	**Cavity 12**	**Cavity 13**	**Total**
Seal with NPS	F_t (N)	-0.60	-0.53	-0.46	-0.41	-0.39	-0.39	-7.88
	Percent	7.6%	6.7%	5.8%	5.2%	5.0%	5.0%	1
Seal with PS	F_t (N)	1.78	1.56	1.35	1.14	0.95	0.77	31.72
	Percent	5.6%	4.9%	4.3%	3.6%	3.0%	2.4%	1
Seal with PSSW	F_t (N)	-0.11	-0.10	-0.09	-0.08	-0.08	-0.03	-1.56
	Percent	7.1%	6.4%	5.8%	5.1%	5.1%	1.9%	1

5. Conclusions

The paper investigated the effects of swirl brakes on rotordynamic performance for three types of seals. Feasibility of swirl brakes on improving performance of damper seals was focused. Evaluation of effective damping and development of pressure distribution were discussed.

The swirl brakes suppress the spiral flow of fluid and enhance the stability of the three typical seals. For labyrinth seal the swirl brakes reverse the sign of effective damping at low frequency and improve seal stability performance in a wide frequency range. Similarly, the swirl brakes also improve the stability performance of fully partitioned damper seal and hole-pattern seal by increasing the low frequency effective damping and reducing their crossover frequency. Installing swirl brakes at damper seals entrance is encouraged.

Nomenclature

C : damping coefficient
C_{eff} : effective damping
h_1 : minimum radial clearance
h_2 : maximum radial clearance
F_r : radial force
F_t : tangential force
F_x : force in X direction

F_y : force in Y direction
k : stiffness coefficient
δ : eccentric
θ : circumferential coordinate
ω : rotor rotational speed
Ω : rotor whirl frequency

6. References

[1] Untaroiu A, Migliorini P, Wood H G, Allaire P E, Kocur J A, 2009, Hole-pattern seals: a three dimensional CFD approach for computing rotordynamic coefficient and leakage characteristics, ASME 2009 International Mechanical Engineering Congress and Exposition, Lake Buena Vista, Florida, USA, IMECE2009-11558, pp. 981-990.
[2] Migliorini P J, Untaroiu A, Wood H G, 2014, A numerical study on the influence of hole depth on the static and dynamic performance of hole-pattern seals, *J. Tribol.*, **137**(1), pp. 011702.
[3] Childs D W, Vance J M, 1997, Annular gas seals and rotordynamics of compressors and turbine, Proceedings of the 26th Turbomachinery Symposium, pp. 201-220.
[4] Moore J J, Walker S T, Kuzdzal M J, 2002, Rotordynamic stability measurement during full-load, full-pressure testing of a 6000 psi re-injection centrifugal compressor, Proceedings of the Thirty-first Turbomachinery Symposium, pp. 29-38.
[5] Benckert H, Wachter J, 1980, Flow induced spring coefficients of labyrinth seal for applications in rotordynamics, Proceedings of the Rotordynamic Instability Problems in High-Performance Turbomachinery Workshop, Texas A&M University, NASA CP-2133, pp. 189-212.
[6] Childs D W, Ramsey C, 1991, Seal-rotordynamic-coefficient test results for a model SSME ATD-HPFTP turbine interstage seal with and without a swirl brake, *J. Tribol.*, **113**(1), pp. 198-203.
[7] Childs D W, Baskharone E, Ramsey C, 1991, Test results for rotordynamic coefficients of the SSME HPOTP turbine interstage seal with two swirl brakes, *J. Tribol.*, **113**(3), pp. 577-583.
[8] Sivo J M, Acosta A J, Brennen C E, Caughey T K, 1995, The influence of swirl brakes on the rotordynamic forces generated by discharge-to-suction leakage flows in centrifugal pumps, *J. Fluids Eng.*, **117**(1), pp. 104-108.
[9] Kwanka K, 1997, Rotordynamic impact of swirl brakes on labyrinth seals with smooth or honeycomb stators, ASME 1997 International Gas Turbine and Aeroengine Congress and Exhibition, Orlando, Florida, USA, 97-GT-232, pp. V004T14A042.
[10] Kimura T, Kawasaki S, Shimagaki M, Uchiumi M, 2011, Effects of swirl brakes on the leakage flow between the casing and the shroud of a centrifugal impeller, ASME-JSME-KSME 2011 Joint Fluids Engineering Conference, Hamamatsu, Japan, AJK2011-06048, pp. 241-248.
[11] Nielsen K K, Childs D W, Myllerup C M, 2001, Experimental and theoretical comparison of two swirl brake designs, *J. Turbomach.*, **123**(2), pp. 353-358.
[12] Brown P D, Childs D W, 2012, Measurement versus predictions of rotordynamic coefficients of a hole-pattern gas seal with negative preswirl, *J. Eng. Gas Turbines Power*, **134**(12), pp. 122503.
[13] Childs D W, Mclean J E, Zhang M, Arthur S P, 2014, Rotordynamic performance of a negative-swirl brake for a tooth-on-stator labyrinth seal, ASME Turbo Expo 2014: Turbine Technical Conference and Exposition, Düsseldorf, Germany, GT2014-25577, pp. V07AT31A011.
[14] Ertas B H, Delgado A, Vannini G, 2012, Rotordynamic force coefficients for three types of annular gas seals with inlet preswirl and high differential pressure ratio, *J. Eng. Gas Turbines Power*, **134**(4), pp. 042503.
[15] Smalley A J, Camatti M, Childs D W, Hollingsworth J R, Vannini G, Carter J J, 2004, Dynamic characteristics of the diverging taper honeycomb-stator seal, *J. Turbomach.*, **128**(4), pp. 717-724.
[16] Childs D W, Shin Y S, Seifert B, 2008, A design to improve the effective damping characteristics of hole-pattern-stator annular gas seals, *J. Eng. Gas Turbines Power*, **130**(1), pp. 012505.
[17] Vannarsdall M, Childs D W, 2014, Static and rotordynamic characteristics for a new hole-pattern annular gas seal design incorporating larger diameter holes, *J. Eng. Gas Turbines Power*, **136**(2), pp. 022507.
[18] Hirano T, Guo Z, Kirk R G, 2005, Application of computational fluid dynamics analysis for rotating machinery—part ii: labyrinth seal analysis, *J. Eng. Gas Turbines Power*, **127**(4), pp. 820-

IOP Conf. Series: Materials Science and Engineering **187** (2017) 012022 doi:10.1088/1757-899X/187/1/012022

826.
[19] Chochua G, Soulas T A, 2006, Numerical modeling of rotordynamic coefficients for deliberately roughened stator gas annular seals, *J. Tribol.*, **129**(2), pp. 424-429.
[20] Untaroiu A, Untaroiu C D, Wood H G, Allaire P E, 2012, Numerical modeling of fluid-induced rotordynamic forces in seals with large aspect ratios, *J. Eng. Gas Turbines Power*, **135**(1), pp. 012501.
[21] Pugachev A O, Kleinhans U, Gaszner M, 2012, Prediction of rotordynamic coefficients for short labyrinth gas seals using computational fluid dynamics, *J. Eng. Gas Turbines Power*, **134**(6), pp. 062501.
[22] Picardo A, Childs D W, 2004, Rotordynamic coefficients for a tooth-on-stator labyrinth seal at 70 bar supply pressures: measurements versus theory and comparisons to a hole-pattern stator seal, *J. Eng. Gas Turbines Power*, **127**(4), pp. 843-855.
[23] Childs D W, Wade J, 2004, Rotordynamic-coefficient and leakage characteristics for hole-pattern-stator annular gas seals—measurements versus predictions, *J. Tribol.*, **126**(2), pp. 326-333.

Acknowledgments
This work was supported by the National Natural Science Foundation of China (No. 51275088 and 51575105).

IOP Conf. Series: Materials Science and Engineering **187** (2017) 012023 doi:10.1088/1757-899X/187/1/012023

Constitutive modelling of polycarbonate at low, moderate and high strain rates

Haitao Wang, Yun Zhang*, Huamin Zhou, Zhigao Huang

State Key Laboratory of Material Processing and Die & Mold Technology, Huazhong University of Science and Technology, Wuhan 430074, Hubei, China

*Corresponding author: marblezy@hust.edu.cn

Abstract. The objective of this paper is to experimentally investigate and predict the mechanical behavior of polycarbonate over a wide range of strain rate from 10^{-4} to 5000 s^{-1}. Considering the influences of the glass (α) transition and the second (β) transition on the mechanical behavior of polycarbonate, a new constitutive model is proposed, and it is decomposed into the α and β components. The α component dominates the low and moderate rate deformations, and the β component is related to the high rate deformation. In comparison with the experimental results, the model can accurately predict the mechanical behavior of polycarbonate at low, moderate and high strain rates.

1. Introduction

Polycarbonate has been extensively used as aircraft windows, eyeglasses and helmets, due to its low density, high impact strength and good transparency [1-3]. The mechanical behavior of polycarbonate is strongly dependent on strain rate, and a large number of studies on its mechanical behavior at low and high strain rates during the last few decades have been conducted. Its low rate (< 10^0 s^{-1}) deformation behavior is readily captured with a MTS 810 servo-hydraulic machine [4] or an Instron servo-hydraulic testing machine [1], and its high rate deformation (>10^2 s^{-1}) behavior can be obtained with a split Hopkinson bar (SHP) [1-3, 5, 6]. Experimental results of polycarbonate indicate the bilinear relationship between the yield stress and the logarithm of the strain rate, and its yield stress at high rates increases more dramatically than that at low rates [1, 7]. However, its moderate rate (10^0~10^2 s^{-1}) behavior is rarely investigated. Actually, polycarbonate experiences the low, moderate and high strain rate deformations in many impact accidents. Consequently, it is significant to investigate the low, moderate and high strain rate deformation behavior of polycarbonate.

With respect to the deformation mechanism of polycarbonate, the glass (α) transition and the secondary (β) transition of polycarbonate can be utilized to expound it. The Dynamic mechanical analysis (DMA) results and the bilinear relationship between the yield stress and the logarithm of the strain rate indicate that the α transition dominates the mechanical behavior of polycarbonate at low strain rates [1, 8], due to the non-activated β transition; the temperature of the β transition increases with an increasing strain rate, and the β transition is observed to be activated at high strain rates, which will have a dramatic influence on the mechanical behavior of polycarbonate in this case [1, 2, 9]. However, the deformation mechanism of polycarbonate at moderate strain rates is still in its infancy.

For the elastic–viscoplastic deformation behavior of polymers, a large number of constitutive models have been developed during the last few decades. Boyce, Parks and Argon [10] proposed a now-famous physical model called BPA for glassy polymers in 1988, and the BPA model employed two parallel resistances to describe the deformation of polymers, including an entropic resistance for strain hardening and an intermolecular resistance for strain softening. Subsequently, based on the BPA model, a large number of modified models were developed, such as the eight-chain Arruda-Boyce model [11], the full-chain model [12], the neo-Hookean approximation model [13] and the Mulliken-Boyce model [1]. Among these models, the Mulliken-Boyce model [1] expounded the deformation mechanism of polycarbonate with two rate-activated processes, namely, the α and β transitions, and the model was demonstrated to have an ability of accurately predicting the deformation of polycarbonate at low and high strain rates. However, there are a large number of material coefficients in these models, and some coefficients are difficult to solve, which makes these models difficult to calibrate and employ in practice.

There are also other models to predict the mechanical behavior of polymers, such as the G'Sell-Jonas [14], the DSGZ model [15] and the ZWT model[16]. The G'Sell-Jonas model [14] can capture initial elasticity, yield and strain hardening deformation features of polymeric materials, while it can not capture strain softening behavior. The DSGZ model [15, 17] compensates the shortcoming of the G'Sell-Jonas model, and it can accurately predict the large deformation of polycarbonate (PC), polymethyl-methacrylate (PMMA) and acrylonitrile-butadiene-styrene (ABS) at low strain rates for various temperatures, while it can not accurately simulate the deformation of polymeric materials at high strain rates. The ZWT model [16] employs a nonlinear spring and two Maxwell elements in parallel, and it can capture the deformation feature of PMMA at low and high strain rates, while the strain hardening deformation of PMMA has not been predicted with the ZWT model. To sum up, these phenomenological models can not expound the deformation mechanism of polymers, and they can not predict the large deformation of polymers from low to high strain rates.

The objective of this paper is to develop a new constitutive model to simulate the deformation of polycarbonate at low, moderate and high strain rates for room temperature. The new model is based on the Mulliken-Boyce model [1], the G'Sell-Jonas model [14] and the DSGZ model [15], and it allows for the deformation mechanism of polycarbonate at low, moderate and high strain rates. In addition, the new model is validated to have an ability to accurately predict the large deformation and the yield stress of polycarbonate from low to high strain rates for room temperature.

2. Experimental

In this study, the material was polycarbonate (EXL1414) from SABIC Innovative Plastics Company. An all-electric injection molding machine (JSW110AC) was utilized to produce cylindrical specimens. The specimens at low strain rates were 12.20 mm in diameter and 6.20 in length, and the specimens at moderate and high strain rates were 5.14 mm in diameter and 5.20 in length. Compressive tests at low, moderate and high strain rates were carried out with a Shimadzu universal testing machine, a Gleeble 3500 thermo-mechanical simulator and a split Hopkinson pressure bar (SHPB), respectively. The low rate tests were conducted at 10^{-4}, 10^{-3} and $10^{-2}\,s^{-1}$, the moderate rate tests were conducted at 1 and 10 s^{-1}, and the high rate tests were conducted at 1600, 3000 and 5000 s^{-1}. All of the tests were conducted at room temperature. Prior to testing, all of the tested specimens were kept at room temperature for three days to eliminate the residual stress. Tests under the same conditions were repeated at least three times.

3. Constitutive model

According to the literatures [1, 18], the glass (α) transition of polycarbonate is related to low strain rates or high temperatures, while the secondary (β) transition of polycarbonate is associated with high strain rates or low temperatures. The activated temperature of the β transition of polycarbonate is quite low under the low strain rate loading. The temperature of the β transition increases dramatically with an increasing strain rate, and it will shift to room temperature under the high strain rate loading. The bilinear relationship between the yield stress and the logarithm of strain rate indicates that the yield stress of polycarbonate at low strain rates is dominated by α transition, and that the yield stress at high strain rates is influenced by the α and β transitions. Consequently, the α transition dominates the mechanical behavior of polycarbonate at low strain rates, due to the non-activated β transition, whereas both the α and β transitions control the mechanical behavior at high strain rates, due to the activated β transition.

Considering the influences of the α and β transitions on the mechanical behavior of polycarbonate, a new constitutive model is proposed to accurately simulate the deformation of polycarbonate at low, moderate and high strain rates. The new model is on the basis of three previous models, including the Mulliken-Boyce model [1], the G'Sell-Jonas model [14] and the DSGZ model [15], as shown in Eqs. (1), (2) and (3),

$$\sigma = \begin{cases} K_\alpha \alpha(\varepsilon)\dot{\varepsilon}^{M_\alpha}, & \dot{\varepsilon} < 10^2 \text{ s}^{-1} \\ K_\beta \alpha(\varepsilon)\beta(\varepsilon)\dot{\varepsilon}^{M_\beta}, & \dot{\varepsilon} \geq 10^2 \text{ s}^{-1} \end{cases} \tag{1}$$

where,

$$\alpha(\varepsilon) = [\exp(-\alpha_1\varepsilon) + \varepsilon^{\alpha_2}][1 - \exp(C\varepsilon)]\exp(\alpha_3\varepsilon) \tag{2}$$

$$\beta(\varepsilon) = [\exp(-\beta_1\varepsilon) + \varepsilon^{\beta_2}]\exp(\beta_3\varepsilon^2) \tag{3}$$

The new model is decomposed into the α and β components, as shown in Eq. (1). The α component dominates the low and moderate rate deformations owing to the non-activated β transition or the insignificant β-contribution, and the β component considers the contribution of the β transition under the high stain rate loading owing to the activated β transition. There are eleven material coefficients in this model, six (K_α, α_1, α_2, α_3, M_α and C) related to the α transition and five (K_β, β_1, β_2, β_3 and M_β) related to the β transition. With regard to the deformation at low or moderate strain rates, $1-\exp(C\varepsilon)$ and $\exp(-\alpha_1\varepsilon)$ describe the initial nonlinear elastic deformation and the strain softening deformation, $\dot{\varepsilon}^{M_\alpha}$ shows the rate dependence of the deformation behavior, ε^{α_2} and $\exp(\alpha_3\varepsilon)$ exhibit the strain hardening deformation; with regard to the deformation at high strain rates, $\exp(-\beta_1\varepsilon)$ describes the strain softening deformation, $\dot{\varepsilon}^{M_\beta}$ shows the rate dependence of the deformation behavior, and ε^{β_2} and $\exp(\beta_3\varepsilon^2)$ denote the strain hardening deformation.

Table 1. Material coefficients of polycarbonate.

Material coefficients	Value
α_1	4.378
α_2	0.918
α_3	0.276
M_α	0.0056
K_α	$89.28\ \text{MPa} \cdot \text{ms}^{M_\alpha}$
C	−17.62
β_1	12.88
β_2	0.0545
β_3	5.09
M_β	0.13
K_β	$79.26\ \text{MPa} \cdot \text{ms}^{M_\beta}$

The new model has been implemented in ABAQUS/Explicit through a user subroutine VUMAT, to simulate the deformation of polycarbonate at low, moderate and high strain rates for room temperature. The corresponding coefficients were calculated and listed in Table 1. The Poisson's ratio and the density were set as 0.3 and 1200 kg/m^3.

4. Experimental results vs. model precisions

Figure 1 shows the experimental results of polycarbonate from low to high strain rates. It is readily observed that all of the mechanical behavior of polycarbonate at low, moderate and high strain rates is decomposed into four parts: initial elasticity, yield, strain softening and strain hardening. The yield stress is found to increase with an increasing strain rate, and there is a small discrepancy between the increasing rate of the stress at low strain rates and that at moderate strain rates, whereas the increasing rate of the stress at high strain rates is much larger than that at low and moderate strain rates, which can be expounded by the two rate-activated processes, namely, the α and β transitions. The α transition dominates the low and moderate strain rate deformation, because the β transition has not been activated or the influence of the β transition on the mechanical behavior is insignificant under these conditions. However, the mechanical behavior of polycarbonate at high strain rates is different from that at low and moderate strain rates, and the stress at high strain rates is significantly influenced by the β transition, due to the activated β transition. Hence, the stress at high strain rates increases rapidly than that at low and moderate strain rates.

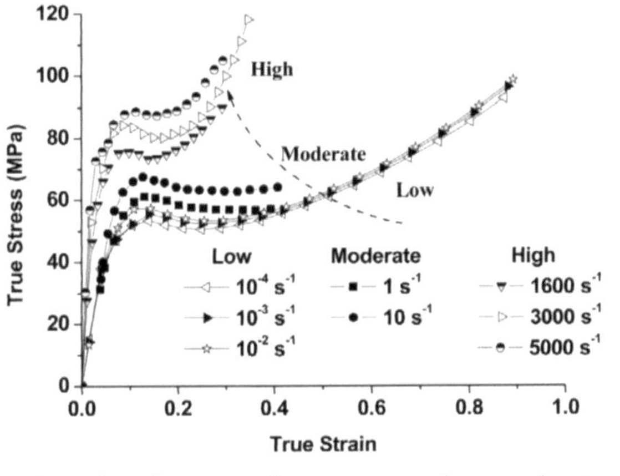

Figure 1. Experimental results of stress-strain responses at low, moderate and high strain rates.

Figures 2-5 show experimental and predicted stress-strain responses at low, moderate and high strain rates. It is observed that the model can capture the typical deformation features of polycarbonate over a wide range of strain rates, including initial elasticity, yield, strain softening and strain hardening. The model can show the same deformation feature as the experiment that the increasing rate of the stress at high strain rates is much larger than that at low and moderate strain rates (see Figure 5). This is attributed to the fact that the model considers the contribution of the β transition to the mechanical performance of polycarbonate at high strain rates. Hence, the model can accurately predict stress-strain responses at high strain rates. In addition, the model is validated that it can accurately simulate the large deformation of polycarbonate at low, moderate and high strain rates for room temperature.

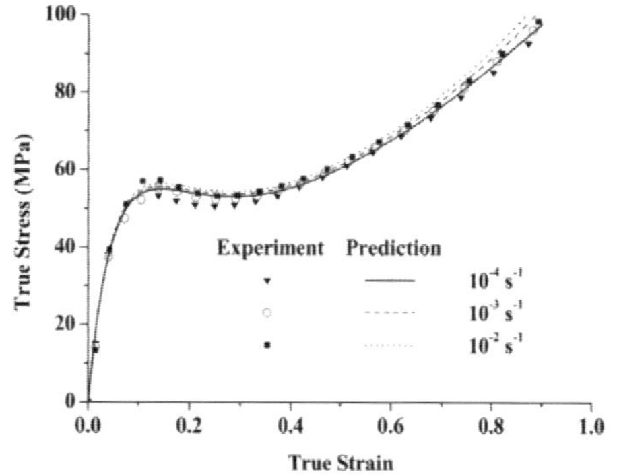

Figure 2. Experimental and predicted results of stress-strain responses at low strain rates.

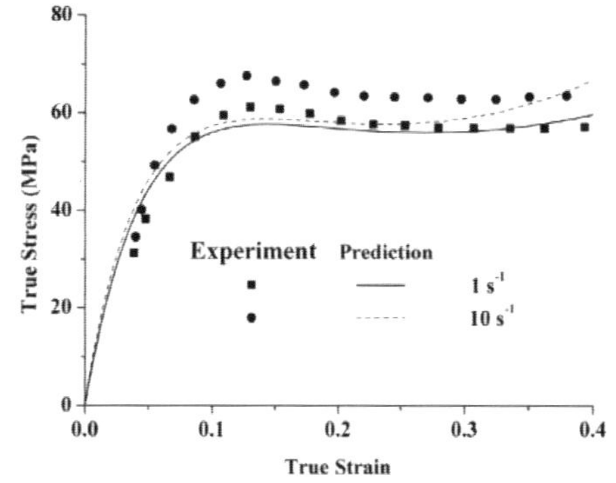

Figure 3. Experimental and predicted results of stress-strain responses at moderate strain rates.

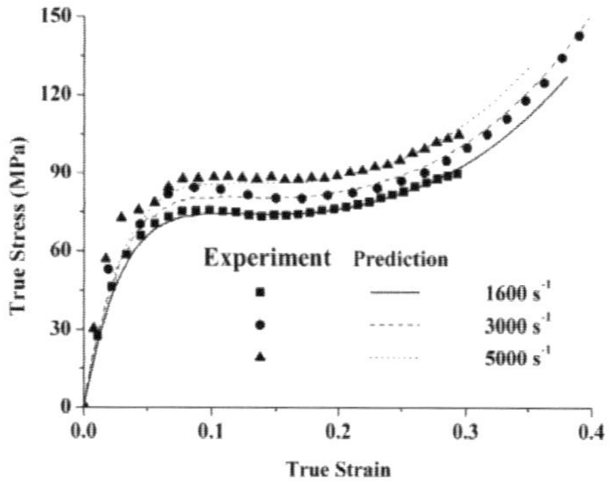

Figure 4. Experimental and predicted results of stress-strain responses at high strain rates.

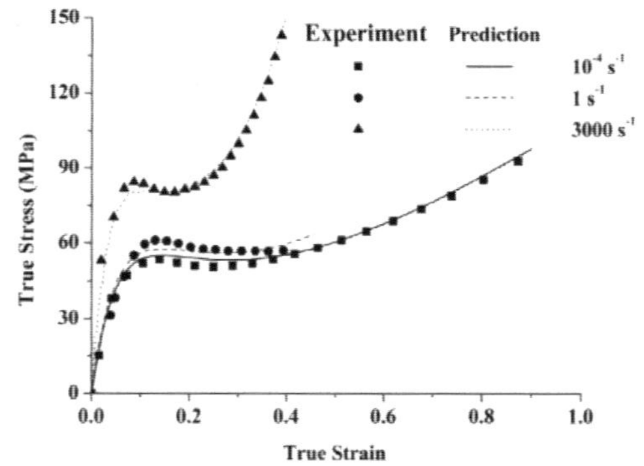

Figure 5. Experimental and predicted results of stress-strain responses at low, moderate and high strain rates.

Table 2. Errors between experimental and predicted true yield stresses.

Strain rate (s^{-1})	Experimental true yield stress (MPa)	Predicted true yield stress (MPa)	Error (%)
10^{-4}	53.47	55.07	2.99
10^{-3}	55.46	55.78	0.58
10^{-2}	57.75	56.51	−2.15
1	61.47	57.58	−6.33
10	67.53	58.7	−13.08
1600	75.85	74.35	−1.98
3000	84.42	80.89	−4.18
5000	88.79	86.16	−2.96

Comparison between the experimental and predicted yield stresses under all of the service conditions is listed in Table 2. The error between the experimental and predicted yield stresses is utilized to evaluate our model's accuracy, which is expresses as

$$error = \frac{\sigma_{P,y} - \sigma_{E,y}}{\sigma_{E,y}} \tag{4}$$

where $\sigma_{E,y}$ and $\sigma_{P,y}$ are the experimental yield stress and the predicted yield stress. Seen from Table 2, the errors under all of the service conditions are less than 14 %, which indicates that there is only a small discrepancy between the experimental and predicted yield stress. Therefore, the model can accurately predict the yield stress over a wide range of strain rate from 10^{-4} to 5000 s^{-1}. In conjunction with the experimental results and the model predictions shown in Figures 2-5 and Table 2, it is validated that the model can accurately predict the stress-strain responses of polycarbonate at low, moderate and high strain rates.

5. Conclusions

The mechanical behavior of polycarbonate at low, moderate and high strain rates has been experimentally investigated and predicted. A new constitutive model was proposed to predict its mechanical behavior and to expound its deformation mechanism. The model allows for the influences of the α and β transitions on the mechanical behavior of polycarbonate, and it is decomposed into the α and β components. The α component dominates the low and moderate rate deformations, and the β component dramatically influences the high rate deformation, due to the activated β transition. Compared with the experimental results, the model predictions can characterize initial elasticity, yield, strain softening and strain hardening. In addition, the model predictions can represent the same experimental deformation feature that the increasing rate of the stress at high strain rates is much larger than that at low and moderate strain rates, and the model can accurately the yield stress and the large deformation of polycarbonate at low, moderate and high strain rates.

6. References

[1] Mulliken AD and Boyce MC 2006 *Int. J. Solids Struct.* **43** p 1331-1356
[2] Safari KH, Zamani J, Ferreira FJ, and Guedes RM 2013 *Polym. Eng. Sci.* **53** p 752-761
[3] Fu S, Wang Y, and Wang Y 2009 *Polym. Test.* **28** p 724-729
[4] Cao K, Wang Y, and Wang Y 2014 *Int. J. Solids Struct.* **51** p 2539-2548
[5] Richeton J, Ahzi S, Vecchio KS, Jiang FC, and Makradi A 2007 *Int. J. Solids Struct.* **44** p 7938-7954
[6] Cao K, Ma X, Zhang B, Wang Y, and Wang Y 2010 *Mater. Sci. Eng.: A* **527** p 4056-4061
[7] Wang H, Zhou H, Huang Z, Zhang Y, and Zhao X 2016 *Mech. Time-Depend. Mater.* p 1-21
[8] Siviour CR, Walley SM, Proud WG, and Field JE 2005 *Polymer* **46** p 12546-12555
[9] Safari KH, Zamani J, Guedes RM, and Ferreira FJ 2015 *Mech. Time-Depend. Mater.* p 1-20
[10] Boyce MC, Parks DM, and Argon AS 1988 *Mech. Mater.* **7** p 15-33
[11] Arruda EM and Boyce MC 1993 *J. Mech. Phys. Solids* **41** p 389-412
[12] Wu PD and Van der Giessen E 1995 *Int. J. Plast.* **11** p 211-235
[13] Tervoort T, Smit R, Brekelmans W, and Govaert LE 1997 *Mech. Time-Depend. Mater.* **1** p 269-291
[14] G'sell C and Jonas JJ 1979 *J. Mater. Sci.* **14** p 583-591
[15] Duan Y, Saigal A, Greif R, and Zimmerman MA 2001 *Polym. Eng. Sci.* **41** p 1322-1328
[16] Wang J, Xu Y, and Zhang W 2014 *Compos. Struct.* **108** p 21-30
[17] Duan Y, Saigal A, Greif R, and Zimmerman MA 2002 *Polym. Eng. Sci.* **42** p 395-402
[18] Richeton J, Ahzi S, Daridon L, and Rémond Y 2005 *Polymer* **46** p 6035-6043

Numerical Simulation of Sugarcane under Wind Load

Fuwei Yuan[1], Wang Yang[1,2,a], Jian Yang [1,2], Huahai Meng[1], Liangjin Wu[1], Huitao Nong[1] and Wenfeng Guo[1]

[1]College of Mechanical Engineering, Guangxi University, Nanning, China
[2]Collaborative Innovation Center for Sugar Industry of Guangxi, Nanning, China

[a]E-mail: yanghope@163.com

Abstract. For difficulties in establishing complex coupling dynamics model of sugarcane-soil-wind system and the dynamic response process of sugarcane by wind effect is unknown problems, establishment of complex coupling dynamics model of sugarcane-soil-wind system was discovered. ALE (Arbitrary Lagrange-Euler) method and soil more hierarchical modeling technology were used in the model. And dynamic response process of sugarcane under wind load was analyzed. Result shows that the modeling method is feasible which can be used on the research of sugarcane lodging. Periodic decay reciprocating vibration occurs in the process.

1. Introduction

At present, the study of sugarcane lodging was mainly concentrated upon different varieties of sugarcane, different depth of implant, hilling up with earth on the process and others' aspects, which researched by sugarcane planting contrast tests. However, the sugarcane lodging experiments effected by multiple factors and comprehensive kinetic mechanism have been made rarely. The region which affected badly by the typhoon, the monsoon lack the effective lodging-resistant measure so that sugarcane lodging badly. The study of the dynamic response process when sugarcanes under wind load and the kinetic mechanism of sugarcane lodging are the foundations of the study of sugarcane lodging. But sugarcane lodging is influenced by many factors. There's tight coupling between this factors and the coupling relationship in between is complicated. In the process of sugarcane lodging, soils produce large deformation and rupture so that it is hard to learn about the dynamic response process when sugarcanes under wind load and the kinetic mechanism of sugarcane lodging when we use general physical test and the finite element method. Accordingly, in this paper, ALE (Arbitrary Lagrange-Euler) method and soil more hierarchical modeling technology were used, to study the numerical simulation of establishment of complex coupling dynamics model of sugarcane-soil-wind system and build the numerical simulation of the dynamic response process of sugarcane by wind effect. It provides the basic theoretics as required for study of sugarcane lodging.

2. The establishment of the numerical simulation model

2.1 The geometric model

The sugarcane-soil-wind dynamic system consists of sugarcane, soil, wind source and air. Among them, the sugarcane is composed of leaves, leaf sheath, stem and roots. As shown in figure 1, due to the complexity of structure of the leaves, the leaf sheath tightly wrap around the stem and meanwhile thanks to the tininess and abundance, the roots have the anchoring effect, as a result, once the sugarcane lodges, its roots near the stem will fasten the soil into a approximately cylindrical soil plate, as shown in figure 2. Therefore, in order to facilitate modeling, according to the structures of leaves and leaf sheath, and the condition where the leaves are subject to the wind, as well as the anchoring effect that the roots exert on the soil, this paper simplifies the sugarcane as follows:

(1) Give the same treatment to the thickness and materials of each part of the leaves, and leave out the too soft part of the leaves' end.

(2) Remove the leaf sheath, connect the leaves and the stem directly, besides, deal with the stem as a regular eight-angular prism made of homogeneous materials.

(3) Replace the anchoring effect existing between the sugarcane roots and the soil with the cylindrical soil plate.

Figure 1. Sugarcane

Figure 2. Cylindrical soil disc

The mesh quality directly affects the accuracy of the results of numerical simulations, thus, in this paper, to obtain good mesh quality in mesh generation, the wind source, the air and the soil are modeled into a rectangular shape. Additionally, as the property of the arable layer's soil changes greatly with the depth changing [1], soil modeling should be conducted in accordance with different layers. At the same time, to ensure the deformation of the soil's surface can be observed when the ALE simulates the soil, a thin layer of air will be built on the surface. What's more, the size of the win source-air-soil geometric model will depend on the scope in which the sugarcane is subject to the win, and the deformation and rupture of the soil. Since the scope where the stem is impacted by the wind is rather small, the model will rule out the air near central part of the stem. This paper adopts the geometric model of the sugarcane-soil-wind coupled system jointly established based on Pro/E and ANSYS/LS-DYNA as shown in figure 3.

2.2 Mesh generation and loading

Above all, the mesh size at the place where materials are coupled should be as the same as possible, otherwise, the penetrating phenomena could happen in the process of coupling [2], consequently, the leaves and the upper part of the stem may couple with the wind, while the plate with the soil. So the meshes of the leaves and the stem should match with that of the air, while the mesh of the plate shall be line with that of the soil. Moreover, to improve the mesh quality, the meshes of the leaves and stem are generated by the sweep method; those of the wind source, the air and the soil are done by the mapped method; and that of the soil plate is by the free method. To guarantee the material exchanges and transportation between the wind and the air, between the soil and the thin air layer, and between different soil layers, the interface nodes are converted into public node. Furthermore, in order to simulate infinite air and soil domains and keep gravitational wave from influencing the calculation results, the boundary surface between the air and the soil is defined as the no reflection boundary. Lastly, the other parameters of the win source can find their definitions in references [2-3]. And the established cutaway view of the finite element model of the sugarcane-soil-wind coupled system could be seen in figure 4.

2.3 The material model

When the sugar cane-soil-wind coupling system modeling, each component adopts hexahedral SOLID 164 unit. Pouring air, air and thin layer chromatography is defined as the blank material, density of 1.29 kg/m³, by the pressure of -1 Pa, dynamic viscosity coefficient of 1.7456×10^{-5}Pa·s, specific heat ratio γ of 1.4. Become a thin layer of air state equation, the air and the linear polynomial equation of state [2-3]. Sugarcane leaves, stems are defined as elastic material, its material parameters measured by physical experiment. Soil disc material parameters take average material parameters of the soil. Soil is defined as the FHWA soil material, the modeling of the soil material parameters measured by physical test and application of reverse technology [4].

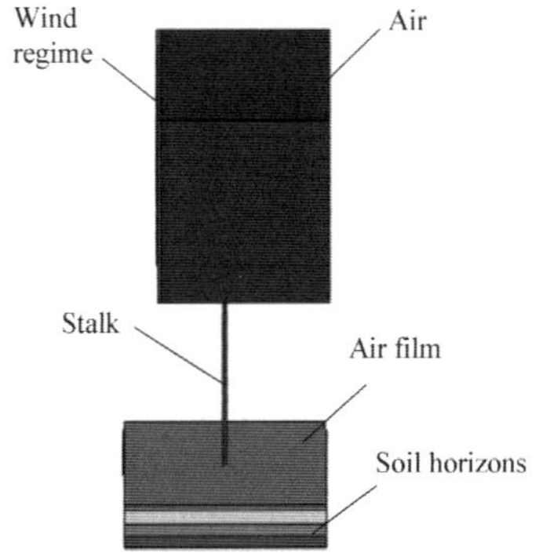

Figure 3. Geometric model of the sugarcane-soil-wind complex coupling system

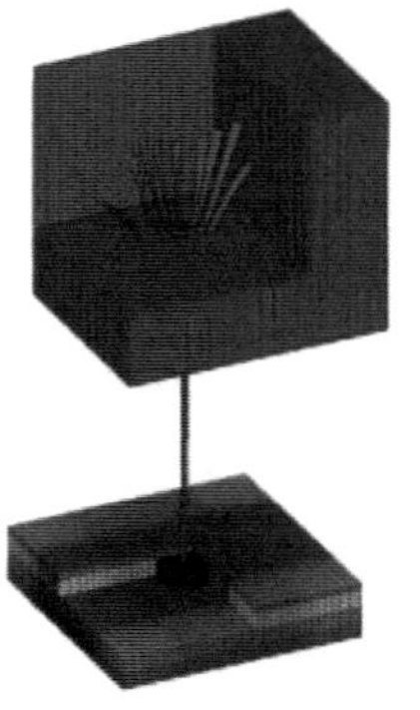

Figure 4. Section view of finite element model of sugarcane-soil-wind coupling system

2.4 The main control parameters

Pouring, air, thin air and soil layer of solid elements ELFORM algorithm parameters are set to 11 (multi-material ALE), AET algorithm parameter is set to become an entity unit 4 (include units). ALE_MULTI - MATERIAL_GROUP for pouring and air with keywords, thin layer between air and soil, the exchange of material between layer and layer of the soil and transport. Wind and sugar cane coupling, coupling between the soil and soil plate adopts way of penalty function, coupled integral number is set to 4 to reduce leakage.

3. Validation of modeling method

Based on the field of physical test methods, through the system of physical experiment results and the comparison of the results of numerical simulation, Validation using the ALE method and soil more hierarchical modeling technology to construct a system of sugar cane - soil - wind coupling numerical simulation model method is feasible. In order to make the field experiment of physical experiment and numerical simulation system conditions are consistent, field experiment uses the air blower generated are, at the same time, the determination of the related parameters in the field experiment in sugarcane and soil, establish verification with sugar cane - soil - wind composite system numerical simulation model, sugar cane and the related parameters of the soil determination method in reference [4-5].

3.1 Test equipment and venues

The main testing equipment: strong industrial fan (model: FS850, Foshan Shunde Leliu de sheng electrical appliance factory, impeller diameter: 850 mm, power: 480 w, the wind speed stability of the gear), digital anemometer (model: GM8902, Shenzhen poly alum source technology co., LTD., resolution: 0.01 m/s), vibration tester (type: DH5938, Jiangsu Donghua testing technology co., LTD.), acceleration sensor (DH131), laptop, etc. Testing ground for sugarcane planting base in Guangxi University, 17 sugarcane varieties for gardens, test time is in November 2014.

3.2 Test method

Field physical test, first choose from sugarcane planting base via a stem straight, no insect pest, the growth of normal sugar cane as a test sugar cane, after clearance test near the sugar cane sugar cane and other sundry, and eliminate the environment influence on test results. At the same time, the clearance test sugarcane leaves and cut off the leaves on the part of the terminal is too soft. And to eliminate the influence of natural wind to test, test in the natural wind speed is low (not more than 0.2 m/s) of time. Test at the same time, fan motor axis aligned with sugarcane stalks, fan out the wind and sugarcane leaf surface parallel to the wind. Sugar cane by wind function test system diagram as shown in figure 5, the mark point height H = 110 mm, the distance between the fan and sugar cane L = 500 mm, H_0 = 230 mm. Test on site are shown in figure 6.

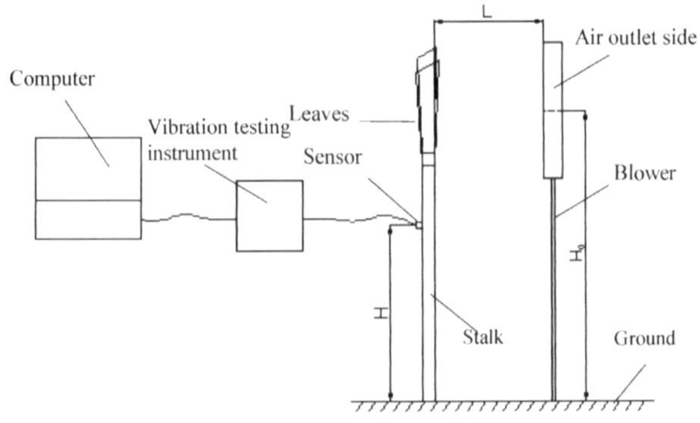

Figure 5. Diagram of sugarcane testing system

Figure 6. The test site

3.3 Material parameters and relative geometric dimensions of the model

According to the test methods of reference [4-5], we measured sugarcane and the corresponding soil material parameters, the average wind speed of the fan, soil and soil anchor plate size roots formation. The results of the material parameters for the modeling of sugarcane are shown in Table 1, the material parameters of the soil model for the test of sugarcane are shown in table 2.

The diameter of the soil plate is 21.8cm, the height of the soil plate is 15.2cm.The average wind speed of the fan is 5.5m/s. The wind regime was applied by the keywords "BOUNDARY_PRESCRIBED_MOTION_OPTION", and the size of the wind regime is the same with the size of fan. For the test of sugarcane stalk length and average diameter were 219.7cm and 2.998cm.The geometrical dimensions of the leaves are shown in table 3.A simplified model of the leaves as shown in Figure 7, leaf and stem leaf angle and torsion angle as shown in figure 8.

Table1. Physical parameters of sugarcane

Name	Average density $/g \cdot cm^{-3}$	Average elastic modulus /MPa
leaf	0.769	1460.3
Stalk	1.051	1250.2
root hair	0.709	649.7

Table 2. Parameters of soil material

Layers No.	Moisture content /%	density $/g \cdot cm^{-3}$	Cohesion /kPa	internal friction angle /°	Elastic modulus /MPa
1	25.76	2.08	13.49	23.06	0.99
2	24.45	1.87	15.08	22.85	1.08
3	24.18	1.97	24.23	22.19	1.30
4	23.34	1.98	28.92	21.67	1.25

3.4 Test results and analysis

Figure 9 is the acceleration curve of the marked point, among them, the solid line is the acceleration curves of physical test, the dashed line is the acceleration curves of numerical simulation test. From figure 9 we can see, the results of numerical simulation are in agreement with the results of physical experiment. Figure 10 is a numerical simulation screenshot of the deformation process of sugarcane, from figure 10 we can see, under the action of the wind, firstly, sugarcane produces large bending deformation, then the partial rebounds, and lets the stalk reciprocating swing. This phenomenon is consistent with the observed in physical experiment. It is proved that the method is feasible for the construction of the numerical simulation model of sugarcane soil air coupling system, and the model is high precision.

Table 3. Size of sugarcane's leaves

Leaves' No.	average value /mm						*The angle between the leaf and stem* θ/°	*Torsion angle of leaf* α/°
	Tip distance h_1	Leaf Length l_2	Front leaf width h_2	Maximum leaf width h_3	Front point to Maximum point l_1	vane thickness w		
1	414	726	23	55	709	0.26	35	23
2	265	701	22	56	693	0.26	23	65
3	121	654	24	56	649	0.25	18	68
4	28	583	21	58	556	0.26	13	71
5	0	525	23	58	498	0.27	1	89

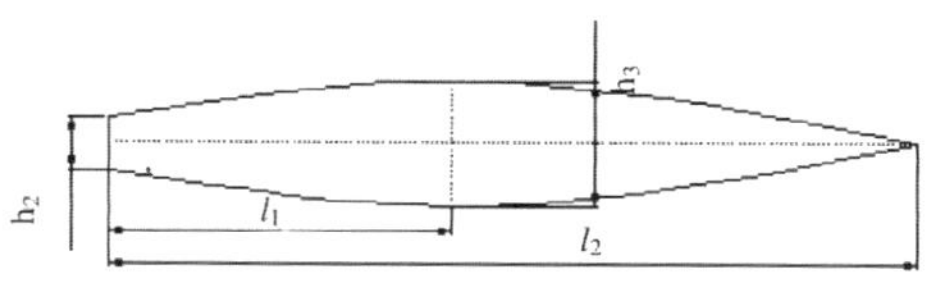

Figure7. Sugarcane leaf

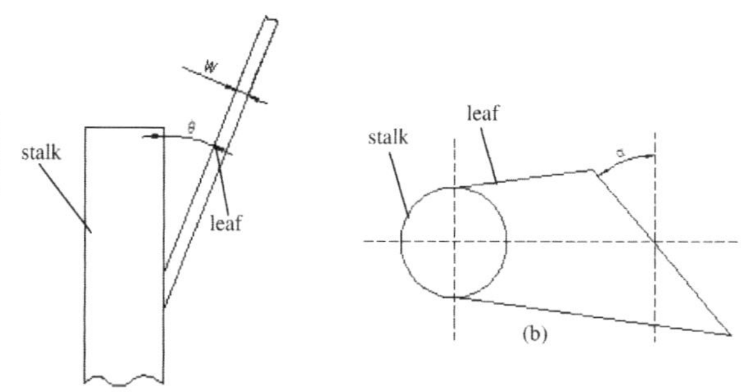

Figure8. Geometric parameters of leaf

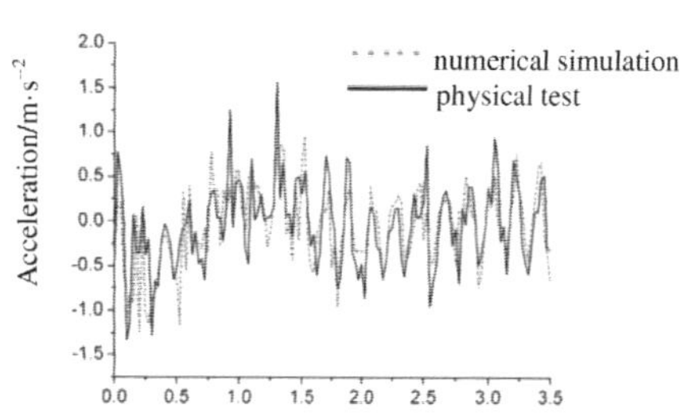

Figure9. Acceleration curve of the marker

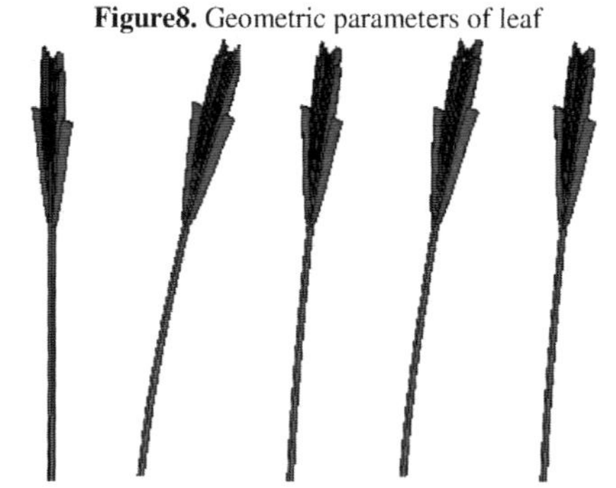

(a) t=0.00s (b) t=0.93s (c) t=1.83s(d) t=2.73s(e) t=3.58s

Figure 10. Deformation of sugarcane

4. Analysis on the process of stress change of sugarcane

Figure 11 is the numerical simulation screenshot of the process of stress change of sugarcane in numerical simulation test. From figure 11 we can see, when t=0.1s,sugarcane started by the effect of wind, the place of leaves and stem connection first appeared greater stress, and the stems came into being slight deformation. With the wind continued to function of sugarcane, the deformation of stems increased gradually, the stress of stem roots also increased gradually, when t=0.925s,the deformation of stems reached the maximum, the stress of stem roots also reached the maximum(2.49MPa).At the same time, after the deformation of stems reached the maximum, its rebounded, during the rebound process, the stress of stem roots gradually reduced, when t=1.825s, the springback of stems reached maximum, stem root stress is reduced to the minimum value, the maximum stress occurs at the junction of the leaves and stems, the value is 1.22MPa.Becouse of the reciprocating swing of the stems, the stress also changed periodically.

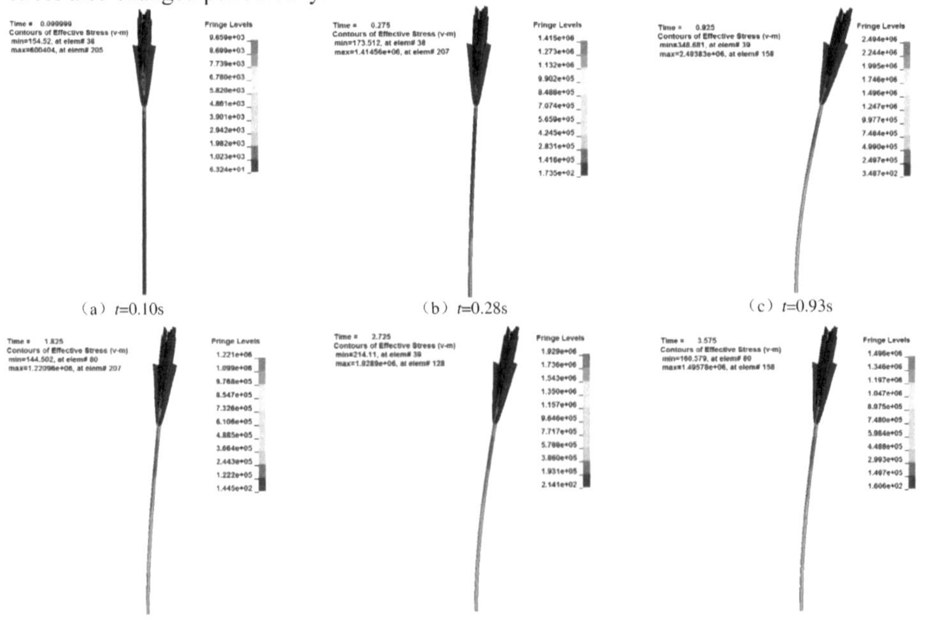

（a） t=0.10s （b） t=0.28s （c） t=0.93s
（d） t=1.83s （e） t=2.73s （f） t=3.58s

Figure 11. Change process of sugarcane's stress

5. Conclusion

(1)The method of constructing the numerical simulation model of sugarcane soil wind coupling dynamic system is feasible which can be used on the research of sugarcane lodging, and the accuracy is higher by using the multi material ALE method and the soil layer modeling technique.
(2)Under the action of wind, cane to produce periodic attenuation of the reciprocating vibration, the stem stress present attenuation change periodically.

6. References

[1] Yang W., Zhang S., Yang J., et al(2015).Experimental study on soil hardness of cultivated horizon at cassava cultivated farm in harvesting period. *Agricultural Mechanization Research*, 37(7), pp.176-180.
[2] Yang W., Mo X.M., Yang J., et al.2015 International Conference on Advances in Mechanical Engineering and Industrial Informatics.Wind-structure system modeling based on multi material ALE method. (Conference paper).
[3] Li Y.C., Shi D.Y., Zhao Y. (2008).ANSYS 11.0/LS-DYNA basic theory and engineering practice. *China Water Resources and Hydropower Publishing House*.(Book)
[4] Yang W., Zhang S., Chen K.Y., et al(2016).Dynamics simulation model of cassava tuber lifting system based on soil layering. *Agricultural Mechanization Research*,38(8),pp.51-55.
[5] Liu Q.T. (2004). Sugarcane cutting mechanism. *South China Agricultural University*.(Thesis)

Acknowledgments
This work was supported by a grant from the National Natural Science Foundation of China (Grant No.51565003), Guangxi Natural Science Foundation (Grant No.2014GXNSFBA118279), College Scientific Research Foundation of Guangxi Education Department (Grant No.YB2014002), Young Doctoral Research Foundation of Guangxi University (Grant No.XBZ160111) and Guangxi Training Programs of Innovation and Entrepreneurship for Undergraduates (Grant No.201510593277).

Research on Superplasticity of 5A90-O Aluminum Alloy Sheet

Zhi-qiang Li*, Wen-lin LU, Feng Zhang, Fengjiao Hu, Yong Wang

Beijing Research Institute of Mechanical&Electrical Technology, Beijing 100083China

*lzq@rdcai.com

Abstract. The influence of temperature and strain rate on the 5A90-O had been researched by costant strain-rate tensile method. The result shows that, although 5A90-O is not an ideal superplastic material, under certain conditions, with a large extension rate, the maximum elongation is193.6%. The strain rate has a significant impact on the flow stress and tensile strength under T=375℃~500℃. With the decrease of strain rate, the load is reduced. Another important factor that affects the flow stress of 5A90 is temperature. Under the same strain rate, the flow stress of 5A90 is lower with the temperature increase. We choose the Backofen law as the constitutive equation of 5A90, the best condition of 5A90 surperplastic deformation is T=400℃, $\varepsilon = 0.005s^{-1}$.

1. Introduction

Before 2020, the Tiangong-2 space station will be established preliminarily by China. From 2016, the spacecraft will be launched frequently. In order to improve the load capacity, the structural material which has light weight and high-strength had been researched. In this paper, our research target is 5A90-O Aluminum alloy, which is in Al-Li series. It has the advantage of low density, highly rigidity and excellent heat resistance. But use the traditional technology (cold stamping) to process a 5A90 thin shell will produce the following phenomena: low-plasticity and high-resilience, complex part is hard to form, at the same time, the costs of cold-stamping-mold will increase[1]. In aerospace field, if you want to form aluminum alloy thin-wall parts, the superplastic forming has been used normally. Because, when use superplastic forming process to aluminum alloy sheet, aluminum alloy has better plasticity, complex shape can be formed at one time, and under the high temperature, Aluminum alloy has almost no rebound spring-back, at the same time, it has simple mold structure. So, in this research, we used the high temperature one-way tensile test to find the best super-plastic parameters for 5A90, including temperature, strain rate and constitutive equation [2-3].

2. Experimental condition

Normally, if we want to find the parameters of superplastic material or superplastic materials with non ideal conditions, we need to do the same strain rate high temperature one-way tensile tests to get it. Then treat experiment data, we can get the elongation of 5A90, stress strain curve, and the Backofen constitutive equation, which include strain-rate sensitivity exponent m and the corresponding constant K [4].

In this paper, the material was provided by South Aluminum, 5A90 aluminum alloy chemical composition is shown in Table.1, and the original thickness of sheet is 1.4 mm.

Table.1 Chemical composition of 5A90 alloy in wt%

Mg	Li	Zr	Cu	Fe	Si	Al
5.29	2.12	0.12	0.03	<0.2	<0.15	Bal.

The dimension of tensile specimen was shown in fig.1.Tensile test is carried out in TG2000-20 universal testing machine. Besides it use retrofitted temperature control unit and the load-displacement sensor to achieve the same strain rate deformation. Finally, we can get the load-displacement curve.

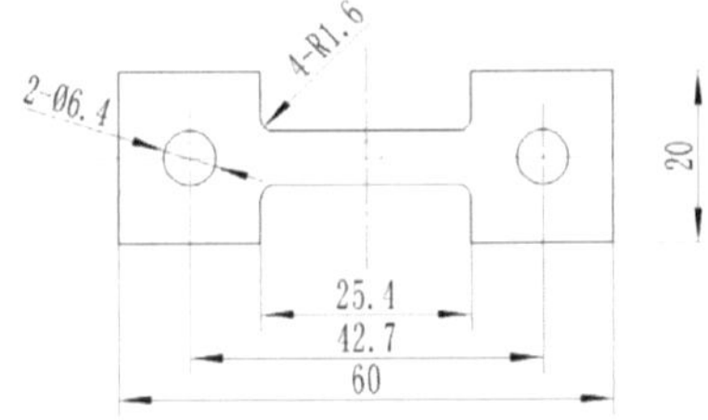

Fig.1 The dimension of tensile specimen

Through consulting sources, we had selected six temperature points:375,400,425,450,475,500℃, and five strain rate points: ε=0.00025s^{-1}, ε=0.0005s^{-1}, $\varepsilon = 0.001s^{-1}$, $\varepsilon = 0.005s^{-1}$, $\varepsilon = 0.01s^{-1}$.

3. Experimental results analysis and discussion

3.1. The elongation of 5A90 aluminum alloy

One-way tensile tests were carried out at different temperatures and strain rates as is stated above. The result of elongation had shown in fig.2. When the temperature is 375℃, the elongation of 5A06 is low, when the temperature is low, grain boundary sliding is difficult to conduct, which caused the low elongation. When the temperature is 400℃, all of the elongation can exceed 100%. The largest elongation is 193.6%. When the temperature is over 450℃, the temperature is too high, grain growth quickly, softening is less than hardening, the material appeared to shrink neck earlier, then break quickly.

When the maximum extension rate occurs, T=400℃, $\varepsilon = 0.01s^{-1}$, so this combination of temperature and strain rate is the best process parameters.

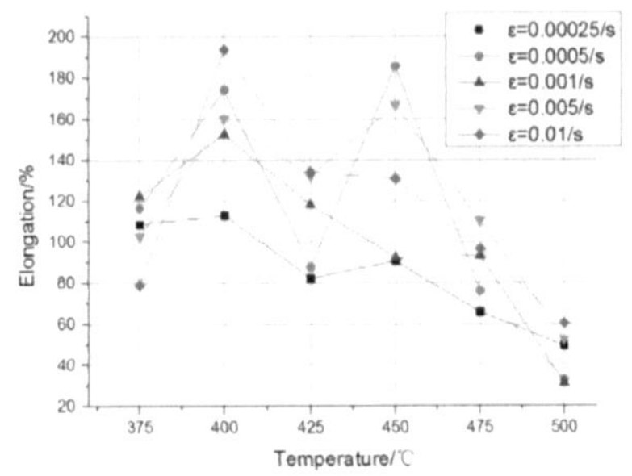

Fig.2 5A90 Al-Li alloy superplasitc one-way tensile elongation under different deformation condition

3.2. Mechanical property

In fig.3, 5A90 alloy flow curves under different strain rate at the same temperature had been shown. It can be seen that the load reaches its maximum value quickly, and then it slows down, finally 5A90 breaks. The load gradually reduces with the decrease of strain rate.

IOP Conf. Series: Materials Science and Engineering **187** (2017) 012025 doi:10.1088/1757-899X/187/1/012025

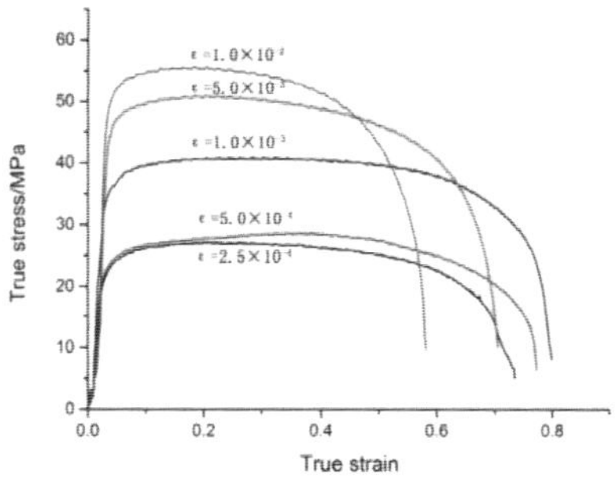

(a) 375℃

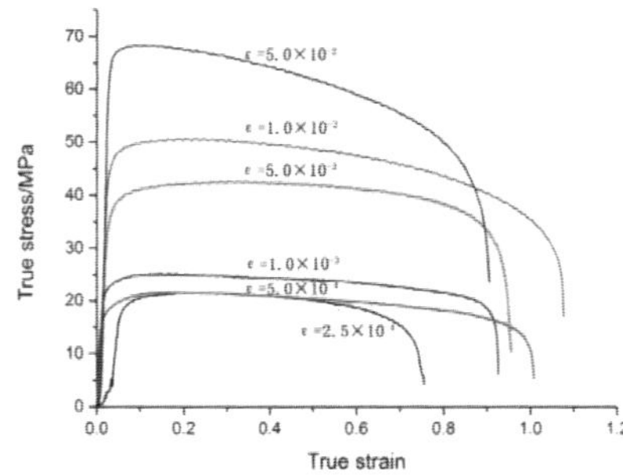

(b) 400℃

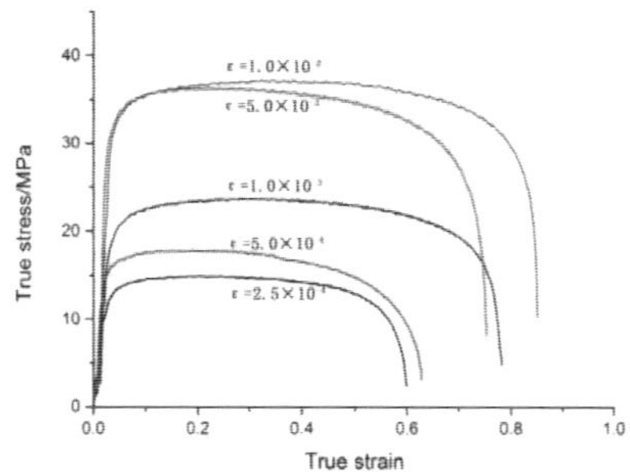

(c) 425℃

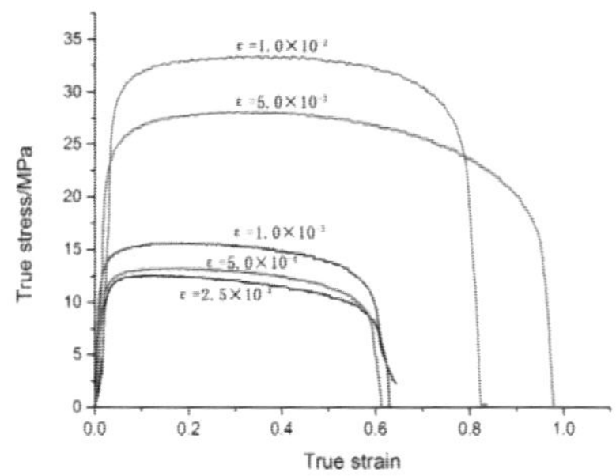

(d) 450℃

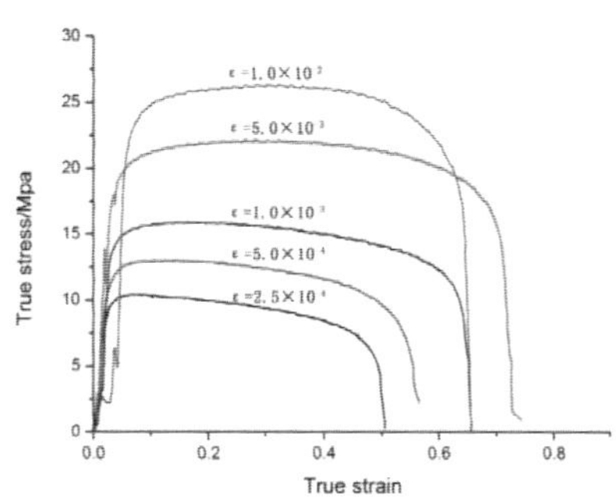

(e) 475℃

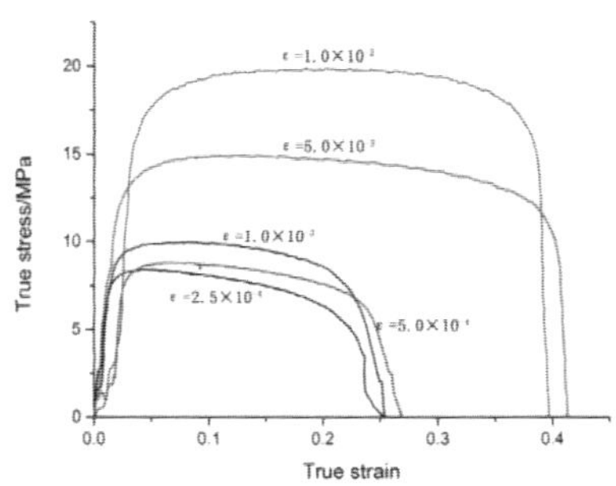

(f) 500℃

Fig.3 5A90 alloy flow curves under different strain rate at the same temperature

The relationship between dislocation velocity and shear stress about 5A90 had shown in equation (1). In this equation, V_0 is initial dislocation velocity; V is final dislocation velocity; A is the activation energy of material; τ is shear stress; T is the temperature of deformation. Through it, we can know that the flow stress is reduced inevitably when there shear stress is down. The movement velocity of dislocation decreased rapidly when the strain rate dipped to 0.0001/s. From the macro view, it is shown that the flow stress decreases rapidly. So the 5A90 Al-Li alloy is a kind of positive strain rate sensitive material.

$$V = V_0 \exp\left(\frac{A}{\tau T}\right) \quad (1)$$

On high temperature, the flow stress of 5A90 aluminum alloy increases rapidly and then tends to be stable when under different deformation conditions. But in this stage, the curves shows like serrate. The result of this phenomenon is that the dislocation density is increasing all the time when the material in the process of deformation. So the resistance of dislocation movement has increased which the load has augmented on the macro. It shows that the work hardening has appeared. At the same time, the dislocation and the energy density increased, when the accumulated energy is over the activation energy of recovery and re-crystallization, the dynamic recovery or dynamic re-crystallization will happen. It is a softening process. Because the hardening and softening coordinated all the time, in the end, the serration been emerged.

The temperature is another important parameter in superplastic forming, so the effects of temperature on the mechanical properties under high temperature needed to study.

Fig.4 shows the curves for 5A90 aluminum alloy, which were stretched at different temperature points when the strain rate$\dot{\varepsilon}$=$0.00025s^{-1}, 0.0005s^{-1}, 0.001s^{-1}, 0.005s^{-1}, 0.01s^{-1}$. We can know that under the same strain rate, the flow stress of 5A90 is lower when the temperature increased. The reason is when the temperature is high, the molecular motion is active, so the kinetic energy is improved, atomic vibration amplitude is larger, the activity of dislocation and vacancy is active, more and more slip system have been started. At the same time, dynamic recovery and re-crystallization has occurred, it can bring to a certain degree of softening. These factors work together to reduce critical shear stress, so the flow stress of 5A90 is reduced.

It follows that, under the high temperature, the deformation of 5A90 Al-Li alloy is a coordinate process which is hardening and softening. The work hardening is mainly derived from the deformation, the softening is mainly derived from the increased dislocation movement and dynamic recovery and re-crystallization. When the temperature is high, the mainly process is softening, on the contrary, is work hardening. On the macro, it shows that when the temperature increased, the flow stress reduced.

3.3. *The constitutive eauation*

In this paper, we had built the constitutive equation of the 5A90 aluminum alloy based on Backofen. When plastic deformation occurs, the flow stress has mainly influenced by strain rate, strain, temperature and the size of grain, it has been shown in formula (2).But the temperature is invariant under superplastic deformation and since the material is 5A90, the relationship about follow stress σ, strain ε and strain rate $\dot{\varepsilon}$ can simplify as formula (3):

$$\sigma = f(\dot{\varepsilon}, \varepsilon, T, d) \tag{2}$$

$$\sigma = K\dot{\varepsilon}^m\varepsilon^n \tag{3}$$

In formula (3), K is a constant about material; m is strain rate sensitivity index; n is strain hardening exponent, but in superplastic, generally, the strain hardening does not appear, so its value is 0.

Finally, the formula about σ and $\dot{\varepsilon}$ can express as formula (4):

$$\sigma = K\dot{\varepsilon}^m \tag{4}$$

Generally, if you want to get the *m* about a material, you can through the same strain rate tensile method, constant velocity tensile method, velocity mutation tensile method and so on. In this experiment, the constant velocity tensile method is been picked.

Firstly, gain the stress strain curve of 5A90. Then select a stable section of the curve, converted it into $\ln\sigma^{-1}$—ε. The slope of the curve is *m*. The formula is (5).

$$m = \frac{d\ln\sigma^{-1}}{d\varepsilon} = -\frac{d\ln\sigma}{d\varepsilon} \tag{5}$$

After calculate, we can get the value of m, they have been shown in fig.4(a). With the temperature increased, the value of m reduced at first, and then increased slowly. Temperature has an important effect on m. In the condition of T=375℃, m=0.3932, material deformation is more uniform.

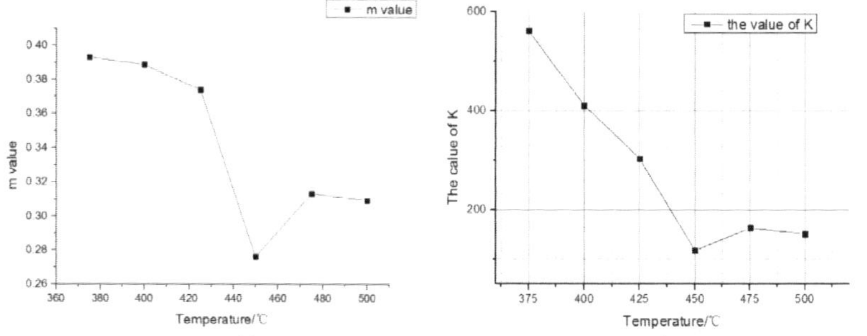

(a) (b)

Fig.4 The value of m and K

(a) Strain rate sensitivity index changing with temperature

(b) The value of K changing with temperature

When we get the value of m, we can through formula (4) to find the value of *K*, it has shown in fig.4 (b). With the temperature increased, the value of K reduced firstly, and then increased slowly. So T=400℃is the best temperature of deformation after take some factors in the practical production into account.

4. Conclusion

1. The best temperature of deformation for 5A90 Al-Li alloy is 400℃, under this temperature points, the maximum elongation is 193.6%.

2. The strain rate has a significant impact on the flow stress and tensile strength under T=375℃ ~500℃. With the decrease of strain rate, the load is reduced.

3. Another important factor that affects the flow stress of 5A90 is temperature. Under the same strain rate, the flow stress of 5A90 is lower with the temperature increase.

4. We choose the Backofen law as the constitutive equation of 5A90, the best condition of 5A90 surperplastic deformation is T=400℃, $\varepsilon = 0.005s^{-1}$.

5. References

[1] TAI Qing-an, LI Zhi-hua, SUN Li-qun.et al. The Aircraft Engine Application and Prospect of Plastic Forming Technology[J]. Aviation manufacturing Technology, 2014(7): 34-39.

[2] KaiFeng Zhang, GuoFeng Wang etc. Advanced material molding technology. Beijing: science press,2012, pp.202-207

[3] Jarrar Firas S,Hector Jrlouis G, Khraisheh,et al. New Approach to Gas Pressure Profile Prediction for High Temperature AA5083 Sheet Forming[J]. Journal of Materials Processing Technology,2010,210(6-7):825-834.

[4] Bao-peng BI.YONG Wang. Research on Superplasticity of Supplied 5A06 Aluminium alloy sheet[C].The 12th international conferenceon superplasticity, n advanced materials.(ICSAM2015)

[5] JIN cheng , HE shi-yu , ZHOU Guang-tao , etal . Experimental and numerical investigations of void damage in aluminum alloy welds under thermal cycling condition [J].Materials Science and Engineering(A), 2009, 499: 171-176.

Acknowledgments

The support of International Science & Technology Cooperation Program of China, Protect No.2013DFR70860, is acknowleged.

IOP Conf. Series: Materials Science and Engineering **187** (2017) 012026 doi:10.1088/1757-899X/187/1/012026

Numerical Investigation on Trapezoidal Cavity Receiver Used In LFR with Water Flow in Absorber Tubes

Rohit Duggal[1, *] **and Ravindra Jilte**[2]

[1] Research Scholar, School of Mechanical Engineering, Lovely Professional University, Jalandhar, Punjab 144411, India
[2] Professor, School of Mechanical Engineering, Lovely Professional University, Jalandhar, Punjab 144411, India

[*]Corresponding author: rohitduggal195@gmail.com

Abstract. In the present study, numerical three dimensional model of trapezoidal cavity used in LFR was analysed. Results are presented in the form of Thermal losses occurring from the receiver operating with an absorber tube temperature from 350-550 K in step of 50 K and emissivity varied from 0.5-1.0. Effect of wind blowing below lower glass plate (cavity aperture) were also analysed considering the heat transfer coefficient from 5 to 25 W/m^{2}K. At lower absorber temperature (350 K) convective losses is found to be 43% of the total heat loss whereas radiative losses accounted 57%. For higher absorber temperature radiative losses are dominant (77%) and convective losses are reduced to 23%. The air temperature gradient in the horizontal direction (parallel to lower glass plate) is found to be negligible whereas it is varied significantly in vertical direction (normal to lower glass plate). The average cavity air temperature is observed to be 480 K for low wind flow (h=5 W/m^{2}K) and it reduces to 360 K for h=25 W/m^{2}K. This has resulted in increased convective losses (27% higher).

1. Introduction

Among the various types of Concentrating Solar Power (CSP) plants, Linear Fresnel Reflector (LFR) technology is considered prospective due to low cost and structural simplicity. Moreover, merits related to cost associated with manufacturing, operation and maintenance is less as compared to Parabolic Trough Collector (PTC) [1-3]. The advantages associated with Fresnel Lens mainly small volume, light weight, low cost mass production and increasing the energy density has paved the way for its application in solar thermal field [4]. An LFR CSP consists of three main components: concentrator, receiver and absorber. It consists of linear array of mirrors that concentrate solar radiation on to a downward facing receiver containing pipe absorbers. These pipes contain a particular Heat Transfer Fluid (HTF) which absorbs energy and applies to power generation. Generally, receiver in the form of inverted trapezoidal shape is preferred [5] (insulated upper and side surfaces, absorber tubes with steam beneath top face and lower glass plate to allow solar flux onto absorber tubes). The performance of the LFR is affected by the thermal losses in the receivers (both radiative and convective) occurring from high temperature absorber tube. Several studies have been attempted to study the heat transfer from the LFR receiver. Pye et al. [6] have studied unsteady flow patterns in trapezoidal cavity receiver. Unsteadiness in the flow structure do not significantly vary overall heat transfer losses. Reynolds et al. [5] analyzed the heat losses from the absorber of trapezoidal receiver. 8% higher efficiency was obtained with round pipes as compared to rectangular absorber pipes. Further,

Singh et al. [7] investigated that thermal efficiency is influenced by surface coating and concentration ratio. The thermal efficiency correlations were developed for absorber surface coating as a function of concentration ratio and entry HTF temperature. Furthermore, deviation in correlated and experimental data was reported for rectangular and round absorber pipe. In addition, Singh et al. [8] investigated heat transfer coefficient for single and double glass cover as well as absorber surface coating. It was concluded that heat losses in nickel black coating was 30% less as compared to matt black absorber coating. Moreover, in case of single glass cover, varying the absorber temperature reduces heat losses for round tubes unlike double layer glass cover where difference in heat losses for both (round and rectangular) absorbers is negligible. Further, the correlations between overall heat transfer coefficient and absorber surface temperature were presented and reported that experimental and correlated data agree to great extent. Facao and Oliveira [9] divided the pipes into two halves and cavity was modelled. The numerical study was carried out considering surface radiation, cavity convection and wall conduction thereby observing least global heat loss coefficient for 45mm cavity depth. Reddy et al. [10] investigated heat transfer and presented Nusselt number correlations for different parameters of trapezoidal cavity receiver considering isothermal top and bottom glass cover. The side walls of cavity were modelled with 0.5 W/m2K as heat loss coefficient and 0.3 as internal emissivity, however, bottom heat loss due to forced convection leading from blowing wind was neglected. Sahoo et al. [11] numerically investigated the thermal losses in trapezoidal cavity with eight absorber tubes. Inside the cavity, isotherms and flow patterns were analyzed ensuring the gaps between absorber tubes and non-dimensionalization was adopted for reporting the results. This study reported different correlations for Nusselt number excluding radiation and conduction parameter. Further Sahoo et al. [12] applied the heat transfer model to investigate the methodology of multiphase flow in absorber tubes. The variation in parameters for single phase region (bulk fluid temperature, pressure drop and heat transfer coefficient) and for two phase region (dryness fraction, pressure loss and local boiling point) was presented. Moghimi et al. [2] carried out computational study to optimize the cavity for lesser heat transfer losses. The LFR cavity was modelled considering four round pipes and walls were assumed as insulated. The fluid inside the cavity was considered as incompressible and density was chosen as a function of temperature. Discrete Ordinate Method was chosen to simulate radiation heat loss and optimized for lesser side wind loss.

Based on the aforementioned studies, it was observed that in numerical investigations actual HTF flow inside the absorber tubes were not considered. The present study focuses on three dimensional numerical investigations on trapezoidal cavity receiver considering both airside as well as HTF side heat losses. However, only the airside results are included in the present paper.

2. Receiver description

A symmetric schematic sketch of cavity located at focal point of LFR system is shown in figure 1. The Heat Transfer Fluid (HTF) is taken as water and air is taken inside the cavity. A closed three-dimensional trapezoidal receiver cavity is considered containing eight tubes having a small gap in between allowing thermal expansion. The tubes are of Nominal Pipe Size (NPS) 1 stainless steel (SS) 304 material configuration. There is a glass covering at bottom of receiver cavity to allow solar radiations inside the cavity. The outer covering of other three sides is insulated to minimize the heat losses. The reflected rays from linear array of mirrors arranged at the bottom falls on the absorber tubes thereby increasing the temperature of HTF flowing.

3. Numerical analysis

3.1. Model description

In the present study, three-dimensional model is adopted to predict the heat loss from the receiver cavity. For the purpose of generating mesh as shown in figure 2, the Meshing package of ANSYS 15 Workbench is used. The accuracy and convergence of the solution for the computational domain is checked for skewness and aspect ratio, which should not exceed certain value (0.85) as reported by

Bakker [13]. The maximum and minimum values of skewness (max. 0.68 and min. 0.032) and aspect ratio (max. 5.86 and min. 1.10) for present study with the average element quality as 0.811 provides the quality mesh for computational domain.

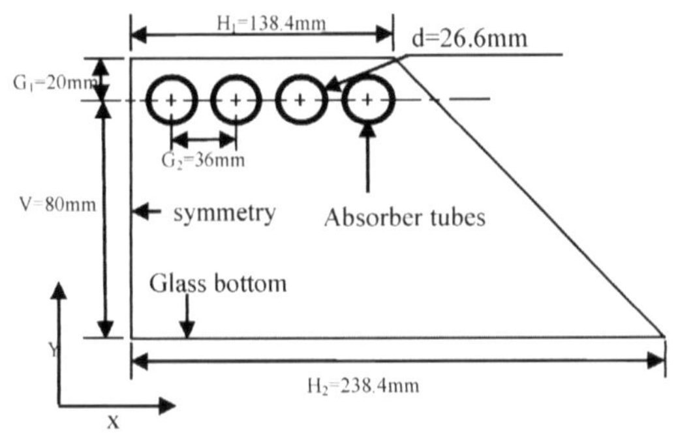

Figure 1. Sketch of receiver cavity.

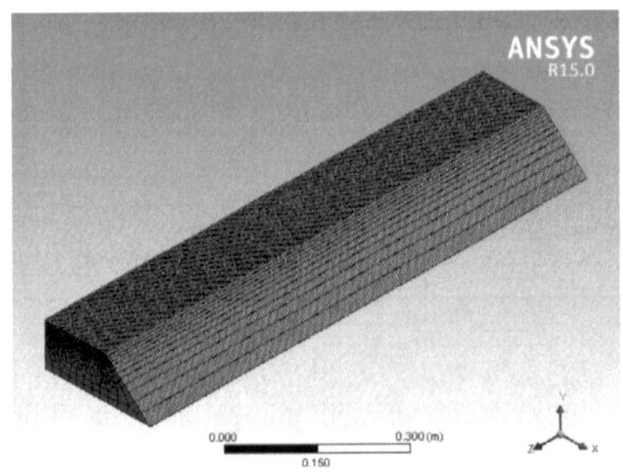

Figure 2. Receiver cavity mesh.

The values obtained from numerical simulations are a result of simultaneous solution of the system of flow and heat transfer equations describing mass, momentum, energy. The equations can be described as below [14]:

Continuity equation

$$\frac{\partial(\rho u)}{\partial x}+\frac{\partial(\rho v)}{\partial x}=0 \quad (1)$$

Momentum equation

$$\left(u\frac{\partial u}{\partial x}+v\frac{\partial u}{\partial y}\right)=-\frac{1}{\rho}\frac{\partial P}{\partial x}+\vartheta\left(\frac{\partial^2 u}{\partial x^2}+\frac{\partial^2 u}{\partial y^2}\right)+g\beta(T-T_h) \quad (2)$$

Energy equation

$$\left(u\frac{\partial T}{\partial x}+v\frac{\partial T}{\partial y}\right)=\alpha\left(\frac{\partial^2 T}{\partial x^2}+\frac{\partial^2 T}{\partial y^2}\right) \quad (3)$$

The current work has 350-550 K of temperature range therefore the Boussinesq approximation is invalid. Since, in this temperature range, the product of temperature difference between wall surface and air with the coefficient of thermal expansion of air (0.142-0.58) is higher than 0.1 for Boussinesq approximation to be applied [15]. Hence non-Boussinesq approximation, i.e. the ideal gas characteristics are chosen for cavity air and above equation is used to solve steady, laminar natural convection model. The surface to surface (S2S) radiation model is coupled with natural convection model. S2S model assumes that the working fluid is not participated in absorption, emission and scattering of radiation. The main assumption of S2S radiation model is that the surfaces are grey and diffuse. However, it should be noted that model assumes heat transfer is affected between two surfaces depending upon their size, orientation and separation distance. View factors are used to account the influence of these variables [16].

3.2. Boundary conditions

Different boundary conditions are applied to the cavity receiver. No slip conditions are assumed at walls. Fluids are taken to be incompressible and Newtonian. Since, tubes are continuously subjected to solar radiations therefore tubes are assumed to be at uniform constant temperature after reaching stagnation conditions. Therefore, isothermal boundary conditions are chosen for absorber tubes. The outer covering of cavity receiver is insulated so as to reduce heat losses. Thus, adiabatic conditions are assumed for outer covering. There is radiative heat loss among the walls of receiver cavity due to the high temperature. Moreover, convection losses are significant with bottom glass cover to surrounding air. Hence, both convection and radiation boundary conditions are applied to glass covering. Mainly, top wall, side wall, bottom wall and absorber tube emissivity as 0.1, 0.1, 0.9 and 0.49-0.9 respectively. External emissivity of bottom wall and heat transfer coefficient as 0.9 and 5.0-25.0 W/m^{2}K. Temperature of absorber tube is varied from 350-550K.

3.3. Numerical procedure

The simulation for turbulent incompressible flow are carried out using FLUENT 15 software package. The standard (k-Ɛ) 2-eq model is chosen to model turbulent conditions, since it is most widely used engineering model which includes sub-model for buoyancy [16]. Second order upwind scheme is used for discretization of pressure velocity coupling equations. For residuals of continuity and momentum equations, a convergence criterion of 10^{-3} was imposed whereas 10^{-6} as energy equation residuals. The convergence is determined by the residual levels and also by monitoring relevant integrated quantity like heat transfer coefficient. The solutions are obtained only after satisfying convergence criterion.

3.4. Grid independence study

To lower the dependency of grid on the obtained solutions different grids were studied. Since, greater the number of elements, lesser is the deviation in solution. Therefore, different number of elements is chosen for the grid to check the solution dependency with respect to the grid. The geometry is divided into quadrilateral and hexahedral unstructured mesh elements ranging from 71,050 to 3,35,274. Table 1 below shows the variation of total heat transfer with respect to the grid size. It is further seen that the variation in heat transfer solution for elements 71,050 and 1,11,600 is negligible (<1%). Hence, 71,050 elements are sufficient to model the geometry to obtain reliable results. Thereby, reducing the computational time and cost on a greater extent.

3.5. Validation and verification of numerical procedure

In order to validate computational results, the simulation values obtained for total heat loss are compared with experimental data obtained by Sahoo et. al. [12] using their cavity lab test setup. It can be seen from figure 3, that the results obtained from simulated data are well in conformation with experimental data.

Table 1. Cavity grid independence study.

Number of elements	Total heat loss (W)	% change in heat loss
71050	106.822	
111600	106.828	0.06
268128	107.830	0.93
335274	107.409	-0.39

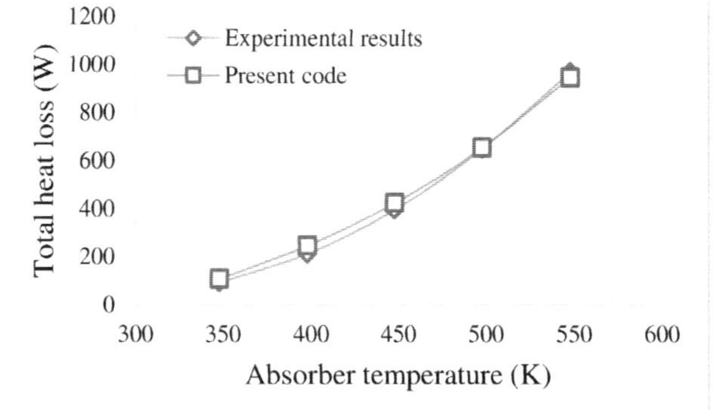

Figure 3. Comparison of Total heat loss (W) with Sahoo et. al. [12] and present numerical simulation.

4. Results

4.1. Isotherm contours

Isotherms contours are plotted in figure 4 and 5 for various cavity parameters. It can be seen from isotherm contours in figure 4 that temperature variations along horizontal direction is almost negligible and hence uniform profile is seen. However, natural convection is taking place due to the temperature gradient along vertical direction. Thus, rising the air from centre of cavity and settling down from the side surfaces. Further, as the absorber tube temperature is increased the heat affected zone with higher average temperature shifts towards bottom glass cover thereby increasing the vertical temperature gradient. figure 4 (a) shows the approximate average temperature 375.5 K which increases as absorber temperature increases up to 487.5 K. However, the percentage vertical temperature gradient is reduced from 9.09% to 7.14% as temperature varies from 350 K to 550 K respectively this is due to the increased losses at higher temperatures.

It can be observed from figure 5 that as the heat transfer coefficient near glass surface is increased there are significant convective losses. The amount of cooling of air that is taking place near the glass surface as a result of ambient air conditions can be observed easily by varying the heat transfer coefficient. It can be seen that for h=5 W/m^{2}K almost all the entrapped air (90% approx.) is at higher average temperature (480 K approx.). However, the amount of entrapped air with higher temperature is reduced from 90% to approximately 40% with increased value of heat transfer coefficient (h=5 to 25 W/m^{2}K). The average temperature is also reduced from 480 K to 360 K as the value of heat transfer coefficient is increased (h=5 to 25W/m^{2}K). It can be concluded from above results that with lesser value of average temperature more heat is being swept away due to blowing wind. Thus, ambient environment conditions significantly affect the heat loss characteristics of the cavity.

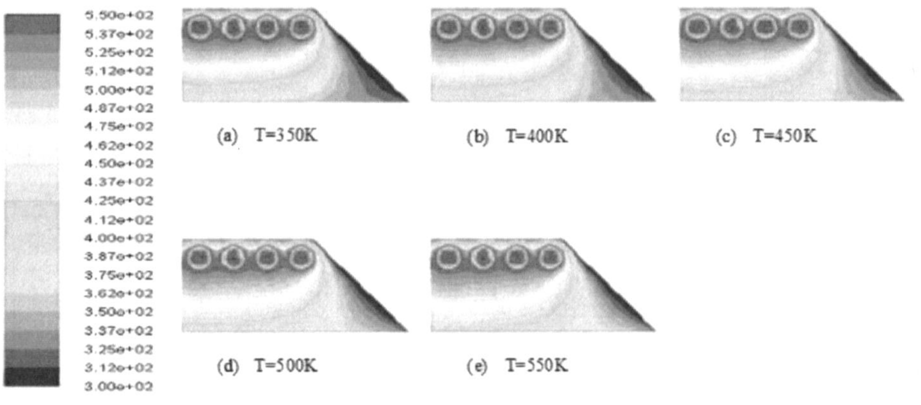

Figure 4. Cavity isotherm contours for m=0.1kg/s, h=5W/m^{2}K, Ɛ=0.49 at different absorber temperatures.

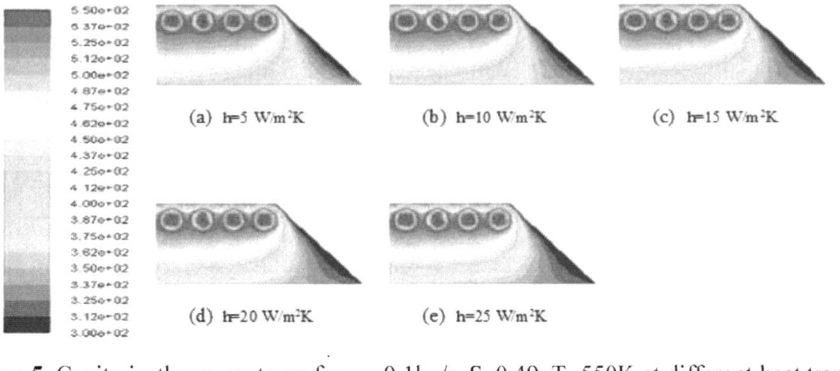

Figure 5. Cavity isotherm contours for m=0.1kg/s, Ɛ=0.49, T=550K at different heat transfer coefficient.

4.2. Cavity heat exchange

4.2.1. Effect of absorber temperature on heat loss characteristics. Figure 6 shows the variation of heat transfer taking place inside the cavity as a function of absorber surface temperature. It is observed that as the wall temperature of absorber tubes increase heat loss is increased for m=0.1kg/s, h=5W/m^{2}K and Ɛ=0.49. Radiative heat loss for temperature 400 K is 244.57 W and for temperature 450 K is 423.32 W showing 87.21% increase in value. The percentage change is decreased from 87.21% to 52.33% as the absorber temperature is increased to 550 K. The convective heat loss amounts to 90.57 W for 400 K and 135 W for 450 K absorber temperature. Thus, there is an increase of 49% in convective losses. This percentage increase changes from 49% to 24.3% as temperature reaches to 550 K. Thus, it is clear that radiative heat losses steep more as compared to convective losses. This is due to the higher temperature of absorber tubes emitting radiation and contributing to radiation losses. It is further observed that at higher absorber temperature (550 K) radiative heat losses are dominant, i.e. 76.45% of the total heat loss for the receiver cavity.

4.2.2. Effect of Rayleigh number. The variation of Nusselt number with respect to Rayleigh number can be observed in figure 7. It is clearly obtained from the graph that Nusselt number is an increasing function of Rayleigh number. As Rayleigh number is increased so does the Nusselt number.

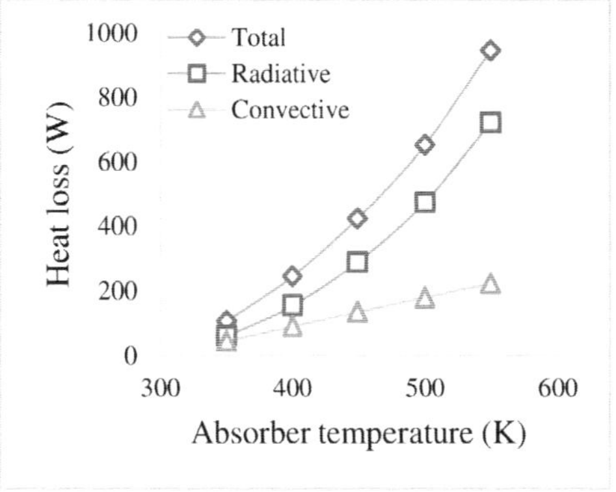

Figure 6. Heat loss characteristics (total, radiative and convective) for m=0.1kg/s, h=5W/m^{2}K and Ɛ=0.49.

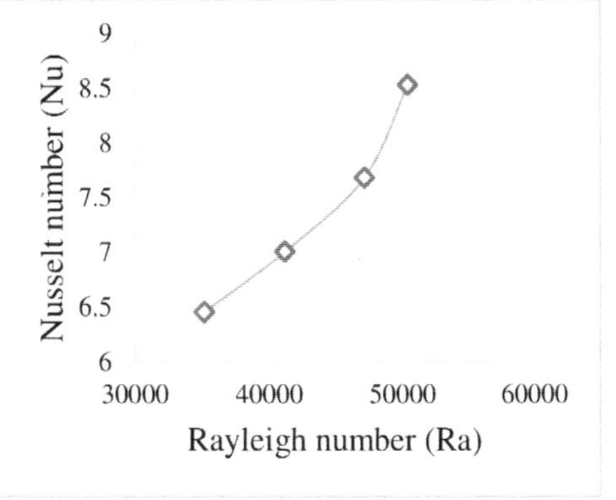

Figure 7. Variation of Nusselt number (Nu) for m=0.1kg/s, h=5W/m^{2}K, Ɛ=0.49.

4.2.3. Effect of absorber emissivity. The heat loss characteristics as a function of absorber tube emissivity can be obtained from figure 8. It is observed that convective heat losses are more or less uniform as well as negligible. The radiation losses are dominant as the emissivity values are varied. Radiative heat exchange is increased with increasing values of absorber tube emissivity.

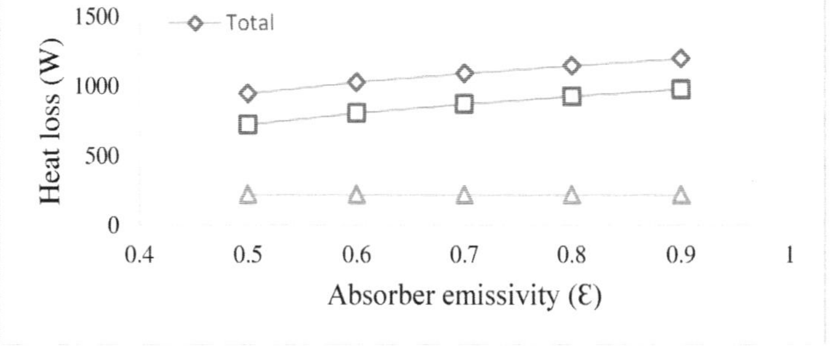

Figure 8. Heat loss characteristics (total, radiative and convective) for m=0.1kg/s, h=5W/m^{2}K, T=550.

5. Conclusions

In the present study, numerical three dimensional model of trapezoidal cavity used in LFR was analysed considering actual HTF flow in absorber tubes. Relative dominance of convective and radiative losses was evaluated. At lower absorber temperature (350 K) convective losses is found to be 43% of the total heat loss whereas radiative losses accounted 57%. For higher absorber temperature (550 K) radiative losses are dominant (77%) and convective losses are reduced to 23%. The cavity air temperature is used to calculate air temperature gradient for studying the effect of wind blow on lower glass plate as well as varied absorber tube temperatures. The air temperature gradient in the horizontal direction (parallel to lower glass plate) is found to be negligible whereas it is varied significantly in vertical direction (normal to lower glass plate). The average cavity air temperature is observed to be 480 K for low wind flow (h=5 W/m^{2}K) and it reduces to 360 K for h=25 W/m^{2}K. This has resulted in increased convective losses (27% higher).

6. References

[1] Mills, D.R., Morrison, G.L., 2000. Compact linear fresnel reflector solar thermal power plant. Sol. Energy 68, 263–283.

[2] Moghimi, M.A., Craig, K.J., Meyer, J.P., 2015. Optimization of a trapezoidal cavity absorber for the linear Fresnel reflector. Sol. Energy 119, 343–361.

[3] El Gharbia N, Derbalb H, Bouaichaouia S, Saida N. A comparative study between parabolic trough collector and Linear Fresnel Reflector technologies. Energy Proc 2011;6:565–72.

[4] Xie, W.T., Dai, Y.J., Wang, R.Z., 2011a. Numerical and experimental analysis of a point focus solar collector using high concentration imaging PMMA Fresnel lens. Energy Conversion and Management 52 (6), 2417–2426.

[5] Reynolds, D.J., Jance, M.J., Behnia, M., Morrison, G.L., 2004. An experimental and computational study of the heat loss characteristics of a trapezoidal cavity absorber. Sol. Energy 76, 229–234.

[6] Pye, J.D., Morrison, G., Behnia, M., 2003a. Transient modeling of cavity receiver heat transfer for the compact linear Fresnel reflector. In: ANZESES Destination Renewable 69-77.

[7] Singh, P.L., Sarviya, R.M., Bhagoria, J.L., 2010a. Thermal performance of linear Fresnel reflecting solar concentrator. Appl. Energy 87, 541–550.

[8] Singh, P.L., Sarviya, R.M., Bhagoria, J.L., 2010b. Heat loss study of trapezoidal cavity absorbers for linear solar concentrating collector. Energy Convers. Manage. 51, 329–337.

[9] Facao, J., Oliveira, A.C., 2011. Numerical simulation of trapezoidal cavity receiver for a linear Fresnel solar collector concentrator. Renew. Energy 36, 90–96.

[10] Reddy, K.S., Natarajan, S.K., Mallick, T.K., 2012. Heat losses characteristics of trapezoidal cavity for solar linear concentrating system. Appl. Energy 93, 523–531.

[11] Sahoo, S.S., Singh, S., Banerjee, R., 2012. Analysis of heat losses from a trapezoidal cavity used for linear Fresnel reflector system. Sol. Energy 86, 1313–1322.

[12] Sahoo, S.S., Singh, S., Banerjee, R., 2013. Steady state hydrothermal analysis of the absorber tubes used in linear Fresnel reflector solar thermal system. Sol. Energy 87, 84–95.

[13] Bakker, A., 2013. Lecture 7 – Meshing, Applied Computational Fluid Dynamics

[14] Jiji, J. L. M. (2006). *Heat Convection*. Springer-Verlag

[15] Gray, D. D., & Giorgini, A. (1976). The validity of the Boussinesq approximation for liquids and gases. International Journal of Heat and Mass Transfer, 19, 545-551

[16] ANSYS, 2013b. Design Exploration User Guide, version 15, ANSYS Incorporated

Research of Customized Aortic Stent Graft Manufacture

Lei Zhang[1, 2, 3], Xin Chen[1, 2, 3, *], Muhan Liu[1, 2, 3]

[1] Department of Mechanical Engineering, Biomanufacturing Center, Tsinghua University, Beijing 100084, P.R.China
[2] Biomanufacturing and Rapid Forming Technology Key Laboratory of Beijing, Beijing 100084, P.R.China
[3] Key Laboratory for Advanced Materials Processing Technology, Ministry of Education, Beijing 100084, P.R.China

[*]stoneszhang@mail.tsinghua.edu.cn

Abstract. Thoracic descending aorta diseases include aortic dissection and aortic aneurysm, of which the natural mortality rate is extremely high. At present, endovascular aneurysm repair (EVAR) has been widely used as an effective means for the treatment of descending aortic disease. Most of the existing coating stents are standard design, which are unable to meet the size or structure of different patients. As a result, failure of treatment would be caused by dimensional discrepancy between stent and vessels, which could lead to internal leakage or rupture of blood vessels. Therefore, based on rapid prototyping sacrificial core – coating forming (RPSC-CF), a customized aortic stent graft manufactured technique has been proposed in this study. The aortic stent graft consists of film and metallic stent, so polyether polyurethane (PU) and nickel-titanium (NiTi) shape memory alloy with good biocompatibility were chosen. To minimum film thickness without degrading performance, effect of different dip coating conditions on the thickness of film were studied. To make the NiTi alloy exhibit super-elasticity at body temperature (37℃), influence of different heat treatment conditions on austenite transformation temperature (Af) and mechanical properties were studied. The results show that the customized stent grafts could meet the demand of personalized therapy, and have good performance in blasting pressure and radial support force, laying the foundation for further animal experiment and clinical experiment.

1. Introduction

Aortic dissection is a kind of aortic aneurysm, which occurs when a tear in the endothelial causes blood flowing into the space between the mesothelial and endothelial, forcing the layers apart. Aortic dissection is an extremely sever lesion among vascular diseases. Without treated in time, the natural mortality rate could reach 36 to 72% within 48 hours after diagnosis [1].

The thoracic aorta consists of the thoracic ascending aorta and the descending thoracic aorta. According to whether the ascending aorta is involved, Stanford classification [2] divide aortic dissection into type A and type B. For type B, endovascular aneurysm repair (EVAR) has been widely used as an effective means for the treatment of descending aortic disease.

Aortic stent graft consists of film and metallic stent. Film material contains polyester, polytetrafluoroethylene and other biopolymers. As the implantation method of aortic stent graft mostly is self-expanding, metal part requires the super-elasticity of NiTi alloy shape memory alloy to unfold.

When insert the self-expandable metallic stent(SEMS), compress and fold the SEMS into a protective sheath, and implant the protective sheath into the location of lesion by the delivery system. Then, withdraw the protective sheath and the stent will gradually unfold by itself because of the super-elasticity of NiTi alloy. The stent will hold a structure to support the vessel and prevent further expansion of the aneurysm. As is shown in Figure 1.

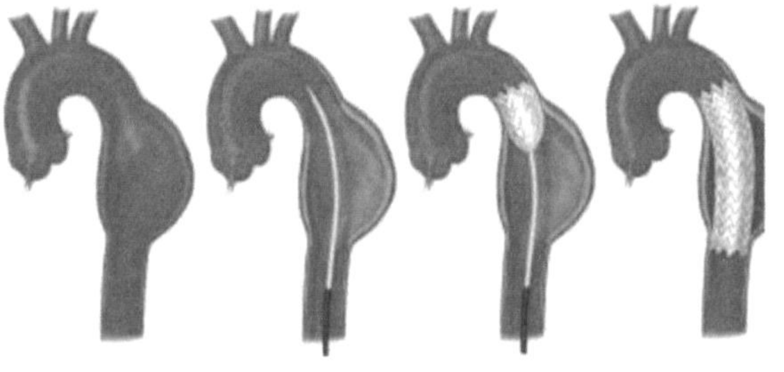

Figure 1. Implantation of aortic stent graft.

At present, size and structure of aorta stent graft are uniform. However, there are differences in the size or shape of each person's blood vessels. Mismatch stents may cause internal leakage or rupture of blood vessels, resulting in failure treatment. Therefore, we proposed a customized aortic stent graft manufactured process based on 3D printing. According to the real structure of patient's descending thoracic aorta, we can manufacture the personalized and suitable stent graft within a short time, which can avoid treatment failures cause by dimensional discrepancy. The stent manufactured by this process has good compliance and mechanical properties. The method will improve the success rate of endovascular aneurysm repair, represents one of the development directions of tissue engineering and medical 3D printing technology, has a good clinical application prospect.

2. Materials

2.1. Film materials

Existing aortic stent graft materials are mainly of polytetrafluoroethylene (PTFE) [3] and polyethylene terephthalate (PET) [4]. Polyurethane (PU) has become the preferred material for the replacement of PTFE and pet because of its good biocompatibility and easy formability. It has reported that [5], there is no difference about complications and patency between PU and PTFE after long term implantation in human body.

Polyurethane is polymerized by polyol (polyether and polyester), isocyanate and chain extender (diol and amine), containing -NHCOO- (carbamate) in backbone. According to the different types of polyol, polyurethane material can be divided into polyether polyurethane and polyester polyurethane. The hydrolysis resistance of polyether type polyurethane is very well, and it has a better biocompatibility than polyester polyurethane [5].

The polyether polyurethane, produced by Zhejiang Huafeng Chemistry Co., Ltd, has been used to the film of aortic stent graft. The material has been certified by the U.S. Food and Drug Administration (FDA), which has good biocompatibility and resistance to hydrolysis.

2.2. Metal materials

NiTi shape memory alloy has special practical value and widely application in medicine for its excellent super-elasticity, shape memory, and good biocompatibility [6, 7].

NiTi alloys exhibit super-elasticity (SE), which occurs at a narrow temperature range just above its transformation temperature; in this case, the material is in the austenitic state, exhibiting enormous elasticity, some 10~30 times that of ordinary metal.

2.3. Sacrificial materials
Sacrificial core material should be consistent with soluble in water and insoluble in organic solvents, non-toxic, good formability and other characteristics. Currently available materials include sugar and polyvinyl alcohol (PVA), etc. Polyvinyl alcohol, with low melting point, good biocompatibility, resistance to organic solvents and molding performance, is the ideal sacrificial core material [8].

3. Method
Additive Manufacturing (AM) [9], also known as 3D printing, is an advanced rapid prototyping method. Complex models can be manufactured by one-time molding. In this paper, AM technique was used as a part of the manufacturing process of stent graft.

A RP sacrificial core – coating forming (RPSC-CF) technique based on 3D printing was proposed by Zhang Lei [10], and multi-branched blood scaffolds with multi-layered wall were manufactured with the technique. Based on RPSC-CF, a customized aortic stent graft molding technique has been proposed in this research.

Step one, customized aortic stent was designed according to patient's computed tomography(CT). Step two, water-soluble and non-toxic sacrificial core was fabricated by fused deposition modeling (FDM), which is one type of 3D printing technique. Step three, bio-polymer materials was coated on the surface of the sacrificial core layer by layer with dip coating process. Step four, NiTi alloy were weaved on the stent, then bio-polymer materials were coated again to make film and metal stent integral. Finally, water-soluble sacrificial core was dissolute in water and customized aortic stent graft was manufactured.

4. Results

4.1. Design of customized aortic stent graft

4.1.1. Compute Tomography of Stanford B Type Patient. The CT image of the 82 years old male patient was provided by the Beijing Huaxin Hospital. According to stratification scan imaging data, the patient's thoracic aorta three-dimensional model was restored by Mimics10.01. After eliminating noise from other tissues and smoothing the surface, the thoracic aorta model is shown as figure 2.

Figure 2. Aorta three-dimensional model restored by Mimics10.01.

4.1.2. Structure Design of customized aortic stent graft. The aortic stent graft in this article is prepared for Stanford B. The diameter of the proximal end of the descending aorta is larger, and the distal end of the descending aorta is smaller. Moreover, due to individual differences, the diameter of the thoracic descending aorta presents irregular changes along the axial direction. Generally, the diameter of descending aorta near the heart is around 20cm, while elders' descending aorta near the heart may increase to 30cm due to loss of blood elasticity.

In order to accurately obtain the size information of the descending aorta, diameter is taken along the axial direction of the aorta by every 5mm, and then the three-dimension model of the stent is established by SolidWorks2012. Meanwhile, to make the stent adaptable and flexible to the structure of blood vessels, "corrugation" is designed to avoid the wrinkle caused by bend of vessels. At the same time, in order to make the stent effectively fixed in the disease area, the diameter of the main part of the stent should be greater than the diameter of the vessel 15%, the port diameter should be greater than 25% of the diameter of the vessel. The stent designed by us is shown as Figure 3.

4.1.3. Sacrificial Core Manufactured by FDM. In this paper, PVA wire with printing temperature of 180-210℃, produced by Shenzhen Esun Industrial Co., Ltd, has been used to be the material of 3D printing. The sacrificial core manufactured by FDM is shown as figure 4.

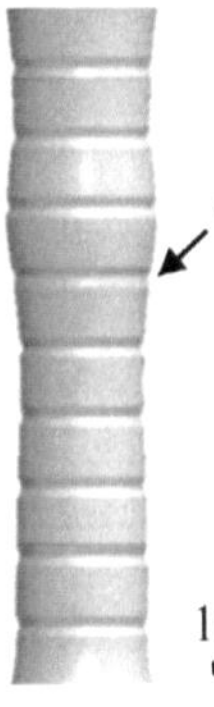

Figure 3. The sacrificial core designed according to the CT images of patient.

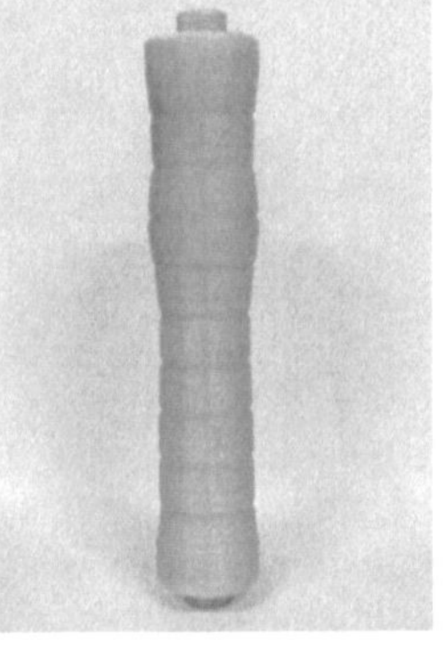

Figure 4. Sacrificial water solubility core manufactured by FDM.

4.2. molding process of film

4.2.1. Biocompatibility test of coated materials. Biocompatibility test is a criterion to evaluate whether a material is suitable to be biological materials. In this paper, hemolysis test and platelet adhesion test have been done to compare the biocompatibility of PU and PTFE.

Hemolysis test is to detect the broken red blood cells in the blood. The aim of the experiment is to test the influence of materials on the red blood cells. The national standard requires that hemolysis rate of the biomaterials to be below 5%. Results are shown as table1.

Although the hemolytic performance of PTFE is slightly better than that of PU, 0.46% of hemolysis rate is still far below the national standard. The hemolysis rate of the two materials are both substantially lower than 5%, which indicate that the damage degree of the two materials on the red blood cells is slight. The result shows that both the two materials have good hemolytic performance.

Table 1. Hemolysis rate of PU and PTFE

Materials	PU	PTFE	National Standard
Hemolytic Ratio（%）	0.46	0.03	5

Platelet adhesion test is to evaluate the blood compatibility of biomaterials. This test is especially important for materials to be implanted in the blood, because the adhesion of platelets on the surface of the material is an important step in the process of thrombosis.

By means of SEM, there is almost no adhesion of platelet on the surface of PU, however, a little platelet adhesion appears on the surface of PTFE. No large amount of aggregation on the surface of the two materials appear, presenting well in platelet adhesion test. As is shown in figure 5.

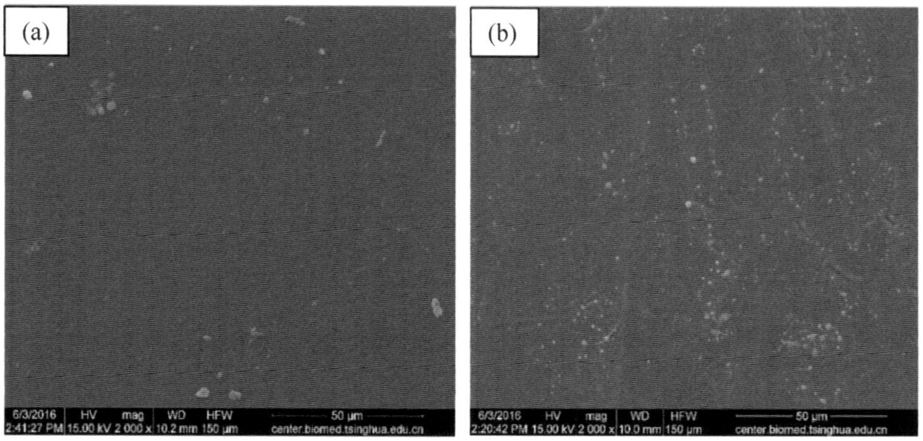

Figure 5. After Platelet adhesion test, surface of PU (a) and PTFE (b) were observed by SEM.

4.2.2. film-making process. The PU film was fabricated by dipping coating technique, which was proposed by Jenaer Glaswerk Schott & Gen in 1939 [11]. For flat substrates, film thickness [12, 13] is affected by the solution concentration, pulling speed, dip coating times etc.

Effect of solution concentration on film thickness. The mass concentrations (w/v) of polyurethane and tetrahydrofuran (THF) solution are 7.5%, 10%, 12.5%, 15%. PVA sacrificial cores with 10mm diameter are dip coated at a pulling speed of 8mm/s. After 5 times dip coating, dissolve the core in water and measure the film thickness. The result shows that the film thickness is exponentially related to the solution concentration, as is shown in figure 6.a.

Effect of pulling speed on film thickness. The PVA sacrificial cores are dip coated at the speed of 2mm/s, 5mm/s, 8mm/s and 10mm/s, respectively. Mass concentration of PU and THF is 12.5%. After 5 times dip coating, measure the film thickness. The film thickness exhibits a logarithmic relationship with the pulling speed, as is shown in figure 6.b.

Effect of dip coating times on film thickness. The PVA sacrificial cores are dip coated at the speed of 8mm/s and the mass concentration is 12.5%. After different dip coating times, dissolve the core and measure the film thickness. The relationship between film thickness and dip coating times presents exponential, as is shown in figure 6.b.

4.2.3. Optimum conditions of dip coating . Under the premise of ensuring film performance, thickness should be reduced as far as possible. Meanwhile, considering the stent should be manufacture in a short time, less dipping coating times and higher solution concentration as well as higher pulling speed should be adopted. However, excessive concentration increases the viscosity of the solution, which is not conducive to the volatilization of the solution. Furthermore, high speed pulling decrease the stability of dip coating. So pulling speed of 8mm/s, mass concentration of 12.5% and 4 times dip coating has been chosen to be the optimum dip coating conditions.

IOP Conf. Series: Materials Science and Engineering **187** (2017) 012027 doi:10.1088/1757-899X/187/1/012027

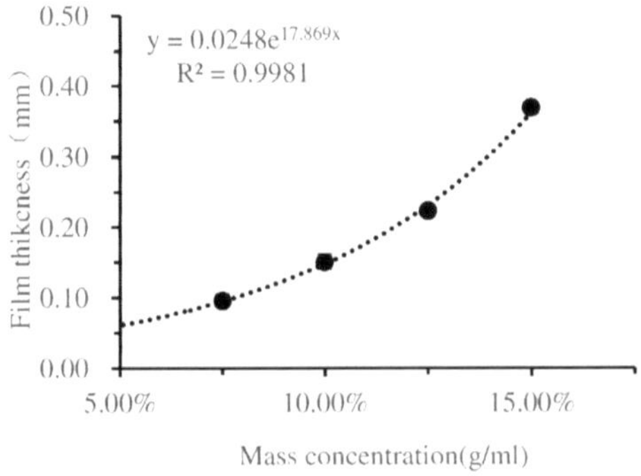

(a)

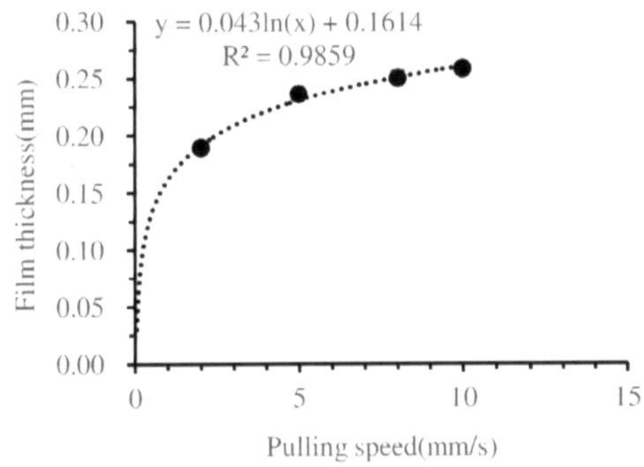

(b)

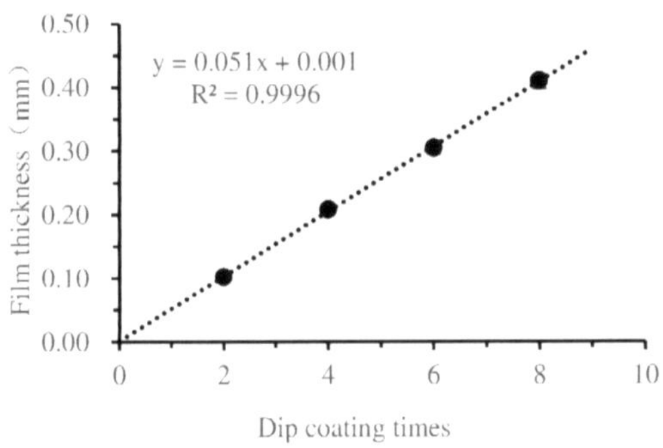

(c)

Figure 6. The influence of mass concentration, pulling speed and dip coating times on film thickness.

4.3. Nickel-titanium alloy and shaping process

4.3.1. shaping process of NiTi alloy. Self-expanding NiTi alloy aortic stent grafts rely on the super-elasticity to unfold at the diseased region, to support the blood vessels and prevent further expansion of aortic aneurysm. The super-elasticity appears when NiTi alloy is austenite. So heat treatment of NiTi alloy mainly has two purposes: on the one hand, shaping NiTi alloy; on the other hand, regulating the austenite transition temperature Af to ensure that NiTi has super-elasticity at body temperature (37℃). Obviously, the heat treatment conditions of NiTi alloys are essential to the performance of aortic stent graft.

4.3.2. Heat treatment conditions of NiTi alloy. In this paper, 0.30mm diameter NiTi alloy wire with austenite phase transition temperature (Af) of 12℃ has been used for the study.

Table 2. Element content of NiTi alloy.

Elements	Ni	C	Co	Cu	Cr	Nb	Fe	H	Ti
Wt %	55.65	0.0043	<0.005	<0.005	<0.005	<0.005	<0.005	<0.001	44.3~44.5

The heat treatment temperature of NiTi alloy is generally around 460~500℃. It's difficult to shape the alloy at low temperature. But excessive aging temperature will cause coarse grain, reducing the mechanical performance of NiTi alloy. Meanwhile, long aging treatment time will increase the austenite phase transition temperature Af. According to the TTT curve [14] of Ti-50.8%Ni alloy, aging treatment condittions of 5min, 10min, 15min, 20min, 25min at 460 C, 480 C and 500 C respectively have been chosen. After that, the stress-strain curve and the phase transition temperature were measured.

4.3.3. Stress-strain cycle curve. Effect of aging time on properties of NiTi alloy. At the same aging temperature, the residual strain increases with the increase of aging time. When the aging time is more than 10mins, the residual strain is more than 3% at all conditions. As is shown in figure7. It shows that NiTi alloy underwent irreversible plastic deformation at a certain extent. As a result, it is not completely recovered the shape before deformation. Therefore, the aging time should not exceed 10min.

4.3.4. Differential Thermal Analysis. Differential scanning calorimetry (DSC) is a thermoanalytical technique which is able to measure phase transition point.

NiTi alloy in the austenitic state has good super-elasticityity, when can provide effective support for blood vessels. So the transition temperature of NiTi should be lower than body temperature 37℃.

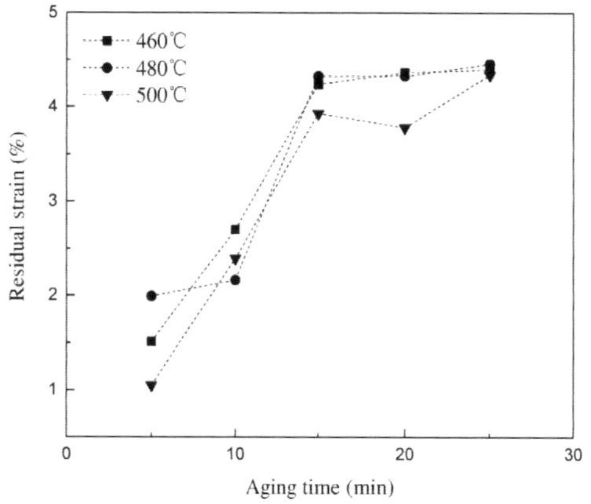

Figure 7. Influence of aging time on Residual Strain.

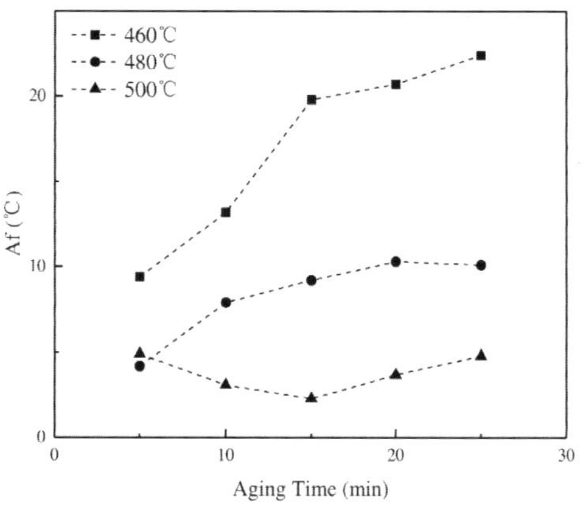

Figure 8. Influence of aging time on Af.

At 460℃ of aging temperature, the transition temperature Af increases rapidly with the increase of aging time. After 25min, the Af is increased from 12℃ to 22.4℃. So aging treatment should be avoided at 460℃. At 480℃ of aging temperature, the Af is also increased with the increase of aging time, but the increase is limited, and Af is controllable. The Af at the aging temperature of 500℃ first decreases then increases, also is controllable. As is shown in figure 8.

The support force of NiTi alloy increases with the decrease of transition temperature Af. NiTi alloys with very low transition temperature have a strong extrapolation, which could injure the blood vessels. Therefore, the transition temperature should be controlled at a reasonable range. So 480℃ is the optimum aging temperature.

4.3.5. Optimum aging treatment conditions of NiTi alloy. Aging treatment temperature should away from 460℃ to avoid rapid increase of transition temperature, and the aging time should not exceed 10min to avoid large residual strain. So aging time should stay within 5~10min at 480℃ or 500℃. But low transition temperature brings strong extrapolation to NiTi alloy. Therefore, 480℃,10min is the optimum aging treatment conditions.

The shaping mold of nickel-titanium alloy is shown as figure 9. Taking into account the thermal inertia of stainless steel mold, the appropriate heat treatment time is increased by 15min. So the optimum aging treatment conditions is 480℃, 25min. After aging treatment, the shape of NiTi alloy is shown in figure 10.

4.4. Aortic stent gragt
The NiTi alloy stent after the shaping treatment was wound on the sacrificible core coated with polyurethane. After twice dipping coating, the sacrificible core was dissolved in water and the customized aortic stent graft remained, as is shown in figure 11.

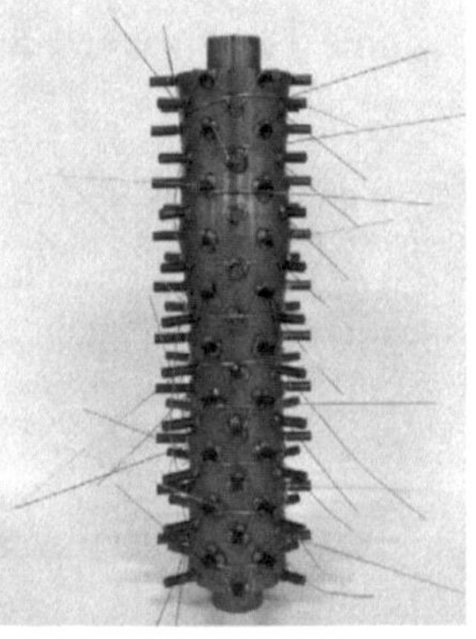

Figure 9. Shaping mold

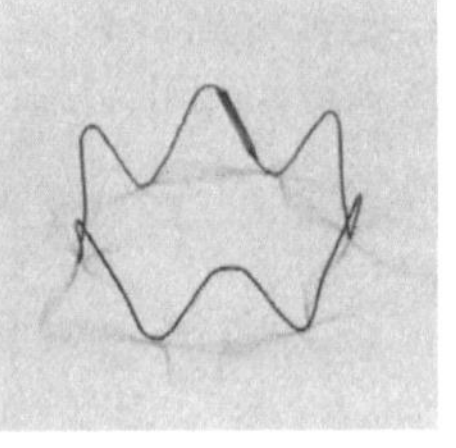

Figure 10. NiTi alloy after aging treatment.

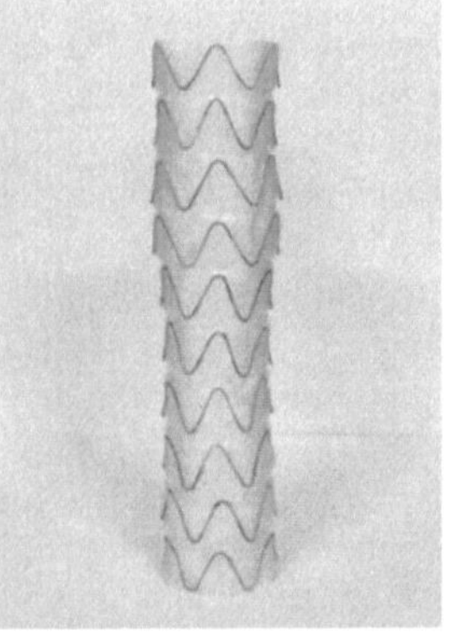

Figure 11. Customized Aortic Stent Graft

4.5. performance test

4.5.1. radial support force. There is no uniform standard to measure the radial support force of the aortic stent. In this paper, bracing force at the moment when the stent is compressed 1/2 diameter, is defined as radial support, as is shown in figure 12. The radial support force of whole stent is 2.1N, and the single ring of the stent is 0.3N. So the stent graft can support an appropriate radical support force.

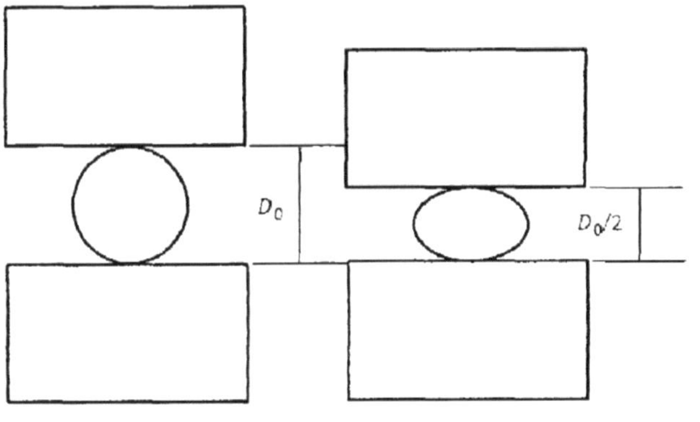

Figure 12. Test method of radial support force

4.5.2. Burst pressure. The endovascular pressure sustained by stent graft is essential to EVAR. Generally, the human normal systolic blood pressure rang is 12.0~18.7kPa, and the diastolic blood pressure range is 8.0~12.0kPa. The burst pressure of customized aortic stent graft is over 100kPa, which is safe enough for implantation into the descending thoracic aorta.

5. Conclusion
In this article, customized aortic stent graft was successfully manufactured by RP sacrificial core – coating forming technique. The stent graft has appropriate radical support force to support the vessels, and enough burst pressure to endure the blood press. Results show that, manufacturing technique of

customized aortic stent graft could meet the demand of personalized therapy, and improve the success rate of EVAR. Moreover, this study lying the foundation for further animal experiment and clinical experiment.

6. References

[1] Anagnostopoulos, C. E.Prabhakar, M. J. & Kittle, C. F. 1972. Aortic dissections and dissecting aneurysms. *The American journal of cardiology*, **30**, 263-273.

[2] Daily, P. O., Trueblood, H. W., Stinson, E. B., Wuerflein, R. D. & Shumway, N. E. 1970. Management of acute aortic dissections. *The Annals of thoracic surgery*, **10**, 237-247.

[3] Xue, L., & Greisler, H. P. 2003. Biomaterials in the development and future of vascular grafts. *Journal of vascular surgery*, **37(2)**, 472-480.

[4] Nerem, R. M., & Seliktar, D. 2001. Vascular tissue engineering. *Annual review of biomedical engineering*, **3(1)**, 225-243.

[5] BAO Junjie, LIU Dubao, LI Bing and XU Gewen. 2007. *Research progress of medical polyurethane materials [J]*, **64**, 72-79.

[6] Cutright, D. E., Bhaskar, S. N., Perez, B., Johnson, R. M. & Cowan, G. S. 1973. Tissue reaction to nitinol wire alloy. *Oral Surgery, Oral Medicine, Oral Pathology*, **35**, 578-584.

[7] Castleman, L. S., Motzkin, S. M., Alicandri, F. P., Bonawit, V. L. & Johnson, A. A. 1976. Biocompatibility of nitinol alloy as an implant material. *Journal of Biomedical Materials Research*, **10**, 695-731.

[8] Holloway, J. L., Spiller, K. L., Lowman, A. M. & Palmese, G. R. 2011. Analysis of the in vitro swelling behavior of poly (vinyl alcohol) hydrogels in osmotic pressure solution for soft tissue replacement. *Acta biomaterialia*, **7**, 2477-2482.

[9] Mellor, S., Hao, L., & Zhang, D. 2014. Additive manufacturing: A framework for implementation. *International Journal of Production Economics*, **149**, 194-201.

[10] ZHANG Lei, 2006, Fabrication and fundamental research of multi-branched blood vessel scaffolds with multi-layered wall[D]. Beijing: Tsinghua University.

[11] Geffcken, W., and Berger, E. 1939. Verfahren zur änderung des reflexionsvermögens optischer gläser. *Deutsches Reichspatent, assigned to Jenaer Glaswerk Schott & Gen., Jena*, **736**, 411.

[12] Landau, L. 2012. Dragging of; 2 Liquid by a Moving Plato. *Dynamics of Curved Fronts*, 141.

[13] GAO Feng and WU Junsen, 2000. Model deduction and verification of coating thickness in pumping process. *Metalware*. **26**, 6-10.

[14] YU Feng and LIAN Cheng, 2004. M, Biomedical nickel titanium alloy, *Science Press.4*

Acknowledgments

The study is supported by Beijing municipal health and Family Planning Commission (2016-2-4121). In this paper, I would like to express my gratitude to Li Lei, Tang Feng and Liu Sibo of Beijing Huaxin hospital. They gave me the guidance about thoracic descending aorta diseases, and provided the blood for the experiments. Meanwhile, thanks Professor Gong Qianming for providing the heat treatment equipment.

IOP Conf. Series: Materials Science and Engineering **187** (2017) 012028 doi:10.1088/1757-899X/187/1/012028

Modeling and Analysis of Non-Uniform Honeycomb Structures Based on Topology Optimization

Xin Jin[1], Guoxi Li[1], Meng Zhang[1] and Jingzhong Gong[2]

[1]College of Mechatronics Engineering and Automation, National University of Defense Technology, Changsha 410073, China

[2]School of Mechanical Engineering, Hunan International Economics University, Changsha 410205, China

anmajin@163.com

Abstract. Honeycomb structure infilling is an important way to achieve lightweight. Focusing on the deficiency of the non-optimized macro-material distribution in uniform honeycomb structures, a modeling method of non-uniform honeycomb structures based on topology optimization was proposed. The loaded component was topology-optimized and the density results were mapped to the relative density matrix of cells. The rapid and automatic modeling of non-uniform honeycomb structures was realized with the using of User-Defined Features and the cyclic definitions of the reference datums. The simulation results show that the mechanic performance of non-uniform honeycomb structures is better than uniform honeycomb structures and the efficiency of the proposed method is validated.

1. Introduction

In the structural design of aerospace, rail transportation and other industry products, lightweight can not only reduce energy consumption in the production and operation, but also bring higher manoeuvrability and better dynamic and static performance, which are key indicators related to the product competitiveness. In recent years, with the rapid development of material preparation and forming technology, honeycomb structure emerges as a kind of novel multifunctional lightweight structure [1]. It exhibits properties of high strength and rigidity except for weight reduction, realize the shock resistance [2], heat transfer and insulation [3], sound absorption [4], and meet the requirements of multifunction [5]. Traditional honeycomb structure is a periodic structure, with the relative density distribution uniform, and the preparation technologies mature [6]. Although the microstructure is optimized on the cell scale, the uniform structure of the macroscale is not optimized. The uniform distribution of the structure doesn't considers the optimal load transfer path; the lightweight efficiency is not sufficient.

Topology optimization is an important way to achieve lightweight when satisfying performance. The essence of this method is to seek the structural layout with the optimal load transfer path under the action of external force and constraint [7]. Topology optimization plays a significant role in the structural design of automobile, aerospace, industrial equipment and other fields [8, 9]. At present, the most mature commercial application is solid isotropic material penalty model (SIMP) proposed by Rozvany [10]. The density of each element in the design space is used as the design variable and the optimal design result is the density distribution between 1 and 0. We can map the optimal density distribution of topology optimization results to the relative density of honeycomb cells, resulting in non-uniform honeycomb structure, and realize the optimization design in the macro-scale and meso-scale, to improve the efficiency of lightweight.

There are two ways to model the honeycomb structure. First method is to use CAD platform for interactive modeling, and the problem is that: due to the non-uniform characteristics of the honeycomb cells, it is not easy to directly model with periodic operations of array or mirror. The manual modeling has the difficulties of high cost. Second method is to use rapid prototyping processing software, such as Magics. This kind of software is to fill the model with fixed units based on STL format. The main problem of the scheme is that: it only provides function of uniform-cell filling. Modeling object is STL format, which is not conducive to the modification of structures. If the structure is generated with defects, it must return to the CAD software. Therefore, it is important to develop the auxiliary application for CAD software by programming to realize rapid and automatic modeling.

2. Structural topology optimization

For the topology optimization of nonuniform honeycomb structure, the reference structure is the honeycomb structure with uniform density. The maximum stiffness at the same weight reduction is achieved by the topology optimization. It adapts to the general definition of the minimum compliance problem [11]: under the given material volume constraint, the optimal distribution of the material in the design domain is sought, so that achieve the structural topology optimization of the maximum stiffness. Therefore, the minimum compliance optimization model is established, the relative density of elements is used as the design variable, and the maximum stiffness is obtained under the constraint of volume.

The expression for the compliance is given as:

$$C = F^T U = U^T K U \tag{1}$$

where K is the overall stiffness matrix, which can be expressed as a function of element density. U is the structural displacement matrix. F is the external force. Constraints include volume constraint, element density range constraints, and other constraints. The volume constraint represents that material distribution is arranged in a fixed volume fraction the same as the uniform honeycomb structure. Element density range is $x_{e\min} \le x_e \le x_{e\max}$. Taking into account that extremely low density means the cell wall is too thin and extremely high density means the cell center hole is too small, which is both not conducive to manufacturing, so the minimum density and maximum density is given as 0.2 and 0.9. Other constraints include other requirements to be considered in the design and manufacturing, and they are expressed in $g(x_e) \le 0$. The mathematical model of topology optimization is obtained:

$$\begin{aligned} &\min C(x_e) = F^T U = U^T K U = \sum_{i=1}^{m} u_i^T k_i^{(e)} u_i \\ &s.t.\ F = KU; \\ &\quad K = K(D(x_e)) = K(x_e); \\ &\quad \sum_{i=1}^{m} x_{ei} \cdot v_i = V; \\ &\quad x_{e\min} \le x_e \le x_{e\max}; \\ &\quad g(x_e) \le 0; \end{aligned} \tag{2}$$

Sensitivity analysis is carried out as the key step to solve the optimization problem. The sensitivity of the objective function to the design variables is obtained [11]:

$$\frac{\partial C(x_e)}{\partial x_e} = -U^T \frac{\partial K(x_e)}{\partial x_e} U \tag{3}$$

Based on the sensitivity of the objective function, the OC algorithm is used to update the design variables. The iteration formula for the design variables based on the optimization criterion method is given [12]:

$$x_e^{k+1} = \begin{cases} \max(x_{e\min}, x_e - m) & if \left(B_e^k\right)^\eta x_e^k \le \max(x_{e\min}, x_e - m) \\ \min(x_{e\max}, x_e + m) & if \ \min(x_{e\max}, x_e + m) \le \left(B_e^k\right)^\eta x_e^k \\ \left(B_e^k\right)^\eta x_e^k & \text{otherwise} \end{cases} \quad (4)$$

where k is used to represent the number of iterations, $\eta = 0.5$ is the damping coefficient to ensure the stability and convergence of the numerical calculation, and $m = 0.2$ to limit the maximum variation of the density variable for each iteration. B_e^k is the optimal criterion:

$$B_e^k = -\frac{\partial C}{\partial x_e} / \lambda \frac{\partial V}{\partial x_e} \quad (5)$$

where Lagrange multiplier λ must satisfy the volume constraints in the optimization model, which can be searched by bisection method. After the determination of optimal criterion, design variables start iteration until convergence.

3. Cellular density mapping

Aiming to the specific object as shown in figure 1:

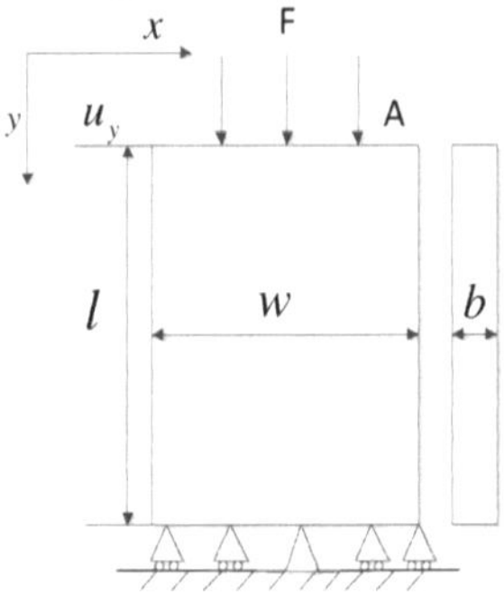

Figure 1. Geometry object ready for design.

It is a solid plate design domain with $l = 120mm$, $w = 60mm$ and $b = 10mm$ under pressure $F = 60N$ on the top. Because the load is on the direction of the cell wall, refer to the mechanical properties analysis carried out by Wang [13] for the typical honeycomb structures including hexagonal honeycomb, square honeycomb and triangle honeycomb, square honeycomb is choose to construct nonuniform honeycomb structure. Based on the elements density matrix obtained by the optimal solution, the densities of the finite elements are calculated by averaging, and the density distribution matrix of honeycomb cells is mapped out. The specific process, as shown in figure 2, is divided into the following three steps:

(1) Determination of some parameters. The length of the design domain is *length*, the width of the design domain is *width*, the dimension of one single cell is a, the dimension of one single element is r, and the density distribution matrix of elements is x_e.

(2) Determination of the arrangement of cells. With the origin O as the starting point for arrangement, the number of cells in x direction is $m = ceil(width / a)$, and y direction is $n = ceil(length / a)$.

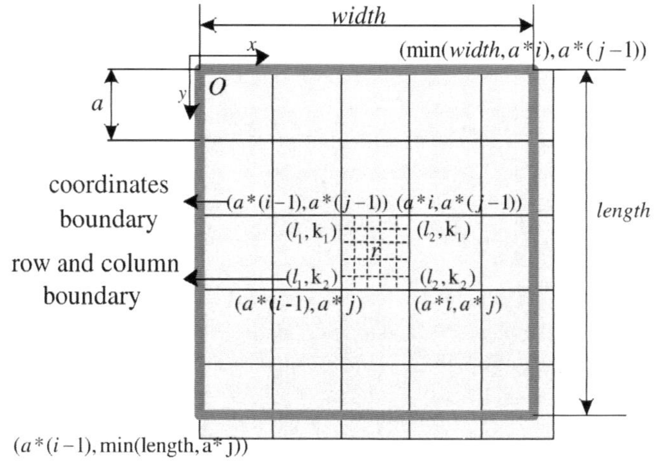

Figure 2. Illustration of map from elements to cells.

(3) Solution for the relative density matrix of cells. The boundary coordinates of the cell in row j and column i are calculated as:

$$\begin{aligned} x_1 &= a*(i-1), x_2 = \min(width, a*i), i = 1,2,\cdots,m \\ y_1 &= a*(j-1), y_2 = \min(length, a*j), j = 1,2,\cdots,n \end{aligned} \tag{6}$$

The number of rows and columns the cell locates in the density matrix of elements are given as:

$$\begin{aligned} &row\ boundary: k_1 = y_1 / r + 1, k_2 = y_2 / r \\ &column\ boundary: l_1 = x_1 / r + 1, l_2 = x_2 / r \end{aligned} \tag{7}$$

Then the relative density of the cell can be calculated with average method in the boundary of rows and columns:

$$x(j,i) = \frac{\sum_{l=l_1}^{l_2} \sum_{k=k_1}^{k_2} x_e(k,l)}{(l_2 - l_1 + 1) * (k_2 - k_1 + 1)} \tag{8}$$

4. Nonuniform honeycomb modeling

With the arrangement and density matrix of cells as input, modeling of nonuniform honeycomb structure is carried out. The PTC Creo is choose as structural modeling CAD software. User-Defined Feature (UDF) is used to simplify the modeling process. With the support of Pro/Toolkit library of Creo, we can develop the automatic modeling auxiliary application for nonuniform honeycomb structure.

The general flow of nonuniform honeycomb structure modeling is shown in figure 3. Firstly, we need to establish parametric model and references of the honeycomb cell features, and so on to complete the creation of cell UDF. The schematic diagram of square cell UDF is shown in figure 4. The key parameters and references in the diagram are shown in table 1,which constitute a parameterized UDF

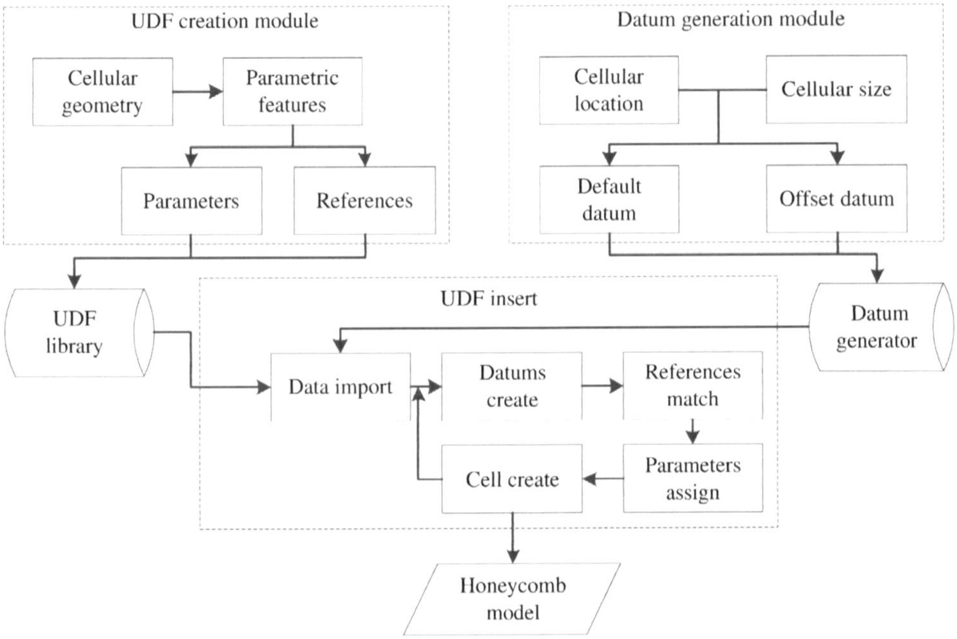

Figure 3. Flow chart of non-uniform honeycomb structures modeling.

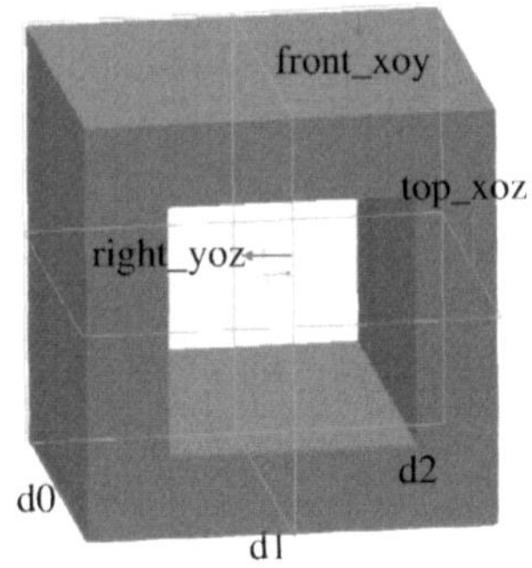

Figure 4. UDF of square cells.

Table 1. Parameters and references of square cells.

Parameters	Description	References	Description
d0	Out-of-plane thickness	front_xoy	Front view datum plane
d1	Cell length	right_yoz	Right view datum plane
d2	Cell hole dimension	top_xoz	Top view datum plane

The relationship between cell size parameters and cell density is satisfied as below, which can determine the basic parameters of each cell:

$$\frac{d1^2 - d2^2}{d1^2} = x \Rightarrow d2 = d1 * \sqrt{1-x} \tag{9}$$

Secondly, the function of generating datums with location and size information is encapsulated as datums generation module. The type of datums includes default datum plane and offset datum plane. Default datum plane is the standard orthogonal plane in the default coordinate system, which is the reference of the initial cell. Offset datum planes are referred by other cells. Based on the default datum, offset value and arrangement number are determined, where offset value is the cell length d1,and arrangement of the number in x and y direction is specified in Sec 3.

Finally, combine cell UDF with datums generation module, and import data of design domain size, cellular dimensions and cellular density matrix. With these data, datums can be generated, cellular references can be matched and values of cell parameters are assigned, so as to circular definite location and geometry of each cell. And nonuniform honeycomb CAD model is obtained.

5. Results and discussion

The object is shown in figure 1 and the Young's modulus and Poisson's ratio of cell wall material are $E = 7.2e10Pa$ and $\nu = 0.3$. Optimization results and nonuniform honeycomb structure model are shown in figure 5 and figure 6. The model via density mapping and CAD modeling is complete without defects, and it is consistent with the density distribution of topology optimization.

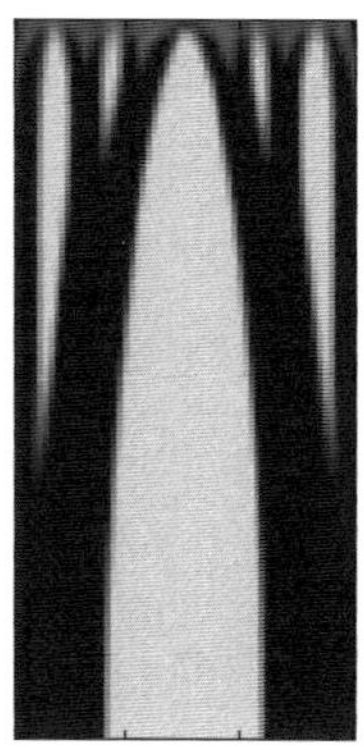

Figure 5. Optimization results.

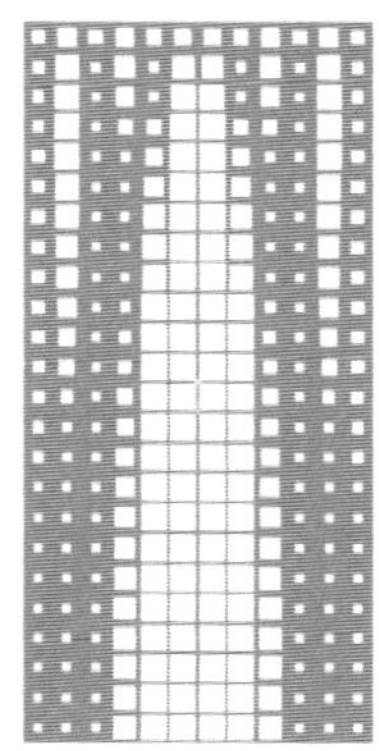

Figure 6. Honeycomb structure.

With ANSYS as the simulation tools, the compare of compressive stiffness of uniform and nonuniform honeycomb structures is carried out by FEA simulation. Compressive stiffness can be expressed as $k = F / u_y$, where u_y is the average displacement in y direction on the top, and F is the pressure. Deformations of uniform and nonuniform honeycomb structures are shown in figure 7 and figure 8.

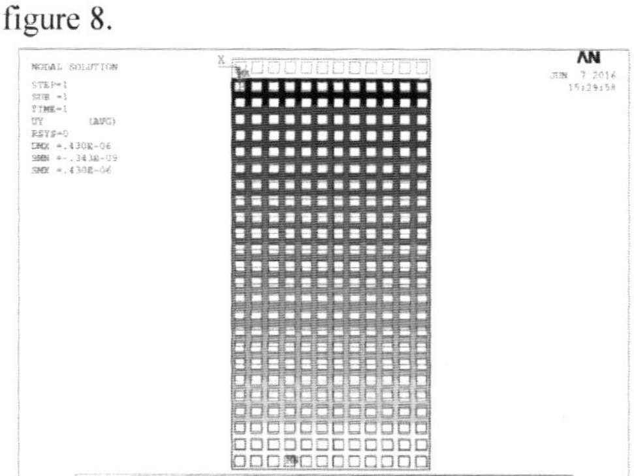

Figure 7. Deformation of uniform honeycomb.

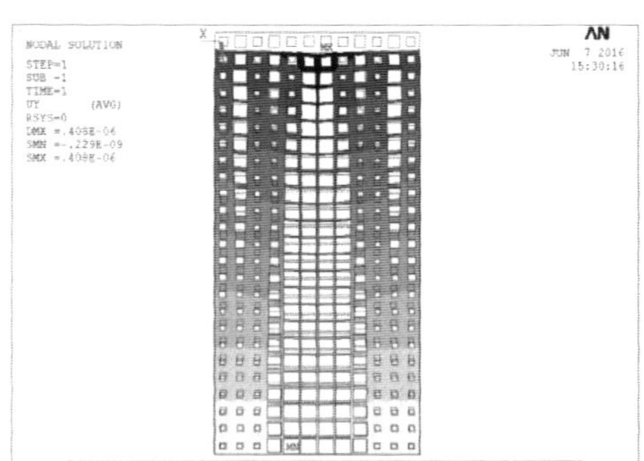

Figure 8. Deformation of nonuniform honeycomb.

Compressive stiffness of uniform honeycomb structure is calculated as $145197N / mm$, and nonuniform honeycomb structure is $164383N / mm$. At the same weight reduction, stiffness of optimized nonuniform honeycomb structure is higher than uniform structure and the extent is 13.2%. The efficiency of the proposed modeling method is validated by FEA simulation.

6. Conclusion

(1)In this paper, a modeling method of non-uniform honeycomb structures based on topology optimization is proposed. Loaded structure is topology optimized, and the optimized density results are mapped to honeycomb cellular relative density distribution matrix. The rapid and automatic modeling of non-uniform honeycomb structures is realized with the using of User-Defined Features and the cyclic definitions of the reference datums.

(2) After density mapping and CAD modeling, the non-uniform honeycomb structure model is complete with no defects, which can reflect the density distribution of topology optimization well.

(3) Paper analyzes and compares the compressive stiffness of uniform honeycomb structure and non-uniform honeycomb structure. The results show that performance of optimized non-uniform honeycomb structure is better, and the effectiveness of uniform honeycomb structure modeling method based on topology optimization is verified.

7. References

[1] Wadley H N G 2006 Multifunctional periodic cellular metals. *Phil Trans R Soc A*. 364 31-68

[2] Elnasri I, Pattofatto S, et al 2007 Shock enhancement of cellular structures under impact loading: Part I Experiments. *J Mech Phys Solids*. 55(12) 2652-2671

[3] Tian J, Kim T, Lu T J, et al 2004 The effects of topology upon fluid-flow and heat-transfer within cellular copper structures. *Int J Heat Mass Transf.* 47 3171-3186

[4] Wang J, Lu T J, Woodhouse J, et al 2005 Sound transmission through lightweight double-leaf partitions: theoretical modeling. *J Sound Vibr*. 286 817-847

[5] Fan H L, Jin F N, Fang D N 2008 Mechanical properties of hierarchical cellular materials: Part I Analysis. *Compos Sci Technol*. 68 3380-3387

[6] Sundarram SS, Jiang W, Li W 2014 Fabrication of small pore-size nickel foams using electroless plating of solid-state foamed immiscible polymer blends. *J Manuf Sci Eng*. 136 021002

[7] Yan J, Guo X, Cheng G 2016 Multi-scale concurrent material and structural design under mechanical and thermal loads. *Comput Mech*. 57 437-446

[8] Gil Ho Yoon 2010Maximizing the fundamental eigenfrequency of geometrically nonlinear structures by topology optimization based on element connectivity parameterization. *Computers and Structures*. 88 120-133

[9] Soren Halkjær, Ole Sigmund, Jakob S Jensen 2006 Maximizing band gaps in plate structures. *Struct Multidisc Optim*. 32 263-275

[10] Rozvany GIN, Zhou M, Birker T 1992 Generalized shape optimization without homogenization. *Struct Optim*. 4 250-254

[11] Bendsoe M P 2003 *Topology Optimization: Theory, Methods and Applications* (Berlin: Springer) pp 9-24

[12] Erik Andreassen, Anders Clausen, et al 2011 Efficient topology optimization in MATLAB using 88 lines of code. *Struct. Multidiscip. Optim*. 43(1) 1–16

[13] Wang A J, McDowell D L 2004 In-plane stiffness and yield strength of periodic metal honeycombs. *ASME J Eng Mater Technol*. 126 137-156

IOP Conf. Series: Materials Science and Engineering **187** (2017) 012029 doi:10.1088/1757-899X/187/1/012029

Simulation of Temperature Field Distribution for Cutting the Temperated Glass by Ultraviolet Laser

B J Yang[1, *], **Y C He**[1], **F Dai** [2] **and X C Lin**[3]

[1]School of Mechatronic Engineering, Wuyi University, Jiangmen, Guangdong, China;
[2]School of Applied Physics & Materials, Wuyi University, Jiangmen, Guangdong, China;
[3]Zhongshan Hantong Laser Science Co., LTD., Zhongshan, Guangdong, China

[*] kurt.yang@163.com

Abstract. The finite element software ANSYS was adopted to simulate the temperature field distribution for laser cutting tempered glass, and the influence of different process parameters, including laser power, glass thickness and cutting speed, on temperature field distribution was studied in detail. The results show that the laser power has a greater influence on temperature field distribution than other paremeters, and when the laser power gets to 60W, the highest temperature reaches 749℃, which is higher than the glass softening temperature. It reflects the material near the laser spot is melted and the molten slag is removed by the high-energy water beam quickly. Finally, through the water guided laser cutting tempered glass experiment the FEM theoretical analysis was verified.

1. Introdution

Tempered glass is a kind of special glass formed by quenching after heating of flat glass. Due to its high bending resistance, high bearing capacity, no acute rupture and anti-thermal break ability[1,2], it is widely used in aerospace vehicles, weapons & equipment, automotive windshield, electronic products screen, high-rise buildings, doors, windows, furniture and so on[3,4]. However, during the process of being cut by laser, it is very likely to form the great uneven thermal stress in the cutting area on the tempered glass owing to its high hardness and brittleness, which will cause the crack or the fracture ultimately[5,6]. Salman Nisar [7] studied the influence of laser cutting speed on the stress by the finite element software ABAQUS, and the cutting path was also optimized. Tsai Chwan-Huei [8] used the thermal stress to induce the material to separate along the cutting direction while cutting LCD glass by the CO_2 laser, and the formation mechanism of crack was analysed. Wang Xuhuang [9] from Zhejiang University of Technology used CO_2 laser to initialize the initial crack in the liquid crystal glass substrate, and then heated by CO_2 laser as a heating resource and cooled by Ar gas.

In the paper, the finite element software ANSYS was adopted to simulate the 3D temperature field on the tempered glass for laser cutting, and then the influence of the different process parameters, including laser power, glass thickness and cutting speed, on the cutting quality were analysed, which provided a solid theoretical basis for the experiment.

2. Simulation of the temperature field

2.1. Basic assumption

Before the numerical simulation of temperature field, the following assumptions were made[10,11]:

(1) The laser energy is evenly distributed throughout the laser beam, and the energy of the laser beam is only absorbed by the surface layer. Thus, the laser is the surface heat source.

(2) The temperature field is a continuous unsteady heat conduction temperature field.

(3) Considering the reflective properties and brittle hardness of the tempered glass, the average values of the corresponding temperature range are adopted for numerical analysis.

(4) The heat flux in the laser processing period is a constant.

2.2. Physical modeling and meshing

According to characteristics of the laser forming process, a simplefied 3D geometric model was generated. The size of the model is 10 mm x 10 mm, and the physical dimension also assure that the model boundary condition has no influence on temperature calculation. The thickness is equal to that of the tempered glass in the experiment. In the processing area, the grid remeshing technique was adopted to reduce the mesh distortion caused by thermal deformation so as to improve the accuracy of simulation calculation. However, the other majority part of the model was grided nomally in order to improve the computation efficiency. The focused laser beam was considered to be an ideal Gauss Distribution and moved along the X axis at a speed of V. I(x,t) was applied to simulate the absorption of laser energy on the surface of the workpiece, and it is converted into heat energy.

In the paper, the grid meshing was adjusted according to the idea above, as shown in figure 1. The minimum length of the element is setted to 0.2 mm for the processing region, and the maximum length of the element is setted to 1 mm. Finally, the amount of the elements was reduce to 19440 successfully by manual method.

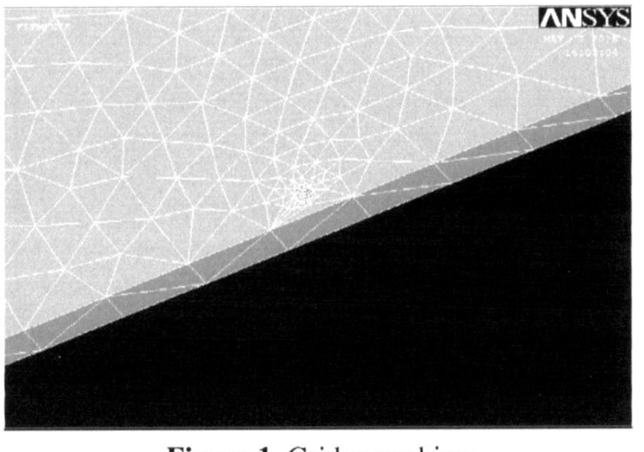

Figure 1. Grid remeshing

2.3. Simulation results

2.3.1. Influence of laser power on temperature field distribution. In this situlation, the glass thickness was selected as 0.5mm, the cutting speed V was setted to 1mm/s, and the laser power was selected as 40W, 50W and 60W successively for the following simulations.

Figure 2(a), (b) and (c) show the temperature field distribution at the same time(2 second) by different laser power 40W, 50W and 60W. It can be seen from the figure 2 that with the increase of the laser power, the temperature of the core cutting area on the surface of the tempered glass is gradually increased. When the laser power is 40W, the maximum temperature of the tempered glass is 528℃, which is lower than the softening temperature of the tempered glass. However, while the laser power reaches 50W, the temperature of laser spot center on tempered glass surface is reaching 623℃, which gets into the softening temperature range of the tempered glass. So it indicates that the material in laser spot center has already partially been melted. Nevertheless, away from the laser spot center, the temperature drops sharply until it gets to the room temperature. More importantly, when the laser power gets to 60W, the highest temperature on the tempered glass surface reaches 749℃, which is well above the softening temperature of the tempered glass. It indicates that the molten pool has already formed in laser spot center, and the material in the molten pool is quickly removed by the high energy water beam. So, 60W laser power is fitable for cutting the tempered glass.

Therefore, by controlling the laser power, the heating temperature on the surface of the tempered glass can be effectively controlled, and the direction of the crack on the tempered glass can also be controlled by changing the temperature field distribution.

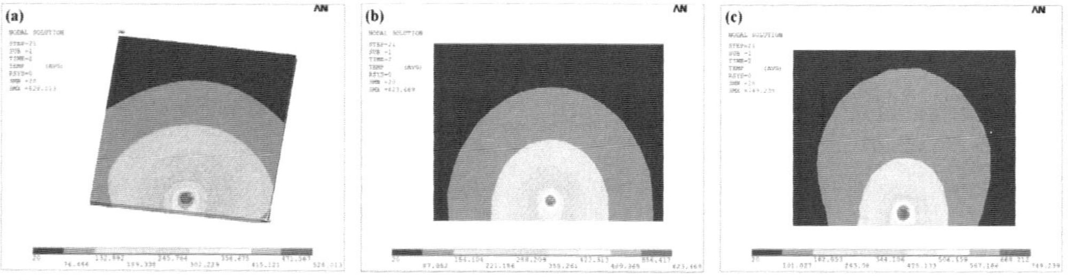

Figure 2. Influence of different laser power on the temperature field distribution (a)40W, (b)50W, (c)60W

2.3.2. Influence of glass thickness on temperature field distribution. For this section, the laser power was setted to 60W, the cutting speed V was setted to 1mm/s, and the glass thickness were selected as 0.5mm, 1.0mm and 1.5mm for the following simulations.

As shown in figure 3, the temperature decreases along the glass thickness direction from the upper surface to the back surface with the continuous laser irradiation, and the maximum value is at the upper surface. In addition, the datas show the thickness has a significant influence on the temperature distribution along the thickness direction. The thicker the tempered glass is, the wider the kerf becomes, and it is harder to be cut off due to the molten pool is fixed with a certain laser power (shown in figure 3). So, 0.5mm thickness is a better choice.

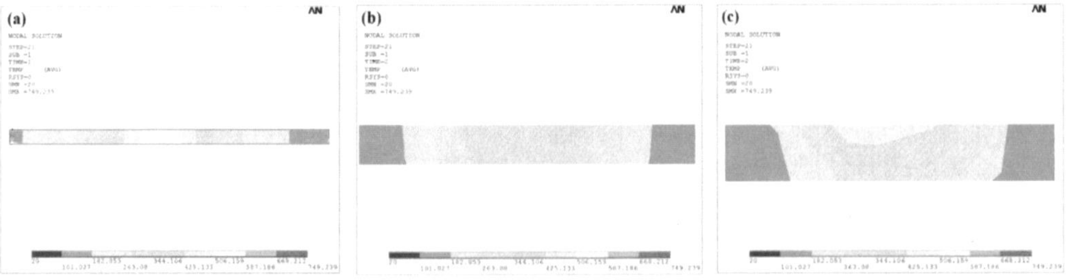

Figure 3. Influence of the glass thickness on temperature field distribution(a) t=0.5mm, (b) t=1.0mm, (c) t=1.5mm

2.3.3. Influence of cutting speed on temperature field distribution. In this part, the laser power was selected as 60W, the glass thickness was selected as 0.5mm, and the cutting speed v were setted to 2mm/s, 4mm/s and 6mm/s respectively for the following simulations.

It is clear from figure 4 that the cutting speed is rather slow, so the time for laser working on the tempered glass gets longer. Thus it makes the effective spot area become larger which directly leads to a more wider kerf on the glass. Moreover, over melting reaction happens in the laser precessing area because the cutting speed is slower than that of material melting speed, and it will also cause a irregular and wider kerf.

However, in figure 4, cutting speed of 6mm/s is a little bit faster, thus the laser beam irradiates on the glass only in a relatively short time, so the laser spot runs over the material ahead but the material has not been well melt yet, which leads to a bad surface roughness, and some slag hanging on the kerf. So, v=4mm/s is suitable.

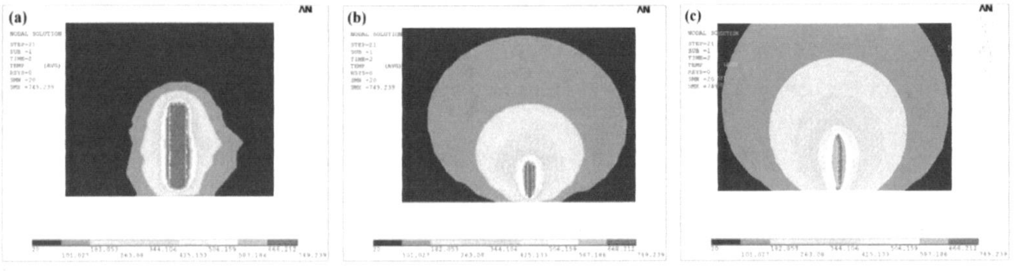

Figure 4. Influence of cutting speed on temperature field distribution (a)v=2mm/s, (b)v=4mm/s, (c)v=6mm/s

4. Experimental Analysis

The laser cutting without transversal crack for the tempered glass experiment was conducted on HT-3P laser processing system (see figure 5) equipped with YPP300 fiber laser and the coupling device of micro water beam and laser beam.

Experimental results show the laser power has a greater influence on temperature field distribution than other parameters and the cutting surface of specimen with 0.5 mm thickness is smooth(see figure 6) while laser power is 60W, cutting speed is 4mm/s, as shown in figure 6. And there is no micro crack near the cutting area. The cutting kerf`s width is about 100 um, as represented in figure 7.

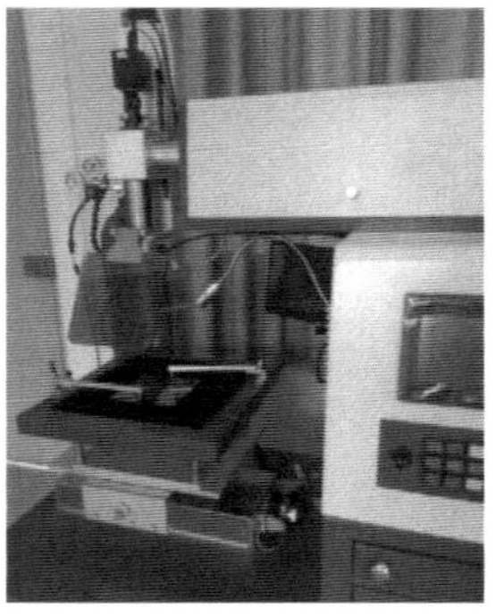

Figure 5. HT-3P laser processing system

Figure 6. Sample(t=0.5mm)

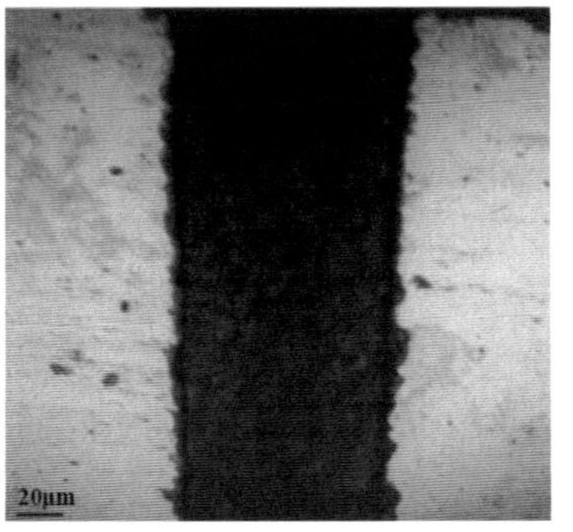

Figure 7. Miroc morphology of the cutting kerf

5. Conclusion

(1) The glass surface temperature gradually increases as the laser power increases. When the laser power get to 60W, the highest temperature reaches 749℃, it reflects the material near the laser spot is melted and the molten slag is removed by the high-energy water beam.
(2) The cutting surface of specimen with 0.5mm thickness is smooth, and no transversal crack appears near the cutting area. The cutting kerf`s width is about 100 um.

6. References

[1] Shao G Z, Wang Q S, Zhao H. Thermal Breakage of Tempered Glass Facade with Down-Flowing Water Film Under Different Heating Rates[J]. Fire Technology, 2016, 52(5), p563-580

[2] Chen J SH, Zhu L Y, Lou X CH. Method discussion of glass tempering[J].Material Review,2012,26(20), p135-138

[3] Chen F , An B J, Yang B ZH. Chemical tempering and Performance Analysis of Thin Glass used for Display panel[J].Glass,2012, 23(5), p45-47

[4] Song L X, Hu X F. Effects of thermal tempering on strenth of window glass of manned spacecraft[J], Chinese Space Science and Technology,1996, 27(4), p43-49

[5] Nielsen J H, Olesen J F, Poulsen P. N., et al. Simulation of residual stresses at holes in tempered glass: a parametric study[J]. Materials and Structures,2010,43(10), p947-961

[6] Roberto D, Ricardo J Z. Analytical model of dynamic crack evolution in tempered and strengthened glass plates[J]. International Journal of Fracture, 2014,190(10), p75-86

[7] Salman N, Sheikh M A, Li L, et al. Effect of Thermal Stresses on Chip-free Diode Laser Cutting of Glass[J]. Optics and Laser Technology. 2009, 41(5), p318-327

[8] Tsai C H, Lin B CH. Laser cutting with controlled crack and pre-bending applied to LCD glass separation[J]. The International Journal of Advanced Manufacturing Technology. 2007,32(12), p11-13

[9] Wang X H, Yao J H, Zhou G B, et al. Research of the technology of laser cutting LCD glass substrates based on thermasl cracking method[J]. Laser technology.2011, 35(4), p472-476

[10] Nielsen J H, Olesen J F, Poulsen P N, Stang H. Simulation of residual stresses at holes in

tempered glass:a parametric study[J]. Materials and Structures, 2010, 43, p947–961
[11] Sheng X J. The simulation and experimental analysis of laser-controlled thermal stress cutting of alumina ceramic[D]. Shanghai, Shanghai Jiaotong University, 2010, p17-19

Acknowledgement
This paper was supported by Innovative Young Talents Project of Guangdong Province (2014KQNCX160), and efforts from Caculation Center of South China University of Technology was also appreciated.

Numerical Study of a Novel Piston-type Synthetic Jet Actuator with a Quick-return Characteristic

Zheng Liu[1], Eriqitai[2] and Liang Hong[2]

[1] Corresponding author, School of Energy and Power Engineering, Beihang University, Beijing, College Road 37#, PRC

[2] School of Energy and Power Engineering, Beihang University, Beijing, College Road 37#, PRC

wangtaoliuzheng@126.com

Abstract. A novel designed piston-type synthetic jet actuator has been developed, which has a quick-return characteristic. When moving to the Top Dead Centre, the piston in high speed route moves faster than the classical one. Compared with classical piston-type SJA, the pressure ratio of the cavity, jet velocity and the jet momentum of the novel SJA increase significantly, which greatly enhances the performance of the actuator. A parametric study has been carried out focusing on the affection in different actuation frequencies and duty cycle. Results show that the performance of the quick-return piston-type SJA is significantly improved.

1. Introduction

Recently, a considerable number of investigations have been carried out in the area of active flow control techniques, and it is a potential field which is beneficial to aerospace, transportation and other fields [1-2]. Synthetic jets have attracted great research interest for the active flow control applications, and the synthetic jet is a jet-like fluid motion formed by the time-periodic, alternative ejection and suction of the fluid through a small orifice bounding a flexible volume cavity. The main structure of the actuator is shown in Figure 1.

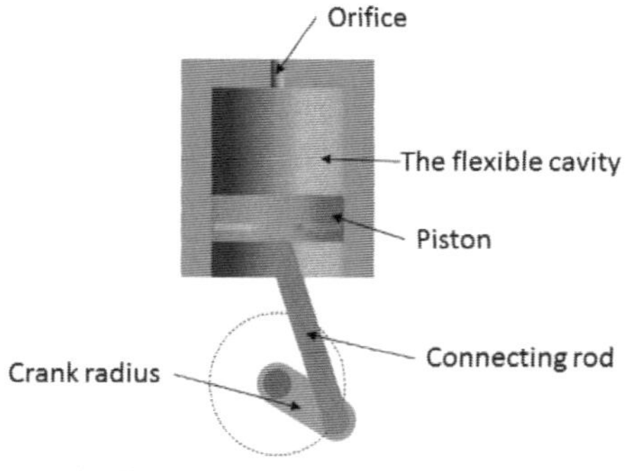

Figure 1. Schematic diagram of synthetic jet actuator driven by piston.

Due to the nature of the zero-net-mass-flux, the synthetic jet is considered as a positive approaching method to the flow control application, involving jet vectoring, flow separation controlling, and boundary layer controlling.

Considerable research has been covered on piston-type SJAs. Georgia Institute of Technology has designed a synthetic jet actuator [3], with the piston diameter 21.2mm, the route 18.4mm, and the volume of the cavity 6.49 cm^3. The max compression ratio was 27.1, and the actuator was set in a normal level of rotate frequency. A conclusion has been drawn that the max velocity of the jet was supersonic in both ejection process and suction process. With the enhancement of the frequency of the piston, the time costed in ejection would be less than that in low level frequency, and the maximum pressure of the cavity would be increased with the minimum of the pressure decreased, the appearance of pressure peak would be put off, which was closer to the t/T=0.5 moment. So, the actuator had to spend more time to breathe more air. The higher the piston frequency was, the more obvious the phenomenon would be. Under the circumstance, the compression ratio (r) was 27.1 with the frequency of the piston 200Hz, the maximum of the pressure up to 6.24atm, the minimum of the pressure 0.22atm, and the peak time t/T=0.5 moment.

From the Figure 2, we could find that the process of suction (the shadow section) cost 71% of a cycle time, leaving about 29% time of the cycle as the ejection process. The reason why the suction time was enlarged and the ejection time was decreased in one period was that the suction was forced by the environment atmosphere's pressure, and the fluid flow wouldn't increase after reaching the critical value. However, ejection was forced by the pressure of the cavity. As a result, the system needed more time to inhale air to meet the requirement of no-mass jet.

Zhang has designed a new piston SJA [4], with the frequency 202Hz, the diameter of the piston 19.5mm's, the 18.6's pressure ration, the maximum pressure was up to the 1.987atm. And with the development of the frequency, the jet velocity would increase, the maximum jet pressure would enlarge in respect with the minimum of the jet pressure decreasing at the same time.

The recently research mostly focus on the performance, including the max jet velocity, the max pressure in the cavity, the total jet mass flow, and the momentum of the jet. Unfortunately, as the things going like the Georgia's, under the high level of the frequency, the jet actuator can't inhale too much air, which doesn't match the mechanical or acoustic resonance phenomenon to get the optimal performance, which is an ideal synthetic jet generator.

This paper proposes a modified mechanical structure, which can enhance the piston's velocity in the ejection process, and the time in outstroke is predicted to be decreased, by putting the piston pin offside the middle line, and shortening the connecting rod length to enhance the quick-return characteristic, aiming at making a significant effect on the ejection velocity and alleviating the shortage of inhaling. The flow-fields are studied by numerical simulation, the effects of actuation frequency and the extent of the quick-return characteristic are also analyzed.

2. Eccentric piston SJA model

2.1 Principle of the quick-return piston-type SJA

The classical piston-type SJA is made up by a crank, a connecting rod, a piston, a cylinder liner, and a jet orifice with the piston pin standing offside the line along which the piston can move away and back. An equation can simulate the motion, (2.1).

$$P_S = \left((r_c + L)^2 - x^2_{offset}\right)^{1/2} - r_c \cos(\theta_c) - \left(L^2 - (r_c \sin(\theta_c) + x_{offset})^2\right)^{1/2} \quad (2.1)$$

Where P_s is the piston location, r_c is the crank radius, L is the connecting rod length, X_{offset} is the piston pin offset, and θ_c is the current crank angle. Here is the draft model in Figure 3.

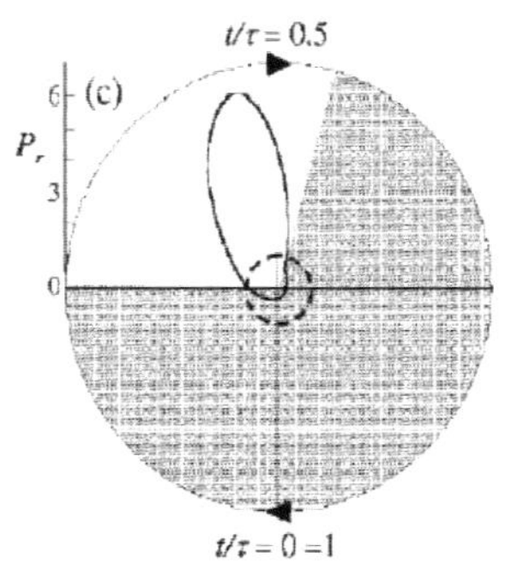

Figure 2. r=27.1, f=200Hz, draft in polar curve.

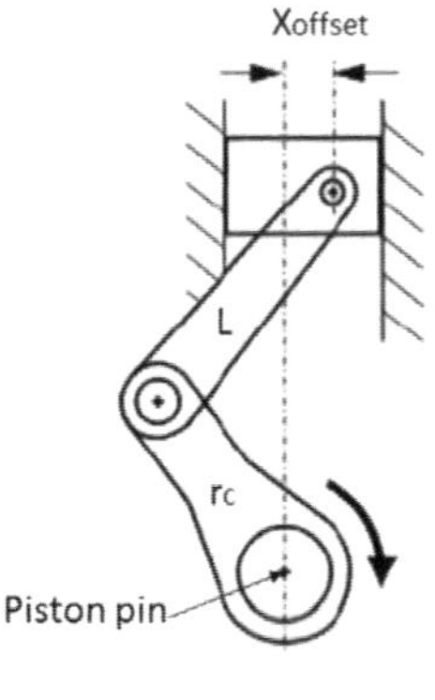

Figure 3. Eccentric piston-type model.

It is because of the quick-return characteristic that using the slider-crank mechanism, such as the feeder, automatic thread rolling machine is popular in the industry. In the automatic feeder machine, the 'giving material' process which would cost a huge amount of energy is beneficial, so the low-speed route is widely used in it. On the contrary, we try to shorten the back route to save time, as a result, the low-speed route is considered. Taking the quick-return structure is actually fit for the requirement, and the requirement in the piston-type SJA happened roughly as same as the feeder machine. In the common SJA, the time costed in ejection is equal to the suction's in terms of the mechanical structure in theory. In order to enhance the ejection velocity, we put the high-speed route into the ejection process, as the same time, the low-speed route is settled in the suction process. Under this condition, the piston velocity in the back route is a bit slower than the original design, but the time we cost in the inhaling route is more, which is beneficial for the suction route, and the cylinder can inhale much more fluid in a cycle. In conclusion, under the circumstance of constant frequency, the ejection speed is enhanced, which is most important to the performance, and much more fluid was inhaled due to the longer suction time. The quick-return mechanical structure is rather fit for the aerodynamic characteristic.

Three mechanical components, piston pin offset, crank radius and connecting rod length, all have an effect on the quick-return characteristic, and through a branch of examples, we can draw conclusions as followings:

1. Quick-return characteristic occurs due to the offset.
2. It is the crank radius rather than the piston pin offset or the connecting rod length that has a huge affection on the route length as well as the maximum velocity. In order to enhance the velocity, the longer crack radius is expected to be considered.
3. The more approximate the length between the piston pin offset and the crack radius was, the more obvious the quick-return characteristic would be, when the connecting rod length is invariable.
4. The more approximate between the whole length of the piston pin offset as well as the crack radius and the connecting rod length was, the more obvious the quick-return characteristic would be.

In order to explain the phenomenon in more details, we shall set several examples. We designed three cases in table 1.

Table 1. Parameters in different quick-return characteristics.

Name	Piston Pin Offset/mm	Crank Radius (r_c)/mm	Connecting Rod Length (L)/mm	Route/mm (L_c)	Frequency/Hz
Case1	5.87	5.4	14	12.2	200
Case2	5	6.01	30	12.2	200

Case3	0	6.1	30	12.2	200

Here is the reason why to set these examples. First of all, according to a number of samples, we find that the wider the piston pin offset is, the earlier the appearance of the jet velocity peak would be. But, the crank radius would be shortened a little respectively considering the condition of the same route length. Besides, with the reduction of the connecting rod length, the quick-return characteristic is expected to be more apparent with the peak appearing earlier.

An explanation has been given in the Figure 4 bellow. But the article doesn't aim at comparing the different crank radius, different offset or different connecting rod length, and the three above examples are totally different from the extent of the quick-return characteristics.

Case 1 is obvious for the quick-return characteristic, the case 2 is a slight quick-return example, and the case 3 is the contrast example. The Figure 4 below reflect piston velocity on the direction of motion. Data about velocity are all counted from the bottom-dead-center (BDC). When the curve moving to the 'piston velocity is equal to 0' straight line again, the piston is on the top-dead-center (TDC). The maximum velocity in case 1 is 47% higher than case 3, and the case 2 is just 4.8% higher than case 3. Besides, the minimum velocity in case 1 is nearly equal to the case 2 and 3. If the velocity in the suction process is slightly smaller than the non-quick-return SJA's, the energy costed in suction process is projected to be saved a little. Furthermore, what we should focus on is the duty ratio, we can easily find that, in the case 1, the suction time is totally enhanced, and it takes 58.3% of a cycle time. What's more, the trend of the curve is fit for the a wide arrange of frequency. From the 100Hz to the 200Hz, the max and the min velocity is expected to increase, and the duty ratio is predicted to be constant.

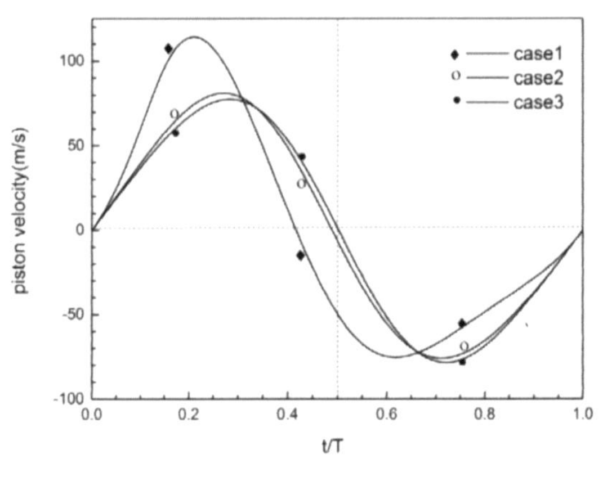

Figure 4. Plot of the piston velocity.

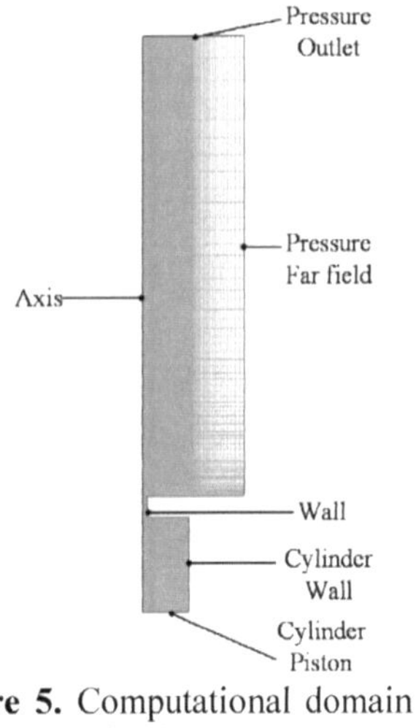

Figure 5. Computational domain and boundary definition.

Another phenomenon appears in the above Figure 4. The peak in case 1 is earlier than case 2 and case 3. A reasonable explanation can be given that the more quick-return characteristic is, the earlier the peak appears.

2.2 Measurement and the parameter definition

The computational domain of the piston-type SJA flow-field is shown in Figure 5. The cylinder wall is set to adiabatic wall, and the environment flow-fields are set as pressure far field and pressure outlet, in which 1atm pressure is given, and a normal temperature air is also used.

A simple cylinder flow channel (L_c=12.2mm in length) with a constant round cross section (*r=9.75mm* in radius) is considered in this paper, the 2D axisymmetric section is also shown in Figure 5. The jet direction domain is 90 mm long and the round cross section radius is 21 mm. The nozzle outlet diameter (d) is 3 mm, the thickness is two times of the d (6mm). The actuation frequency (*f*) is tested every 50 Hz from 100 Hz to 200 Hz, corresponding to the ability of the novel SJA in the experiment. The crank radius and the connecting rod length are flexible. The movement of the piston is simulated by a cylinder dynamic mesh. Specifically the piston wall is defined as a movement wall. The mesh growth of the piston wall is based on smoothing and remeshing methods of mesh deformation. The crank angle step size is 0.1°.

Viscous flow-fields of SJAs are simulated by finite volume method of CFD software Fluent [5]. Conservation equations are solving coupling with the method of density based method. An implicit algorithm with second-order spatial accuracy is used to solve the Navier–Stokes equations for steady flow-fields. The jet flow in the present investigation is unsteady, turbulent, and compressible. Density is calculated based on the ideal gas law. Gravity is neglected.

In order to analysis the performance of the quick-return piston-type SJAs, the seven cases are devised in table 2.

Table 2. Parameters in different cases.

Name	Offset/mm	Crack radius (r_c)/mm	Connecting rod length (L)/mm	Route (L_c)/mm	Frequency/Hz
Case1	5.87	5.4	14	12.2	200
Case2	5	6.04	30	12.2	200
Case3	0	6.08	30	12.2	200
Case4	5.87	5.4	14	12.2	150
Case5	0	6.08	30	12.2	150
Case6	5.87	5.4	30	12.2	100
Case7	0	6.08	30	12.2	100

The table 2 includes diversity of factors, and all of the data are in the same route and pressure ratio (5.8). Case 1 ,4 and 6 are all the obvious forms, Case 2 is the unapparent forms, and Case 3, 5and 7 are all the non-quick-return forms.

3. Comparison of the computational results

3.1 Analysis of the quick-return characteristic impact under 200Hz.

3.1.1 Flow image about the main calculation field under the 200 Hz.

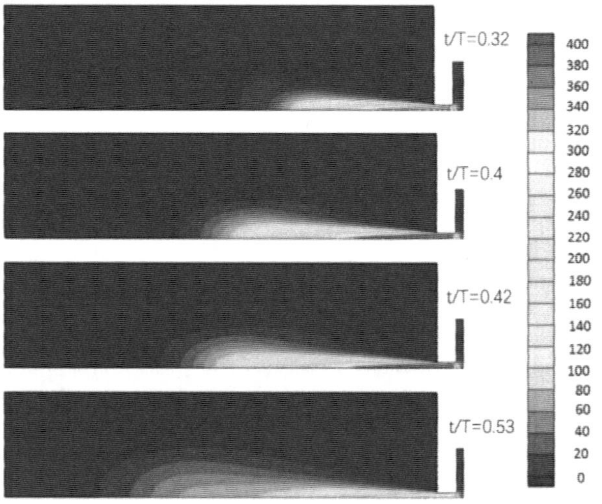

Figure 6. Jet image at different moments.

The picture above contains 4 images in the condition of case 1, they are different from diversity of velocity at diverse moment in a cycle. The first image is at the t/T=0.32 moment, the piston is moving towards the TDP, and the jet flow is gathering. The second image is at the t/T=0.4 moment which is actually under the maximum pressure of the cavity in the ejection process, and this phenomenon is expected to be discussed in the following paragraphs. And the third image is at the t/T=0.42 moment when the piston is at the TDP. The last one is at the t/T=0.53, the piston is moving away the TDP, but the jet flow is still expanding.

3.1.2 Comparison of the velocity in the X axis direction under the 200Hz

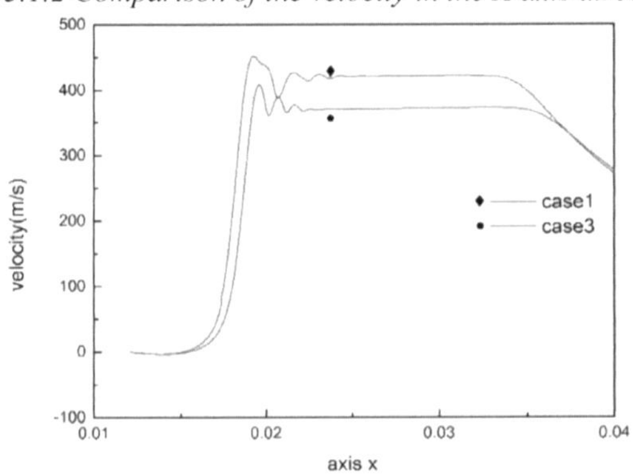

Figure 7. Velocity in the x-axis direction under the 200Hz.

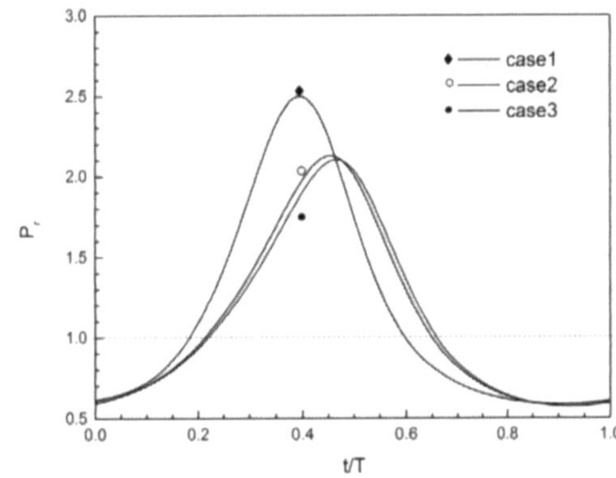

Figure 8. Pressure ratio in 200Hz.

The positive direction of the x-axis is considered as the jet flow direction. A visual survey can be concluded that the jet flow thrown from the nozzle has a slight supersonic, and the weak shock wave shows in the curve in Figure 7. Besides, the velocity in case 1 is higher than that in case 3, but the depth of the centre jet flow has not formed, because the jet haven't expanded.

3.1.3 Comparison of the pressure under the 200Hz. First, what we should know is the definition of the pressure ratio, which is actually the outcome of the environment pressure divided by the pressure of the cavity. Under the frequency of 200Hz, we can see that (**Figure. 8**), the case 1 takes an advanced phase position, and the peak of the pressure in case 2 is slightly ahead than the case 3, the control group. Besides, the peak of case 1 is 2.5, the peak of case 2 and 3 are 2.13 and 2.1, so the apparent quick-return characteristic facility has a priority in pressure ratio.

3.1.4 Comparison of the mass flow under the 200Hz

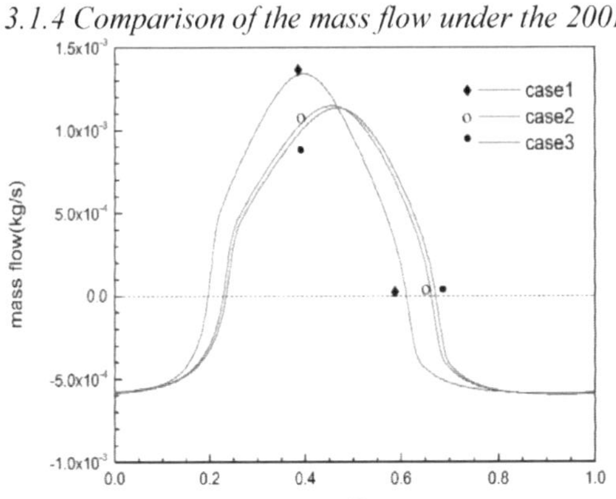

Figure 9. Mass flow in 200Hz.

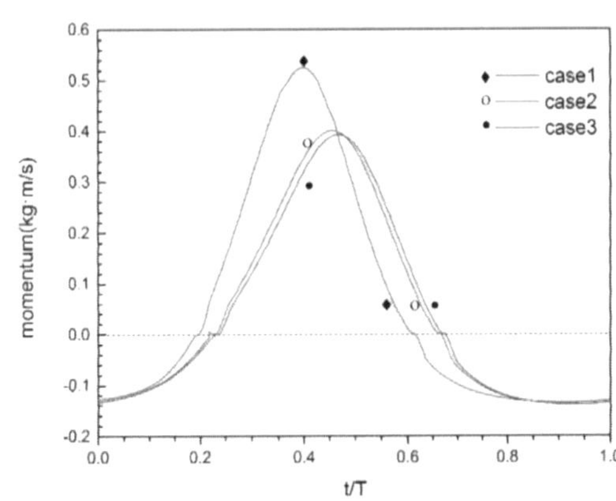

Figure 10. Momentum in 200Hz.

From the Figure 9, we can see the same tendency happening. Due to the obvious quick-return structure, the maximum mass flow in case 1 is obviously higher than case 2 and case 3. Besides, through the integration of the mass of the jet flow in one cycle, we can find that the total mass of the jet flow in case 1 is 8 percent more than case 3, and the case 2 is 0.4 percent more than case 3. In the suction process, both the pressure ratio and the mass flow are at the same level, it is expected not to be altered by the quick-return structure.

What's more, the dash line 0.0 is the boundary between ejection and suction, and the part under dash line is representative for suction. Time costed by suction in case 1 takes 58.5 percent of a cycle, by contrast, time costed by suction in case 2 and case 3 respectively are 56.3% and 56.1%. Considering the mechanical suction time of case 1 is 58.3%, the obvious quick-return characteristic facility performs better.

3.1.5 Comparison of the momentum under the 200Hz. The curve (**Figure. 10**) about the momentum has a slight wave around the line 0.0, because of the manual calculation, which shouldn't be noticed a lot. The trend of the momentum is similar to the former's, and the case 1 have a remarkable priority than the case 3. Through the integration of the ejection momentum, the whole jet momentum in case 1 is 22.3 percent more than case 3, and the case 2 is 1.5 percent more than case 3. So the quick-return piston-type SJA not only has a potential to jet velocity and pressure ratio, but also behaves better in the mass and momentum in a cycle.

4. Comparison of P_r in different frequency

4.1 Comparison of pressure ratio in two SJAs types

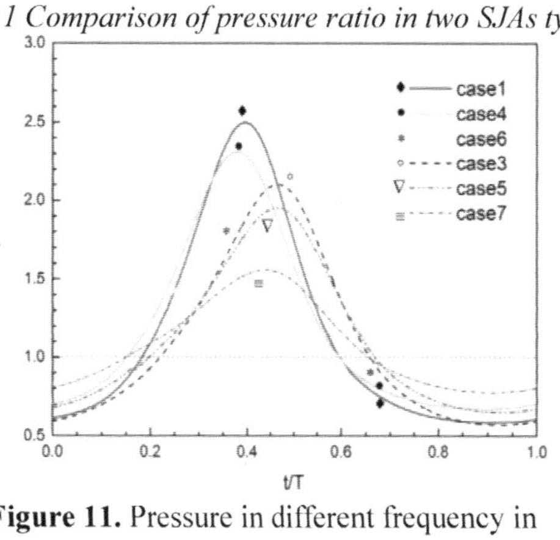

Figure 11. Pressure in different frequency in two types.

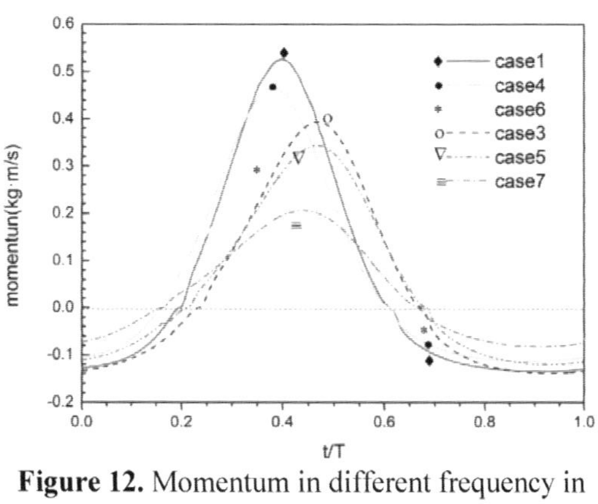

Figure 12. Momentum in different frequency in two types.

There are six curves in Figure11, three full lines are all from the obvious quick-return SJAs, and the contrast group contains three dash curves, which are from the original SJA's. In the suction process, the beginning time of the suction is disparate from the frequency in terms of the same mechanical structure, but they have the same ending time of suction. SJAs can inhale fluid because of the pressure difference. In the suction process, during the piston moving to a special position where the critical pressure difference has just formed, the actuator is projected to inhale forced by the pressure difference, so it is not difficult to explain why piston in different frequency begin inhaling the fluid at the same moment. However, the faster the piston is, the more difficult suction process the piston is in. In order to breathe more fluid, SJAs have to expand the suction time, so the beginning of the ejection time are in diversity.

Another phenomenon can't be overlooked that the intersection points of the full lines and the P_r=1.0 line are look-ahead from the intersection point of the dash lines, because the quick-return piston-type SJA is predicted to meet the peak of velocity ahead in the ejection process.

Furthermore, contrasting with different frequency, the suction time in 100Hz in the non-quick-return SJA is 49.58 percent, it would need 3.4 percent addition time to inhale more fluid under 150Hz or 6.67 percent addition time under 200Hz. Nevertheless, the suction time in 100Hz in the quick-return SJA is 52.56 percent, it would need 2.94 percent addition time to inhale more fluid under 150Hz or 5.94 percent addition time under 200Hz. It needs a slightly less time to inhale fluid for the quick-return facility with the enhancement of the frequency. Last but not least, no matter what the frequency is, the quick-return piston-type SJA has a better performance in terms of the P_r, jet mass and momentum.

4.2 Comparison of momentum in two SJAs types

No matter what the frequency is, the quick-return piston-type SJA has a better performance in momentum in Figure 12. In the 200Hz circumstance, the whole momentum of ejection in a cycle in case 1 is 22.3 percent more than case 3, case 4 is 23.6 percent more than case 5, and case 6 is 24.4 percent more than case 7. Even though with the enhancement of the frequency, effect by the quick-return characteristic has a slight reduction, it is obvious that the obvious quick-return one can make a better influence on the fluent field because of the added momentum.

5. Conclusion

This paper puts forward a novel designed piston-type SJA with a quick-return characteristic structure. According to the requirement of the SJA performance, the rapid traverse is applied to the route, in which the piston is moving to the TDC. Besides, the quick-return piston-type SJA can take more time to inhale more fluid because of the slow-route in which the piston moves to the BDC. Then, two various quick-return characteristic structures and a non-quick-return SJA have been proposed to analyze how the different quick-return characteristic influence the performance.

The quick-return piston-type SJA has a significant influence on the pressure of the cavity, jet velocity, mass flow and momentum, and the more apparent the quick-return characteristic is, the better the performance is. For example, the peak of the pressure ratio for the obvious quick-return SJA is up to the 2.5, and the non-quick-return SJA is just 2.1 under the same frequency condition of 200Hz. Besides, the obvious quick-return SJA have a 22.3 percent more than the non-quick-return one in terms of the jet momentum.

Under the other cases of frequency, the quick-return SJA also performs better than the classical one. Besides, with the development of the frequency, the suction time would increase in the whole cases, but the quick-return structure can reduce the influence of the frequency. For example, the obvious one just need 5.94 percent addition time from 100Hz to 200Hz, but the non-quick-return one need 6.67 percent addition time from 100Hz to 200Hz.

In conclusion, the novel quick-return piston-type SJA have a potential to application, and it can be applied with other advanced improvement measures, such as the auxiliary air system, and so on.

6. Reference

[1] Scott G. Anders, William L. Sellers, Anthony E. Washburn. "Active flow control activities at NASA Langley," AIAA 2004-2623.

[2] C.L.Rumsey, T.B.Gatake, W.L.Sellers. "Summary of the 2004 computational fluid dynamics validation workshop on synthetic jets," AIAA Journal, Vol.44, No.2, 2006.

[3] Crittenden T M, Glezer A. A high-speed, compressible synthetic jet [J]. Physics of Fluids (1994-present), 2006, 18(1): 017107.

[4] Zhang liang. Piston synthetic jet actuator efficiency analysis and improvement research [J]. Journal of propulsion technology, 2015, 36 (11) : 1648-1655.

[5] Patankar S V, "Numerical heat Transfer and Fluid Flow," McGraw Hill, New York, pp 89-109.1980.

IOP Conf. Series: Materials Science and Engineering **187** (2017) 012031 doi:10.1088/1757-899X/187/1/012031

Research on Creep Damage and Life Forecast of Rod Fastening Rotor Based on Damage Mechanics

Liu Qiang[1,2] **and Wang Ai-lun**[1,2]

[1] State Key Laboratory of High Performance Complex Manufacturing, Central South University, Changsha 410083, China

[2] School of Mechanical and Electrical Engineering, Central South University, Changsha 410083, China

LQLJRMS@126.com

Abstract. Rod fastening rotor (RFR) is the core part of gas turbine, the degradation of RFR has great effect on gas turbine's performance. One of the main reasons which cause the RFR performance degradation is creep damage, while few studies have been carried out in terms of it so far. In order to analyze the influence of preload on virtual material parameters, a dynamic model of RFR considering interface contact effect was built. Then equivalent stiffness of RFR and elements was analyzed as well. Furthermore, creep damage of elements under higher stress were analyzed with damage mechanics to get their influence on the total damage. Likewise, RFR were analyzed with damage mechanics to get the connection between the total damage and rupture life. The results showed that connection between the total damage and rupture life was a complicated, non-linear process. Moreover, the rods of turbine and combustion chamber were the biggest influencing factors. The results of this dissertation can be a support for structural design and life prediction of RFR.

1. Introduction

RFR is composed of wheel disks which compressed by rods, so the structure of it is more complicated than integral rotor. RFR is the core part of gas turbine, marine engine, etc. Furthermore, it always works in high temperature and high rotating speed environment for a long time, so performance degradation will generate easily. According to the statistics of Siemens, 58% malfunction of gas turbine is from RFR [1]. Therefore, studying the connection between damage and rupture life has important significance for ensuring gas turbine running safety in a long-term.

RFR works in harsh conditions, so it is easy to generate creep damage. However, the theory for studying damage and rupture life of rotor is almost linear cumulative damage theory which is very inaccurate. At present, Damage Mechanics is applied to many domains, such as fatigue damage, ductile damage, creep damage, etc [2]. 600MW turbine rotor with Damage Mechanics model considering multiaxial stress has been researched in order to get the nonlinear cumulative damage model [3]. Linear cumulative damage theory was combined with Damage Mechanics theory to study rotor stress field with finite element analysis, the result was more accurate [4]. Creep damage of 2Cr11NiMoVNbNB steel with creep experimental data has been studied to get damage rule and relationship between stress and strain when steel occur to fracture [5]. Creep damage and anti-cracking characteristics of turbine rotor has been studied in order to indicate connection between fracture parameters and propagation rate of creep crack [6]. Calculation method of turbine rotor creep life was proposed to calculate unsteady state creep stress and steady state creep stress with finite element program [7]. High-temperature creep of turbine rotor has been analyzed with finite element program to get temperature field, stress field and strain field [8].

The above research objects are all about solid rotor. But RFR includes many elements, so the total creep damage of RFR also contain different creep damage of elements, while few studies have been carried out in terms of it so far. Therefore, building creep damage analyzing model of RFR and indicating connection between element damage and the total damage of RFR has important significance for ensuring gas turbine running safety. This paper built a dynamic model of RFR considering interface contact effect to get equivalent stiffness of RFR. What's more, creep damage of combustion chamber, rods of turbine and the total rotor was analyzed in order to indicate connection among the total damage, element damage and rupture life.

2. Rod fastening rotor model considering interface contact effect

2.1. Virtual Material Method

Virtual Material Method proposes an isotropic virtual material hypothesis-based analytical dynamical modeling method for immovable joint surface in order to enhance the modeling precision of ensemble machine tools. Some analytical expressions of elastic modulus, poisson ratio, thickness and density was put forward for Virtual Material to simulate interface contact effect of joint surface [9].

Analytical expressions of virtual material elastic modulus, poisson ratio, thickness and density will be got as the following [10]:

$$E_c = \frac{2^{0.5D} D \psi^{1-0.5D}}{3\pi\sqrt{\pi}} E' G^{1-D} a_L^{0.5D} (a_c^{-0.5} - a_L^{-0.5}) \tag{1}$$

$$v_c = \frac{(1+\mu')E^*}{G^*_x} - 1 \tag{2}$$

$$h_c = h_1 + h_2 \tag{3}$$

$$\rho_c = \frac{\rho_1 h_1 + \rho_2 h_2}{h_1 + h_2} \tag{4}$$

Where D is fractal dimension, ψ is a parameter decided by D, E' is equivalent elastic modulus, G is fractal length scales, a_L is the largest contact area, a_C is critical contact area, μ' is equivalent poisson ratio, E^* is dimensionless virtual material elastic modulus, G_x^* is dimensionless virtual material shear modulus, ρ_1, ρ_2 are joint wheel disc density, l_1, l_2 are joint wheel disc thickness.

2.2. Contact model between wheel discs of rod fastening rotor

According to the real project, material of RFR is GH4169 whose density is $7850 kg/m^3$, elastic modulus is $205 GPa$, yield stress is $950 MPa$, poisson ratio is 0.3.

If material surface roughness is $0.8 \mu m$, joint surface virtual material thickness $h_1 = h_2 \approx 0.5 mm$ [11]. The thickness of virtual material layers can be got as:

$$h = h_1 + h_2 = 1 mm$$

The virtual material thickness between wheel discs is $1 mm$, as shown in Figure 1.

Figure 1. Virtual material layer

By the above equation (1)(2)(3)(4), material parameters of virtual material with different preload can be calculated, as shown in Table.1.

Table 1. Material parameters of virtual material with different preload

Preload of each rod (kN)	Elastic modulus (GPa)	Poisson ratio	Density (kg/m^3)
1000	93.792	0.270	7800

2.3. Research on equivalent bending stiffness of rod fastening rotor

The study object of this paper is 300MW gas turbine rotor designed by Dong Fang Turbine Company Limited. Figure 2 shows the model of RFR. RFR is composed of compressor, combustion chamber and turbine. Compressor and turbine consist of 17 wheel discs and 12 tied rods. Turbine consists of 4 wheel discs and 12 tied rods. Combustion chamber is between the compressor and turbine. The wheel discs on compressor and turbine are connected with 12 tied rods on which the pre-load is applied, many contact interfaces exist between wheel discs. Physical parameters of RFR are showed in table 2.

IOP Conf. Series: Materials Science and Engineering **187** (2017) 012031 doi:10.1088/1757-899X/187/1/012031

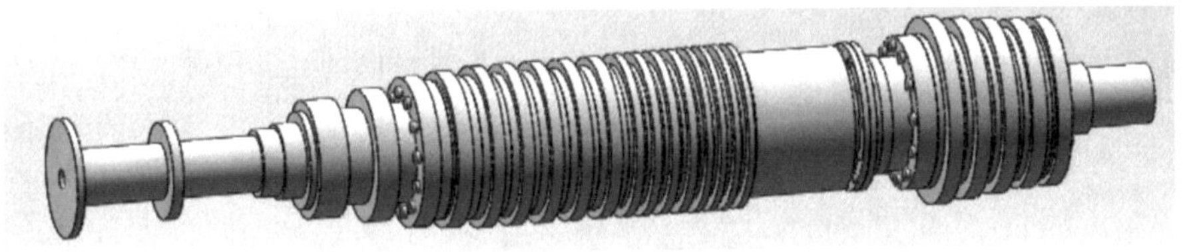

Figure 2. Model of RFR

Table 2. Physical parameters of RFR

Physical parameters	Value(mm)
Total length of the rotor	11582
Rod length of compressor	3802
Rod length of turbine	1611
Rod diameter	80
Length of Combustor	1905
The thickness of wheel of compressor	310
The thickness ofwheel of turbine	304
Wheel diameter	1424

RFR is composed of compressor, combustion chamber and turbine, and its total stiffness can be cascaded with the three parts. So the total stiffness can be got as the following:

$$K_{rot} = (\frac{1}{K_{com}} + \frac{1}{K_{tur}} + \frac{1}{K_{axl}})^{-1} \tag{5}$$

Compressor and turbine are consist of wheel discs, virtual material layers and rods. Wheel discs are cascaded with virtual material layers and they are all parallel with rods. Therefore, the stiffness of compressor and turbine can also be got:

$$K_{com}(K_{tur}) = (\frac{n_d}{K_d} + \frac{n_c}{K_c})^{-1} + K_r \tag{6}$$

Analytical expressions of the part and total stiffness is shown as the following [12]:

(1) Twisting stiffness of wheel discs

The twisting moment of wheel discs is M and strain energy is U_d, the equation of twisting stiffness is:

$$K_d = \frac{M^2}{2U_d} \tag{7}$$

(2) Twisting stiffness of virtual material layers

With calculation of E_c, the stiffness of virtual material layers can be got as the following:

$$K_c = \frac{E_c I_{Pc}}{l_c} \tag{8}$$

(3) Twisting stiffness of rods

The distribution of rods is shown in Figure 3, so the equation of twisting stiffness is:

$$K_{rt} = 3n_r(\frac{r_p}{l_r})^2 K_{rb1} \tag{9}$$

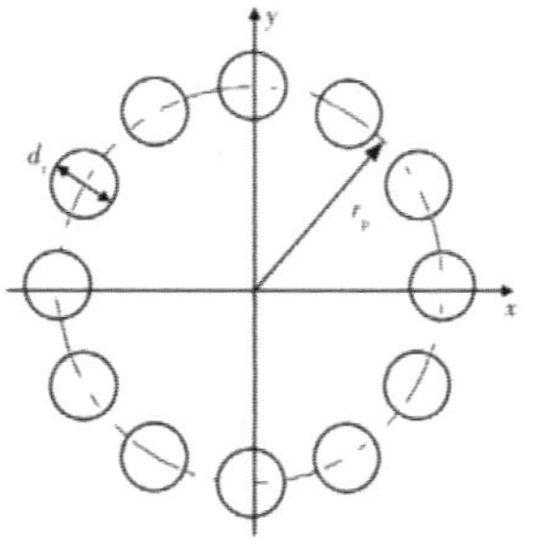

Figure 3. Distribution of rods

(4) Twisting stiffness of combustion chamber

The twisting stiffness combustion chamber is given as:

$$K_{axl} = \frac{EI_{Paxl}}{l_{axl}} \tag{10}$$

Where $K_{rot}, K_{com}, K_{tur}, K_{axl}$ are twisting stiffness of RFR,compressor, turbine and combustion chamber. K_d, K_c, K_r are twisting stiffness of wheel discs, virtual material layers and rods of compressor and turbine. K_{a1}, K_{rb1} are axial stiffness and twisting stiffness of each rod. n_d, n_c, n_r are numbers of wheel discs, virtual material layers and rods. l_d, l_r, l_{axl} are length of wheel discs, rods of turbine and combustion chamber. r_p is pitch radius of rods.

By the equations(5)~(10), some stiffness of RFR can be calculated, as shown in table 3.

Table 3. Twisting stiffness of RFR

Position	Wheel disks	Rods (turbine)	Rods (compressor)	Virtual material layers	Combustion chamber	RFR
Stiffness (N/m)	4.32E11	1.23E6	9.39E4	1.04E13	5.82E10	6.28E9

3. Research on creep damage of rod fastening rotor

3.1 Creep damage cumulative model

Creep damage cumulative model was proposed by Lermaitre[13,14], as the following equation:

$$D_c = 1-(1-R_V(1+\alpha)(\frac{\sigma_{eq}}{\lambda})^r t)^{\frac{1}{1+\alpha}} \tag{11}$$

Where α, λ, r are material parameters.

Multiaxia factor is defined as the following:

$$R_v = \frac{2}{3}(1+\upsilon)+3(1-2\upsilon)(\frac{\sigma_H}{\sigma_{eq}})^2 \tag{12}$$

Where σ_H is hydrostatic stress, σ_{eq} is equivalent stress, ν is poisson ratio.

If element has no damage, $t=0, D_c=0$. If element is ruptured completely, $t=t_r, D_c=1$. Integrate equation (11), t_r and D_c can be got as the following:

$$t_r = \frac{1}{R_V(\alpha+1)}(\frac{\sigma_{eq}}{\lambda})^{-r} \tag{13}$$

$$D_c = 1-(1-\frac{t}{t_r})^{\frac{1}{\alpha+1}} \tag{14}$$

The equation (13) and (14) is no-linear creep damage cumulative model which considering influence of equivalent stress. Therefore, it can calculate creep damage with multiaxial stress.

3.2. Element creep damage combined model in series or parallel

With equilibrium condition and equivalent strain principle[15], if $D=1-\frac{E'}{E}$, then $\sigma = E(1-D)\varepsilon$.

Damage parameter D can be defined as[16]:

$$D = 1-\frac{K'}{K} \tag{15}$$

(1) If elements are in parallel, then:

Total displacement is $L_p = l_1 = \cdots = l_i$, total stiffness is $K_p = \sum_{i=1}^{n} k_i$, total force is $F_p = \sum_{i=1}^{n} F_i = \sum_{i=1}^{n} k_i(1-d_i)l$.

According to the above equations , total damage parameter in parallel is:

$$D_p = \frac{\sum_{i=1}^{n} k_i d_i}{K_p} \tag{16}$$

(2) If elements are in series, then:

Total displacement is $L_s = \sum_{i=1}^{n} l_i$, total stiffness is $\frac{1}{K_s} = \sum_{i=1}^{n} \frac{1}{k_i}$, total force is $F_s = F_1 = \cdots F_i = K(1-D)L = k_i(1-d_i)l$.

According to the above equations, total damage parameter in series is:

$$D_s = 1 - \frac{1}{K_s \sum_{i=1}^{n} \frac{1}{k_i(1-d_i)}} \tag{17}$$

3.3. Research on creep damage of rod fastening rotor

According to finite element analysis of RFR, the position of maximal stress is combustion chamber and rods of turbine, therefore, creep damage will generate more easily. Accordingly, this paper will research on creep damage of RFR caused by combustion chamber and rods of turbine.Eequivalent stress and creep strain of RFR are shown in Figure 4 and Figure 5.

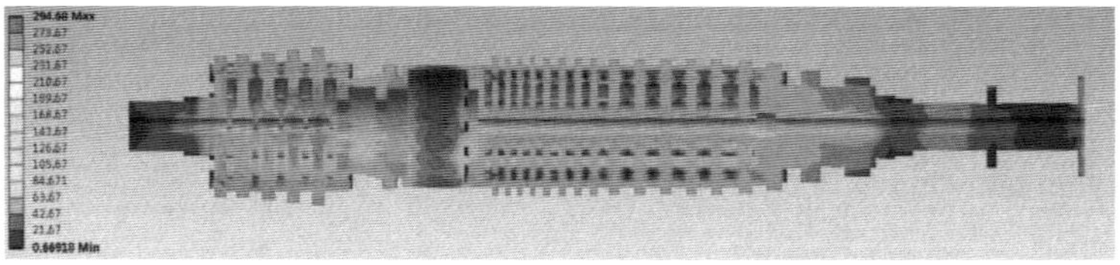

Figure 4. Equivalent stress of rotor

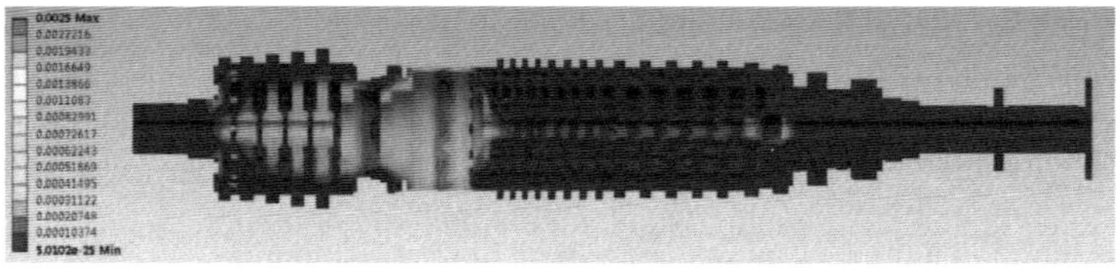

Figure 5. Equivalent creep strain of rotor

The creep damage of combustion chamber, rod of turbine and the total rotor can be calculated with the above equations.

(1) Creep damage of rods of turbine and combustion chamber

According equations (11)~(14), creep damage can be got as the following.

Damage of Rods is : $R_V = 1.46$, $t_r = \frac{1}{10.264}(\frac{\sigma_{eq}}{1575.48})^{-8.596}$, $D_c = 1-(1-\frac{t}{t_r})^{0.1724}$.

Damage of combustion chamber is: $R_V = 1.41$, $t_r = \frac{1}{8.178}(\frac{\sigma_{eq}}{1575.48})^{-8.569}$, $D_c = 1-(1-\frac{t}{t_r})^{0.1724}$

(2) Creep damage of the total rotor when creep damage of rods is homogeneous

Only the creep damage of combustion chamber and rods of turbine is considered, furthermore, the creep damage of rods is homogeneous. As turbine and combustion chamber are in series, therefore, the creep damage of turbine can be calculated as the following:

IOP Conf. Series: Materials Science and Engineering **187** (2017) 012031 doi:10.1088/1757-899X/187/1/012031

$$D_{tur} = D_{turr} = \frac{\sum_{i=1}^{n} k_{turri} d_{turri}}{K_{tur}} \tag{18}$$

According to above analysis, the total creep damage of rotor can be calculated as the following:

$$D_{rot} = 1 - \frac{1}{K_{rot}[\frac{1}{K_{tur}(1-D_{tur})} + \frac{1}{K_{axl}(1-D_{axl})}]} \tag{19}$$

It is supposed that the creep damage of rods is homogeneous, Compare the nonlinear creep damage of combustion chamber, rods and the total RFR with Miner linear damage cumulative model[17], damage curves are shown in Figure 6.

Conclusion from the creep damage curves:

① When $t/t_r \leq 0.9$, creep damage increase equably as t/t_r increase; when $t/t_r \geq 0.9$, creep damage increase rapidly as t/t_r increase; when $t/t_r = 1$, creep damage reach 1, it means that RFR is fractured thoroughly.

② The total creep damage of RFR is decided by creep damage of combustion chamber and turbine,but bigger than them.

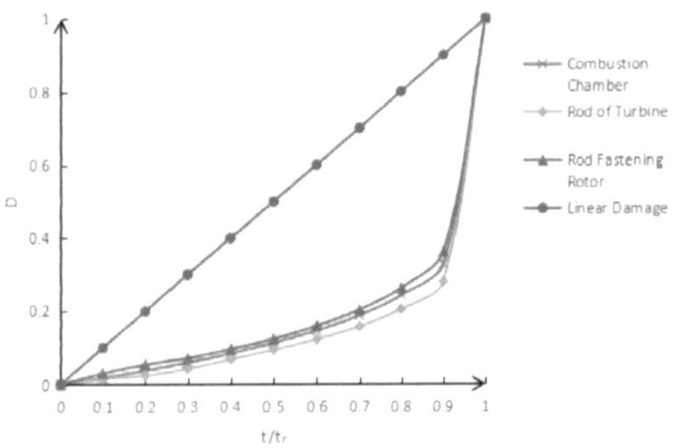

Figure 6. Comparison of creep damage curves

4. Conclusion

In this paper, nonlinear creep damage model was proposed for indicating connection between the total creep damage and rupture life against the disadvantages of current creep damage theory and life prediction. Furthermore, a dynamic model of RFR which has virtual material layer between wheel disks was built in order to calculate equivalent stiffness of RFR.

Results show that:

(1) The elastic modulus and poisson ratio of virtual material layer between wheel disks increase as preload increase, but the density is not change.

(2) According to the analysis of creep damage with nonlinear creep damage model, damage parameter D increase nonlinearly as cycle index. RFR will fracture thoroughly when D=1.

(3) The total creep damage of RFR is decided by combustion chamber and rods of turbine, and the total creep damage is lager than creep damage of combustion chamber and turbine.

5. References

[1] David k Hall. Performance Limits of Turbomachines[D]. Massachusetts Institute of Technology,2010:102-113.

[2] Assimina A.Pelegri.Evolution of interlayer and intralayer cracks under compressive fatigue in composites[J] .Trans ASME .Journal of Engineering Materials and Technology ,1999, 121(4):430-435 .

[3] Jing Jian-ping, XIA Song-bo, SUN Yi. Nonlinear continuum damage mechanics model on high temperature creep analysis[J].Journal of Propulsion Technology, 2001,02:139-142

[4] Wang Li-bin, JING Jian-ping, SUN Yi. Application of Damage Mechanics on High Temperature Creep Analysis of Turbine[J]. Turbine Technology, 2002,42(2):88-90.

[5] Mao Xue-ping, LIU Zong-de, YANG Kun. Study of Creep Damage Behaviors for 2Cr11NiMoVNbNB[J]. Steel Proceedings of the CSEE,2003,01:160-163.

[6] Wang You-sheng, ZHANG Jian-ying. Research on Creep Damage and Resisting Rupture Property of a Steam Turbine Rotor Material[J]. Chinese Journal of Applied Mechanics,1989,01:72-77+114.

[7] Shi Jin-yuan, SUN Qing, YANG Yu. Creep Life Design and Evaluation of Large Steam Turbine Parts[J]. Proceeding s of the CSEE.2002， 22(3):103-106.

[8] Wu Qiong, WANG Wei-zhe, ZHANG Jun-hui. High-temperature Creep Strength Analysis for IP Rotor of an Ultra-supercritical Steam Turbine[J]. Journal of Chinese Society of Power Engineering,2015,01:25-29.

[9] Tian Hongliang，Li Bin，Liu Hongqi，etal. A new method of virtual material hypothesis-based dynamic modeling on fixed joint interface in machine tools[J]. ELSERVIER International Journal of Machine Tools & Manufacture，2011，51(3)：239-249.

[10] Tian Hong-liang. Dynamic Modeling on Fixed Joint Interface in Mechanical Structure[D]. Huazhong University of Science and Technology.

[11] Fishkis M, Lin J C. Formation and Evolution of a Subsurface Layer in a Metalworking Process[J]. Wear, 1997,2006(1-2):156-170.

[12] Li Pu, YUAN Qi, GAO Jin. Investigation on Stiffness of Circumferential Distributed Tie-rod Rotor with Curvic Couplings of Trochal Disk[J]. Journal of Aerospace Power,2013,07:1618-1623.

[13] Lemaitre J. A Course on Damage Mechanics. Spring-Verlag,1992.

[14] Lemaitre J .Application of damage concepts to predict creep-fatigue failures[J] .Transaction of the ASME, Journal of Engineering Materials and Technology .1979 , 101:284～ 292.

[15] Yu Tian-qing, QIAN Ji-cheng.Damage Theory and Application[M]. Beijing:National Defend Industry Press,1993.

[16] Wu Bo, LI Hui. The Mechanic Method for Damage Analysis of Structures[J].Earth Quake Engineering and Engineering Vibration,1997,01:15-23.

[17] Jing Jian-ping, MENG Guang. Fatigue-creep Damage Analysis of a Steam Turbine Rotor by Nonlinear Continuous Damage Mechanics Model[J]. Proceeding of the CSEE,2003,09:167-172.

Acknowledgments

This research is supported by the National Basic Research Program of China (973 program: 2013CB035706).

Design and Implementation of Multifunctional Automatic Drilling End Effector

Zhanxi Wang*, Xiansheng Qin, Jing Bai, Xiaoqun Tan and Jing Li

School of Mechanical Engineering, Northwestern Polytechnical University, Xi'an 710072, China

Email: *zxwang@nwpu.edu.cn

Abstract. In order to realize the automatic drilling in aircraft assembly, a drilling end effector is designed by integrating the pressure unit, drilling unit, measurement unit, control system and frame structure. In order to reduce the hole deviation, this paper proposes a vertical normal adjustment program based on 4 laser distance sensors. The actual normal direction of workpiece surface can be calculated through the sensors measurements, and then robot posture is adjusted to realize the hole deviation correction. A base detection method is proposed to detect and locate the hole automatically by using the camera and the reference hole. The experiment results show that the position accuracy of the system is less than 0.3mm, and the normal precision is less than 0.5°. The drilling end effector and robot can greatly improve the efficiency of the aircraft parts and assembly quality, and reduce the product development cycle.

1. Introduction

In the assembly of aircraft structures, all internal and external parts are fixed by solid rivets or fasteners [1]. The quality of holes determines the final quality of the aircraft, manufacturing costs and production cycle. Much of the assembly processes involved in creating high-level structures use conventional manual drilling due to the high complexity of the parts to be processed. Such manual drilling is frequently associated with risk of rework, reduction in the process capability range, and structural impairment, resulting in extra costs [2].

Recently, industrial robotic applications are becoming widely used in the aviation sector [3] because of its lower investment, increasing of automation degree, the table working performance and the good accessibility. From the beginning of twenty-first Century, American GEMCOR, EI (ElectroimPact), Italy COMAU, German BROETJE-Automation is committed to the design and development of drilling robot system, and their drilling robot system production has been used in aircraft manufacturing enterprises, such as F-16, F-22vertical wall, F-35 aircraft wing panel and A380 wing pane [4, 5]. S H Bi developed the drilling robot system, which uses industrial camera to establish the relationship between the workpiece and the robot coordinate system [6].

The end effector is one of the key components of the drilling robot system. The performance of drilling end effector will directly affect the accuracy and efficiency of drilling robot system. In this paper, a drilling end effector is designed to realize automatic drilling of aircraft wing beam, which integrates a variety of function, like as compaction, base detection, vertical normal adjustment, drill and countersink.

2. The structure and principle of drilling end effector

According to the requirements of automatic drilling process for aircraft assembly, the design for automatic drilling end effector actuator includes the following functions: (1) the automatic clamping function; (2) the precision drilling function; (3) high precision countersink function, which can be used for drilling countersink hole of the spindle with the requirements; (4) release and grasp knife function, which can realize the automatic tool changing with the pneumatic actuator and robot; (5) the base detection function, which provide reference measurement data for drilling system; (6) the vertical normal adjustment function, provide deviation data between tool axis and hole position vertical normal for drilling robot system.

Based on the function and space layout, the overall structure of drilling end effector are detailedly designed. The function modules are set as frame structure, pressing unit, drilling unit, robot flange, measurement unit, integrated pneumatic signal unit and integrated electric control unit. The specific structure of drilling end effector is presented in Figure 1.

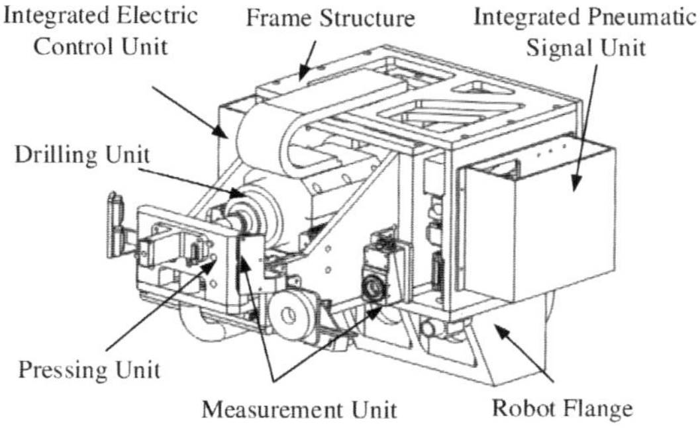

Figure 1. The specific structure of drilling end effector

2.1. Design for Pressing Unit

With standard robotic manipulators, the system provides high degrees of flexibility to enable functionality in large-volume processes, including those involving the manufacture of parts with complex geometries. However, this flexibility causes poor stiffness of the robotic arm, and this relatively lesser rigidity affects the relationship between the end-effector and the work piece, but this low degrees of mechanical stiffness in robotic arms could be compensated by special features fixtures.

The main challenge to be overcome in making increased use of industrial robots in aircraft assembly lines is the positioning accuracy required for these processes. During the entire drilling cycle, the pressure force must be controlled in order to reduce the tangential forces' potential to spoil the hole quality. For instance, the sliding of the drilling tool on the surface during the operation can seriously degrade both the position and other characteristics of the hole.

The whole structure of the pressing unit is shown in Figure 2, which is mainly composed of a pressure nose, a pressure foot, a pair of earrings, a ball bearing, a compression cylinder, a guide rail slide block and a vacuum bypass.

The pressure nose is fixed on the pressure foot, which is connected with the compression cylinder by a pair of earrings and two ball bearings. The ball bearing can be completely connected in a certain processing and assembly error by use the self-aligning function. The pressure foot is connected with the bottom flame plate of the end effector by the high precision guide rail and slide block, and ensures the rigidity of the other direction except the moving direction of pressure foot. The pressure foot end is provided with a chip discharge hole, which is connected with the vacuum bypass.

2.2. Design for Drilling Unit

As shown in Figure 3, the structure of drilling unit includes integrated sliding table, synchronous wheel, servo motor, spindle clamping seat, electric spindle, knife and handle. The servo motor could actuate the integrated sliding table to achieve linear moving by using of synchronous wheel. The feed movement of electric spindle can be realized through the spindle clamping seat and sliding block. The countersink sensor is mounted on the spindle clamping seat and is used for real-time monitoring the depth of drill and countersink.

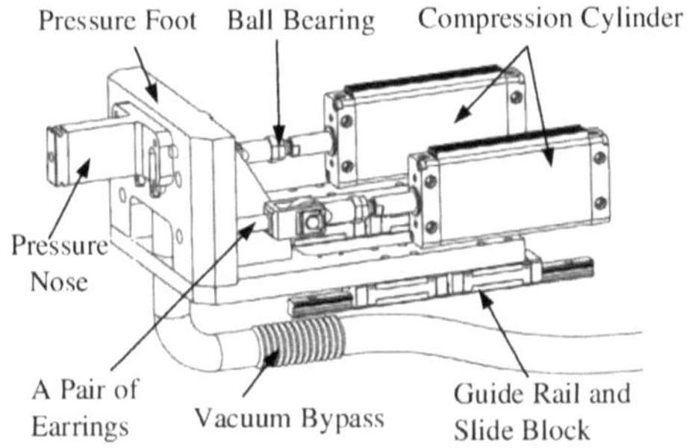

Figure 2. The integral structure of pressing unit

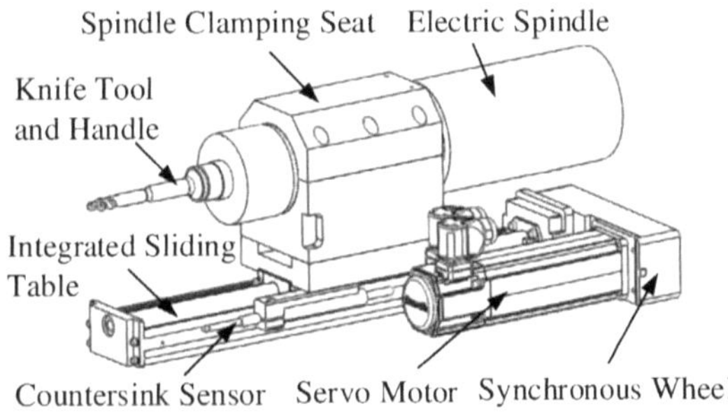

Figure 3. The integral structure of drilling unit

The servo motor is installed on the left side of integrated sliding table, which will makes the drilling unit structure is more compact and reasonable. The actuator power is transferred through the synchronous belt wheel between servo motor and integrated sliding table. The electric spindle rotates knife tool and handle to complete high quality drilling, while the depth of hole and the countersink is critically controlled by countersink sensor which has high measuring precision and response speed. The electric spindle adopts the way of circulating oil to ensure the continuous operation at high speed.

2.3. Principle of measurement unit-- the vertical normal adjustment

The vertical normal adjustment technology is using four laser displacement sensors installed on the drilling end effector to detect and adjust the knife tool posture, which can maintain the deviation between tool axial direction and vertical normal of work piece in a normal range. This technology is one of the key technologies to improve the quality of making hole.

In this paper, the four point vertical adjustment algorithm is used to calculate the intersection angle between tool axis and hole position vertical normal. The measurement principle is shown in Figure 4.

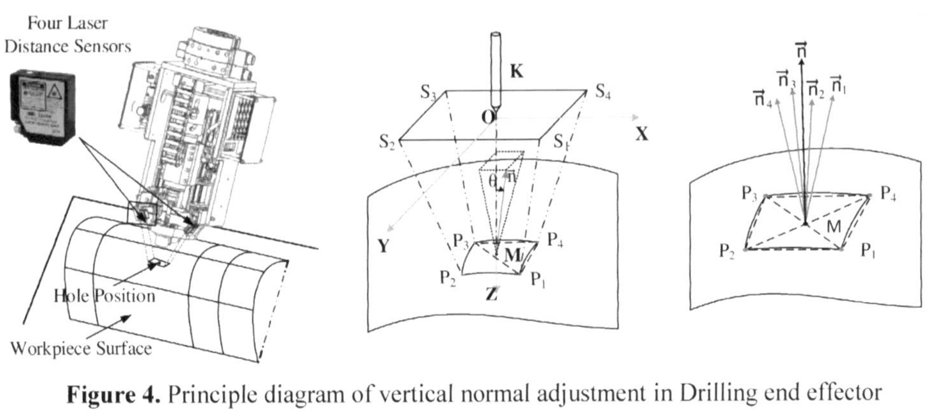

Figure 4. Principle diagram of vertical normal adjustment in Drilling end effector

Four laser distance sensors is installed on the both sides of drilling end effector, and the projection direction of the laser distance sensor and the actual position in reference coordinate system are already calibrated as premise work [7]. In Figure 4, the position of four laser distance sensors are respectively presented S_1, S_2, S_3 and S_4. P_1, P_2, P_3 and P_4 are the light spots on the surface of workpiece emitted by the laser beam of four laser displacement sensors. Before the drilling-hole process, four laser distance sensors are used to measure the distance the light spots and sensors. By using the distance data of four sensors, the coordinate value of P_1, P_2, P_3 and P_4 in O-XYZ coordinate system can be calculated respectively. Since any three points of P_1, P_2, P_3 and P_4 are not collinear, four points can be formed around four micro processing plane ($P_1P_2P_3$, $P_1P_2P_4$, $P_1P_3P_4$ and $P_2P_3P_4$). Four normal vectors ($\vec{n}_1$, $\vec{n}_2$, $\vec{n}_3$ and $\vec{n}_4$) of four micro plane can be deduced with normal calculation mathematical model and coordinate value of three point. By using the specific algorithm, four normal vectors are "weighted average" to approach the actual normal vector of the machining point M.

In the vertical normal adjustment process, the measured data of laser distance sensors are assumed to be l_1, l_2, l_3 and l_4. Then, the coordinate value (x_1, y_1, z_1), (x_2, y_2, z_2), (x_3, y_3, z_3) and (x_4, y_4, z_4) of P_1, P_2, P_3 and P_4 in O-XYZ coordinate system can be written as:

$$\begin{cases} x_i = (l_i - l_i^0)\cos\alpha_i \cos\beta_i + \vartriangle x_i^0 \\ y_i = (l_i - l_i^0)\cos\alpha_i \sin\beta_i + \vartriangle y_i^0 \\ z_i = (l_i - l_i^0)\sin\alpha_i \end{cases} \quad i=1, 2, 3, 4 \tag{1}$$

Where l_i are the measured data of laser distance sensors, l_i^0, $\vartriangle x_i^0$ and $\vartriangle y_i^0$ are the constant value, which are determined in premise calibration procedure. α_i are angles between the laser beam direction of four laser distance sensors and the reference plane of the tool coordinate system. β_i are angles between the projection line of the laser beam on the reference plane and the Coordinate axes Y. α_i and β_i are already deduced in premise calibration procedure, too.

In each micro plane of $P_1P_2P_3$, $P_1P_2P_4$, $P_1P_3P_4$ and $P_2P_3P_4$, the normal vector two of micro plane can be cross multiplied with two vectors composed by three points:

$$\vec{n_1} = \frac{\overrightarrow{P_1P_2} \times \overrightarrow{P_1P_3}}{\left|\overrightarrow{P_1P_2} \times \overrightarrow{P_1P_3}\right|}, \ \vec{n_2} = \frac{\overrightarrow{P_1P_2} \times \overrightarrow{P_1P_4}}{\left|\overrightarrow{P_1P_2} \times \overrightarrow{P_1P_4}\right|}, \ \vec{n_3} = \frac{\overrightarrow{P_1P_3} \times \overrightarrow{P_1P_4}}{\left|\overrightarrow{P_1P_3} \times \overrightarrow{P_1P_4}\right|}, \ \vec{n_4} = \frac{\overrightarrow{P_2P_3} \times \overrightarrow{P_2P_4}}{\left|\overrightarrow{P_2P_3} \times \overrightarrow{P_2P_4}\right|} \tag{2}$$

The weighted average of four normal vector is obtained as the actual normal vector of the machining point M:

$$\vec{n} = \frac{1}{4}(\vec{n_1} + \vec{n_2} + \vec{n_3} + \vec{n_4}) \tag{3}$$

Then the angle θ between the tool axial direction and actual normal vector $\vec{n}$ is:

$$\theta = arc\cos([1\ \ 0\ \ 0] \cdot n) = arc\cos(n(1,1)) \tag{4}$$

If $\theta < 0.5^{\circ}$, then there is no need to implement the vertical normal adjustment program. If $\theta \geq 0.5^{\circ}$, the vertical normal adjustment diagram will be steadily carried out until meeting the requirements.

2.4. *Principle of measurement unit-- the base detection*

In the drilling process, the position of drilling holes is located with a high precision industrial camera to identify the reference hole on the workpiece. When the robot arrive at the designated location, there is a deviation between theoretical robot coordinate system and the actual robot coordinate system. On the otherwise, the installation of processing parts may also produce deviation the workpiece and the theory position, as shown in Figure 5.

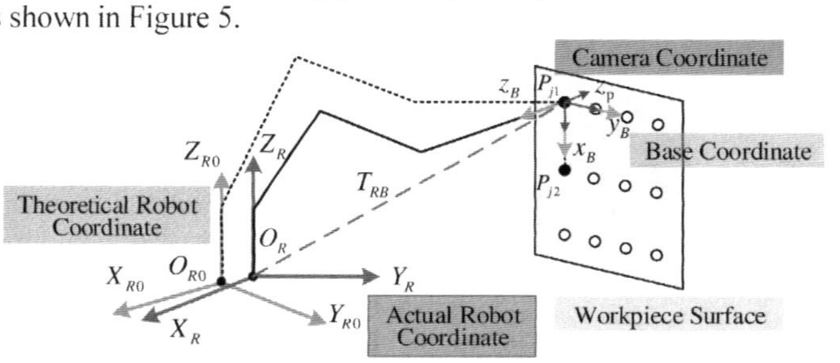

Figure 5. Principle diagram of base detection in Drilling end effector

These reasons will lead that the actual position relationship between the robot and processing parts is different with the theoretical relationship. Therefore, the base detection program is necessarily implemented to establish the base coordinate system and determine the actual position between parts workpiece and drilling end effector.

The original point of camera coordinate system is define as $POS_{Rp}(x, y, z, A, B, C)$. (x, y, z) means the coordinate value of POS_{Rp} in actual robot coordinate system, while A, B, C are the angles of Euler transform from the base coordinate system $O_R - X_R Y_R Z_R$ to the camera coordinate system $P_{j1} - x_p y_p z_p$. Then, the homogeneous matrix of Euler transform can be written as:

$$T_{Rp} = Trans(x, y, z) \cdot Rot(z, A) \cdot Rot(y, B) \cdot Rot(x, C)$$
$$= \begin{bmatrix} 1 & 0 & 0 & x \\ 0 & 1 & 0 & y \\ 0 & 0 & 1 & z \\ 0 & 0 & 0 & 1 \end{bmatrix} \cdot \begin{bmatrix} \cos A & -\sin A & 0 & 0 \\ \sin A & \cos A & 0 & 0 \\ 0 & 0 & 1 & 0 \\ 0 & 0 & 0 & 1 \end{bmatrix} \cdot \begin{bmatrix} \cos B & 0 & \sin B & 0 \\ 0 & 1 & 0 & 0 \\ -\sin B & 0 & \cos B & 0 \\ 0 & 0 & 0 & 1 \end{bmatrix} \cdot \begin{bmatrix} 1 & 0 & 0 & 0 \\ 0 & \cos C & -\sin C & 0 \\ 0 & \sin C & \cos C & 0 \\ 0 & 0 & 0 & 1 \end{bmatrix} \quad (5)$$
$$= \begin{bmatrix} \cos A \cos B & \cos A \sin B \sin C - \sin A \cos C & \cos A \sin B \cos C + \sin A \sin C & x \\ \sin A \cos B & \sin A \sin B \sin C + \cos A \cos C & \sin A \sin B \cos C - \cos A \sin C & y \\ -\sin B & \cos B \sin C & \cos B \cos C & z \\ 0 & 0 & 0 & 1 \end{bmatrix}$$

Two base holes are pre-made before the base detection, then the drilling end effector is approach to base hole P_{j1} by operating the robot arm and made the circle center of hole P_{j1} into the camera vision center. At this moment, the base detection program will record point coordinates value P_{pj1} of P_{j1} in the camera coordinate system $x_p y_p z_p$. According to the same process, point coordinates value P_{pj2} of P_{j2} in the camera coordinate system $x_p y_p z_p$ will be accurately identified. The base coordinate system $x_B y_B z_B$ will be established on the basis of P_{pj1} , P_{pj2} and T_{Rp} .

3. The experiment research

In this experiment, we have developed a Drilling End Effector system which is installed on Kuka KR-500 robot as the experiment platform, as shown in Figure 6.

Figure 6. The experiment platform of Drilling end effector

The measurements and calculation results of vertical normal adjustment program are shown in table 1, the angle error shows that the method to meet the accuracy requirements (less than 0.5 deg).

The robot and drilling end effector can realize auto-drilling for Aluminium laminated material, and the efficiency can reach 6 holes / min. The parameters of holes can meet the drilling requirements, which greatly improve the drilling efficiency and quality of aircraft components assembly. The results

IOP Conf. Series: Materials Science and Engineering **187** (2017) 012032 doi:10.1088/1757-899X/187/1/012032

of drilling hole experiment and precision measurement are shown in Figure 7. The position error of holes is less than 0.3mm; the diameter precision can reach H9; the vector normal errors of all the holes are less than 0.5°, and the holes' surface roughness is less than 1.6μm.

Table 1. The results of vertical normal adjustment program

Num.	Laser senor 1	Laser senor 2	Laser senor 3	Laser senor 4	Calculated angle (Deg.)	Actual Value (Deg.)	Error (Deg.)
1	135.90	138.60	136.50	134.30	0.106	0.063	0.044
2	144.23	146.93	144.83	142.63	0.086	0.062	0.025
3	144.07	146.77	144.67	142.47	0.098	0.073	0.025
4	135.74	138.44	136.34	134.14	0.112	0.103	0.009
5	139.74	142.44	140.34	138.14	0.050	0.063	0.112
6	135.64	138.34	136.24	134.04	0.097	0.061	0.036

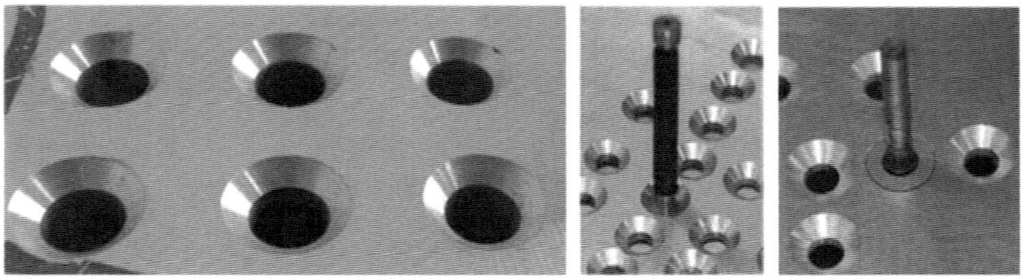

(a) (b) (c)

Figure 7. The experiment and precision measurement: (a) countersink specimens; (b) diameter detection; (c) countersink depth detection

4. Conclusion

A drilling end effector is designed for the robot automatic drilling system. Through the research of pressing unit, drilling unit and measurement unit, the effective integration of each functional module of drilling end effector is realized. The vertical normal adjustment is proposed to significantly improve the vertical accuracy of the robot automatic drilling system and the base detection program can be realize the automatic orientation of base coordinate system. The experiment research illustrate that drilling end effector and robot can greatly improve the efficiency of the aircraft parts and assembly quality, and reduce the product development cycle.

5. References

[1] Webb, P., N. Jayaweera and G. Lowe 2007 *Assembly Automation* **27**(4) 343
[2] Bo Yong, X.G.-k. and Xiao Qing-dong 2009 *Aeronautical Manufacturing Technology* **24** 61
[3] Olabi A, Béarée R, Gibaru O and Damak M 2010 *Control Engineering Practice* **18**(5) 471
[4] Atkinson J , Hartmann J and Jones S P 2007 SAE AeroTech Congress & Exhibition (Los Angeles, CA, USA) 28
[5] DeVlieg R, Feikert E, One-up assembly with robots[J]. Training, 2008. **2013**: p. 09-30.
[6] Bi S S, and Liang J 2011 *The International Journal of Advanced Manufacturing Technology*,. **54**(5) 767.
[7] Wang Z C, Qin X S, Bai J and Wang W L *Machinery Design & Manufacture* 2014 **6** 160.

Acknowledgments
This work is partially supported by the National Natural Science Foundation of China (Grant No. 51505380), Shaanxi Science and Technology Innovation Project (Grant No. 2016KTZDGY06-01).

IOP Conf. Series: Materials Science and Engineering **187** (2017) 012033 doi:10.1088/1757-899X/187/1/012033

The Influence of Depth of Cut, Feed Rate, and Step-over on Dimensional Accuracy in Subtractive Rapid Prototyping of Polycarbonate Material

T J Suteja

Department of Manufacturing Engineering, University of Surabaya, Raya Kalirungkut, Surabaya, Indonesia

jayasuteja@staff.ubaya.ac.id

Abstract. Subtractive rapid prototyping is fast and automatic three dimensions physical modelling that uses computer aided design model as the input. The dimensional accuracy of the result of the subtractive rapid prototyping is influenced by its process parameters. The aim of this research is to study and then develop a model that shows the influence of depth of cut, feed rate, and step-over on the vertical length error, horizontal length error, and depth error in subtractive rapid prototyping of polycarbonate material. This research implements response surface methodology to develop the model and then followed by the residual tests to evaluate the developed model. The result shows that the increase of the feed rate and the step-over will increase the horizontal dimension error. The most influenced factor on the horizontal dimension error is the step-over. Meanwhile, the vertical dimension error will be affected mostly by the step-over. Last, the depth error is influenced by the feed rate, the step-over, and the depth of cut. The depth of cut is the most critical factor that increases the depth error. The developed models give an insight on how several process parameters of rapid prototyping will influence the dimensional accuracy of a polycarbonate material. Based on the model, efficient resources utilization can be achieved.

1. Introduction

Polycarbonate material is a strong, tough, and transparent thermoplastic material that is used in various applications including in tooling. As an example, a polycarbonate can be implemented as a complex shape chocolate mould. In order to produce a complex shape mould, a subtractive rapid prototyping is implemented as it is typically suitable for polycarbonate material to produce parts that require specific finishes.

Subtractive rapid prototyping is fast and automatic three dimensions physical modelling that uses Computer Aided Design model as the input. The physical modelling process is performed by implementing milling process to cut the raw material with tool that rotates in very high speed. According to Toh, C.K., the milling process with 10 mm tool diameter that is rotated in 10.000 rpm is called high speed milling [1]. Therefore, the subtractive rapid prototyping can be categorised as a high speed milling.

One of the important considerations in manufacturing a mold by using subtractive rapid prototyping is the dimensional accuracy of the mold. However, no research found in the literature that investigates the influence of high speed milling process parameters on the dimensional accuracy of the polycarbonate material. Most of the previous studies found in the literature investigate the dimensional accuracy of part produced by additive rapid prototyping [2-6]. According to Ippolito, R., et. al., one of the additive rapid prototyping processes which is Stereolithography process can achieve the average dimensional accuracy until 0,08 mm [3]. This research also describes that the highest dimensional accuracy for additive rapid prototyping process is achieved by using Fused Deposition Modelling. The achieved dimensional accuracy of this process is 0,03 mm. However, this process will generate higher surface roughness compared to other processes.

Nieminem, I., et. al., investigate the possibility to use high speed milling to fabricate a thin fin of polycarbonate material by changing the depth of cut and step-over [7]. However, Nieminem, I., et. al. did not investigate the influence of these parameters on the dimensional accuracy of the polycarbonate material.

Based on the studies conducted by Oktem, H., et al., The, J.S., et. al., the dimensional accuracy of a milling process are affected by interpolation type, tool holder type, controller of the machine, computer aided manufacturing software, physical and mechanical characteristics of the material, physical and mechanical characteristics of the tool, vibration, depth of cut, step-over, feed rate, cutting speed, and cut type. [8-12].

The aim of this research is to study and develop a model that shows the influence of depth of cut, feed rate, and step-over on the dimensional accuracy of polycarbonate material in subtractive rapid prototyping. These three process parameters are selected because these parameters are the dominant factors that influence the material rate removal and the dimensional accuracy simultaneously.

2. Research methodology

This research implements Response Surface Methodology to develop the dimensional accuracy model. To validate the developed model, three residual tests, which are independence test, constant variance test, and normality test are conducted.

In this research, the subtractive rapid prototyping process is implemented on the polycarbonate material to achieve the shape shown in figure 1. The subtractive rapid prototyping machine used in this research is Roland MDX 40. The machine is assisted by CAM Modela Player 4.0 software to generate the tool path. The required input to generate the tool path is a three dimension model in STL format. The software uses zigzag cut type to move the cutting tool. The cutting tool used in this research is carbide solid square end mill with 5 mm diameter.

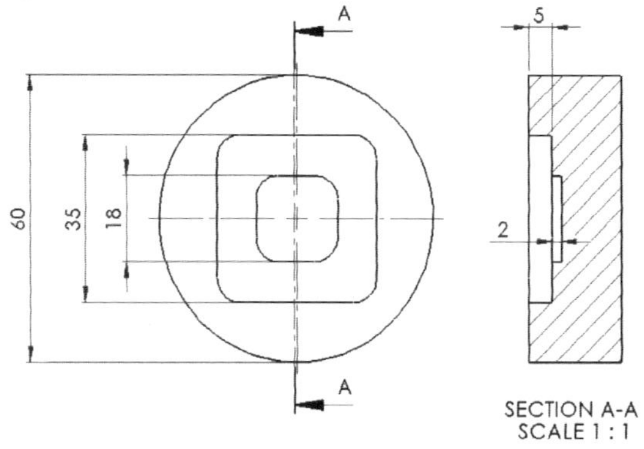

Figure 1. The designed shape.

Roughing and finishing processes are applied in order to create the shape above. Each process requires different parameter values. Table 1 shows the parameter value of the subtractive rapid prototyping for the roughing process. For finishing process, three levels of value for depth of cut, feed rate, and step-over are determined. The value of each level for each parameter is determined based on the machine specification, literature study, and the preliminary experiment as shown in table 2. Meanwhile, the spindle and the entry speed for finishing process are 10000 rpm and 4 mm/s consecutively. In this research, no coolant is used in performing the subtractive rapid prototyping.

IOP Conf. Series: Materials Science and Engineering **187** (2017) 012033 doi:10.1088/1757-899X/187/1/012033

Table 1. Parameter value for roughing.

Feed Rate	: 12 mm/s
Entry Speed	: 4 mm/s
Spindle Speed	: 8500 rpm
Depth of Cut	: 0.37 mm
Step-over	: 1 mm

Table 2. Parameter value for finishing.

	Low	Middle	High
Depth of Cut (mm)	0.1	0.235	0.37
Feed Rate (mm/s)	12	14.5	17
Step-over (mm)	0.3	0.65	1

This research takes several assumptions, which are the polycarbonate material is always homogeneous, the cutting temperature is always constant, and the tool wear occurs after performing three roughing and finishing processes.

The response investigate in this research is the dimensional accuracy. The measurement of the dimensional accuracy is conducted at Industrial Metrology Laboratory of University of Surabaya. The dimensional accuracy is determined based on the difference between the actual and the designed length of horizontal edge (paralel to the feed direction), the difference between the actual and the designed length of vertical edge (paralel to the step over direction), and the difference between the actual and the designed depth of the hole. The actual lenght and the depth are measured by using a caliper with 0.01 μm of accuracy. After the measurement process, the measured data is analyzed by using MINITAB release 14 software.

3. Result and discussion

The first step in implementing the response surface methodology is designing and conducting the first order experiment. The experimental design and the result of the first order experiment are shown in table 3. Based on the experiment results, the first order models of horizontal length error, vertical length error, and depth error are developed.

Table 3. Experiment results.

Std Order	Run Order	Feed Rate (mm/s)	Step-over (mm)	Depth of Cut (mm)	Hor. Length Error (mm)	Ver. Length Error (mm)	Depth Error (mm)
11	1	14.50	0.65	0.235	0.05	0.57	0.00
13	2	14.50	0.65	0.235	0.05	0.61	-0.01
2	3	17.00	0.30	0.100	0.04	0.28	-0.01
12	4	14.50	0.65	0.235	0.03	0.67	-0.04
7	5	12.00	1.00	0.370	0.06	1.00	0.02
9	6	14.50	0.65	0.235	0.04	0.72	0.01
3	7	12.00	1.00	0.100	0.09	1.03	0.00
5	8	12.00	0.30	0.370	0.05	0.36	0.01
10	9	14.50	0.65	0.235	0.05	0.73	0.00
6	10	17.00	0.30	0.370	0.03	0.29	0.03
1	11	12.00	0.30	0.100	0.04	0.40	0.01
4	12	17.00	1.00	0.100	0.09	1.07	0.03
8	13	17.00	1.00	0.370	0.09	1.02	0.02

The mathematical model to predict the horizontal length error values is shown in equation (1). Figure 2 and 3 shows that the residual of the horizontal length error model is to be normally and independently distributed. Furthermore, based on the regresion equation analysis, the residual of the horizontal length error model is identical. The residual tests show that the first order model of horizontal length error is satisfy all the assumptions and can be used as the best prediction of horizontal length error model and no second order experiment is required to be conducted. This model shows that the increase of the feed rate and the step over will increase the horizontal length error value. In the opposite, the increase of the depth of cut will decrease the horizontal length error value. According to the model, the horizontal length error is mostly affected by the step-over.

$$E_{hor} = 0.0144289 + 0.0005\ x\ F + 0.0607143\ x\ S - 0.0277778 \quad (1)$$

where:
E_{hor} is horizontal length error (mm)
F is feed rate (mm/s)
S is step-over (mm)
D is depth of cut (mm).

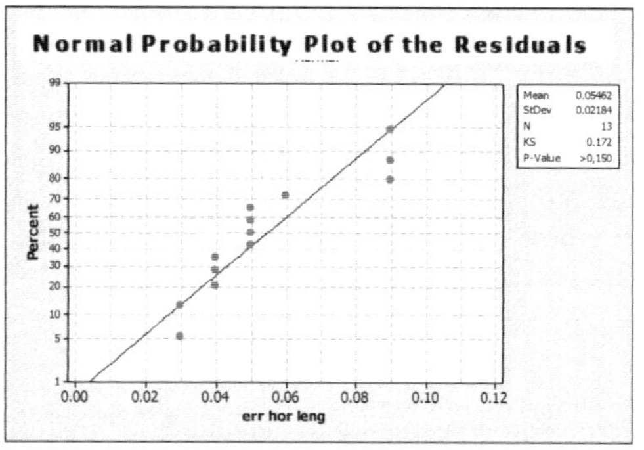

Figure 2. Normal Probability plot of the residuals for horizontal length error.

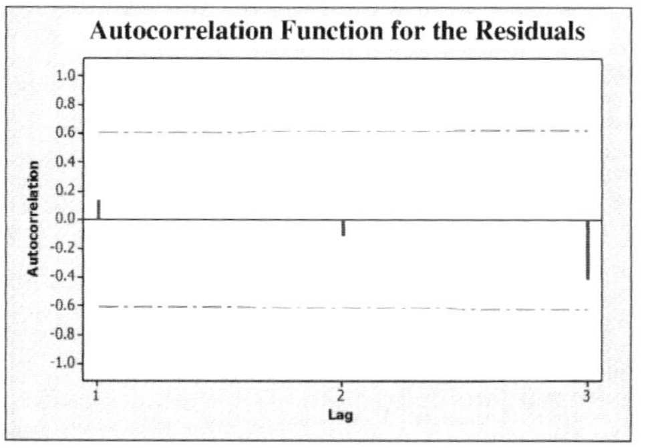

Figure 3. Autocorrelation plot of the residuals for horizontal length error.

Furthermore, the mathematical model to predict the vertical length error values is shown in equation (2). Figure 4 and 5 shows that the residual of the vertical length error model is to be normally and independently distributed. Furthermore, the residual of the vertical length error model is identical based on the regresion equation analysis.

$$E_{ver} = 0.143584 - 0.0065\ x\ F + 0.996429\ x\ S - 0.1018528\ x\ D \quad (2)$$

where:
E_{ver} is vertical length error (mm)
F is feed rate (mm/s)
S is step-over (mm)
D is depth of cut (mm).

This model shows that the increase of the step-over will increase the vertical lenght error value. Meanwhile, the increase of the depth of cut will decrease the vertical length error value. Based on the residual tests, the first order model of vertical lenght error is satisfy all the assumptions and can be used as the best prediction model for vertical length error.

Last, the mathematical model to predict the depth error is developed as shown in equation (3).

$$E_{depth} = -\ 0.0342093 + 0.0015\ x\ F + 0.0107143\ x\ S + 0.0462963 \quad (3)$$

where:
E_{depth} is depth error (mm)
F is feed rate (mm/s)
S is step-over (mm)

D is depth of cut (mm).

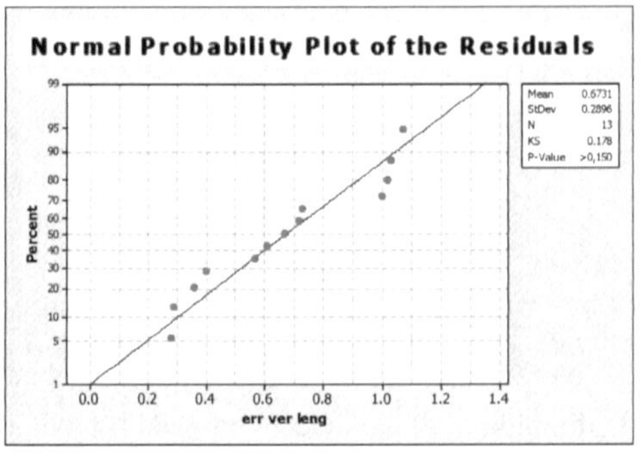

Figure 4. Normal Probability plot of the residuals for vertical length error.

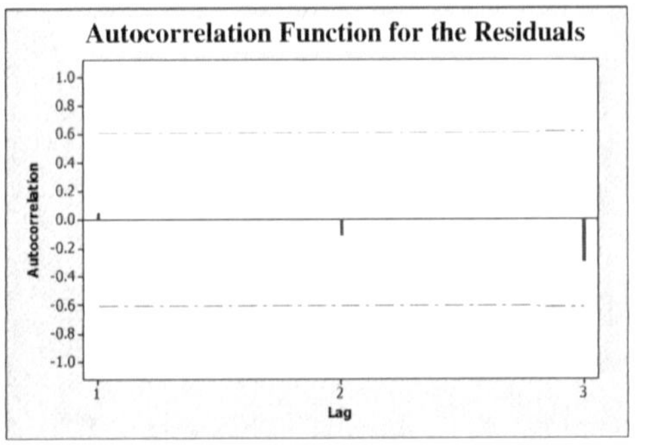

Figure 5. Autocorrelation plot of the residuals for vertical length error.

Figure 6 and 7 shows that the residual of the depth error model is to be normally and independently distributed. Furthermore, based on the regresion equation analysis, the residual of the depth error model is identical. The mathematical model shows that the depth error will increase following the increase of the feed rate, the step over, and the depth of cut. The model also shows that the most influenced factor is the depth of cut. The residual tests shows that the first order model of depth error is satisfy all the assumptions and can be used as the best prediction model for depth error.

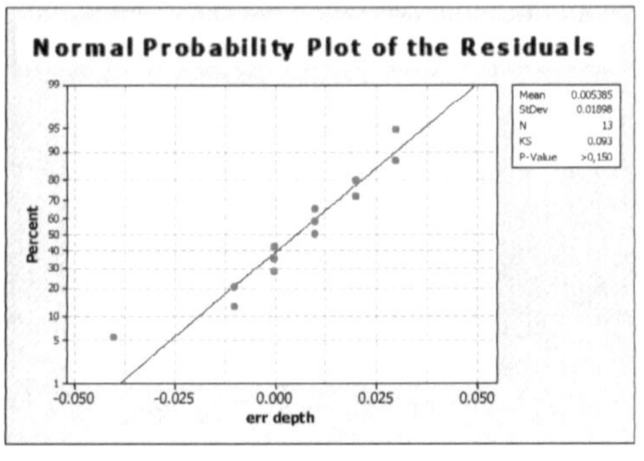

Figure 6. Normal Probability plot of the residuals for depth error.

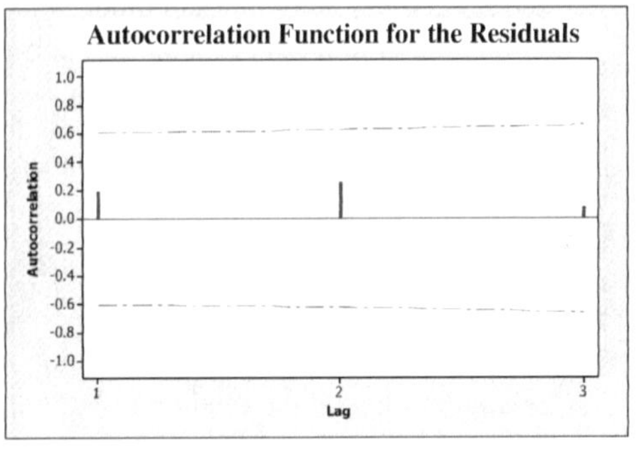

Figure 7. Autocorrelation plot of the residuals for depth error.

4. Conclusions

This result of this research shows that the increase of the step-over will increase the vertical and horizontal length errors. In addition, the increase of the feed rate will increase the vertical length error. Meanwhile, the increase of the depth of cut will decrease the vertical and horizontal length error value. Additionally, the depth error will increase following the increase of all process parameters. This most affected factor of the length error and the depth error are the step-over and the depth of cut consecutively.

The developed error models give an insight on how several process parameters of rapid prototyping will influence the dimensional accuracy of a polycarbonate material. Based on the model, efficient resources utilization can be achieved. This research also provides a preliminary study to optimize the material removal rate and the accuracy in rapid prototyping of the polycarbonate material. For the future work, the optimization of various parameters in rapid prototyping of the polycarbonate material to achieve the highest accuracy, the highest material removal rate, and the minimum surface roughness need to be done. In addition, based on the experiment results, there is a possibility that the length error is affected by the length of the designed shape. In the future, it is also important to investigate this possibility further.

5. References

[1] Toh C K 2004 Surface topography analysis in high speed finish milling inclined hardened steel *Precision Engineering* **28(4)** 386-398
[2] Islam M N, Boswell B and Pramanik 2013 A An investigation of dimensional accuracy of parts produced by three-dimensional printing *Proc. the World Congress on Engineering* vol 1
[3] Ippolito R, Iuliano L and Gatto 1995 A benchmarking of rapid prototyping techniques in terms of dimensional accuracy and surface finish *CIRP Annals-Manufacturing Technology* **44(1)** 157-160
[4] Taft R M, Kondor S and Grant G T 2011 Accuracy of rapid prototype models for head and neck reconstruction *The Journal of prosthetic dentistry* **106(6)** 399-408
[5] Pal D K and and Ravi B 2007 Rapid tooling route selection and evaluation for sand and investment casting *Virtual and Physical Prototyping Journal* **2(4)** 197-207
[6] Brajlih T, Valentan B, Balic J and Drstvensek I 2011 Speed and accuracy evaluation of additive manufacturing machines *Rapid prototyping journal* **17(1)** 64-75
[7] Nieminen I, Paro J and Kauppinen V 1996 High-speed milling of advanced materials *Journal of Materials Processing Technology* **56(1-4)** 24-36
[8] Oktem H, Erzurumlu T and Kurtaran H 2005 Application of response surface methodology in the optimization of cutting conditions for surface roughness *Journal of Materials Processing Technology* **170(1-2)** 11-16
[9] The J S and Candra S 2008 Optimasi proses pemesinan milling fitur pocket material baja karbon rendah menggunakan response surface methodology *Jurnal Ilmiah Teknik Mesin Universitas Kristen Petra* **10(1)** 1-7
[10] Ma W, He G, Zhu L and Guo L 2016 Tool deflection error compensation in five-axis ball-end milling of sculptured surface *The International Journal of Advanced Manufacturing Technology* **84(5-8)** 1421-1430
[11] Zeroudi N and Fontaine M 2015 Prediction of tool deflection and tool path compensation in ball-end milling *Journal of Intelligent Manufacturing* **26(3)** 425-445
[12] Shimana K, Kondo E, Karashima H, Nakao M and Yamashita S 2016 An approach to real-time compensation of machining error using deflection of tool estimated from cutting forces in end-milling process *Journal of Advanced Mechanical Design, Systems, and Manufacturing* **10(2)** 1-10

IOP Conf. Series: Materials Science and Engineering **187** (2017) 012034 doi:10.1088/1757-899X/187/1/012034

Kinematic Structures Description of Bionic Hand Based on VF Set

Xiancan Liu[1, a], Pengying Bai[1], Min Luo[1], Meng Gao[1] Qiang Zhan[2]

[1] Beijing Institute of Mechanical Equipment, Yong Ding Road No.50, Hai Dian District, Beijing, China

[2] Robotics Institute, BeiHang University, Beijing, China

[a]liuxiancan@126.com

Abstract. This paper presents a method for describing kinematic structure of bionic hand based on VF (virtual finger) set. At first, a 20 DOFs (degrees of freedom) human hand kinematic model is built, which is expressed by five fingers' kinematic chains consisting of kinematic pairs and symbols that represent geometric relationships of kinematic pairs' axes. Based on the concept of VF, the hand fingers are divided into two types: VF_{AA} having adduction/abduction motion and VF_{FE} having flexion/extension motion. The concept of VF set comprising VF_{AA}s and VF_{FE}s is defined, human hand and six basic grasp postures are described by VF set. Then, the structures corresponding VF_{AA} and VF_{FE} are given according to active and passive forms of finger joints, and VF_{FE} Structure-Base comprising 20 conventional structures is built. Based on VF set and the structures of VF_{AA} and VF_{FE}, VF sets and kinematic structures of several classic bionic hands are given.

1. Introduction

Human hand is a dexterous and complex tool, which can perform different grasp tasks in our daily life, and it is very important for human being. Therefore, many bionic hands used in rehabilitation or other areas have been designed from views of bionics including anatomy skeleton and grasp characteristics of human hand. In the process of a bionic hand design, how to determine kinematic structure of bionic hand is a key part

The classification of human hand grasp has been studied in anthropology, rehabilitation and medical science. In order to capture human hand versatility for designing prosthetic hands, Schlesinger developed a classic taxonomy of human hand grasp consisting of cylindrical grasp, precision grasp, hook prehension, tip grasp, spherical grasp, and lateral hip [1]. Napier [2]presented a prehensile classification based on function and anatomy structure of human hand, and grasp tasks are divided into power grasp and precision grasp. Based on the concept of virtual finger, Iberall [3] proposed a classification of human hand grasp including pad opposition, palm opposition and side opposition.

For human hand grasp description, there have been some researches about force or contact relation of human hand and grasped object, grasp force imposed on object, virtual fingers and opposition space etc. Arbib[4] presented the concept of VF, which is an abstract representation with a function unit for a collection of individual fingers and hand surfaces applying an opposition force. A single finger or a palm can be a VF, and several fingers grouped together to apply force or torque opposing other VFs or task torques can also be a VF. Iberall [5]proposed the term opposition to describe three basic directions where the human hand can apply forces, which includes pad opposition that occurs between

IOP Conf. Series: Materials Science and Engineering **187** (2017) 012034 doi:10.1088/1757-899X/187/1/012034

hand surfaces along a direction generally parallel to the palm, palm opposition that occurs between hand surfaces along a direction generally perpendicular to the palm, and side opposition that occurs between hand surfaces along a direction generally transverse to the palm. The opposition space can be described by both physical terms such as amount and orientation of force vectors and abstract terms such as types and numbers of oppositions, virtual-to-real finger mappings and VF lengths. Liu and Zhan [6] presented a method to describe and analyze the human hand grasp postures so as to indicate which fingers should act during grasping and the required movements of those fingers.

2. Kinematic structure of bionic hands based on VF Set

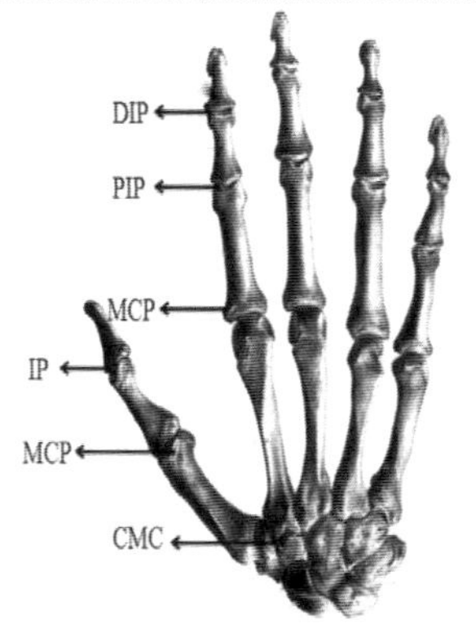

Figure 1. Human hand skeleton.

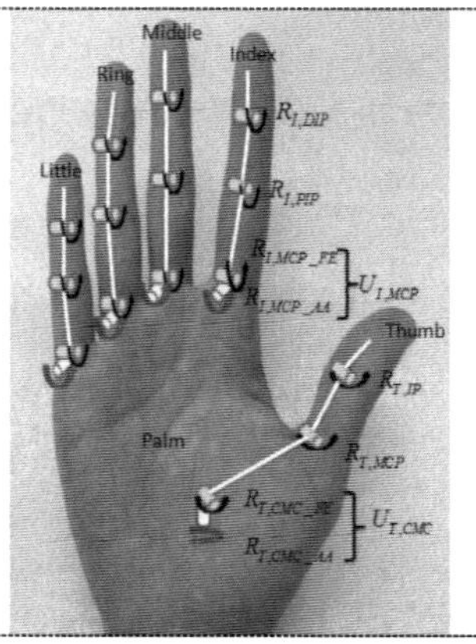

Figure 2. Simplified kinematic model of human hand.

Human hand is a complex and dexterous tool with multi-DOFs, and it can perform various grasp postures in our daily life. Based on human hand skeleton (Fig.1), a kinematic model with 20 DOFs is presented, which is shown in Fig.2. We do not consider the CMC (carometacarpal) joints of four fingers because of having little effect on grasp. In the kinematic model, the thumb has 4 DOFs, 2 DOFs for adduction/abduction and flexion/extension motions in the CMC joint, two for flexion/extension in the MCP (metacarpophalangeal) joint and the IP (interphalangeal) joint, respectively. Each of four fingers has 4 DOFs, 2 DOFs for adduction/abduction and flexion/extension motions in the MCP joint, and the remaining two in the DIP (distal interphaalangeal) joint and PIP joint, respectively. All joints of the simplified human hand model are divided into two types: revolute (proxomal interphalangeal) joints and universal joints. The revolute joints perform flexion/extension movement of fingers, and universal joints perform flexion/extension and adduction/abduction movements of fingers. The DIP and PIP joints of four fingers belong to revolute joints, and the MCP and IP joints of the thumb also belong to revolute joints. The MCP joints of four fingers and the CMC joint of thumb belong to universal joints. In order to conveniently analyze, it is regarded a universal joint as two revolute joints, axes of which are orthogonal and intersecting. Therefore, based on the method proposed by Yangtingli[7] , the kinematic structure of human hand is expressed as follows:

$$
\begin{aligned}
& TC\left\{R_{T,CMC-AA} \perp R_{T,CMC-FE} \parallel R_{T,MCP} \parallel R_{T,IP}\right\}, IC\left\{R_{I,MCP-AA} \perp R_{I,MCP-FE} \parallel R_{I,PIP} \parallel R_{I,DIP}\right\} \\
& MC\left\{R_{M,MCP-AA} \perp R_{M,MCP-FE} \parallel R_{M,PIP} \parallel R_{M,DIP}\right\}, RC\left\{R_{R,MCP-AA} \perp R_{R,MCP-FE} \parallel R_{R,PIP} \parallel R_{R,DIP}\right\} \\
& LC\left\{R_{L,MCP-AA} \perp R_{L,MCP-FE} \parallel R_{L,PIP} \parallel R_{L,DIP}\right\}
\end{aligned}
\qquad (1)
$$

Where, "$\perp$" represents that axis of one kinematic pair is perpendicular to axis of the other one. "$\parallel$" represents that one kinematic pair is parallel to the other one. The thumb chain is abbreviated to TC. IC, MC, RC, and LC are defined in the same way.

IOP Conf. Series: Materials Science and Engineering **187** (2017) 012034 doi:10.1088/1757-899X/187/1/012034

Virtual Finger (VF) is an abstract representation, a function unit that comprises at least one real physical finger. Therefore, a VF can represent thumb, index finger and all four fingers, and it applies a force or torque opposing other VFs or task torques. Anatomical joint configurations, range of motion and finger size are described as kinematic components of VFs such as lengths, widths, and orientations[5]. Human hand grasp movement is performed by combination motion of some virtual fingers. Based on revolute and universal joints of fingers, motions of each finger can be divided into two types: flexion/extension motion and adduction/abduction motion. The virtual finger having flexion/extension motion is defined as VF_{FE}, and the virtual finger having adduction/abduction motion is defined as VF_{AA}. If one finger has flexion/extension and adduction/abduction motions, it is considered that the finger belongs to not only VF_{FE} but also VF_{AA}. A set that contains different types of VFs is named VF set, which can be used to express any grasp posture of human hand bionic hands. For human hand, it can be described by VF set, as follows:

$$S_{human\ hand} = \{VF_{thumb-FE}, VF_{thumb-AA}, VF_{index-FE}, VF_{index-AA}, VF_{middle-FE}, VF_{middle-AA}, VF_{ring-FE}, VF_{ring-AA}, VF_{little-FE}, VF_{little-AA}\} \quad (2)$$

3. Kinematic structure of bionic hands based on VF set

The two types of VFs describe what motion a finger has, but how are all joints of a designed bionic hand finger distributed? In this paper, a VF kinematic structure is expressed by revolute joints and symbols that represent the geometric relationship of active and passive finger joints.

3.1. Structures of VF_{AA}

A VF_{AA} has usually one DOF making the virtual finger adduction/abduction motion. The adduction/abduction DOFs locate in the CMC joint of the thumb and the MCP joints of four fingers. In this paper, the kinematic structure of VF_{AA} is represented by a kinematic pair, and is shown in Fig.2. Therefore, while these joints are active, VF_{AA} structures for five fingers are given as follows:

$$\{R_{T,CMC-AA}\}, \{R_{I,MCP-AA}\}, \{R_{M,MCP-AA}\}, \{R_{R,MCP-AA}\}, \{R_{L,MCP-AA}\} \quad (3)$$

The VF_{AA} structure for the thumb is the most important among five VF_{AA} structures, because it makes the thumb to be opposite other four fingers. In addition, the CMC joint of the thumb can be passive, and the adduction/abduction of thumb is achieved by manual assistance. Therefore, VF_{AA} structure for the thumb is expressed as $\tilde{R}_{T,CMC-AA}$ in which"~" represents that the CMC joint is passive.

3.2. Structures of VF_{FE}

The flexion/extension joints of VFFE can be active or passive, so there are various VFFE kinematic structures of bionic hand finger based on different active and passive ways that can including all actuated, under-actuated and motion-coupled. When some fingers having the synchronous motion as a VFFE, in whose structure the MCP joints of fingers can be connected. We give a VFFE Structure-Base including different VFFE structures, as shown Table.1.

Table 1. VF_{FE} Structure-Base.

Order number	DOF	Fingers	VF_{FE} structures
1	3	thumb	$\{R_{T,CMC-FE} \parallel R_{T,MCP} \parallel R_{T,IP}\}$
2	2	thumb	$\{R_{T,CMC-FE} \parallel R_{T,MCP} \parallel \tilde{R}_{T,IP}\}$
3	1	thumb	$\{R_{T,CMC-FE} \parallel \tilde{R}_{T,MCP} \parallel \tilde{R}_{T,IP}\}$

4	3	index finger	$\{R_{I,MCP-FE} \parallel R_{I,PIP} \parallel R_{I,DIP}\}$
5	2	Index finger	$\{R_{I,MCP-FE} \parallel R_{I,PIP} \parallel \tilde{R}_{I,DIP}\}$
6	1	Index finger	$\{R_{I,MCP-FE} \parallel \tilde{R}_{I,PIP} \parallel \tilde{R}_{I,DIP}\}$
7	3	middle finger	$\{R_{M,MCP-FE} \parallel R_{M,PIP} \parallel R_{M,DIP}\}$
8	2	middle finger	$\{R_{M,MCP-FE} \parallel R_{M,PIP} \parallel \tilde{R}_{M,DIP}\}$
9	1	middle finger	$\{R_{M,MCP-FE} \parallel \tilde{R}_{M,PIP} \parallel \tilde{R}_{M,DIP}\}$
10	3	ring finger	$\{R_{R,MCP-FE} \parallel R_{R,PIP} \parallel R_{R,DIP}\}$
11	2	ring finger	$\{R_{R,MCP-FE} \parallel R_{R,PIP} \parallel \tilde{R}_{R,DIP}\}$
12	1	ring finger	$\{R_{R,MCP-FE} \parallel \tilde{R}_{R,PIP} \parallel \tilde{R}_{R,DIP}\}$
13	3	little finger	$\{R_{L,MCP-FE} \parallel R_{L,PIP} \parallel R_{L,DIP}\}$
14	2	little finger	$\{R_{L,MCP-FE} \parallel R_{L,PIP} \parallel \tilde{R}_{L,DIP}\}$
15	1	little finger	$\{R_{L,MCP-FE} \parallel \tilde{R}_{L,PIP} \parallel \tilde{R}_{L,DIP}\}$
16	1	index finger, middle finger	$\{\not{R}_{I,MCP-FE} \parallel \tilde{R}_{I,PIP} \parallel \tilde{R}_{I,DIP}\}$ $\{\tilde{\not{R}}_{M,MCP-FE} \parallel \tilde{R}_{M,PIP} \parallel \tilde{R}_{M,DIP}\}$
17	1	middle finger, ring finger	$\{\not{R}_{M,MCP-FE} \parallel \tilde{R}_{M,PIP} \parallel \tilde{R}_{M,DIP}\}$ $\{\tilde{\not{R}}_{R,MCP-FE} \parallel \tilde{R}_{R,PIP} \parallel \tilde{R}_{R,DIP}\}$
18	1	ring finger, little finger	$\{\not{R}_{R,MCP-FE} \parallel \tilde{R}_{R,PIP} \parallel \tilde{R}_{R,DIP}\}$ $\{\tilde{\not{R}}_{L,MCP-FE} \parallel \tilde{R}_{L,PIP} \parallel \tilde{R}_{L,DIP}\}$
19	1	middle finger,ring finger, little finger	$\{\not{R}_{M,MCP-FE} \parallel \tilde{R}_{M,PIP} \parallel \tilde{R}_{M,DIP}\}$ $\{\tilde{\not{R}}_{R,MCP-FE} \parallel \tilde{R}_{R,PIP} \parallel \tilde{R}_{R,DIP}\}$ $\{\tilde{\not{R}}_{L,MCP-FE} \parallel \tilde{R}_{L,PIP} \parallel \tilde{R}_{L,DIP}\}$

20	1	index finger, middle finger, ring finger,little finger	$\left\{ \not{R}_{I,MCP-FE} \parallel \tilde{R}_{I,PIP} \parallel \tilde{R}_{I,DIP} \right\}$ $\left\{ \tilde{\not{R}}_{M,MCP-FE} \parallel \tilde{R}_{M,PIP} \parallel \tilde{R}_{M,DIP} \right\}$ $\left\{ \tilde{\not{R}}_{R,MCP-FE} \parallel \tilde{R}_{R,PIP} \parallel \tilde{R}_{R,DIP} \right\}$ $\left\{ \tilde{\not{R}}_{L,MCP-FE} \parallel \tilde{R}_{L,PIP} \parallel \tilde{R}_{L,DIP} \right\}$

[a] "~" represents that the joint is passive and actuated by under-actuated or motion-coupled way.
[b] "—" represents that coaxial axes of some MCP joints that have the synchronous motion are connected, and the whole VF_{FE} structure is drove by one motor.

[c] " $\tilde{\not{R}}$ " represents that the joint is passive and its coaxial axis is connected.

The conventional 20 VF_{FE} structures are given in this paper. However, VF_{FE} structure-base is opening and some other VF_{FE} structures can be added. All actuated VF_{FE} structures (1, 4, 7, 10, 13) are shown in Fig.3. In passive way, 2 DOFs (2, 5, 8, 11, 14) and 1 DOF (3, 6, 9, 12, 15) VF_{FE} structures are shown in Fig.4, respectively. The passive VF_{FE} structure (20) whose coaxial axes are connected is shown in Figuer.5.

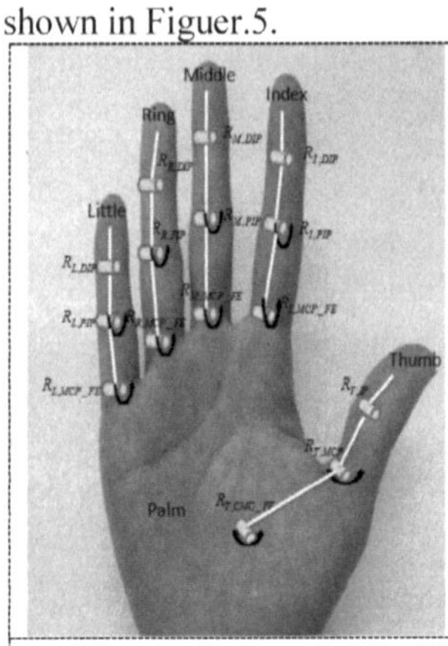

Figure 3. Two DOFs VF_{FE} structures.

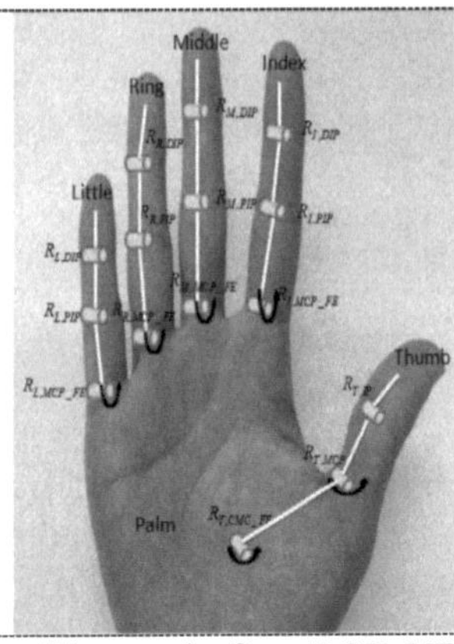

Figure 4. One DOF VF_{FE} structures.

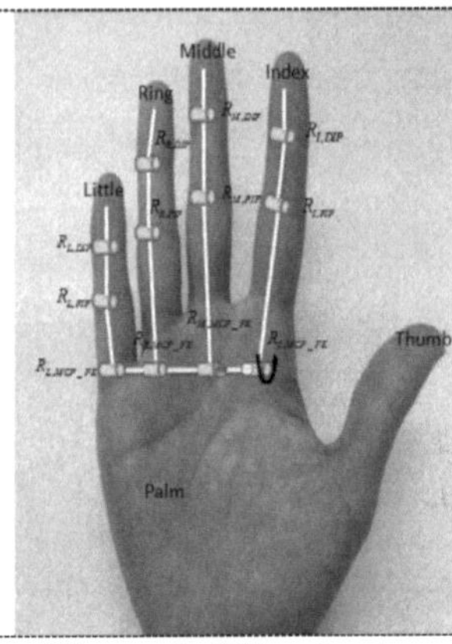

Figure 5. VF_{FE} structure of NO.20.

3.3. Structures of classical bionic hands

A finger has usually adduction/abduction and flexion/extension motions, so it has not only VF_{AA} structure but also VF_{FE} structure, whose axes are perpendicular in the CMC and the MCP joints. Bionic hands are made up of fingers, so structures of bionic hands are consisted of VF_{AA} and VF_{FE} structures. By analyzing grasp functions and mechanism design, kinematic structures of several DLR Hand and i-LIMB Hand are given.

- Kinematic structure of DLR Hand is given as follows:

$$TC\left\{ R_{T,CMC-AA} \perp R_{T,CMC-FE} \parallel R_{T,MCP} \parallel \tilde{R}_{T,IP} \right\}, IC\left\{ R_{I,MCP-AA} \perp R_{I,MCP-FE} \parallel R_{I,PIP} \parallel \tilde{R}_{I,DIP} \right\}$$
$$MC\left\{ R_{M,MCP-AA} \perp R_{M,MCP-FE} \parallel R_{M,PIP} \parallel \tilde{R}_{M,DIP} \right\}, RC\left\{ R_{R,MCP-AA} \perp R_{R,MCP-FE} \parallel R_{R,PIP} \parallel \tilde{R}_{R,DIP} \right\} \quad (4)$$

- Kinematic structure of i-LIMB Hand is given as follows:

$$
\begin{aligned}
&TC\left\{\tilde{R}_{T.CMC-AA} \perp R_{T.CMC-FE} \parallel \tilde{R}_{T.MCP\ and\ IP}\right\}, IC\left\{R_{I.MCP-FE} \parallel \tilde{R}_{I.PIP\ and\ DIP}\right\} \\
&MC\left\{R_{M.MCP-FE} \parallel \tilde{R}_{M.PIP\ and\ DIP}\right\}, RC\left\{R_{R.MCP-FE} \parallel \tilde{R}_{R.PIP\ and\ DIP}\right\}, LC\left\{R_{L.MCP-FE} \parallel \tilde{R}_{L.PIP\ and\ DIP}\right\}
\end{aligned} \quad (5)
$$

4. Conclusion

This paper presents a method of determining kinematic structure of bionic hand based on set mapping. The designers can directly obtain kinematic structure of bionic hand to design according to required grasp functions. The method is general for bionic hands such as prosthetic hands and dexterous robotic hands. In addition, it is easily understood and used by designers who can choose appropriate VF_{AA} and VF_{FE} structures according to the actual situation.

Based on the concept of virtual finger (VF), human hand fingers are divided into two types: VF_{AA} and VF_{FE}. It helps us understand human hand finger movement characteristic. Six grasp postures in this paper are usually used in our daily life. Of course, some grasp postures can be performed by less fingers or different fingers. For example, the best case of spherical grasp is that five fingers take part in grasp. In actual situation, even if two fingers in opposition directions can perform spherical grasp. However, it is less steady than five fingers for heavy object grasped. The VF set comprising VFs is used to describe not only six basic grasp postures but also any grasp posture. The VF set can also describe grasp postures of bionic hands and hominine hands. The description method makes people understand earsily the inside structure of a bionic hand, and it shows that the relation of grasp function and the bionic hand structure.

In future works, some special VF_{FE} structures will be added into Structure-Base of VF_{FE}. In this paper the VF set shows which fingers take part in and what motion types the fingers in grasping. However, VF sets of some grasp postures such as tip grasp and lateral hip can have the same VF set. Therefore, the VF set will describe grasp postures in detail, and contact areas of the fingers in grasping will be expressed.

5. References

[1] Kyberd P and Pons J 2003 *Proc. Int. Conf. on Robotics & Automation* vol 3 pp3231-3236
[2] Napier J 1956 *Bone Joint Surg* 38B: 902-913
[3] Iberall T 1986 *Opposition Space as A Structuring Concept for The Analysis of Skilled Hand Movements* (Berlin: Springer-Verlag)
[4] Arbib M and Iberall T 1985 *Coordinated Control for Movements of the Hand* (Berlin: Springer-Verlag)
[5] Iberall T 1997 *The International Journal of Robotics Research* vol 16(3) pp285-299
[6] Liu X C and Zhan Q 2012 *Medical Engineering & Physics* vol 35(7) pp1020-1027
[7] Yang T L 2004 *Topology Structure Design of Robot Mechanism* (Beijing: China Machine Press)

IOP Conf. Series: Materials Science and Engineering **187** (2017) 012035 doi:10.1088/1757-899X/187/1/012035

Noise optimization of a regenerative automotive fuel pump

J F Wang*, **H H Feng, X L Mou and Y X Huang**

School of Mechanical Engineering, Beijing Institute of Technology, Beijing, 100081 China

E-mail: wjfbit@163.com

Abstract. The regenerative pump used in automotive is facing a noise problem. To understand the mechanism in detail, Computational Fluid Dynamics (CFD) and Computational Acoustic Analysis (CAA) together were used to understand the fluid and acoustic characteristics of the fuel pump using ANSYS-CFX 15.0 and LMS Virtual. Lab Rev12, respectively. The CFD model and acoustical model were validated by mass flow rate test and sound pressure test, respectively. Comparing the computational and experimental results shows that sound pressure levels at the observer position are consistent at high frequencies, especially at blade passing frequency. After validating the models, several numerical models were analyzed in the study for noise improvement. It is observed that for configuration having greater number of impeller blades, noise level was significantly improved at blade passing frequency, when compared to that of the original model.

1. Introduction

Regenerative pumps have found applications in many industrial areas which require high heads at low flow rates including automotive and aerospace fuel pumping, booster systems, water supply, agricultural industries, shipping and mining, and chemical and food processing systems[1]. The fluid and NVH characteristics of regenerative fuel pump need more attention simultaneously. Both vibrations and noise can affect the regenerative pump performance and its life. In regenerative pumps, the sources of vibrations and noise may lie in hydraulic or mechanical aspects, but under normal operating condition, the blade passing frequency is the most usual excitation of vibrations and noise. This frequency results from the nonuniformity of the flow at the impeller outlet which is caused by the effects of the rotor-stator interaction.

Momentum exchange theory was used to explain fluid mechanism within regenerative pumps[2], and was improved by the following researchers[3-6]. Meanwhile, the fluid performance of regenerative pumps was improved by optimizing different parameters numerically[7-9] and experimentally[10-12]. Recent years, noise improvement has gradually got more attention[13,14]. Hence, it is important to conduct in-depth research on the noise performance of regenerative pump.

Therefore, the main purpose of this study is to explore the effects of factors such as the blade numbers on the noise performance of a regenerative automotive fuel pump. To achieve this objective, the CFD and CAA models were considered for the regenerative pump. For verifying the simulation results, the fluid and noise performance tests were applied. Accuracy of the simulations was assessed by comparisons to experimental data.

2. Regenerative fuel pump geometry and grid generation

The regenerative fuel pump-tank system consists of a fuel tank and a fuel pump module, as plotted in **Figure 1**. For CFD simulation, we just considered the flow domain in the fuel pump module, as shown in **Figure 2**. More detailed specifications are shown in **Table 1**, and sectional view of the impeller in **Figure 3**. Simplification was considered in the outlet domain.

For establishing grid independency, analysis was carried out using the meshing program ICEM (ANSYS Inc.), with number of elements restricted to approximately 1 million, 2 million and 4 million elements containing hybrid hexahedral/tetrahedral grids, and the grids were refined at the wall region where the flow properties change rapidly, the y+ value of the first grid point from the wall was adjusted to be less than 2. The mass flow rates of the three schemes and experiment are shown in **Table 2**. Since the variation in the mass flow rate values was less than 1.2% for models containing elements greater than 1 million, mesh size of approximately 1 million was adopted to save the computational time, and relative discrepancy between experimental and computational results was 1.9%.

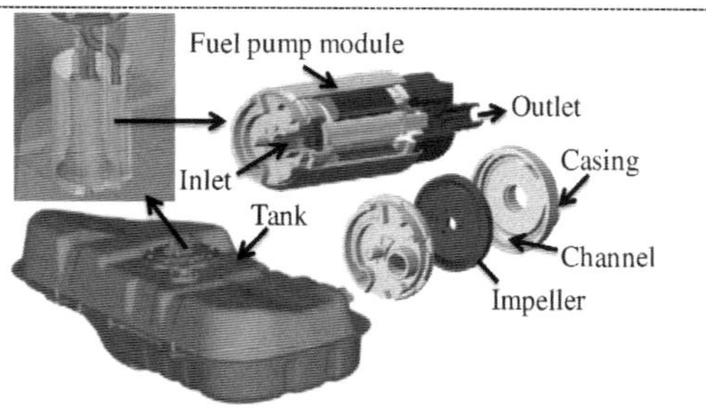

Figure 1. Regenerative fuel pump-tank system.

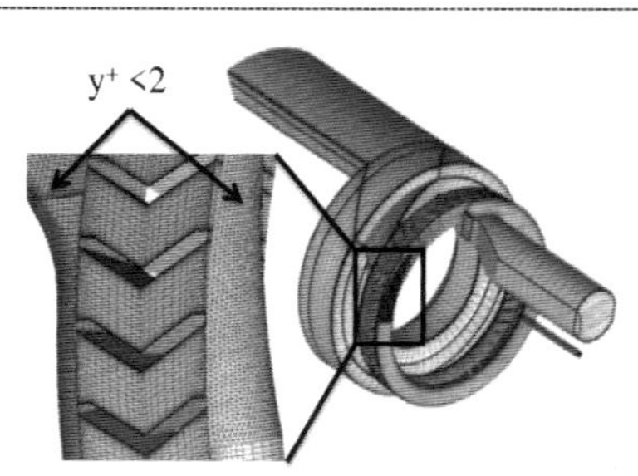

Figure 2. CFD computational domain and grids.

Table 1. Design specifications for the regenerative fuel pump.

Impeller thickness (mm)	3.8
Impeller diameter (mm)	32.8
Blade thickness b (mm)	0.4
Number of baldes Z (EA)	47
Chevron blade angle α (°)	30
Design mass flow rate (g/s)	20.5
Rotational speed (rpm)	6000
Pressure rise (kPa)	400

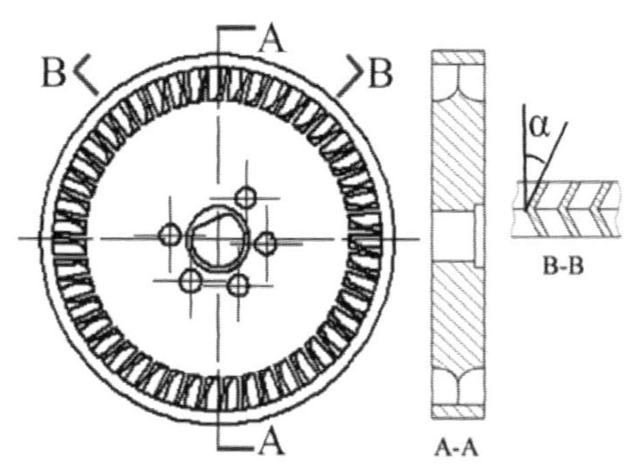

Figure 3. Sectional view of the impeller.

Table 2. Grid independency test.

Number of elements	1092018	2122816	4137091	Experimental result
Mass flow rate (g/s)	20.11	19.99	19.85	20.5

3. Computational models set up

3.1. CFD model

Three-dimensional RANS equations were solved in ANSYS CFX 15.0 to obtain the fluid performance. Gasoline was considered as the working fluid. The total pressure 0 kPa and 400 kPa were assigned at the inlet and outlet domain respectively, and the rotating speed, 6000 rpm, was set to the impeller domain. No slip wall condition was specified at the wall boundaries of the blades and the casing. The turbulence is simulated using the k-ω-based shear stress transport (SST) turbulence model. A high-resolution scheme with second-order accuracy in space was used for the numerical scheme of the convection terms in the governing equations.

The interface between stationary and rotating domains was set to frozen rotor method in steady state computations, and to transient rotor stator method in unsteady state computations, respectively. The time step and coefficient loops for time scale control were set to 2.12766×10^{-5} s and 10 times respectively in order to reduce all maximum residuals to a value below 10^{-5}. This provided 10 time steps per impeller blade passage. Since the nature of flow is unsteady, it requires numerical analysis until the transient fluctuations of the flow field become time periodic as judged by the pressure fluctuations. In the present analysis this has been achieved after 5 complete rotations of the impeller. The pressure at the monitor points were recorded corresponding to each rotation of the impeller by time step advancement. Since the nature of flow is unsteady, it requires numerical analysis until the transient fluctuations of the flow field become time periodic as judged by the pressure fluctuations. In the present analysis this has been achieved after 5 complete rotations of the impeller. The pressure at the monitor points were recorded corresponding to each rotation of the impeller by time step advancement, as shown in **Figure 4**. The data extracted from the fifth revolution will be used for the following acoustic analysis.

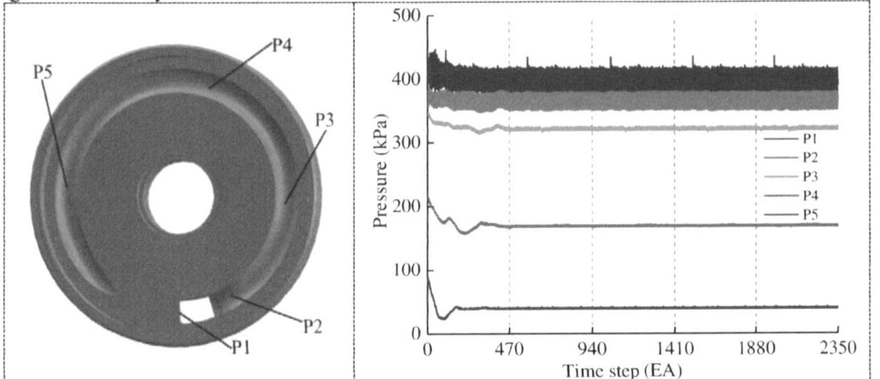

Figure 4. Monitor points and corresponding pressure fluctuations.

3.2. CAA model

In this work, acoustic analysis was implemented using finite/boundary element method, vibration-acoustic coupling method and Automatically Matched Layer (AML) method in LMS Virtual. Lab Rev12. In this process, the pressure fluctuations on the inner wall of the fuel pump extracted from the unsteady flow were used for acoustic analysis.

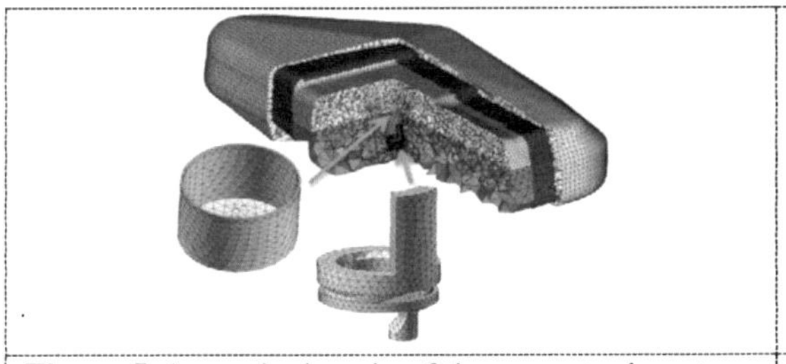

Figure 5. Acoustic domain of the pump-tank system.

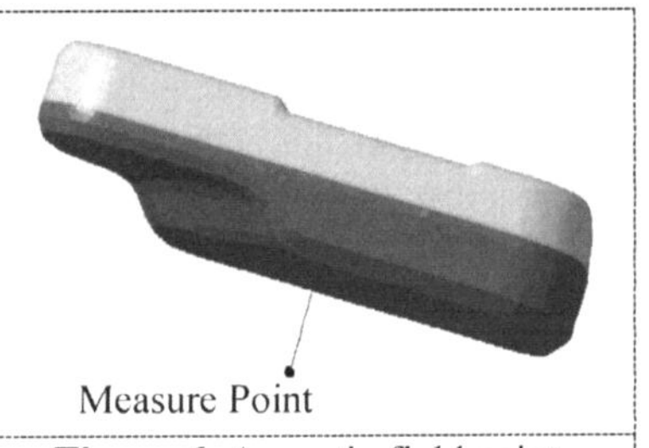

Figure 6. Acoustic field point.

Figure 5 represents the whole acoustic domain for the acoustic analysis, which consists of pump boundary, oil pool boundary, air, gasoline and fuel tank domains. A tetrahedral grid system was constructed in the acoustic domain for acoustic analysis with approximately 136061 grid points to satisfy the target element length. An observer point is set to messure the sound pressure, which is 200mm below the fuel tank, corresponding to the measure position, as shown in **Figure 6**.

4. Validation of the models

For validating the accuracy of the computational model, acoustic experiment of the regenerative fuel pump was implemented in the semi-anechoic room. The fuel pump was tested under installation condition, as shown in **Figure 7**. The 40AK type microphone and 26AK type preamplifier together were used to measure the sound pressure, located 200mm below the fuel tank. The LMS SCADAS III system was used for data acquisition.

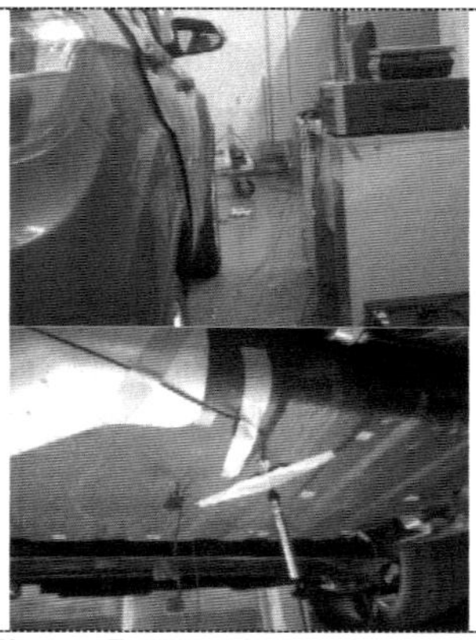

Figure 7. Experimental set up.

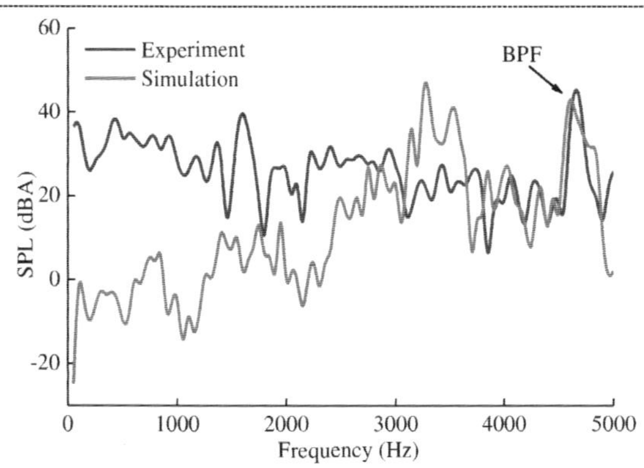

Figure 8. SPL comparison between simulation and experiment.

Figure 8 represents the sound pressure levels comparison between computational and experimental results at the monitor point. As we just considered fluid induced noise, the blade passing frequency noise was the interest we had in the fuel pump. Mechanical noise caused by rotor misalignment and electromagnetic noise caused by DC motor commutators were not considered, which related to the basic frequency and harmonic frequencies corresponding to low and medium frequency range. So the simulational results were lower than experimental results under 3000Hz. Between 3000 and 4000 Hz, sound pressure levels were higher in simulation than that in experiment, the probable reason was that some structures such as feed tubes, flange and damping layer were not considered. Similar trend between simulation and experiment was observed after 4000 Hz, especially at the blade passing frequency(BPF) where SPLs were 45.38 dB and 47 dB respectively, with a relative discrepancy about 3.4%.

5. Geometry modification

Effects of chevron blade angle α, blade thickness b and blade number Z on noise performance of the fuel pump were studied. Different parameters are shown in **Table 3**, where Z=47, b=0.4 and α=30° are the original parameters.

Table 3. Different parameter modifications.

Parameter	Blade number Z	chevron angle α	blade thickness b
	42	25°	0.35 mm
Value	47	30°	0.4 mm
	52	35°	0.45 mm

Sound pressure level at the measure point was the criteria to evaluate different parameter modifications. **Figure 9**, **Figure 10** and **Figure 11** show the comparisons of the 1/3 octave band SPL spectra between numerical results of different parameters. It can be seen that around the BPF and higher frequencies, chevron blade angle and blade thickness have no much influence on the sound pressure levels. On the contrary, the modification with 52 blades decreases the SPL significantly at high frequencies, especially at BPF. **Figure 12** represents the comparison of the whole frequency band SPL spectra between 52 blades model and original model, value of SPL is decreased by 20.2 dB at BPF. At the same time, efficiency of the fuel pump only decreases from 56% to 53%.

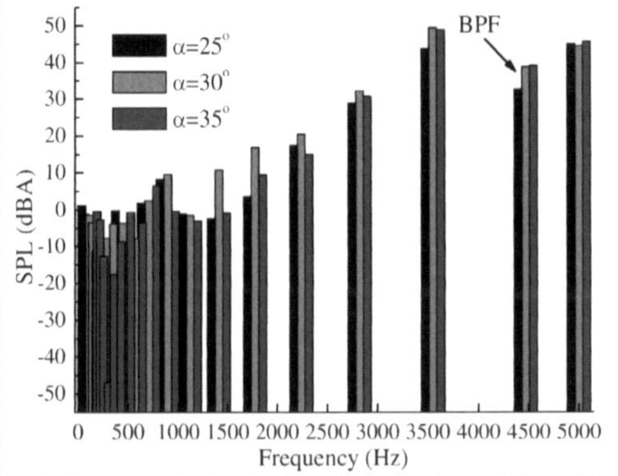

Figure 9. 1/3 octave band SPL spectra for different chevron blade angles.

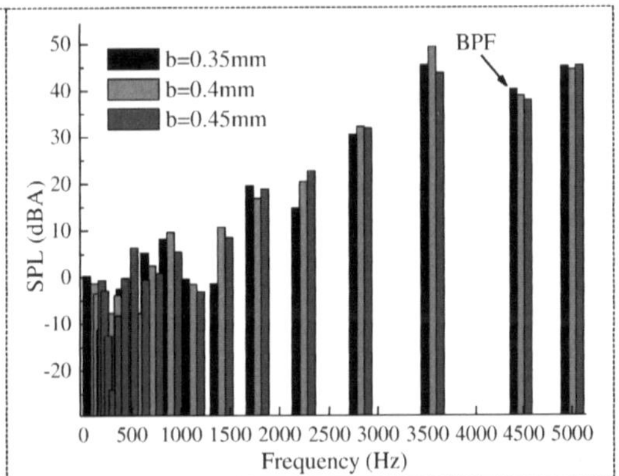

Figure 10. 1/3 octave band SPL spectra for different blade thicknesses.

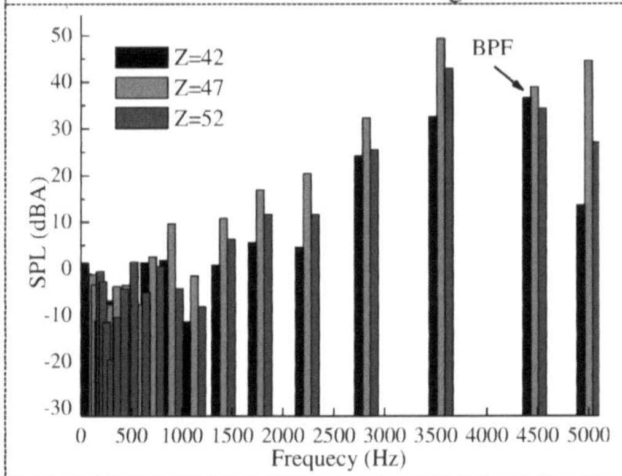

Figure 11. 1/3 octave band SPL spectra for different blade numbers.

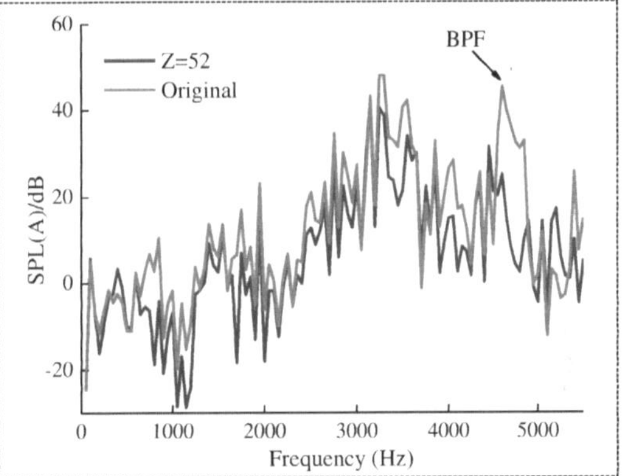

Figure 12. Whole frequency band SPL spectra for modified and original models.

6. Conclusion

The numerical analysis using CFD and CAA helped understand the fluid and noise performance of a regenerative pump. The CFD results of the mass flow rate only had a relative discrepancy of 1.9% compared to experiment, and the calculated sound pressure levels agreed well with experimental results at high frequencies, especially at blade passing frequency only a relative discrepancy of 3.4% existed. Three parameters were selected to study the effects on the noise level of the regenerative fuel pump. The modification with 52 blades can improve the noise level, and decreases the SPL by 20.2 dB at BPF, only decreases the efficiency from 56% to 53%.

7. References

[1] Y Teshome, E Dribsa 2007 *M.S. thesis* University of Addis Ababa, Addis Ababa, Ethiopia
[2] M Badami 1997 *SAE Technical Paper Series* No. 971074 45-55
[3] J W Song, A Engeda, M K Chung 2003 *Proc. IMechE* A *J. Power and Energy* **217** 311-321
[4] M M Raheel 2003 *Ph.D. thesis* Michigan State University East Lansing MI USA
[5] I S Yoo, M R Park, M K Chung 2005 *Proc. IMechE* A *J. Power and Energy* **219** 567-581

[6] F J Quail, M Stickland, A Baumgartner 2011 *Trans. ASME* **133** 093001-1-093001-7
[7] K V Karanth, N Y Sharma 2014 *Proc. 3rd World Conf. on Applied Sciences, Engineering and Technology* 344-349
[8] F J Quail, T Scanlon, M Stickland 2011 *Int. J. Numerical Methods for Heat & Fluid Flow* **21** 95-111
[9] F J Quail T Scanlon, A Baumgartner 2012 *European J. Mechanics* B/Fluids **31** 181-187
[10] M Hübel, B Blättel, W Strohl 1995 *SAE Technical Paper Series* No. 950077 131-139
[11] W C Choi1, I S Yoo, M R Park, M K Chung 2013 *J. Power and Energy* **227** 585-592
[12] T A-M Mekhail, O M Dahab, M F Sadik, M M El-Gendi and H S Abdel-Mohsen 2015 *Open J. Fluid Dynamics*, **5** 224-237
[13] T A R Rossetti, M P D Passo, F B Geremias 2013 *SAE Technical Paper Series* No. 2013-36-0625
[14] M W Heo, T W Seo, H S Shim, K Y Kim 2016 *J. Mechanical Science and Technology* **30** 1197-1208

Acknowledgements
This research was financially supported by the National Natural Science Foundation of China (Grant No. 51006010) and a joint China-UK research programme "111" (No. B12022).

IOP Conf. Series: Materials Science and Engineering **187** (2017) 012036 doi:10.1088/1757-899X/187/1/012036

Experimental Study of Material Removal Rate in Electrical Discharge Turning of Titanium Alloy (Ti-6al-4v)

Y M Puri[1] and Vikas Gohil[1]

[1]CAD-CAM Centre, Department of Mechanical Engineering, Visvesvaraya National Institute of Technology, Nagpur 440010, MS, India

2vikasgohil@gmail.com

Abstract. Electrical discharge turning (EDT) is a new machining process in which an external axis is added to a conventional EDM machine in order to produce precise cylindrical forms on hard and difficult to machine materials. By feeding a pre shaped tool electrode against a rotating work piece, axially symmetrical pats can be produce. The machining performance of EDT process is influenced by its machining parameters, which directly affect the quality of the machined component. This paper presents an experimental study on the effects of EDM parameters namely pulse-on time, peak current, gap voltage, spindle speed and flushing pressure on material removal rate (MRR) in electrical discharge turning of titanium alloy Ti-6Al-4V. This has been done by means of the Taguchi's design of experiment technique. A mathematical model has been developed for MRR by regression analysis and factor effects were analyzed using analysis of variance (ANOVA). Signal-to-noise ratio analysis is used to find the optimal condition.

1. Introduction

Electrical discharge machining (EDM) is largely disseminated among several industrial sectors and is no longer considered an unconventional machining process [1]. EDM machines accounts for about 7 percent of all machine tool sales in the world [2]. This machining method uses a series of discrete electrical sparks between a tool electrode and a work piece immersed in a dielectric fluid and subjected to an electric voltage. Literally thousands of high temperature electrical discharges per second are generated and each electrical discharge produces a tiny crater by melting and vaporization, thus work material is removed and desired shape of the tool is produced into the work piece [3]. The hardness and strength of the difficult-to-machine work-materials are no longer the dominating factors that affect the tool wear and hinder the machining process. This makes the EDM process particularly suitable for machining hard, difficult-to- cut materials [4]. The EDT process is being extended from wire electrical discharge turning (WEDT) in order to achieve a high degree of accuracy, surface finish and cylindricity in straight turning as well as in form turning of hard-to-machine material having high slenderness ratio. One of the main applications of the die-sinking EDM machine is in turning aerospace honey-comb seals where internal and external seals are machined to a burr free finish with internal or external diameters from 300mm-1400mm [5]. The concept and part machined using EDT method is shown in fig.1 and fig. 2 respectively.

The idea of employing EDM to machine cylindrical parts for manufacturing small diameter pins has been reported by Dr. Masuzawa at university of Tokyo [6]. The small pins can be utilized as a tool for micro EDM application [7]. Soni and Chakraverty conducted a comparative study of output

IOP Conf. Series: Materials Science and Engineering **187** (2017) 012036 doi:10.1088/1757-899X/187/1/012036

performance characteristics between stationary and rotary tool electrode. Their findings show improvement in material removal due to better flushing action and sparking efficiency in rotary EDM of titanium. However, this results in high surface roughness [8]. Mohan et al. perform the experimental study on Al-SiC metal matrix composite material. They showed that the rotary tool improves the material removal rate and reduce the surface roughness [9]. Y.H. Guu et al. study the influence of workpiece rotation during electrical discharge machining of AISI D2 tool steel using copper electrode. Experimental findings indicate that the surface finish improves at higher rotational speed [10]. E. Uhlmann et al. conducted a comparative experimental study of EDT, electric discharge grinding and wire electric discharge grinding to analyze the machining effect, resulted from high rotational speed of the electrodes. The experimental results make clear that the MRR growing with higher circumferential speed and discharge energy, but it causes increase in roughness of machined surface [11]. Matoorian et al. established optimum process condition for electrical discharge turning process. They adopted Taguchi design method to study the influence of process parameters namely, voltage, power, pulse-on time, spindle speed and servo upon MRR while machining HSS-1.3255. Based on experimental findings, power and spindle speed were found most significant factors, that is by increasing the value of these two parameter MRR increases significantly. However pulse-on time and servo have significant and reciprocal effect on it [12]. Teimouri and Baseri introduced magnetic field in rotary EDM. Results indicate that the applying of rotational magnetic field near machining zone lead to better flushing condition and subsequently improves the material removal rate and surface finish [13]. Song et al. developed a new strip EDM turning process which performs the same function as wire in wire electric discharge turning. Experimental findings indicate that strip ED'M produces nearly 74% more MRR than wire electrical discharge turning. In addition, the machined surface was found cusp free and nearly four times smoother than WEDM turned surface [14]. Janardhan et al. used pulse train data analysis to study the effect of process parameters on material removal, surface roughness and roundness [15]. Recently Aravind Krishnan and Samuel has attempted to optimize process parameters by considering the material removal and surface roughness as output parameters in wire electrical discharge turning process [16].

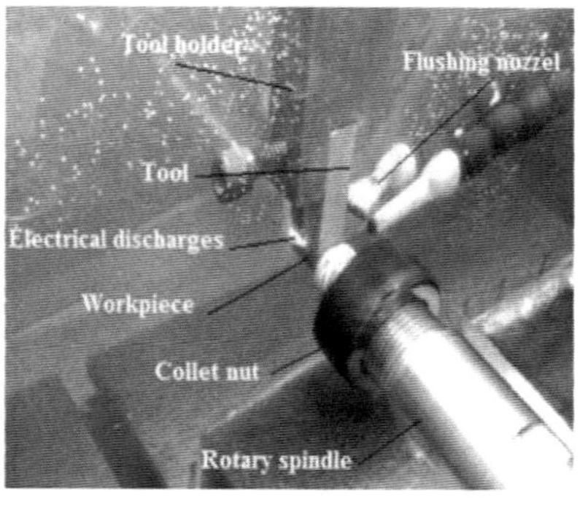

Figure 1. Concept of EDT process.

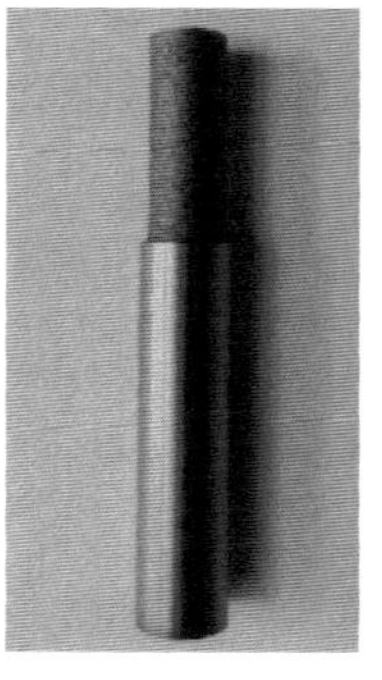

Figure 2. A cylindrical EDT part.

From the above literature it is found that several researchers have contributed to the development of EDT process. However, because of its stochastic nature the full potential utilization of this process is not completely solved. So the need has been felt towards the highlighting this process and explores the possible ways to adjust its parameters to enhance the process performance. This paper presents a study on straight turning of titanium alloy using preshaped solid copper electrode. This has been done by determining significant factors using ANOVA technique. Regression analysis than conducted to establish a relationship between factors and response. Finally signal to noise ratio (S/N) analysis is conducted to find the optimal settings and factor levels. Therefore optimal process parameters are verified through the confirmation experiment.

2. Experimental setup and equipments

In this research all experiments were conducted on Electronica 500×300 ZNC EDM machine. The dielectric fluid used for experiments was industrial grade EDM oil. The experiments are aimed at considering effects of several controllable factors on MRR. In these experiments, electrolytic copper tool electrode was used. The material selected for carrying out the experiments was cylindrical bar (10 mm dia. and 75 mm length) of titanium alloy. This material is a difficult-to-machine material and is extensively used in aerospace, automobiles, chemical and biomedical applications. To determine the material removal rates the specimen was weighted before (w_i) and after (w_f) machining using a precision self calibrating digital balance (AND, GR-200 e = 1mg.). The material removal rate was specified using the following equation (1):

$$MRR(mg/\min.) = \frac{W_i - W_f}{t} \times 1000 \qquad (1)$$

Table 1. Experimental layout using an L18 orthogonal array.

Run	Ton	IP	V	RPM	FP	MRR	S/N ratio
1	5	5	40	40	0	4.7425	13.5201
2	5	20	60	80	0.25	16.632	24.4188
3	5	35	80	120	0.5	13.052	22.3135
4	10	5	40	80	0.25	9.023	19.1070
5	10	20	60	120	0.5	24.706	27.8560
6	10	35	80	40	0	19.428	25.7685
7	15	5	60	40	0.5	7.17	17.103
8	15	20	80	40	0	17.588	24.9043
9	15	35	40	120	0.25	40.308	32.1078
10	5	5	80	120	0.25	0.344	9.2688
11	5	20	40	40	0.5	14.65	23.3167
12	5	35	60	80	0	19.456	25.7810
13	10	5	60	120	0	1.042	0.3573
14	10	20	80	40	0.25	13.96	30.2114
15	10	35	40	80	0.5	32.402	30.2114
16	15	5	80	80	0.5	5.088	14.1309
17	15	20	40	120	0	26.866	28.5840
18	15	35	60	40	0.25	28.7255	29.1653

2.1. Design of experiments

The accuracy and effectiveness of an experimental program depends on careful planning and execution of the experimental procedure. With a view to achieve the aforesaid objective, Taguchi orthogonal array (OA) $L_{18}(2^1 \times 3^7)$ was adopted for design of experiments. Thus 18 experiments were conducted at parameter levels shown in table 1. Note that the experiments were run in random order and the machining time for each experiment was fixed to 20 minute. Peak Current, pulse on-time, voltage, spindle rotation speed and flushing pressure each at three levels are adopted as a factor (variable parameters) which varies during the experiments according to the design of experiment. The commercial software MINITAB 16 was used to design and analysis the experiments. Table 1 shows the assigned orthogonal L_{18} array along with the data acquired for material removal rate. In this research only the main effect of factors are of interest and their interaction are excluded from the data analysis.

3. Data analysis

The first step in data analysis of this study is to summarize test results for each experiment performed by the use of Taguchi design of experiments technique. Figure 3 exhibits the main effect plot of pulse-on time, peak current, gap voltage, spindle speed and flushing pressure on the material removal rate. Figure 3 a-b shows the effect of applied pulse-on time and current on material removal rate. It indicates that both the parameters has most significant effect on MRR and shows direct proportion to it. That is by increasing the value of these two parameters MRR Increases significantly. Gap voltage as indicated in Figure 3 (c) is reciprocally proportional to the MRR; that is higher material removal can be obtain at the lower voltage value. The Figure 3 d-e presents the effect of spindle rotational speed and flushing pressure on MRR. As indicated none of the parameter has a significant effect on material removal. Analysis of Variance (ANOVA) was used for discussing and interpreting data. It is numerously employed by researchers because it covers the shortcomings of graphical assessment. Before any presumptions can be made based on ANOVA table, the assumptions through ANOVA have to be checked (normality error, variance consistency etc.). Figure 4 shows the normal plot of residuals for MRR. A normal probability plot is just a graph of cumulative distribution of the residuals on normal probability paper. Anderson-Darling (AD) statistic is used here to check the normal distribution of the residuals. As shown in Figure 4 the p-value calculated based on AD statistic is higher than the α-level of confidence (0.05), the error normality is considered to be valid. Table 2 indicates ANOVA table for MRR. It can be seen from table that current, pulse-on time and voltage have most significant impact on material removal rate at confidence level of 95%. So the P-values which are less than 0.05 indicate that null hypothesis should be rejected and thus the effect of respective parameter is significant.

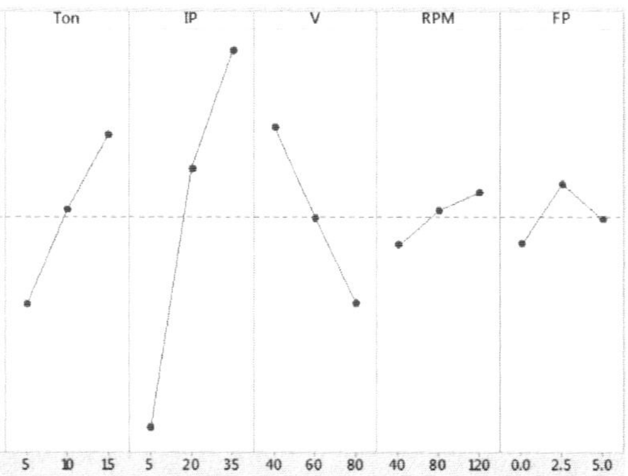

Figure 3. Mean effect plot of factors for MRR.

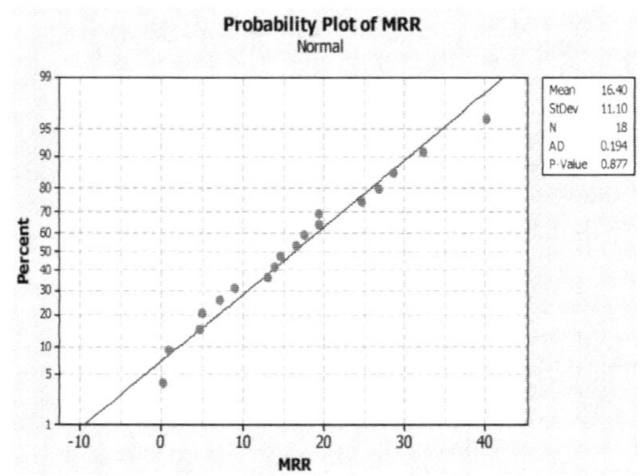

Figure 4. Normal probability plot of residual for MRR.

Table 2. Analysis of Variance for Ra, using Adjusted SS for Tests.

Source	DF	Seq SS	Aj SS	Adj MS	F	P
Ton	2	270.68	270.68	135.34	10.21	0.008
IP	2	1386.26	1386.26	693.13	52.28	0.000
V	2	285.60	285.60	142.80	10.77	0.007
RPM	2	26.74	26.74	13.37	1.01	0.412
FP	2	33.34	33.34	16.67	1.26	0.342
Error	7	92.80	92.80	13.26		
Total	17	2095.43				
S= 2.64	R-sq.= 95.57%		R-sq (adj)= 89.24%			

Equation (2) presents the relationship between factors and response, which is the outcome of multiple linear regression analysis.

$$MRR = 3.96 + 0.948\ Ton + 0.700\ IP - 0.244\ V \tag{2}$$

Table 3 indicates that the above mention regression model is the best one in comparison with the others that can be used with these parameters and parameter levels by R^2(adj.) test. The R^2 value indicates that the model can reasonably explain 91.1% of the variance in MRR. R^2(adj.) is 89.24%, which accounts for the number of predictors in the model. Both value show that the data are fitted well. S/N ratio analysis is utilized to obtain the optimal condition. The type of problem which is dealt with here is "higher the better" [17] For such type of problem η is calculated as:

$$(HB)\ \eta = -10\log\left[\frac{1}{n}\sum_{i=1}^{n}\left(\frac{1}{y_i^2}\right)\right] \tag{3}$$

Table 4 shows the mean S/N ratios obtained for each parameter level. The purpose of the confirmation experiments is to validate the conclusions drawn during analysis. The predicted S/N ratio using the optimal level of the machining factors can be calculated as Equation (4). The results of confirmation test are presented in table 5.

$$\eta_{opt} = \eta_m + \sum_{j=1}^{k}(\eta_j - \eta_m) \tag{4}$$

Table 3. Table of coefficients for regression analysis for MRR.

Predictor	Coef.	SE coef.	T	P
Constant	3.956	5.153	0.77	0.458
Ton	0.9478	0.2279	4.16	0.001
IP	0.6997	0.0759	9.21	0.000
V	-0.2438	0.0569	-4.28	0.001
RPM	0.0367	0.0284	1.29	0.221
FP	2.649	4.599	0.58	0.572
S= 3.94	R-sq.=91.1%	R-sq (adj)= 87.4%		

Table 4. S/N values (dB) for MRR.

Factor level	Ton	IP	V	RPM	FP
1	16.680	9.160	24.475	21.963	19.819
2	21.033	25.330	20.782	23.092	19.738
3	24.334	27.558	16.791	16.992	22.490
Delta	7.654	18.398	7.683	6.101	2.752
Rank	3	1	2	4	5

Table 5. Results of the experimental confirmation for MRR.

Initial machining parameter		Optimal machining parameter	
		Prediction	Experiment
Level	A2B2C2D2E2	A3B3C1D2E3	A3B3C1D2E3
MRR	16.952	57.884	57.651
S/N ratio	24.58	35.25	35.22

4. Conclusion

In this study the influence of machining parameters i.e. pulse on time, current and spindle speed over MRR was investigated. The study has been made for Ti-6Al-4V because of its growing range of applications in industry. Experiments have been designed by Taguchi's design of experiments method. Experimental findings shows that Ton, IP and V has most significant effect upon MRR. A multiple regression equation was derived and presented as Eq. 1 for MRR. An optimum parameter combination for the maximum MRR was obtained by using the S/N ratio analysis. The confirmation test results are indicated in table 5. This table evaluates the predicted and actual value of MMR. Thus, MRR is increased by 3.41 times and S/N ratio increased by 1.43 times. This confirms the success of the experiments by using estimated model and proposed statistical technique.

5. References

[1] Peças P and Henriques E 2009 Intrinsic innovations of die sinking electrical discharge machining technology: estimation of its impact *The International Journal of Advanced Manufacturing Technology.* **44** 880-89.

[2] Leão F N and Pashby I R 2004 A review on the use of environmentally-friendly dielectric fluids in electrical discharge machining, *J Mater Process Technol.* **149(1)** 341-6.

[3] Ramasawmy H and Blunt L 2004 Effect of EDM process parameters on 3D surface topography, *J Mater Process Technol.* **148** 155-64.

[4] Kunieda M, Lauwers B, Rajurkar KP and Schumacher BM 2005 Advancing EDM through fundamental insight into the process, *CIRP Ann: Manuf Techn.* **54** 64-8.

[5] Gohil V and Puri Y M 2015 Turning by electrical discharge machining: A review, *Proceedings of the Institution of Mechanical Engineers, Part B; Journal of Engineering Manufacture*, DOI: 10.1177/0954405415590560.

[6] Masuzawa T and Tönshoff H K 1997 Three-dimensional micromachining by machine tools, *CIRP Annals Manufacturing Technology.* **46** 621-28.

[7] Rajurkar K P and Yu Z Y 2000 3D micro-EDM using cad/cam, *CIRP Annals-Manufacturing Technology.* **49** 127-30.

[8] Soni JS and Chakraverti G 1994 Machining characteristics of titanium with rotary electro-discharge machining, *Wear.* **171** 51-8.

[9] Mohan B, Rajadurai A and Satyanarayana KG 2002 Effect of SiC and rotation of electrode on electric discharge machining of Al–SiC composite, *J Mater Process Technol.* **124** 297-304.

[10] Guu YH and Hocheng H 2001 Effects of workpiece rotation on machinability during electrical-discharge machining, *Mater Manuf Processes.* **16** 91-101.

[11] Uhlmann E, Piltz S and Jerzembeck S 2005 Micro-machining of cylindrical parts by electrical discharge grinding, *J Mater Process Technol.* **160** 15-23.

[12] Matoorian P, Sulaiman S and Ahmad MM 2008 An experimental study for optimization of electrical discharge turning (EDT) process *J Mater Process Technol.* **204** 350-6.

[13] Teimouri R and Baseri H 2012 Effects of magnetic field and rotary tool on EDM performance, *J Manuf Processes.* **14** 316-22.

[14] Song K Y, Park MS and Chu CN 2013 EDM turning using a strip electrode, *J Mater Process Technol* **213** 1495-1500.

[15] Janardhan V and Samuel GL 2010 Pulse train data analysis to investigate the effect of machining parameters on the performance of wire electro discharge turning (WEDT) process, *Int. J. Machine Tool Manuf.* **50** 775-88.

[16] Krishnan S A and Samuel G L 2013 Multi-objective optimization of material removal rate and surface roughness in wire electrical discharge turning, *The International Journal of Advanced Manufacturing Technology*, **67** 2021-2032.

[17] Phadke M S 1995 *Quality engineering using robust design*, Prentice Hall of India, New Delhi, India.

IOP Conf. Series: Materials Science and Engineering **187** (2017) 012037 doi:10.1088/1757-899X/187/1/012037

Energy Conversion for Thermal Comfort and Air Quality Within Car Cabin

Daniel Kristanto* **and Thananchai Leephakpreeda**

School of Manufacturing Systems and Mechanical Engineering, Sirindhorn International Institute of Technology, Thammasat University, Pathum Thani, Thailand

*e-mail: kristanto.dan12@gmail.com

Abstract. Thermal comfort and air quality within a car cabin are required during driving throughout various climates where energy is efficiently consumed to maintain acceptable conditions by air conditioning (AC) unit. This paper proposes an analysis of energy conversion within a car cabin for thermal comfort and air quality. Mathematical models, based on energy and mass balances, are developed to determine process variables of a car cabin. Experimental data from real conditions is compared with simulated results for model validation. There is very good agreement between those results. The proposed models are used to simulate interesting case studies in real circumstances for investigation on trade-off among thermal comfort, air quality, and energy usage.

1. Introduction

Nowadays, people spend a lot of time in cars to travel. It is challenging to automotive engineering design in automatically making passengers comfortable within cars all the time. Weather and ambient conditions mainly determine thermal conditions of an air within a car cabin. In sunny day, air temperature within a car cabin can reach 70°C during parking [1]. Besides temperature, an amount of CO_2 can also affect comfortable conditions of passengers. High concentration of CO_2 makes passengers sleepy and dizzy. With three passengers in a sedan, concentration of CO_2 can reach 3,500 ppm, which is much higher than acceptable level [2].

With those conditions, thermal comfort and air quality are needed to be investigated when a car is driven in long period, especially in a hot climate. Typically, an Air Conditioning (AC) unit is designed to make up the air within a car cabin. Its capability is implemented to maintain temperature, humidity, and CO_2 concentration under comfortable conditions. In that case, the AC unit consumes an amount of energy efficiently.

Recently, researchers have paid attention to analysis on energy conversion for thermal comfort and air quality within a car cabin. For example, mathematical models of car cabins have been developed to describe thermal conditions [1], [3-5]. Most of those works used temperature and humidity to represent thermal conditions within a car cabin. In few cases, air quality of car cabin is indicated by CO_2 concentration [2]. It should be noticed that the previous works analyzed thermal conditions and air quality, separately.

In this work, mathematical models of energy conversion for thermal comfort and air quality are proposed. The usage of Predicted Mean Vote (PMV) index and CO_2 concentration in quantifying thermal comfort and air quality, respectively, is novelty of this work.

2. Experiments

A sedan car is used as an experimental rig for model verification in this study as shown in figure 1. An experiment is performed during night time with ambient temperature of 31 °C and humidity of 70%, respectively. There are 2 passengers inside the car where an AC unit is turned on. The air temperature, humidity, and CO_2 concentration within the care cabin are measured every 30 seconds in 10 minutes. They are recorded by a data acquisition system.

The temperature within the car cabin is measured by a thermistor with measuring range from 0°C to 100°C with an accuracy of ±0.3°C. The relative humidity is measured by a hygrometer with measuring range from 10 to 90 % RH with an accuracy of <3% RH. The CO_2 concentration is measured by a carbon dioxide sensor with measuring range from 0 to 5,000 ppm with an accuracy of ±40 ppm. An anemometer with measuring range 0 to 20 m/s with an accuracy of ±3% is used to measure a wind speed for airflow of a AC unit.

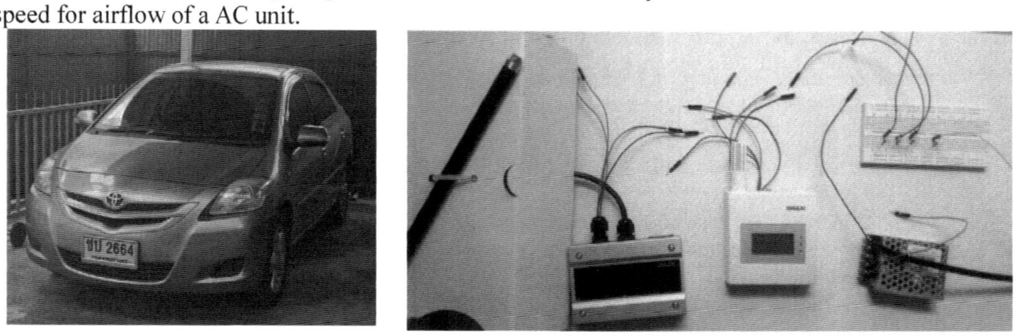

(a) (b)

Figure 1. Experimental rig: (a) sedan car and (b) measurement devices.

3. Methodology

3.1. Mathematical model

Mathematical models are developed to determine the air temperature, humidity, and CO_2 concentration within a car cabin by applying principles of mass and energy balances. Figure 2 shows a schematic diagram of energy and mass transfers across the car cabin from environment.

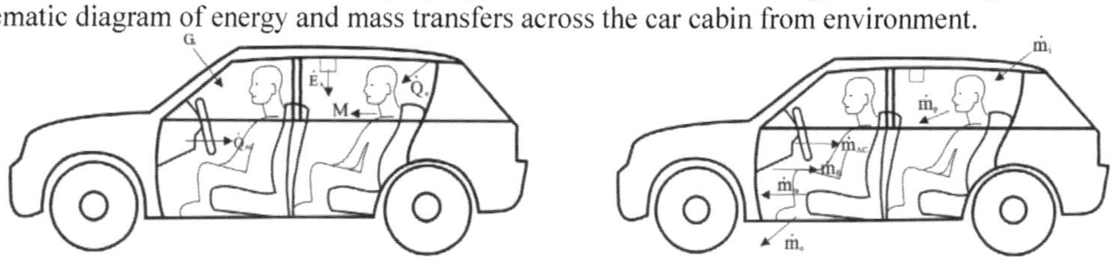

(a) (b)

Figure 2. Energy (a) and mass (b) balance of car cabin.

From figure 2a, the energy equations of a car cabin can be expressed in equation (1)-(2).

$$m_d(Cv_d + w_c Cv_w)\frac{dT_c}{dt} = M + G + \dot{E}_A + \dot{Q}_e - \dot{m}_{AC}(Cv_d + w_c Cv_w)(T_c - T_{AC}) \quad (1)$$

$$\dot{Q}_e = hA(T_e - T_c) + \varepsilon\sigma A\left(T_e^4 - T_c^4\right) \quad (2)$$

where m is the mass (kg), Cv is the specific heat (J/kg°C), w is the absolute humidity (kg/kg), M is the metabolic rate (W), G is the solar radiation of sunlight, which is transmitted to car envelope (W), $\dot{E}$ is the heat generation (W), $\dot{Q}$ is the heat transfer (W), h is the heat transfer coefficient W/m^{2}°C, A is the area of car envelope (m^2), ε is the emissivity, σ is the Stefan-Boltzman constant, $\dot{m}$ is the mass flow rate (kg/s), and T is the temperature (°C). The subscripts, such as d, w, c, A, e, and AC, represent the dry air, water vapor, car cabin, auxiliaries, envelope of car cabin, and AC unit, respectively. The left term of (1) is the internal energy of an air within the car cabin, which consisting of dry air and water vapor. The right hand terms are the metabolic rate of human, radiation heat from windshield, heat generation from auxiliaries, convective and radiative heat transfers between air within car cabin and envelope, and heat extraction of an AC unit, respectively.

From figure 2b, the mass equations are written in terms of the absolute humidity and CO_2 concentration in equation (3)-(5).

$$m_d \frac{dw_c}{dt} = \dot{m}_i w_a - \dot{m}_o w_c + N\dot{m}_p - (1-X)\dot{m}_{AC} w_c + \dot{m}_{AC} w_{AC} \quad (3)$$

$$m_d \frac{dC_c}{dt} = \dot{m}_i C_a - \dot{m}_o C_{cab} + X\dot{m}_{AC} C_a + NC_p \quad (4)$$

$$X = \frac{\dot{m}_{fa}}{\dot{m}_{ra}} \quad (5)$$

where N is the number of passengers, X is the ratio of fresh air to return air in AC unit, and C is the CO_2 concentration (ppm). The subscripts, such as i, a, o, p, fa, and ra, represent infiltration, ambient, exfiltration, passenger, fresh air, and return air, respectively. It should be noticed from equation (3)-(5) that the air temperature, absolute humidity and CO_2 concentration within a car cabin is affected by infiltration, exfiltration, human exhale, and, in particular, functions of a AC unit.

3.2. *PMV index*

The PMV index, which is proposed by Fanger, is used to quantify thermal comfort of the passengers [6]. According to ISO standards 7730, PMV index is the function of air temperature (T_c), air velocity (v), relative humidity (RH_c), mean radiant temperature (MRT), clothing (Clo), and metabolic activity (M), as summarized in equation (6).

$$PMV = f(T_c, v, RH_c, MRT, Clo, M) \quad (6)$$

The value of Clo and M are determined by human clothing and activity. In this work, the value of Clo and M are chosen to be 0.5 for general clothing and 70 W for driving activity, respectively.

The values of PMV index with respect to thermal sensation is shown in figure 3. The acceptable range for PMV index for comfortable environment is from -0.5 to +0.5 [7].

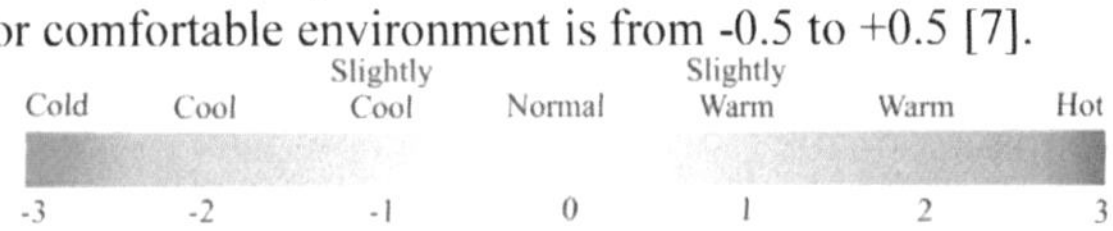

Figure 3. Scales of PMV index for thermal sensation

3.3. *CO_2 concentration*

The CO_2 concentration has unit in ppm (part per million). The standard indicates that the CO_2 concentration must be below 700 ppm over ambient condition to accomplish the air quality [2].

3.4. *Energy conversion*

In achieving acceptable conditions of thermal comfort and air quality, the AC unit consumes an amount of energy to make up the supply air, which is mixed of the return air within the car cabin together with fresh air from environment. The amount of energy for making up air is called cooling load CL, which can be calculated by equation (7)-(8).

$$CL = \dot{m}_{AC} C_v (T_m - T_{AC}) \quad (7)$$

$$T_m = (1 - X)T_c + (X)T_{fa} \quad (8)$$

where T_m is the temperature of mixing air between return air and fresh air (°C) and T_{fa} is the temperature of fresh air (°C).

4. Results and discussion

4.1. Performance of proposed model

The mathematical models of equation (1)-(5) are numerically solved. The simulated results are compared with experimental data. For model validation, several experiments are performed as a driver and a passenger sit in the car when the AC unit operates. An experiment is observed that the air temperature and relative humidity decrease from an initial condition as shown in figure 4a and figure 4b, respectively. After 300 s, the air temperature and humidity reach their steady state conditions. In figure 4c, two persons inside the car cabin cause increase of CO_2 concentration. However, the CO_2 concentration slowly rises due to taking fresh air to the car cabin. There is very good agreement between the simulation and experiment.

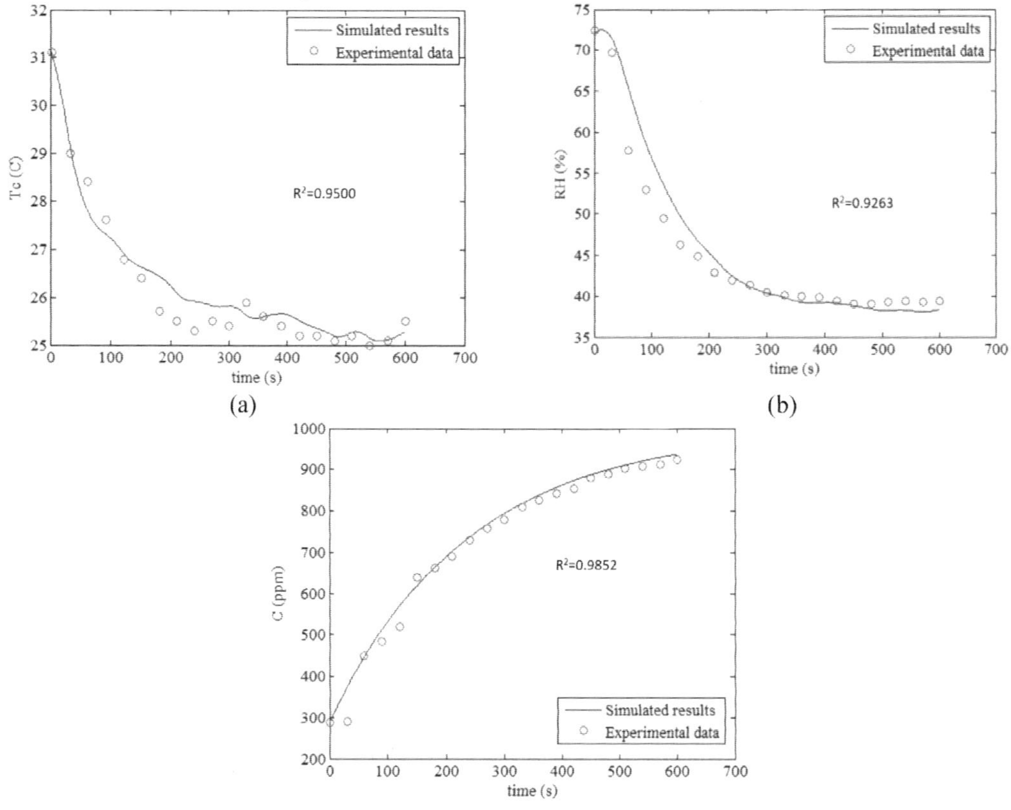

Figure 4. Comparison between simulated results and experimental results of (a) temperature, (b) relative humidity, and (c) CO_2

4.2. Analysis on thermal comfort and air quality

The validated model is used to simulate some case studies. Accordingly, the PMV index, CO_2 concentration, and cooling load of the car cabin are determined. The simulation takes place in the morning with ambient temperature of 32°C and relative humidity of 70%. There are six conditions in the simulation. These conditions are continuing. Each condition is separated by dash lines in figure 5 and figure 6.

4.2.1. Condition A. The car is parked outside without any cover or roof. The PMV index increases because the air temperature get high since solar radiation is transmitted to the envelope of the car cabin. The CO_2 concentration remains during this period. Obviously, there is no cooling load.

4.2.2. Condition B. This condition is similar with the condition A. One driver and one passenger get on the car. Consequently, the air temperature increases. It is shown in figure 6 that the PMV index drastically increases. The CO_2 concentration has the same trend due to CO_2 production from human exhale and lack of fresh air from an AC unit.

4.2.3. Condition C. In this period, the car is driven with speed of 20 m/s. The AC unit is turned on with mass flow rate of 0.0548 kg/s, supply air temperature of 12°C and mixing ratio of 0.4. It is expected that the air temperature decreases. In turn, the PMV index decreases to reach the acceptable level. Also, the fresh air, which is taken into the car cabin, dilutes the CO_2 concentration. The cooling load increases in making up the hot return air and fresh air.

4.2.4. Condition D. Another passenger gets on the car in this time. This minor change in number of passengers causes the heat and CO_2 generations from metabolism of humans. The PMV index and CO_2 concentration slightly increase.

4.2.5. Condition E. To obtain comfortable conditions, the mass flow rate of the AC unit is changed to 0.0948 kg/s. The PMV index rapidly decreases until it reaches -0.1, which is in the acceptable level. More amount of fresh air alleviates the CO_2 concentration to 700 ppm. However, the cooling load of the AC unit is an energy usage of 1,600 W.

4.2.6. Condition F. To reduce the CO_2 concentration, the mixing ratio is changed to be 0.8. The CO_2 concentration decrease significantly to 500 ppm. The air quality is acceptable. The higher cooling load with 1,800 W is traded off than Condition E.

5. Conclusion

Mathematical models of car cabin are developed in this paper. Analysis on energy conversion for thermal comfort and air quality within a car cabin is performed effectively. The model validation is confirmed by a very good agreement between simulated results and experimental data. Some interesting case studies are simulated in real circumstances. The simulations show the dynamic changes of thermal comfort and air quality. The proposed models can be used for design in automatic trade-off among thermal comfort, air quality, and energy usage in making up the air within the acceptable conditions.

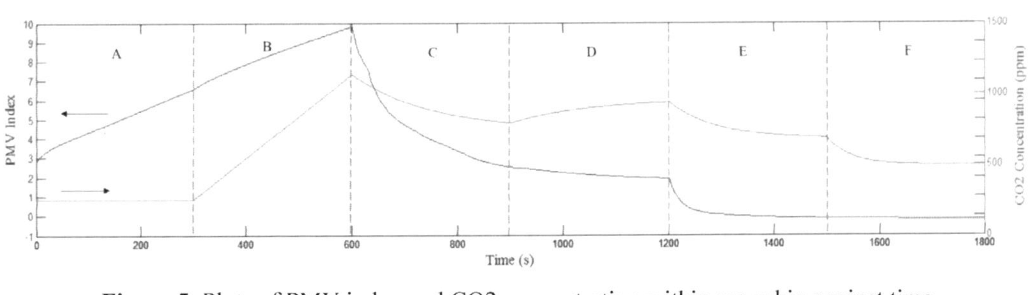

Figure 5. Plots of PMV index and CO2 concentration within car cabin against time.

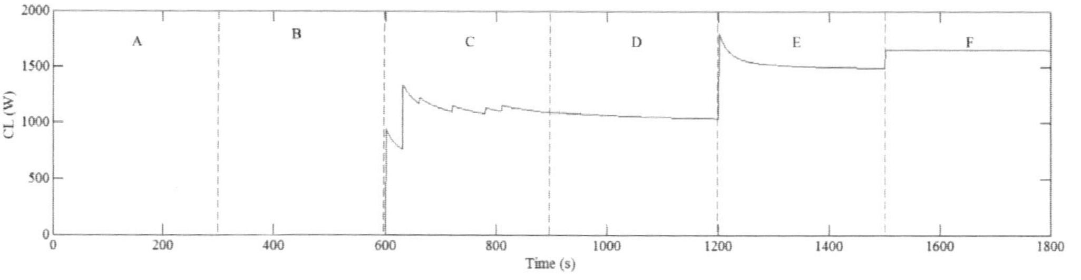

Figure 6. Plots of cooling load of car cabin against time.

6. References

[1] H. Lee, Y. Hwang, I. Song, and K. Jang 2015 Transient thermal model of passenger car's cabin and implementation to saturation cycle with alternative working fluids *J. Energy*, vol. 90, pp. 1859–1868

[2] H. Jung 2013 Modeling CO 2 Concentrations in Vehicle Cabin *SAE Int.* doi:10.4271/2013-01-1497

[3] M. A. Fayazbakhsh and M. Bahrami 2013 Comprehensive Modeling of Vehicle Air Conditioning Loads Using Heat Balance Method *SAE Int.* doi:10.4271/2013-01-1507

[4] D. Marcos, F. J. Pino, C. Bordons, and J. J. Guerra 2014 The development and validation of a thermal model for the cabin of a vehicle *Appl. Therm. Eng.*, vol. 66, no. 1–2, pp. 646–656

[5] B. Torregrosa-Jaime, F. Bjurling, J. M. Corberan, F. Di Sciullo, and J. Paya 2015 Transient thermal model of a vehicle's cabin validated under variable ambient conditions *Appl. Therm. Eng.*, vol. 66, no. 1–2, pp. 646–656, 2014

[6] ISO Standard 7730 1994 Moderate thermal environments–Determination of the PMV and PPD indices and specification of the conditions for thermal comfort

[7] P. Aryal and T. Leephakpreeda 2015 CFD Analysis on thermal comfort and energy consumption effected by partitions in air-conditioned building *International Conference on Alternative Energy in Developing Countries and Emerging Economies*, vol. 79, pp. 183–188.

Acknowledgments

The author wishes to thank Sirindhorn International Institute of Technology for providing Excellent Foreign Student scholarship to do this research.

IOP Conf. Series: Materials Science and Engineering **187** (2017) 012038 doi:10.1088/1757-899X/187/1/012038

Study of Fuze Structure and Reliability Design Based on the Direct Search Method

Zhang Lin and Wang Ning

Air Force Engineering University, Xi'an, Shaanxi, China

Kjddlz2002@163.com

Abstract. Redundant design is one of the important methods to improve the reliability of the system, but mutual coupling of multiple factors is often involved in the design. In my study, Direct Search Method is introduced into the optimum redundancy configuration for design optimization, in which, the reliability, cost, structural weight and other factors can be taken into account simultaneously, and the redundant allocation and reliability design of aircraft critical system are computed. The results show that this method is convenient and workable, and applicable to the redundancy configurations and optimization of various designs upon appropriate modifications. And this method has a good practical value.

1. Introduction

With the development of aerospace technology, the requirement for the system reliability design is getting higher and higher [1]. The fundamental approach is to produce more reliable components by means of appropriate improvement measures. However, in actual production process, a series of problems are involved in the improvement of component reliability such as: the quality of raw materials, manufacturing process and design standards of components. In case of improving component reliability when it reaches a certain value, the cost will increase incredibly, while the effect of this method on the overall reliability of the system is not obvious in such case.

Redundant design improves the reliability of the system by means of configuration and rational management of multiple resources of the system. However, increasing the redundancy will increase resources, cost or weight, because of the increase the complexity and error rate of the system [2], the reliability is reduced on the contrary. So need to consider the impact of multiple factors to get a more perfect redundancy configuration proposal.

2. Reliability allocation problem models

Generally, the reliability allocation can be divided into two categories, one is constraint condition based on reliability index, with the lower limit value being given, taking cost, structural weight and other factors as objective function. The other one is to calculate the upper limit value by taking cost and structural weight as constraint conditions [3], and the highest reliability of the system as objective function.

Because of the constraint of many factors in aircraft design, the reliability allocation is not only a matter of simple mathematical programming, but to specifically assign the reliability index such as importance level, complexity level, technical level and work environment, into each sub-system, determine the reliability index of each sub-systems, and then assign to the overall unit according to its mathematical mode [4], and finally, assign to each component based on the functions and the environment of the overall unit and some other factors.

- Assume the total requirements of system reliability is given, and the subsystems relations, system reliability the components, cost, structural weigh and other factors are known, it is required to design (allocate) the quantity of the components of each subsystem to minimize the total cost or structural weight.
- Assume cost, structural weight and other factors as given as constrain condition, and the subsystems relations, system reliability the components, cost, structural weigh and other factors are known, it is required to design (allocate) the reliability of each subsystem to maximize the system reliability (minimize the unreliability).

Several components may be paralleled on the existing series system when making redundant design. In order to build the series parallel system model which takes the minimum cost as the objective function and the reliability index and quality as the constraint conditions, a cost function must be established first. The cost function shows the relationship between reliability and cost, it is an abstract concept, including the sum of all kinds of costs for manpower, material resources and financial resources which are used to improve unit reliability. It is originally named effort function. It is a monotonically increasing function. Because it is difficult to obtain the statistical data between the cost and reliability of each unit, so an empirical formula is generally unable to be established. For the redundant design of a complex system, in order to make it convenient for the study, it can be assumed that the parallel components in subsystems are identical, and the cost of each component is a constant. So for the general series parallel system, it is as shown in Figure 1.

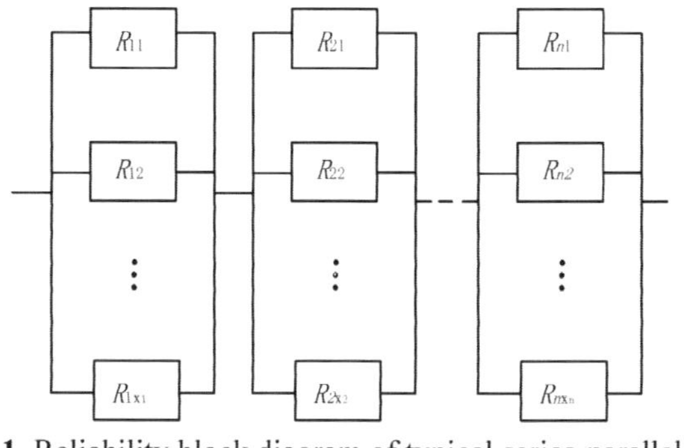

Figure 1. Reliability block diagram of typical series-parallel system

Reliability of typical series-parallel system

$$R_s = \prod_{i=1}^{n}(1 - F_i^{x_i}) \tag{1}$$

The above problems can be summarized as the following mathematical forms:

$$\left.\begin{aligned} &\min f(x) \\ &f(x) = \sum_{i=1}^{n} w_i x_i \\ &R_s = \prod_{i=1}^{n}(1 - F_i^{x_i}) \geq R^* \\ &x_i \geq 0, i = 1, 2, \cdots, n \end{aligned}\right\} \tag{2}$$

$$\left.\begin{aligned} &\max R(x) \\ &f(x)=\sum_{i=1}^{n} w_i x_i \le W^* \\ R_s=&\prod_{i=1}^{n}(1-F_i^{x_i}) \ge R^* \\ &x_i \ge 0, i=1,2,\cdots,n \end{aligned}\right\} \tag{3}$$

Among the formula, n is the number of sub-systems,， F_i is the unreliability of subsystem components,， x_i is the number of parallel elements in the subsystem， W_i is the cost (or structural weight) of the system components, $f(x)$ is the total cost of the system, R_s is the system reliability, R^* is the index required by the system, C^* is the constraint value of the total cost (structural weight) of the system.

Make following assumptions before making redundancy optimal and allocation:

(1) Each subsystem is independent and not repairable in mission phase.

(2) Each subsystem has two states, namely, the working state and the fault state

(3) The reliability of each subsystem is a constant.

There are many kinds of reliability allocation method with constraint conditions, such as dynamic programming method, Lagrange multiplier method and integer programming method [5], but in practice, there is a relatively simple engineering algorithm of Direct Search Method.

3. Algorithm principle of Direct Search Method

The basic idea of Direct Search Method is to find a link (subsystem) and to parallel a redundant part on the link so as to achieve a maximum system reliability increment compared to unit cost, then checks constraint conditions and continuously search in this way, and constantly increase system reliability, until all constraint conditions are satisfied [6]. The approaches to the above problems of the Direct Search Method are as follows:

- Search a certain level. Parallel a redundant part in this level to maximize the system reliability increment compared to unit cost, and search within constraint conditions. Search constantly and continuously until the constraint conditions are satisfied.
- Each time, parallel a redundant part on the series system with largest unreliability, check constraint conditions, and maximize the system reliability by doing a series of tests.

4. Numerical example

Fuze is a key link in the weapon system to realize the effective damage to the target. If its reliability cannot be guaranteed, a military operation will be either hindered due to the failure to complete the scheduled mission because of fuze blind fire, or the outcome of a battle will be even directly affected. Or because of the premature burst or early burst of the fuze, it cannot destroy enemy, but cause casualties and damage equipment of our side.

Generally, mechanical and electrical products work above ground or in indoor environment, environmental force the bear in work process is not large, and failure detection and failure maintenance can be done in work process. Fuze working process is featured by long-term storage, with a storage life of generally 10-20 years, and gradual declining reliability in the long-term storage process, so its function cannot be predicted. IN addition, it is required to pass through launch, flight and missile target intersection and other process, the environmental conditions in working process is very harsh, such as High impact overload and high speed rotation. To ensure the normal operation under high and low temperatures, the components and parts are required to have small size and light weight, good performance, low price, and high reliability and safety. Another feature of the fuze working process is a very short working time, the failure occurred during the course of the work cannot be troubleshooted and maintained manually, and it is also hard to take failure detection of fuze in launch process. In such circumstances, it is very important to improve the reliability of the fuze.

The redundancy technology provides an effective way to improve the reliability, safety, reliability and fault tolerance ability of the system. This study is intended to consider the reliability, cost and weight when making design of redundant system of the fuze, study on the optimal allocation of redundancy, so as to effectively improve the reliability of the system.

The key components of fuze are composed of four elements in series, and their respective weight and reliability data are shown in table 1. Because its work reliability directly related to whether the aircraft can successfully complete the task, and its work is not reversible, so its reliability is required to be higher than 0.99.

Table 1. Unit weight and reliability.

Unit	Weight（kg）	Reliability
1	0.53	0.92
2	0.28	0.94
3	1.12	0.97
4	0.74	0.95

Its reliability logic block diagram is simplified as Figure 2:

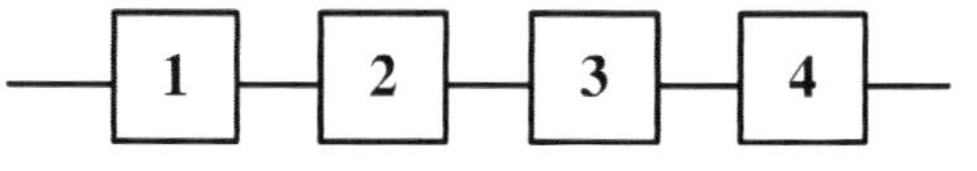

Figure 2. System reliability logic block diagram.

4.1 Computation of initial reliability

In the case of non-redundant design, system reliability is

$$R_s = \prod_{i=1}^{n} R_i = 0.7879$$

It can be seen from the computation result that the reliability of the system does not meet the specified requirements, so redundant design must be done to improve the reliability of the system. Applied redundancy technology can effectively improve the reliability of the system composed of components with relatively low reliability, that is, "exchange the high reliability with the cost of unconventional resources". It is an effective means to improve the reliability and safety of the system.

4.2 Reliability optimization design

The optimization model of the system obtained from the above problems is as follows: when system reliability is or above design reliability R_0, compute $n = \{n_1, n_2, \dots n_k\}$, to compute the minimum value of weight of $\sum_{i=1}^{k} w_i n_i$.

The specific algorithm of Direct Search Method is as follows:

Step one:

(1) Obtain from constraint conditions:

$$1 - q_i^{n_i} > R_0, i = 1, 2, \cdots, k \quad (q_i = 1 - R_i) \tag{4}$$

Making n_i^* as the number of due parallel components at least of the ith subsystem,

IOP Conf. Series: Materials Science and Engineering **187** (2017) 012038 doi:10.1088/1757-899X/187/1/012038

Then get $$n_i^* > \ln(1-R_0)/\ln q_i \tag{5}$$

Because the number of components n_i^* must be a positive integer, so take

$$n_i^* = \ln(1-R_0)/\ln q_i + 1 \tag{6}$$

So the minimum weight of the system is $$W^* = \sum_{i=1}^{k} w_i n_i^* \tag{7}$$

(2) If taking the formula now $$\prod_{i=1}^{k}(1-q_i^{n_i^*}) \geq R_0 \tag{8}$$

Then we get the optimum answer of $\{n_1^*, n_2^*, \ldots, n_k^*\}$. Otherwise transfer to Step Two.

Step Two:

(1) Use the answer form $\{n_1^*, n_2^*, \ldots, n_k^*\}$ to compute $\Delta Q_i = q_i^{n_i^*} - q_i^{n_i^*+1}$ and $\Delta Q_i/W$

(2) Given $$\Delta Q_{i0}/w_{i0} = \max_{1\leq i\leq k}\{\Delta Q_i/w_i\} \tag{9}$$

Add a parallel component on i_0 part, so as to make the number of parallel components as $n_{i0}^* + 1$.

Then compute $$R_r = \left(1-q_i^{n_{i0}^*+1}\right)\prod_{\substack{i=1\\ i\neq i_0}}^{k}\left(1-q_i^{n_i^*}\right) \tag{10}$$

If $R_r \geq R_0$, then get the optimum answer of $$\{n_1^*, n_2^*, \ldots, n_{i0-1}^*, n_{i0}^* + 1, \ldots, n_k^*\} \tag{11}$$

Otherwise repeat the computation of Step Two until the optimum answer is obtained. The detailed computation process is shown in Table 2 below.

Table 2. Computation process.

j	Items	1	2	3	4
0	n_i^0	2	2	2	2
	Q_i	0.0064	0.0036	0.0009	0.0025
	ΔQ_i	0.005888	0.003384	0.000873	0.002375
	$\Delta Q_i / W_i$	0.011109	<u>0.012086</u>	0.000779	0.003209
	R_r		0.9866		
1	n_i^1	2	3	2	2
	Q_i	—	0.000216	—	—
	ΔQ_i	—	0.00020	—	—
	$\Delta Q_i / W_i$	<u>0.011109</u>	0.000714	—	—
	R_r		0.989984		
2	n_i^2	3	3	2	2
	Q_i	—	0.000512	—	—
	R_r		0.995872		

In the Table, $R_r = \prod_{i=1}^{k}(1-q_i^{n_i}) = \prod_{i=1}^{k}(1-Q_i) \approx 1-\sum_{i=1}^{k} Q_i$, the blank square refers to the value without need to recomputed, and the underlined square refers to the number of parallel components of this level + 1.

By computation the optimum answer is obtained $n=$ {3， 3， 2， 2} .On the premise of ensuring reliability , $W^* = \sum_{i=1}^{k} w_i n_i^* = 6.15kg$.

4. Conclusion

The study only discusses the redundancy of single equipment, and the redundant design of multiple equipments can be done with slight modification. In practical work, the reliability model can also be designed into other forms based on actual needs such as series and parallel models and embedded model, which only require corresponding modification on this basis. And tabulation method is used in the computation process, which is simple and intuitive, easy to calculate the answers, so that the reliability allocation can be quick and accurate. With small amount of calculation, it is convenient for engineering and practical use. In addition, the algorithm can be realized by using VC++6.0 language programming on compute. Direct Search Method is a process of mathematical optimization, and it is applicable to large and complex systems such as airplanes and ships. Satisfactory effect can be achieved by using the method into reliability allocation and design.

References

[1] Wang Qi and Cui Wei 2015 Study of Redundancy Reliability Design Based on the Genetic Algorithm Journal of Nanchang Hangkong University 29 pp 16–20

[2] Wang Zhi-qiang and Gu Xing-sheng 2009 Redundancy Design on Flight Controller. Journal of East China University of Science and Technology 35 pp 158–162

[3] Shan Xin and Dong Wen-hong 2006 Direct Search Method of Reliability Allocation for MWS,Journal of Projectiles,Rockets,Missile and Guidance 26 pp 1098–1100

[4] Xiong Jun-jiang and Wang Jia-feng 2004 System Reliability Distribution Optimization Model For Redundancy Design Acta Aeronautica et Astronautica Sinica 25 pp 45–48

[5] Gao Xiang and Wang Shao-song 2003 Reliability Distribution Method of Missile Weapon Systems Tactical Missile Technology 2 pp 30–32

[6] Liu Fei and Zhang Wei-Hua 2005 System reliability optimization allocation based on cost minimization Machinery Design & Manufacture 11 pp 34–36

IOP Conf. Series: Materials Science and Engineering **187** (2017) 012039 doi:10.1088/1757-899X/187/1/012039

Motorcycle Start-stop System based on Intelligent Biometric Voice Recognition

A Winda[1, a], **W R E Byan**[2, b], **Sofyan**[1], **Armansyah**[2], **D L Zariantin**[3] **and B G Josep**[4]

[1] Computer Engineering, Bina Nusantara University, Indonesia
[2] Industrial Engineering, Bina Nusantara University, Indonesia
[3] Mechanical Engineering, Pancasila University, Indonesia
[4] School of Computer Science, Bina Nusantara University, Indonesia

[a]wastuti@binus.edu, [b]eriyandwita@binus.edu

Abstract. Current mechanical key in the motorcycle is prone to bulgary, being stolen or misplaced. Intelligent biometric voice recognition as means to replace this mechanism is proposed as an alternative. The proposed system will decide whether the voice is belong to the user or not and the word utter by the user is 'On' or 'Off'. The decision voice will be sent to Arduino in order to start or stop the engine. The recorded voice is processed in order to get some features which later be used as input to the proposed system. The Mel-Frequency Ceptral Coefficient (MFCC) is adopted as a feature extraction technique. The extracted feature is the used as input to the SVM-based identifier. Experimental results confirm the effectiveness of the proposed intelligent voice recognition and word recognition system. It show that the proposed method produces a good training and testing accuracy, 99.31% and 99.43%, respectively. Moreover, the proposed system shows the performance of false rejection rate (*FRR*) and false acceptance rate (*FAR*) accuracy of 0.18% and 17.58%, respectively. In the intelligent word recognition shows that the training and testing accuracy are 100% and 96.3%, respectively.

1. Introduction

Modern internal combustion engines in motorcycle typically lack the ability to self start, it need additional mechanism to crank the engine. One of the oldest form of manual mechanism to crank the engine usually employs a kickstart lever. Current commercial motorcycle starting engine system can both employ the kickstart lever and a starter motor which is activated by a mechanical switch in a form of pushing a button after turning on an ignition key. The key is also used to turn off the engine simply by turning the key to the off position to cut the electricity supply to the ignition coil. Also, a kill switch or idle stop button sometimes available to disrupt the process of combustion engine and engineered to close it in the fastest way possible. The key is designed to limit the access of the vehicle for unauthorized user. However, this mechanical key is prone to forcible bulgary access, being stolen, misplaced, lost or duplicated. In Indonesia itself, motorcycle theft tends to increase from year 2012 as many as 41,816 cases to 42,508 cases in 2013[1].

Limitations of the key as conventional security system urge some researcher to consider biometric verification process. Biometrics methods used for personal authorization utilize some features such as the face, voice, hand shape, finger print and iris[2]. Finger print pattern recognition is one of an example of biometric technique that can be applied in vehicle ignition authentication where a person finger print is matched with the database where if the recognition is not valid then the control will not ignite the vehicle[3]. Each of biometric method has its own advantages and disadvantages in their application[4]. Voice as one of biometric characteristic is increasingly popular to be used as it is a natural signal which is fast, simple and convenient to produce and not easy to be stolen. Therefore, the ability to access the system should always remain with the user wherever they go, increasing the easiness and safety of the user.

This paper introduces a start-stop system based on intelligent biometric voice recognition. The proposed system is based on the performance biometric which offers an ability to provide positive verification of identity from an individual's voice characteristics to access the start-stop engine mechanism. The proposed system features are extracted from the person voice data and then Support Vector Machines (SVMs) is used to develop models of the voice persons based on the feature extracted from the authorized person voices.

First, the prototype of the voice based automation starting engine control is described. Next, the voice based model used in the proposed system is discussed in detail. Finally, the performance of the proposed intelligent biometric voice recognition for engine start-stop access control is evaluated experimentally.

2. Proposed System

2.1. Proposed System Description

Figure 1 show the proposed system description, the system basically consists of five main components, namely voice recording, voice based recognition system, word based recognition system, Arduino, and two automotive relay. One relay is to turn on a starter motor and the other is to turn off a capacitor discharge ignition (CDI) pulse to an ignition coil. A hand phone microphone used as a voice sensor to record the person voice. The recorded voice then processed in an intelligent voice recognition system which will recognize the person identity based on his/her voice. Once the person is identify then the system will recognize the word utter by the recognize person with an intelligent word recognition system. Recognized word is then used as an input to the Arduino which will activate the automotive relay to activate the engine's starter or turn off ignition pulse.

A personal computer (PC) of 2.1 MHz Intel Core i3 processor equipped with a sound card is used for voice recognition system implementation. The sound card records the voice data based on the sampling frequency of 44 kHz. In this system, all of the voice data processing and speaker recognition algorithms are implemented in the PC using MATLAB and its toolbox. As a result of the intelligent voice and word recognition, a decision signal which will switch on or off the engine is sent through the parallel port of the PC to the Arduino. This decision signal is sent from Arduino to one of the automotive relay. The automotive relay works on 12 volts DC power supply and it is set in normally open (NO) condition.

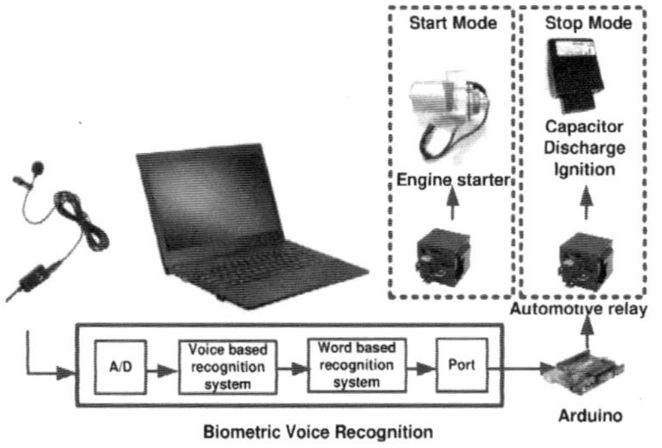

Figure 1. Proposed intelligent biometric voice recognition for motorcycle start-stop system.

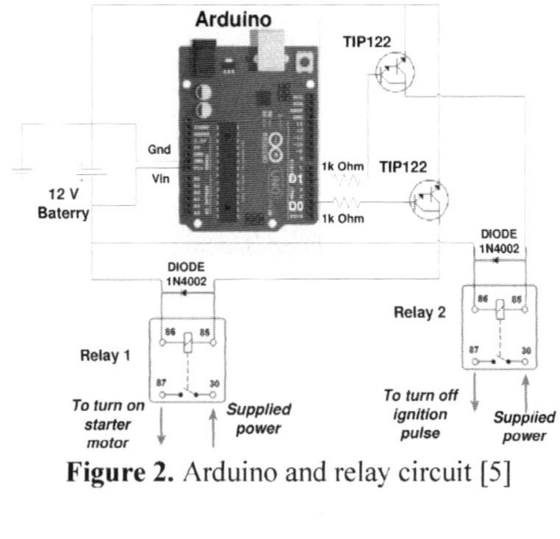

Figure 2. Arduino and relay circuit [5]

In the case a person is recognized by the intelligent voice recognition as an authorized user, the access is granted [6]. In sequence, the word utter by the authorized user is detected by the word recognition system whether it is in 'On' or 'Off'. In case the word is 'On', Arduino sends a signal to the automatic relay no 1 so that the automotive relay is demagnetized. As a result the starter motor is activated to turn the engine on by that authorized person. In case the word is 'Off', Arduino sends a signal to the automatic relay no 2 so that the automotive relay is demagnetized and turn off the ignition pulse which in that case turning the engine off. The system diagram is shown in Figure 2. Therefore, without an 'On' command signal from the intelligent biometric voice recognition system, the circuit for starter motor will always remain at an off condition which means the engine also remains off. Similarly, without an 'Off' command signal from the recognition system, the circuit for ignition pulse will always remain on during engine running.

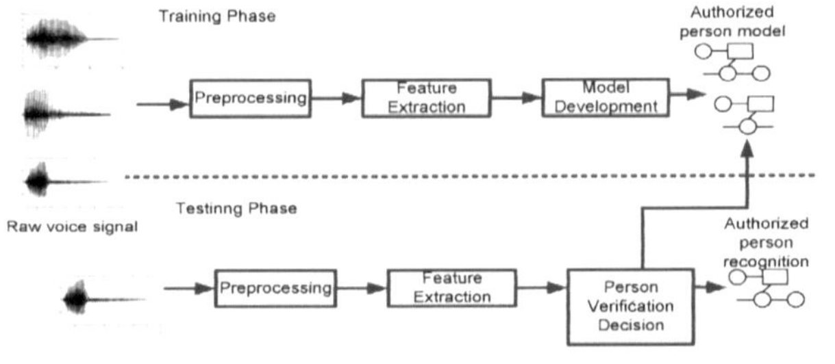

Figure 3. Intelligent biometric voice recognition system

The overall process of the proposed system for intelligent biometric voice recognition is shown in Figure 3. There are two important parts of the voice recognition, namely person identification and speech recognition. On the person identification part, the system will authorize the person. The system will decide whether the person is the authorized person or not. Once the person is authorized by the system, then the system will recognize the word uttered by the authorized person.

2.2. Feature Extraction
As shown in Figure 3, feature extraction is an important process in the proposed system. Features are some quantities, which are extracted from preprocessed voice and can be used to represent the voice signal. In this work, the MFCC technique is applied to the system. The MFCC takes advantages of several well-known properties of the auditory system, since provide relatively good performance and straightforward to be implemented[7]. Futhermore, the MFCC process is explain detail on the other work [8].

2.3. Support vector machine (SVMs) classification
SVMs classification is implemented as pattern classification. The basic idea of SVM is the mapping of non-linearly training data into higher-dimensional feature space through the kernel function. There are three important aspects of SVM, namely discrimination (optimal) hyperplane, optimization via Lagrange multipliers and kernel functions [9][10]. The discussion detail of the SVM was explain in the previous work [8].

3. Results

3.1. Experimental setup
In order to evaluate the effectiveness of the proposed intelligent biometric voice recognition for motorcycle start-stop system, the proposed system is installed in Intelligent Control System Laboratory, Binus-Aso School of Engineering. The voice of 10 (ten) speakers is used in the experiment. Each of the speakers has to utter the word 'On' and 'Off' for 10 times, as shown in Figure

4 and 5. There are three series of experiments. Firstly, speaker recognition experiment, this experiment is used to determine the accuracy of the proposed voice recognition system. In this experiment the

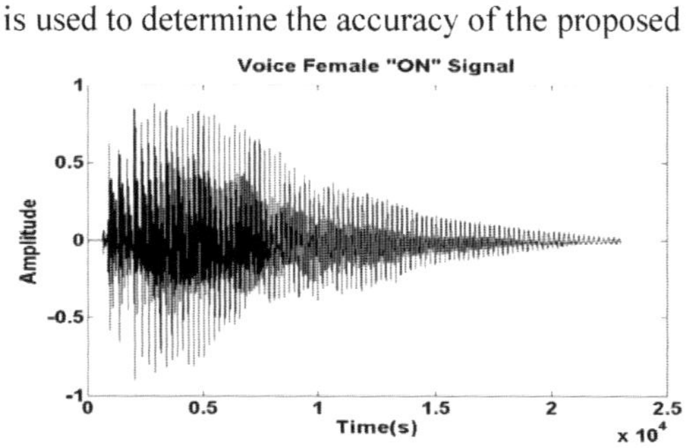

Figure 4. Female 'On' voice sample.

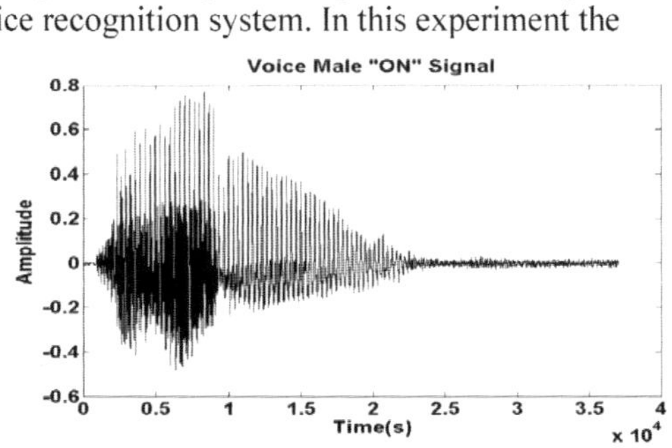

Figure 5. Male 'On' voice sample.

system determines the speaker identity. This experiment consists of two phases, training phase and testing phase. In the training phase, the system is trained with the data in order to develop the model of the recognize speaker, seven (7) speakers are used to train the system and the other three (3) speakers are used to test the system. Therefore, the training data and testing data are contained of 70 words and 30 words of 'On' and 'Off' words, respectively.

The second experiment is the access control system, this experiment is implemented to determine the authorize person in the system. In general, this experiment has four possible decisions; the authorized person is accepted, the authorized person is rejected, the unauthorized person (an impostor) is accepted and the unauthorized person (an impostor) is rejected[3]. The accuracy of the access control system is then specified based on the rate in which the system makes the decision to reject the authorized person and to accept the unauthorized person. The quantities to measure the rate of the access control accuracy to reject the authorized person is then called as a false rejection rate (*FRR*) and that to measure the rate of access control to accept the unauthorized person is called to as a false acceptance rate (*FAR*). Mathematically, both rates are expressed as a percentage using the following simple calculations:

$$FRR = \frac{NFR}{NFA} \times 100\% \quad (1)$$

$$FAR = \frac{NAA}{NIA} \times 100\% \quad (2)$$

where *NFR* is the numbers of false rejections and *NFA* is the number of false acceptance. *NAA* is the number of the authorized person attempts and *NIA* is the number of impostor person attempts. In order to achieve the high security of the proposed system, it is expected that the proposed system will have both low *FRR* and low *FAR*. Furthermore, in the second experiment, one (1) speaker is considered as the authorized person to access the motorcycle and the other three (3) speakers are assumed as outside impostors. Finally, the third experiment is applied to recognize the word uttered by the authorize speaker. Each speaker, who is assumed as authorized person, has to say the word 'On' for 10 times where 5 voice data are used as training data and the other 5 voice data are used as a testing data for each person. This means the text-dependent speaker verification system is used in the proposed system.

In the third experiments, the data is extracted with MFCC feature extraction technique. The MFCC coefficients, the 12 Cepstral coefficients are used. These filters are simulated by integrating the FFT spectrum of 20-ms Hamming-window speech segments in which the frame rate is 10-ms. Figure 6 and 7 shown the 12 MFCC Cepstral coefficient extracted from the voice signal for 'On' and 'Off', respectively. The extracted data, are then become an input to the classification system. The SVMs based multi-class classification is applied to perform classification process using one-against-one method. The RBF kernel function is used. The effectiveness of the proposed method is evaluated based upon the training phase and the testing phased, respectively.

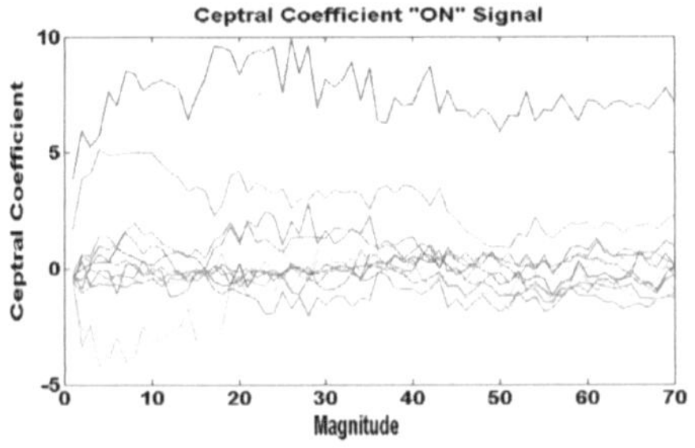

Figure 6. Cepstra Coefficient of the 'On'signal.

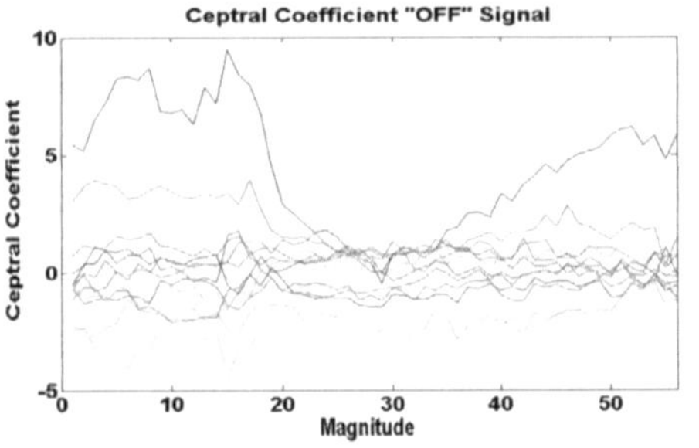

Figure 7. Cepstra Coefficient of the 'Off'signal.

3.2. Experimental Results

In the first experiment, the extracted process resulting 1732 data are extracted from the word 'On' for the training data and 759 data resulting from the extracting process of the testing data in order to evaluate the proposed system. The SVM-base multiclass classification one against one is applied, the Gaussian kernel with a degree of 11 is used.

The training accuracy and testing accuracy of the proposed system achieved 99.31 and 99.43%, respectively as shown in Table 1. Person 3 have the highest testing accuracy of 100%, and person five have the lower testing accuracy of 98.38%.

The second experiment, consist of one authorize person and other three are unauthorized person. The extracted process resulting 128 data are extracted from the word 'On' and 'Off' are used to train the system and 759 data resulting from the extracting process of the testing data in order to evaluate the proposed system. The evaluation result is shown in Table 2. The result shows that the average of *FRR* and *FAR* accuracy are 0.18% and 17.58%, respectively. The result shows that the access control system convenient for the authorized person.

The third experiment, the system recognizes the word utter by the authorize person. The system will decide whether the authorize person speaks the word 'On' and 'Off'. 359 data are used to train the system and 540 data resulting from the extracting process of the testing data in order to evaluate the proposed system. The evaluation result is shown in table 3. The average training and testing accuracy of the proposed system achieved 100% and 96.3%, respectively, as shown in table 3. The accuracy of the testing data consists of accuracy of the word 'On' and 'Off', which accuracy of 95.31% and 97.3 %, respectively.

Table 1. Experiment I results.

Model	Authorized Person	Training Classification Rate (%)	Testing Classification Rate (%)
SVM-Multiclass Classification	Person 1		99.71
One-Against-One	Person 2		99.75
Kernel=Gaussian;	Person 3		100.00
Kernel option=11; c = 1000;	Person 4	99.31	99.66
lambda = 1e-7;	Person 5		98.38
	Person 6		99.43
	Person 7		99.52
average		99.31	99.43

Table 2. Experiment II results.

Authorized person	FRR(%)	Close set FAR(%)	Open Set FAR(%)
Person 1	0.29	0	32.7
Person 2	0.25	0	8.8
Person 3	0	0	11.24

Table 3. Experiment III results.

Model	Authorized Person	Training Classification Rate (%)	Testing Classification Rate (%)	
			'On'	'Off'
SVM-Multiclass Cl. One-Against-One Kernel=Gaussian; Kernel option=11; c = 1000; lambda = 1e-7;	Person 1	100	95.31	97.3

4. Conclusions

This study has documented the development of intelligent biometric voice recognition for motorcycle start-stop engine access control. The proposed system adopted Mel-Frequency Ceptral Coefficient (MFCC) coefficients as the feature of the person voice and used Support Vector Machines to develop authorized voice models based on their words. Experimental results showed that the proposed system produced a good performance, especially it gave a good training and testing accuracy. The system be able to start and stop the motorcycle based on the authorize person voice.

5. References

[1] Sub Direktorat Statistik Politik dan Keamanan 2014 *Statistik Kriminal 2014* (Badan Pusat Statistik Indonesia)

[2] Kung S Y, Mak M W and Lin S H 2004 *Biometric Authentication: Machine Learning Approach* (Prentice Hall)

[3] Omidiora E O, Fakolujo O A, Arulogun O T and Aborisade D O 2011 *A Prototype of a Fingerprint Based Ignition Systems in Vehicles* Eropean Journal of Scientific Research vol. 62, no. 2 p 164-171

[4] Matyáš V and Říha Z 2002 *Biometric Authentication — Security and Usability*. Proceedings of the IFIP TC6/TC11 Sixth Joint Working Conference on Communications and Multimedia Security: Advanced Communications and Multimedia Security p 227-239

[5] Bench H *TIP122 Arduino Relay Tutorial Hendry' Bench*. [Online]. Available: http://henrysbench.capnfatz.com/henrys-bench/arduino-output-devices/tip122-arduino-relay-tutorial/. [Accessed: 29-Aug-2016]

[6] Wahyudi, Astuti W and Mohamed S 2007 *Intelligent Voice-Based Door Access Control System Using Adaptive-Network-based Fuzzy Inference Systems (ANFIS) for Building Security* (*J. Comput. Sci.* vol. 3 no. 5) p 274–280

[7] Mporas I and Ganchev T 2007 Comparison of speech features on the speech recognition task *J. Comput.Sci* vol. 3 no. 8 p 608–616

[8] W. Astuti and E. Byan Wahyu Riyandwita 2016 *Intelligent Automatic Starting Engine Based on Voice Recognition* IEEE SCOReD 2016.

[9] Cristianini N and Taylor J S 2003 *An Introduction to Support Vector Machine and other Kernel based learning method* (Cambrige University Press)

[10] Rifkin R and Klautau A 2004 *In Defense of One-Vs-All Classification* (*J. Mach. Learn. Res.*, vol. 5) p 101–141.

Acknowledgments

We would like to express our thanks to our colleagues and student in BINUS-ASO School of Engineering for their contribution in providing voice data collection.

IOP Conf. Series: Materials Science and Engineering **187** (2017) 012040 doi:10.1088/1757-899X/187/1/012040

Vehicle Maximum Weight Limitation Based on Intelligent Weight Sensor

W Raihan[1, a], R M Tessar[1], C O S Ernest[1], W R E Byan[2, b] and A Winda[1]

[1] Computer Engineering, Bina Nusantara University, Indonesia
[2] Industrial Engineering, Bina Nusantara University, Indonesia

[b]eriyandwita@binus.edu, [a]wastuti@binus.edu

Abstract. Vehicle weight is an important factor to be maintained for transportation safety. A weight limitation system is proposed to make sure the vehicle weight is always below its designation prior the vehicle is being used by the driver. The proposed system is divided into two systems, namely vehicle weight confirmation system and weight warning system. In vehicle weight confirmation system, the weight sensor work for the first time after the ignition switch is turned on. When the weight is under the weight limit, the starter engine can be switched on to start the engine system, otherwise it will be locked. The seconds system, will operated after checking all the door at close position, once the door of the car is closed, the weight warning system will check once again the weight during runing engine condition. The results of these two systems, vehicle weight confirmation system and weight warning system have 100 % accuracy, respectively. These show that the proposed vehicle weight limitation system operate well.

1. Introduction

Safety in driving plays important rules in order to reduce the number of accident in the road [1]. However, safety in driving usually being neglected when there is no enforcing regulates related to that practice. It can be ranging from not wearing a seat belt, over speed and also ignoring the load being carried by the vehicle, regardless it can promote a real danger which can lead to injury or even death of the driver or other people who may involve in the accident [2]. Overload will increase the braking distance of a car and could create a dangerous condition especially for the other vehicle in front of the car if it is not well considered by the driver [3].

Therefore, every vehicle manufacturers has designed a load limit for all of entire vehicle which aims to make the vehicle safe. When an excess weight is being applied to the vehicle, the vehicle will become unstable where the center of gravity of the car can shift away from the original designation and this definitely will harm the driver. Transportation excess burden is one of the causes of traffic accidents in the world [4] . Not only dangerous, it makes the vehicle very unstable, and also reduced the lifetime of the auto parts. The infamous accident that happened in Indonesia that's related to vehicle overload was the Indonesian celeberity Saipul Jamil. He fitted up to 10 people into a 7 seater car and some cargo, which cause the car to be unstable when he suddenly turn and hit the road seperator [5].

Some methods are applied in some countries such as active surveillance, weighing bridge, portable weighing etc. However, these methods are not popular with the public and mostly only being applied to high load transportation such as a truck [6]. This paper proposes a concept of an intelligent weight monitoring system which can be embedded in any kind of vehicles. With this system, weight in a vehicle can be monitored and limited to a safe value following the recommendation of the vehicle maker.

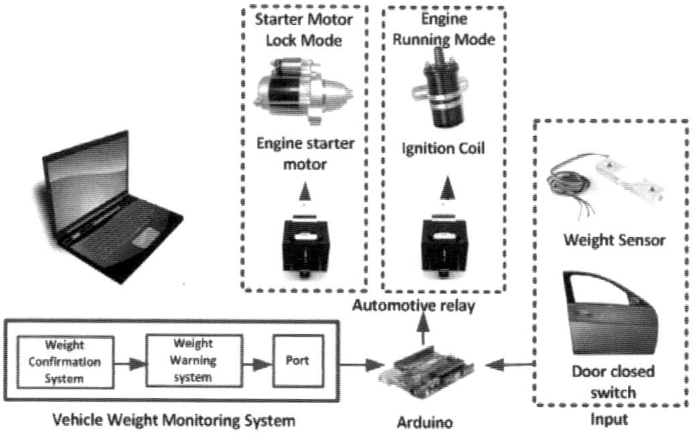

Figure 1. Proposed Vehicle Maximum Weight Limitation Based on Intelligent Weight Sensor.

2. Proposed System

2.1. Proposed System Description

Figure 1 shows the proposed system description, the system basically consists of three main systems, namely vehicle weight limitation system, automotive relay and input for weight monitoring system. In the vehicle weight limitation system, there are two steps involve which are weight confirmation system and weight warning system. In the weight confirmation system, the system will check the total weight of the vehicle and load in the vehicle when ignition key [7] being turned ON. The weight warning system is the next step where the total weight of the vehicle and load inside the vehicle will be checked after all the car doors are closed. This system is added to make sure that all loaded being checked are already all the load inside the vehicle exactly at the moment when the vehicle is ready to be driven.

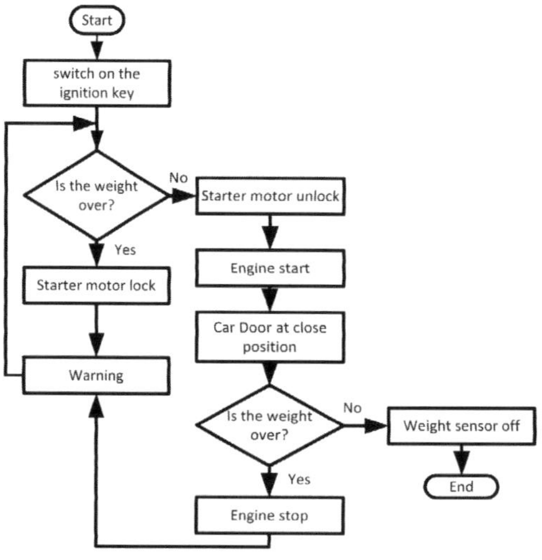

Figure 2. Flowchart of the proposed system.

IOP Conf. Series: Materials Science and Engineering **187** (2017) 012040 doi:10.1088/1757-899X/187/1/012040

The detail flowchart of the proposed system is shown in Figure 2. First the system will check the weight of the vehicle for the first time after the driver switch ON the ignition key. If the weight of the vehicle is overweight, the system will lock the engine's motor starter by cutting off the electrical access to it using an NC (Normally Closed) relay [8] and a warning signal will be sent, otherwise the engine will be ON. If the vehicle is being overloaded, it can only be started if the driver already reducing the weight of the vehicle.

The weight warning system is the next checking sequence for the weight of the vehicle. Furthermore, this second checking is initiated after the all the door of the car is closed. If the weight is below the normal load, then weight monitoring system will be turned off to avoid dynamic load picking and the vehicle can run normally. If the weight is over the weight limit, then the engine will be stopped immediately by using an NC (Normally Closed) relay to cut off the electrical signal to the ignition coil and a warning signal will be given. The driver needs to check and maintain the maximum load so that the vehicle can be operated normally.

In the vehicle weight monitoring system, the weight of the vehicle is measured for the first time once the ignition key is switched on. If the weight of the vehicle is in the normal range, then the starter motor will unlock and the engine will start, otherwise, the starter of the motorcycle will lock, and there will be a warning. In the second system is input for weight warning system, the system is implemented based on the door car position. The system will check all the doors of the car is closed. Once the door is closed, then the weight is measured from the last time, before the sensor is switched off and the engine is continued working.

3. Results

3.1. Experiment Setup

A personal computer (PC) of 2.1 MHz Intel Core i3 processor is used to develop the program of the proposed system. The flowchart of the program developed based of the Arduino [9] software in the computer. The program is then being uploaded to the Arduino to run the system. Weight sensor and car door position switch are used as input to the system. The weight sensor located at pin A2 and A3. The car door position input in pin A4 (input from the switch door sensor). The output of the Arduino is connected to pin D0, D1, D11 and D12 as shown in Figure 3 and 4. Output from the Arduino is the sent to the relay which works on 12 volts DC power supply and it is set in normally closed (NC) condition to lock the starter motor or to turn off the ignition signal. Figure 3.a shows the configuration

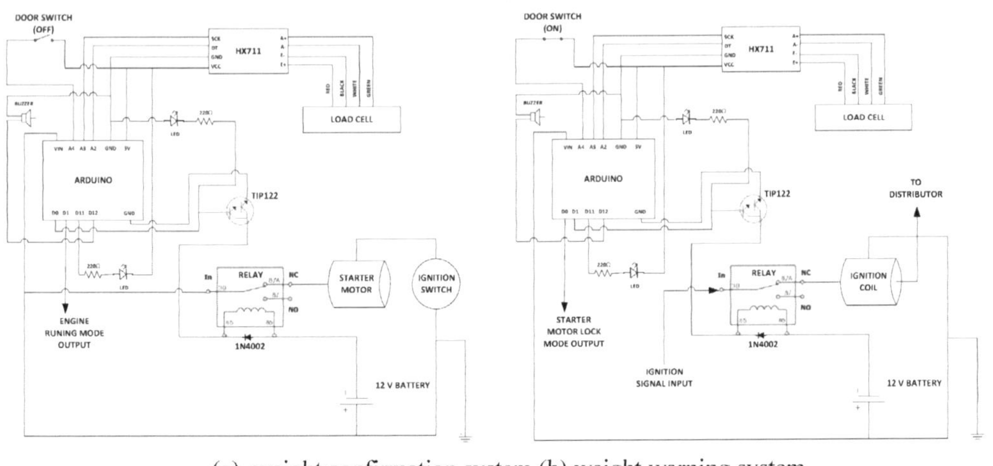

(a) weight confirmation system (b) weight warning system

Figure 3. Arduino and relay circuit [8] for weight confirmation configuration.

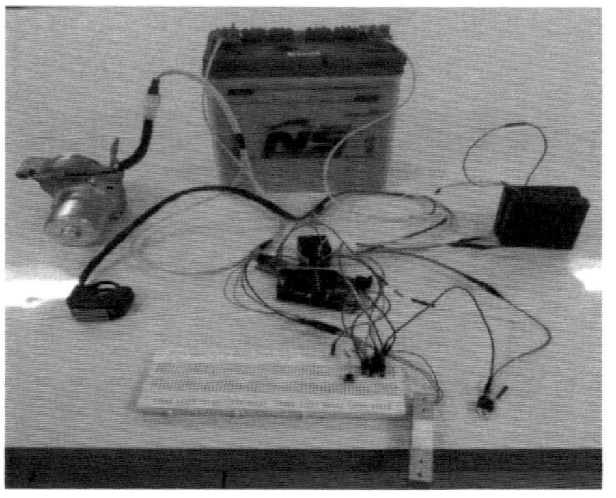

Figure 4. Experimental setup for simulation of the proposed system

for weight confirmation where the first checking is applied. Figure 3.b shows the configuration for a weight warning system of the second step where the door position is closed, once all the door car is closed, then the weight sensor will activate to measure the weight. Once the weight is in the normal range, then the Arduino will switch off the sensor otherwise it will send signals to turn off the engine. Figure 4 shows the experimental setup of the proposed system, the simulation consists of several parts, namely ignition system, ignition switch, NC relay, battery 12 Volts, weight cell sensor [10], circuit simulation and switch ON/OF the door sensor simulation. The input from the weight sensor will activated the electrical circuit which will activate the Arduino. The output from the Arduino will activate the NC (Normally Closed) relay to switch on the electrical signal to the ignition coil.

3.2. Experiment Setup

The weight of the vehicle is measure using weight cell sensor. The result of the weight sensor is used as input to the proposed system which will activate the Arduino. Once the Arduino is activated, then the automotive relay is active which will start the engine through activating starter motor or stop the engine through stopping the ignition pulse. The experiment divided into two parts, firstly when the condition of the weight after the same weight as the first condition and the second condition is when the weight a bit increase from 0 to 900 grams, by an interval of 50 or 100 grams from the previous weight. The weight limit is set to be not more than 800 grams. The result of the first experiment, where it is assumed that the car door is still open is shown in Table 1. Meanwhile, Table 2 shows the result where it is assumed that the engine is already turn on and the all the door already closed. Additional weight is added to check the weight warning system after it is assumed that all of the door is closed.

Table 1. The result of the experiment when the car door is still open.

No.	Weight (gr)	A proposed system designated action (starter motor condition)	Proposed system condition based on experiment
1	0	UNLOCKED	UNLOCKED
2	200	UNLOCKED	UNLOCKED
3	300	UNLOCKED	UNLOCKED
4	400	UNLOCKED	UNLOCKED
5	500	UNLOCKED	UNLOCKED
6	600	UNLOCKED	UNLOCKED
7	700	UNLOCKED	UNLOCKED
8	800	LOCKED	LOCKED
9	900	LOCKED	LOCKED

Table 2. The result of the experiment when there is increasing weight after all the door is closed.

No.	Weight (gr)	Door switch is On (All door are closed) (gr)	The Proposed system designated action (Ignition coil condition)	Proposed system condition based on experiment
1	0	(+50)50	ON	ON
2	200	(+50)250	ON	ON
3	300	(+100)400	ON	ON
4	400	(+100)500	ON	ON
5	500	(+100)600	ON	ON
6	600	(+100)700	ON	ON
7	700	(+100)800	OFF	OFF
8	800	(+100)900	OFF	OFF
9	900	(+0)900	OFF	OFF

The result shows that the proposed system works well for the weight confirmation and weight warning system. The experiment achieved 100% accuracy in the condition after ignition switch is turned on and conditional control whether the starter motor can be turned on (unlocked) or turn off (locked) is aligned with the designated purpose. Furthermore, the experimental results of the proposed system; with additional weight during all the car doors are closed is as shown in Table 2. The result shows that t proposed system achieved 100% accuracy where the signal to turn off the engine will be sent if the weight is above the limit. It shows that the system will work well for weight vehicle monitoring.

4. Conclusions
This paper shows a preliminary study to apply a weight limitation system in a vehicle to make sure that the vehicle is not overloaded, thus increase the safety of the vehicle. Experimental results showed that the proposed system produced a good performance. The application to the real vehicle need to be developed.

5. References
[1] Transport Departement, "Road Safety bulletin," Transportation Departement, 2009. [Online]. Available: http://www.td.gov.hk/en/road_safety/. [Accessed: 01-Oct-2016].
[2] President Republik Indonesia, "PERATURAN PEMERINTAH REPUBLIK INDONESIA NOMOR 55 TAHUN 2012 TENTANG KENDARAAN," 2012.
[3] arrivealive, "Overloading and Road Safety," 2016. [Online]. Available: https://arrivealive.co.za/pages.aspx?u=Overloading-and-Road-Safety. [Accessed: 01-Oct-2016].
[4] F. Edition, "the Damaging Effects of Overloaded Heavy," no. April, 1997.
[5] Heru Margianto, "Kronologi Kecelakaan Saipul Jamil," Kompas. [Online]. Available: http://entertainment.kompas.com/read/2011/09/03/14183685/kronologi.kecelakaan.saipul.jamil. [Accessed: 19-Oct-2016].
[6] FHWA, "Compilation of Existing State Truck Size and Weight Limit Laws," J. Chem. Inf. Model., vol. 53, no. 9, pp. 1689–1699, 2016.
[7] LBCC Automotive Technology, "Ignition System Basic Operation," linnbenton, 2014. [Online]. Available: http://www.linnbenton.edu/auto/perform/ignition.html. [Accessed: 01-Nov-2016].
[8] H. Bench, "TIP122 Arduino Relay Tutorial," Hendry' Bench. [Online]. Available: http://henrysbench.capnfatz.com/henrys-bench/arduino-output-devices/tip122-arduino-relay-tutorial/. [Accessed: 29-Aug-2016].

[9] H. World, P. Computing, K. Sensor, P. Melody, K. Serial, L. C. D. Display, U. S. Motor, and S. U. Advanced, "Arduino Tutorials from www.arduino.cc Table of Content," pp. 1–37, 2013.
[10] S. T. L. C. Czlc, "Datasheet 3138 - S Type Load Cell (0-100kg) - CZL301C," pp. 1–4, 2012.

Acknowledgments
We would like to express our thanks to our colleagues and student in BINUS-ASO School of Engineering for their contribution in providing necessary equipment for the experiment setup.

IOP Conf. Series: Materials Science and Engineering **187** (2017) 012041 doi:10.1088/1757-899X/187/1/012041

Simulation Analysis of Resonant Frequency of Mix Sucker Rod with absorber

Mingming Xing[1,*], Lili Zhou[2] and Xue Kaifeng[1]

[1] School of Mechanical & Vehicle Engineering, Linyi University, China
[2] School of Chemistry & Chemical Engineering, Linyi University, China

*xingmingming2009@126.com

Abstract. Considering the influence of sucker rod absorber (SRA) on longitudinal vibration of sucker rod string (SRS), a new model of mix sucker rod with absorber (MSRWA) in the form of numerical integration is built. In the detail, based on the dynamical theory of continuous systems, the wave equation of instantaneous motion of MSRWA is deduced. The dynamic response is solved out with the mode superposition method. The affecting factors of natural frequency are analyzed, such as the length, diameter, density and elastic modulus of SRS. Then three dimensional curves of natural frequency are obtained. The results are very important for the design and application of SRS.

1. Introduction

The sucker rod pumping system (SRPS) is an important method of the mechanical recovery methods in the domestic and overseas oilfields [1-3]. In order to reduce energy consumption and boost its crude production, the optimal design of SRPS had been done since the nineteen sixties [4, 5]. The resonant frequency of SRPS is foundation of optimal design of SRPS, so the resonant frequency of SRPS has been studied. Two steps in estimating natural frequencies and damping ratios are given by Lu X [6], the first step is from measurement records to the power spectral density function, the second step is from the power spectrum to system parameter estimates. Dong S gave the conclusion that there will be resonance phenomena, when the excited frequency and even multiple of fundamental frequency are equal [7]. Peng Y analyzes the twisting vibration of rotary sucker rod string, a formula of calculating every harmonic vibration inherent frequency is presented [8]. The inherent frequency of longitudinal vibration of single stage rod string is analyzed and calculated from the point of view of vibration theory by Zhu B [9]. The 3-D vibration of sucker rod string in the reciprocating motion is proposed by Chen J, such as horizontal vibration, longitudinal vibration and torsional vibration. The sucker rod string is considered the elastic body whose quality is distributed by using the elastic body vibration theory. The dynamic models for the vibrations of sucker rod string in three directions are established [10]. Although the resonant Frequency of SRPS has been studied for years, the most of aforementioned studies have not been studied sufficiently. In the detail, the formula of natural frequency is not accurate [11], when the free vibration of SRS is researched. When the SRA is installed between polished rod and SRS, the influence of SRA on the SRS's vibration is usually ignored. In the period of oil field development, the cost of oil production has been the focus of the global attention at present [12-13]. However, with application of SRA, the vibration of SRS will be absorbed and the stress of SRS and the peak of dynamic load will be reduced, at the same time, the motor load and power consumption will decrease. Therefore, the resonant frequency of longitudinal vibration of MSRWA is necessary to analyzed, which lays the foundation for the design and application of SRA.

2. Dynamic simulate model of MSRWA

The number stages of SRS assemblage are supposed to be k, and the polished rod displacement is u*. The relative displacement between polished rod and SRS top is u0. Considering the influence of SRA on SRS's longitudinal vibration, a mechanical model of MSRWA is established, as shown in Fig.1.

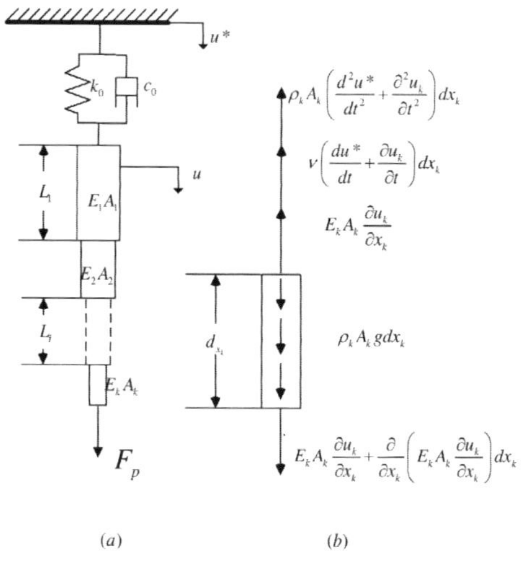

Figure 1. The mechanical model of MSRWA

Based on the mechanical model of longitudinal vibration, a mathematical model is built, as follows

$$\begin{cases} \dfrac{\partial^2 u_k}{\partial t^2} - a_k^2 \dfrac{\partial^2 u_k}{\partial x_k^2} + c_k \dfrac{\partial u_k}{\partial t} = -\dfrac{d^2 u^*}{dt^2} - c_k \dfrac{du^*}{dt} + g \\ E_1 A_1 \dfrac{\partial u_1(0)}{\partial x_1} = k_0 u_0 + c_0 \dot{u}_0 \\ E_k A_k \dfrac{\partial u_k(L_d)}{\partial x_k} = F_p \end{cases} \tag{1}$$

Where,

$$\begin{cases} a_k = \sqrt{\dfrac{E_k}{\rho_k}} \\ c_k = \left[\dfrac{\left(0.20 + 0.39 \dfrac{D_k}{D_t}\right) + \dfrac{2.197 \times 10^4}{25}}{\left(\dfrac{D_{ck}}{D_t} - 0.381\right)^{2.57} \dfrac{D_{ck}^{\ 2} - D_k^{\ 2}}{l_{0k} D_k}} \right] \dfrac{12\pi\mu m_k}{\rho_k A_k} \left(\dfrac{D_k}{D_t - D_k} \right) \end{cases} \tag{2}$$

Where, a_k is the velocity of sound in SRS (m/s); c_k is the friction coefficient (Pa.s); F_p is the load of pump plunger (N); ω is the exaction frequency (rad/s)

3. Free vibration

IOP Conf. Series: Materials Science and Engineering **187** (2017) 012041 doi:10.1088/1757-899X/187/1/012041

Ignoring the influence of the damping of SRA on the SRS's vibration, the equation of free vibration is obtained based on Eq.(1).

$$\begin{cases} \frac{\partial^2 u_1}{\partial t^2} - a_1^2 \frac{\partial^2 u_1}{\partial x_1^2} = 0 \\ \vdots \qquad \cdots \qquad \vdots \\ \frac{\partial^2 u_k}{\partial t^2} - a_k^2 \frac{\partial^2 u_k}{\partial x_k^2} = 0 \end{cases} \tag{3}$$

The equations of boundary and continuous conditions are given, as follows.

When k=1

$$E_1 A_1 \frac{\partial u_1}{\partial x_1}\Big|_{x=0} = k_0 u_0, \qquad E_{rk} A_{rk} \frac{\partial u}{\partial x}\Big|_{x=L} = 0 \tag{4}$$

When k=2, 3, 4…

$$\begin{cases} E_1 A_1 \frac{\partial u_1}{\partial x_1}\Big|_{x_1=0} = k_0 u_0, \qquad u_{k-1}\Big|_{x_{k-1}=L_{k-1}} = u_k\Big|_{x_k=0} \\ E_{(k-1)} A_{(k-1)} \frac{\partial u_{k-1}}{\partial x_{k-1}}\Big|_{x_{k-1}=L_{k-1}} = E_k A_k \frac{\partial u_k}{\partial x_k}\Big|_{x_k=0}, \ E_k A_k \frac{\partial u_k}{\partial x}\Big|_{x_k=L_k} = 0 \end{cases} \tag{5}$$

The vibration model functions are defined, as follows.

$$\begin{cases} \Phi_1(x_1) = B_1 \sin(b_1 x_1) + D_1 \cos(b_1 x_1) & 0 \le x_1 \le L_1 \\ \Phi_2(x_2) = B_2 \sin(b_2 x_2) + D_2 \cos(b_2 x_2) & 0 \le x_2 \le L_2 \\ \Phi_i(x_i) = B_i \sin(b_i x_i) + D_i \cos(b_i x_i) & 0 \le x_i \le L_i \end{cases} \tag{6}$$

Combining Eqs. (4) and (5), the transcendental equation of natural frequency is given, as follows

$$B_k \cos(b_k L_k) - D_k \sin(b_k L_k) = 0 \tag{7}$$

Where,

$$\begin{cases} D_1 = B_1 \frac{E_1 A_1 b_1}{k_0} \\ B_2 = B_1 \frac{E_1 A_1 b_1}{E_2 A_2 b_2} \left[\cos(b_1 L_1) - \frac{E_1 A_1 b_1}{k_0} \sin(b_1 L_1) \right] \\ D_2 = B_1 \left[\sin(b_1 L_1) + \frac{E_1 A_1 b_1}{k_0} \cos(b_1 L_1) \right] \\ B_i = B_1 \frac{E_{i-1} A_{i-1} b_{i-1}}{E_i A_i b_i} \left[\cos(b_{i-1} L_{i-1}) - \frac{D_{i-1}}{B_1} \sin(b_{i-1} L_{i-1}) \right] \\ D_i = B_1 \left[\frac{B_{i-1}}{B_1} \sin(b_{i-1} L_{i-1}) + \frac{D_{i-1}}{B_1} \cos(b_{i-1} L_{i-1}) \right] \end{cases} \tag{8}$$

Then the formula of regular function is obtained, as follow.

$$\begin{cases}\varphi_1(x_1)=B_1\left[\sin(b_1x_1)+\dfrac{E_1A_1b_1}{k_0}\cos(b_1x_1)\right] & 0\le x_1\le L_1\\ \varphi_2(x_2)=B_1\left[\dfrac{E_1A_1b_1}{E_2A_2b_2}\cos(b_1L_1)\sin(b_2x_2)-\dfrac{E_1^2A_1^2b_1^2}{E_2A_2b_2k_0}\sin(b_1L_1)\sin(b_2x_2)\right.\\ \left.+\sin(b_1L_1)\cos(b_2x_2)+\dfrac{E_1A_1b_1}{k_0}\cos(b_1L_1)\cos(b_2x_2)\right] & 0\le x_2\le L_2\\ \varphi_i(x_i)=B_1\left[\dfrac{E_{i-1}A_{i-1}b_{i-1}}{E_iA_ib_i}\cos(b_{i-1}L_{i-1})\sin(b_ix_i)-\dfrac{E_{i-1}A_{i-1}b_{i-1}D_{i-1}}{E_iA_ib_iB_1}\sin(b_{i-1}L_{i-1})\sin(b_ix_i)\right.\\ \left.+\dfrac{B_{i-1}D_i}{B_1^2}\sin(b_{i-1}L_{i-1})\cos(b_ix_i)+\dfrac{D_{i-1}D_i}{B_1^2}\cos(b_{i-1}L_{i-1})\cos(b_ix_i)\right] & 0\le x_i\le L_i\end{cases} \tag{9}$$

Where,

$$\begin{cases}B_1=\sqrt{\dfrac{1}{\rho_1A_1G_1+\rho_2A_2G_2+\rho_iA_iG_i}}\\ G_1=\left[\dfrac{L_1}{2}-\dfrac{1}{4b_1}\sin(2b_1L_1)\right]+\dfrac{D_1^2}{B_1^2}\left[\dfrac{L_1}{2}+\dfrac{1}{4b_1}\sin(2b_1L_1)\right]+\dfrac{D_1}{2b_1B_1}\left[1-\cos(2b_1L_1)\right]\\ G_2=\dfrac{B_2^2}{B_1^2}\left[\dfrac{L_2}{2}-\dfrac{1}{4b_2}\sin(2b_2L_2)\right]+\dfrac{D_2^2}{B_1^2}\left[\dfrac{L_2}{2}+\dfrac{1}{4b_2}\sin(2b_2L_2)\right]+\dfrac{B_2D_2}{2b_2B_1^2}\left[1-\cos(2b_2L_2)\right]\\ G_i=\dfrac{B_i^2}{B_1^2}\left[\dfrac{L_i}{2}-\dfrac{1}{4b_i}\sin(2b_iL_i)\right]+\dfrac{D_i^2}{B_1^2}\left[\dfrac{L_i}{2}+\dfrac{1}{4b_i}\sin(2b_iL_i)\right]+\dfrac{B_iD_i}{2b_iB_1^2}\left[1-\cos(2b_iL_i)\right]\end{cases} \tag{10}$$

The numerical solutions of the time functions in regular coordinate system are solved out with the fourth order Runge-Kutta method. Then the numerical solutions of the time functions in Natural coordinate system are obtained with the coordinate transformation.

4. Characteristic analysis

The basic parameters are as follows: when the length of SRS L_d is 1000m, the length of first stage L_1 is 600m and the length of second stage L_2 is 400m. When the length of SRS L_d is 5000m, the length of first stage L_1 is 3000m and the length of second stage L_2 is 2000m. The Young's modulus of SRS E_1 as well as E_2 is 2.1×10^{11} Pa; the density of rod ρ1 as well as ρ_2 is 7850 kg/m^3; the diameter of first stage D_1 is 22mm and the diameter of second stage D_2 is 19mm.

Table 1. The comparison of natural frequency

older	Theoretical	Current model		New model	
	rad/s	rad/s	Error %	rad/s	Error %
1	8.1	8.1	0	8.1	0
2	24.4	24.4	0	24.4	0
3	40.6	56.9	16.3	40.6	0
4	56.9	67.7	10.8	56.9	0
5	73.1	89.2	16.1	73.1	0
6	89.4	138.1	48.7	89.4	0

When L_d =1000m, L_1=L_2, two stages SRS is transformed into one stages SRS. Based on the current model, new model and the analytic expression of natural frequency, the six natural frequencies are obtained, in Tab.1. The table shows that there is a large error in results of current model, but the

IOP Conf. Series: Materials Science and Engineering **187** (2017) 012041 doi:10.1088/1757-899X/187/1/012041

simulation results of the new model are found to be in good agreement with the simulation results of theoretical value. Therefore, the new model of natural frequency in the paper is advised to use in actual engineering.

When the elasticity modulus of SRS E_1=E_2=2.1×10^{11} Pa, the density of SRS ρ_1=ρ_2=7850kg/m^3, the length of SRS L_d=5000m, the diameter of SRS D_1=0.022m. The combination ratios of length and diameter are changed and three dimensional curves of natural frequency are obtained, in Fig.2.

As you can see in the above figures, the first natural frequency shows large changes when the diameter ratio of SRS is larger than 1.5 or the length ration of SRS is an interval value 0.3-0.7. The first natural frequency is increasing with the diameter ratio increasing. With the length ratio increasing, the first natural frequency is first increasing and then decreasing. With second natural frequency, the length ratio is major factor of natural frequency.

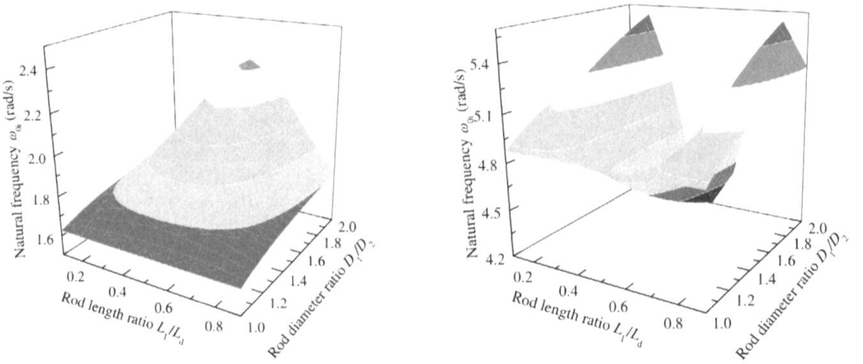

(a) First natural frequency (b) Second natural frequency

Figure 2. The natural frequency of SRS with length and diameter

When the elasticity modulus of SRS E_1 =2.1×10^{11} Pa, the density of SRS ρ_1=7850kg/m^3, the length of SRS L_1=3000m, L_2=2000m, the diameter of SRS D_1=0.022m, D_2=0.019m. The combination ratios of elasticity modulus and density are changed and three dimensional curves of natural frequency are obtained, in Fig.3.

In Fig.3, with elasticity modulus ratio decreasing and density ratio increasing, the natural frequency of SRS is increasing. When the elasticity modulus ratio is smaller areas, the density ratio has a significant impact on natural frequency. When the density ratio is bigger areas, the elasticity modulus ratio has a significant impact on natural frequency. Therefore, the above three-dimensional curves have important implications for the design and application of SRS.

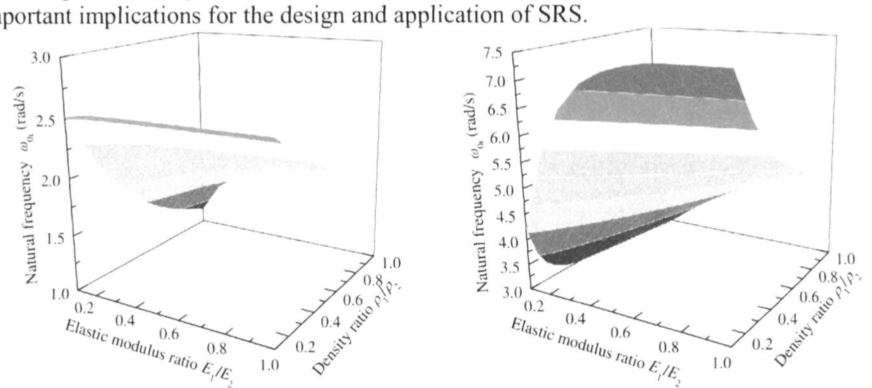

(a) First natural frequency (b) Second natural frequency

Figure 3. The natural frequency of SRS with elasticity modulus and density

5. Conclusions

Considering the influence of SRA on longitudinal vibration of SRS, a simulation model of MSRWA is built. A computational model of natural frequency is corrected against the deficiency of current model. With the free vibration of SRS, a computational model of natural frequency is corrected against the deficiency of current model. With the forced vibration of MSRWA, a mode superposition method is used to find the numerical solution. The affecting factors of natural frequency are analyzed, such as the length, diameter, density and elastic modulus of SRS. Then three dimensional curves of natural frequency are obtained. The above conclusions are very important to the design and application of SRA in practical applications.

6. References

[1] Liu X.F, Qi Y.G. A Modern Approach to the Selection of Sucker Rod Pumping Systems in CBM Wells. Journal of Petroleum Science and Engineering, Vol.76, pp. 100–108, 2011.
[2] Xing Mingming, Dong Shimin. A new simulation model of beam pumping system applied in energy saving and resource consumption reduction. SPE Production & Operations, Vol. 30, pp.130-140, 2015.
[3] Shardakov I.N, Wasserman I.N. Numerical modelling of longitudinal vibrations of a sucker rod string, Journal of Sound and Vibration, Vol. 329, pp. 317-327, 2010.
[4] Gibbs S.G. Predicting the Behavior of Sucker-Rod Pumping System. Journal of Petroleum Technology, Vol. 15 , pp.769-778, 1963.
[5] Gibbs S.G., Neely A.B. Computer Diagnosis of Down-Hole Conditions in Sucker Rod Pumping Wells. SPE Journal Paper, 1165-PA, 1966.
[6] Lu, X, Vandiver, J. K. Damping and Natural Frequency Estimation Using the Least Pth Optimization Technique, Offshore Technology Conference, Houston, Texas, OTC-4283-MS, 1982.
[7] Dong S, Ma D, Huang X. The resonance characteristic and resonance condition of sucker rod string. China Petroleum Machinery, Vol. 29, pp. 22-24. 2001.
[8] Peng Y, Yan W, Wang L. Analysis of Twisting Vibration Eigenfrequencies of Rotary Sucker-rod String. Journal of Xi'an Petroleum Institute, Vol. 18, pp. 57-59, 2003.
[9] Zhu B. The Analysis and Calculation of the Inherent Frequency of Longitudinal Vibration of Rod String. Oil Field Equipment, Vol. 36, pp. 11-14, 2007.
[10] Chen J, Qi Y, Liu X, Liu C. Study on 3-D Vibrations of Sucker Rod String Based on the Elastic Body Vibration Theory. Oil Field Equipment, Vol. 38, pp. 15-19, 2009.
[11] Fernando C, María J. E. Longitudinal vibration of a damped rod. Part I: Complex natural frequencies and mode shapes. International Journal of Mechanical Sciences, Vol.48, pp. 969-975, 2006.
[12] Xing M, Dong S, Tong Z, Tian R, Chen H. Dynamic simulation and efficiency analysis of beam pumping system. Journal of Central South University, Vol.22, pp. 3367-3375, 2015,
[13] Takacs G. and Belhaj H. Latest Technological Advances in Rod Pumping Allow Achieving Efficiencies Higher than with ESP System. Journal of Canadian Petroleum Technology, Vol. 50, pp. 53-58, 2011.

Acknowledgments

This research was financially supported by the Scientific Research Starting Foundation for Linyi University (Grant NO.LYDX2016BS032) and Natural Science Foundation of Shandong Province (Grant NO. ZR2016HB59).

IOP Conf. Series: Materials Science and Engineering **187** (2017) 012042 doi:10.1088/1757-899X/187/1/012042

Contrastive Analysis for Diversion Performance of Three Kinds of Deflector

Chengde Wang[1], Yi Jiang[1], Aolin Zhang[1] and Yongyue Li[1]

[1]Department of Aerospace Engineering, Beijing Institute of Technology, Beijing, 100081, China

bitwcd@163.com

Abstract. Deflector is generally used as deflecting device of Vehicle-mounted Vertically Thermal Launched Missile. To study the features of three different kinds of deflectors, numerical simulation is used to simulate the jet flow around single-faced, double-faced, triple-faced deflector respectively in this dissertation. Pathlines, Mass flow rate through the observation surfaces built around the deflector, regions influenced by jet flow on ground are standards to be compared, characters of them can be concluded from an analysis.

1. Introduction

Vehicle-mounted missile is widely applied because of its excellent concealment and maneuverability, deflector, as an important part of Vehicle-mounted Vertically Thermal Launched system, bearing the impact of jet flow and guiding it to the direction that is favorable to launch [1], reduces the ablation and impact caused by jet flow with high speed and high temperature on launchers. At present, the type of the deflector on Vehicle-mounted missile mainly includes single-faced, double-faced, triple-faced, cone-type and so on [2]. JingSong Chen [3] studied the law of double-faced deflector, but no one compared the diversion performance of them.

Structured grid and unstructured grid are applied to modeling the jet flow field around the different kinds of deflectors, by numerical simulation, comparing and analyzing the result, it is concluded that the characteristics of different deflectors, the results will provide theoretical basis for rational selection.

2. Physical Model

Three-dimensional N-S equation is used for the numerical simulation. The Control Equations [4] are shown as follow:

Continuum equations is shown as follow

$$\partial\rho/\partial t+\nabla\cdot(\rho\vec{u})=0 \tag{1}$$

Where, ρ is flow density, $\vec{u}$ is flow velocity vector.

Momentum equations is shown as follow

$$\partial(\rho u_i)/\partial t+\partial(\rho u_u u_j)/\partial x_j=-\partial p/\partial x_i+\partial\tau_{ij}/\partial x_j \tag{2}$$

Where u_i, u_j are i and j weight flow velocity. p is the flow pressure. x_i, x_j are i ,j direction coordinate. τ_{ij} is the stress tensor. Its expression is shown as follow

$$\tau_{ij} = [\mu(\partial u_i / \partial x_j + \partial u_j / \partial x_i)] - \frac{2}{3}\mu \frac{\partial u_l}{\partial x_l}\delta_{ij} \tag{3}$$

Energy equations is shown as follow

$$\frac{\partial}{\partial t}(\rho H) + \frac{\partial}{\partial x_j}(\rho u_j H) = \frac{\partial p}{\partial t} + \frac{\partial}{\partial x_j}(u_i \tau_{ij} - q_j) \tag{4}$$

H is total enthalpy $H = h + u_i^2 / 2$, $h = C_p T$, C_p is stated specific heat, T is static temperature. q_j is heat flux.

Turbulence Equations is shown as follow

The Realizable k-ε model [5] equation was used. The transportation equation [6] of the turbulence kinetic energy k is

$$\frac{\partial(\rho k)}{\partial t} + \frac{\partial(\rho k u_i)}{\partial x_i} = \frac{\partial}{\partial x_j}\left[\left(\mu + \frac{\mu_t}{\sigma_k}\right)\frac{\partial k}{\partial x_j}\right] + G_K + G_b - \rho\varepsilon - Y_M + S_k \tag{5}$$

The equation of the dissipation rate of the turbulence kinetic energy is

$$\frac{\partial(\rho\varepsilon)}{\partial t} + \frac{\partial(\rho\varepsilon u_i)}{\partial x_i} = \frac{\partial}{\partial x_j}\left[\left(\mu + \frac{\mu_t}{\sigma_\varepsilon}\right)\frac{\partial \varepsilon}{\partial x_j}\right] + \rho C_1 S\varepsilon - \rho C_2 \frac{\varepsilon^2}{k + \sqrt{\upsilon\varepsilon}} + C_{1\varepsilon}\frac{\varepsilon}{k}C_{3\varepsilon}G_b + S_\varepsilon \tag{6}$$

Where $\mu_i = \rho C_\mu k^2 / \varepsilon$

3. Calculation Model

The computational model is shown in **Figure 1. - Figure 3.** , it includes nozzle, missile, deflector, baffle (except triple-faced deflector) and ground.

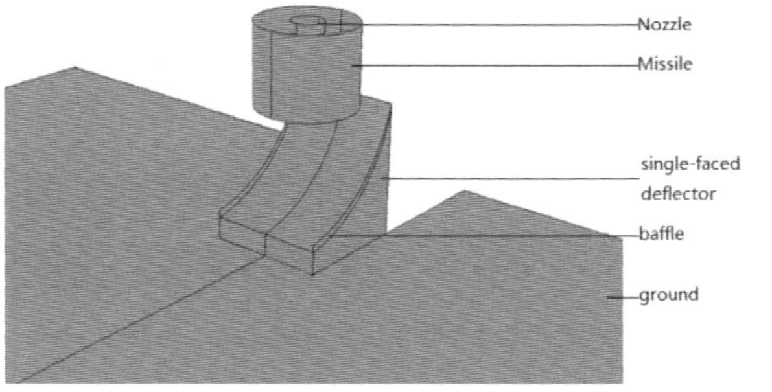

Figure 1. calculation model of single-faced deflector

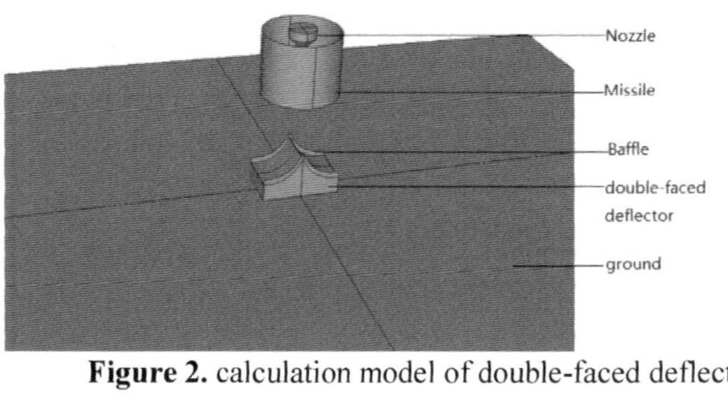

Figure 2. calculation model of double-faced deflector

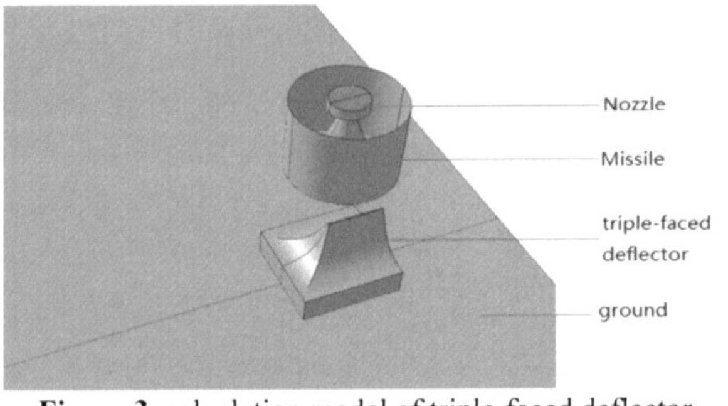

Figure 3. calculation model of triple-faced deflector

The computational domain is shown in **Figure 4.** - **Figure 6.** Based on symmetry, the calculation model of single-faced deflector and triple-faced deflector adopt half model, the calculation model of double-faced deflector uses quart model. Structured grid is used in most regions except for some complicated regions near the deflector. The mesh number of three kinds of deflector is about 2 million.

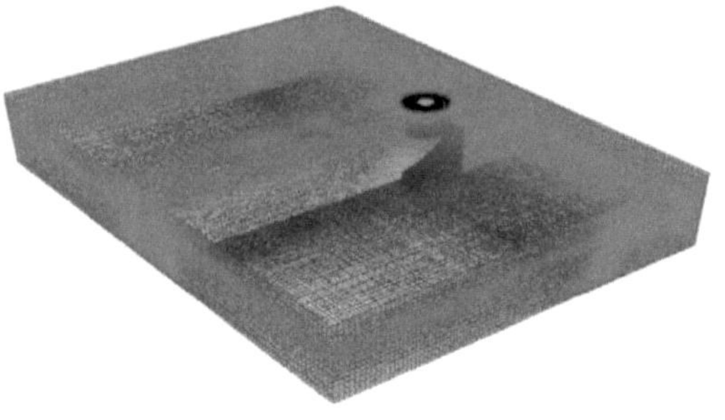

Figure 4. calculation model of single-faced deflector

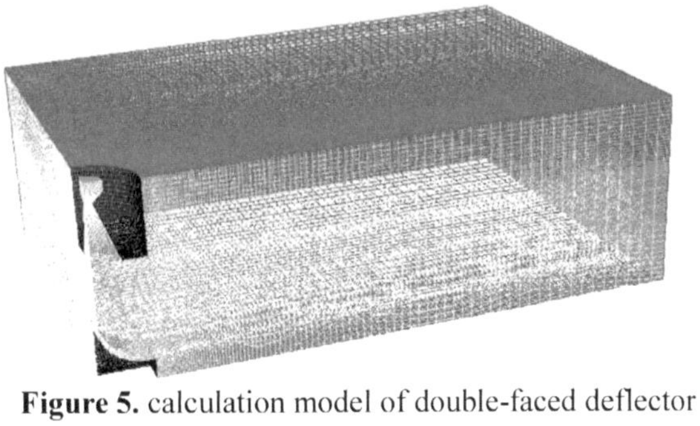

Figure 5. calculation model of double-faced deflector

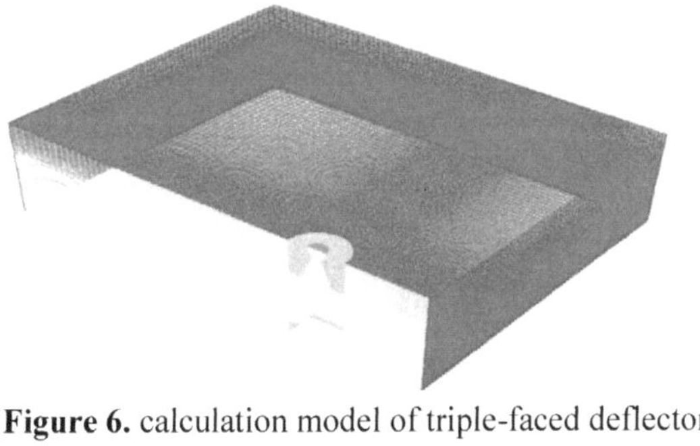

Figure 6. calculation model of triple-faced deflector

IOP Conf. Series: Materials Science and Engineering **187** (2017) 012042 doi:10.1088/1757-899X/187/1/012042

Two components, air and gas, are adopted for calculation, the boundary condition of the inlet of nozzle is pressure inlet, the total combustion chamber pressure is 7 MPa, the total temperature is 3200 K, free boundaries are set to pressure outlet, the pressure is 101325 Pa, and temperature is 300 K. The wall boundary [7] is adiabatic and stationary.

4. Results and Discussions

It is compared by three aspects in this paper, Pathlines, Mass flow rate through the observation surfaces built around the deflector, Regions influenced by jet flow on ground, respectively.

4.1. The pathlines

The pathlines of three kinds of deflector are shown in **Figure 7. - Figure 9.**

Figure 7. Pathlines of single-faced deflector

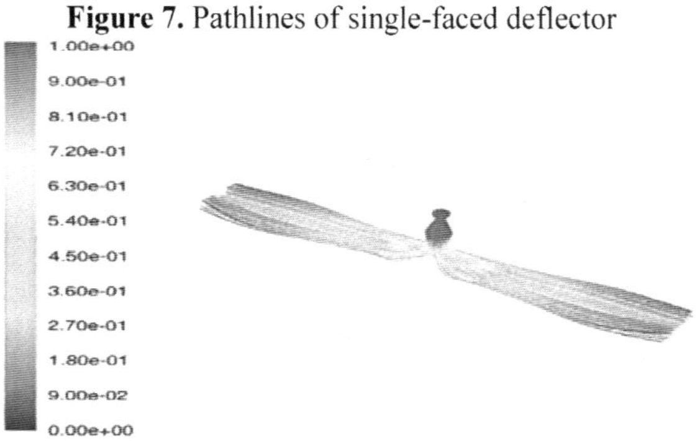

Figure 8. Pathlines of double-faced deflector

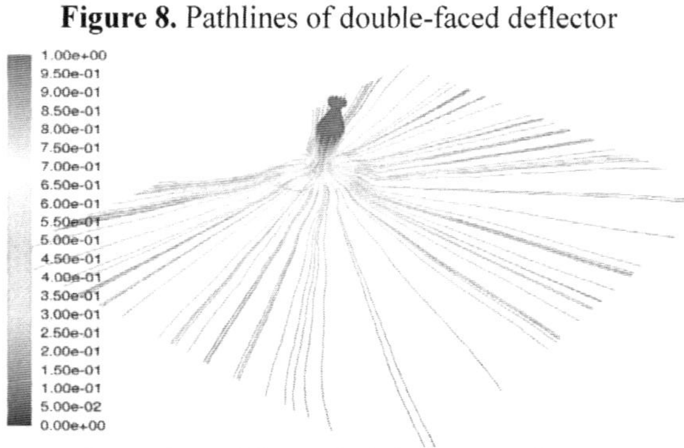

Figure 9. Pathlines of triple-faced deflector

As shown in pictures of pathline ,because of the deflector, jet flow is guided into deflecting side, which, thanks to the baffles, is mainly the back for single-faced deflector, the two sides for double-faced, however ,not only both sides and back ,but also the front for triple-faced without a baffle.

4.2. Mass flow rate through the observation surfaces

Observation surfaces are set around the deflector, which are shown in **Figure 10.** - **Figure 12.**

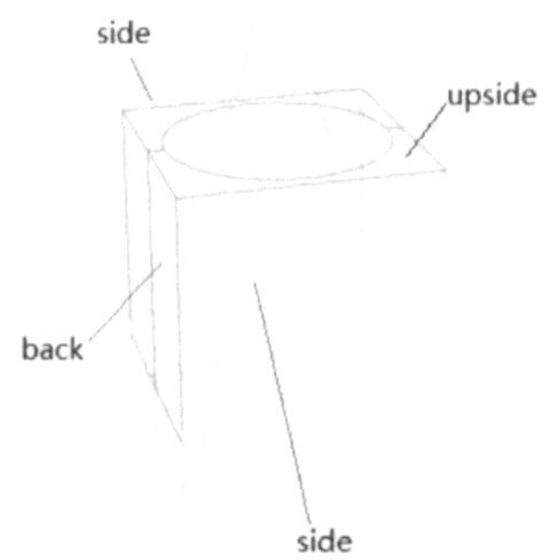

Figure 10. Observation surfaces of single-faced deflector.

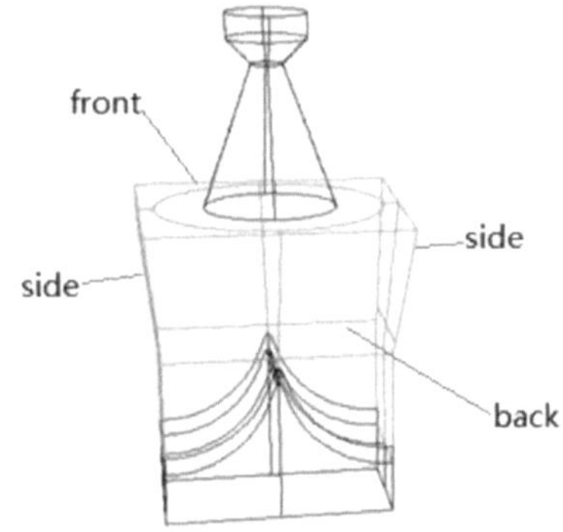

Figure 11. Observation surfaces of double-faced deflector.

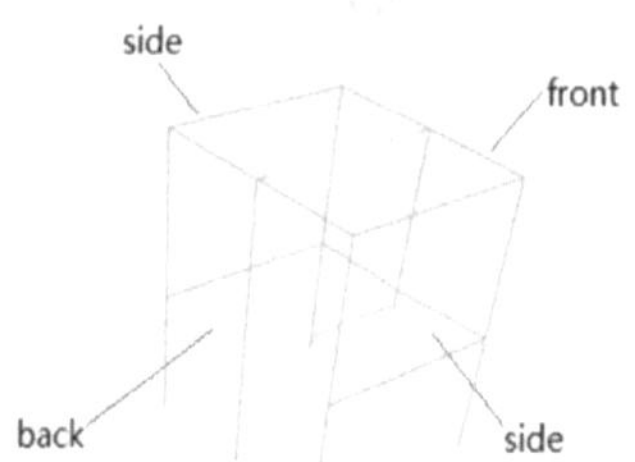

Figure 12. Observation surfaces of triple-faced deflector.

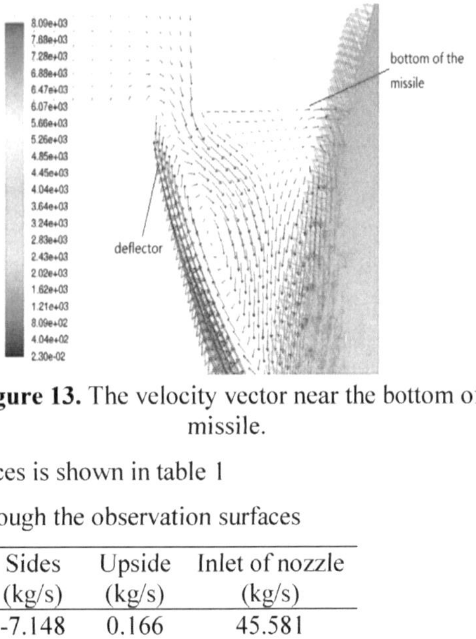

Figure 13. The velocity vector near the bottom of missile.

The mass flow rate through the observation surfaces is shown in table 1

Table 1. The mass flow rate through the observation surfaces

	Front (kg/s)	Back (kg/s)	Sides (kg/s)	Upside (kg/s)	Inlet of nozzle (kg/s)
Single-faced	---	-38.850	-7.148	0.166	45.581
Double-faced	-5.373	-5.373	-35.968	0	47.946
Triple-faced	-4.834	-10.336	-32.939	0	47.705

It is shown that jet flow is mainly deflecting to backside of single-faced deflector, but it appears a little backwashing (the mass flow rate through upside is not zero), it also can be shown in **Figure 13.** While no backwashing in double-faced and triple-faced deflector. Meanwhile the input (inlet of nozzle) equals to output (the sum of different sides) as shown.

4.3. Regions influenced by jet flow on ground

Figure 14. - **Figure 16.** show the temperature contours on the ground (only display the region beyond 1000K).

IOP Conf. Series: Materials Science and Engineering **187** (2017) 012042 doi:10.1088/1757-899X/187/1/012042

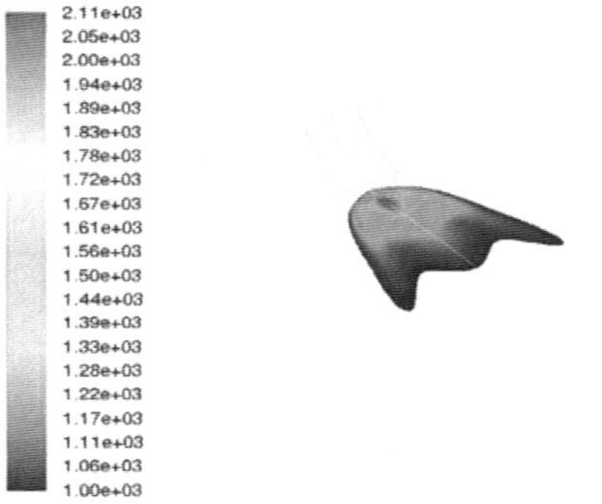

Figure 14. The temperature contours of single-faced deflector on the ground

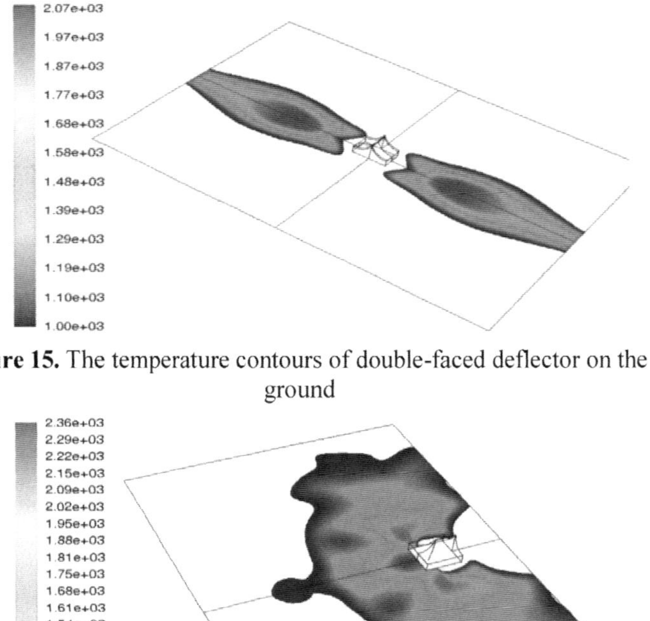

Figure 15. The temperature contours of double-faced deflector on the ground

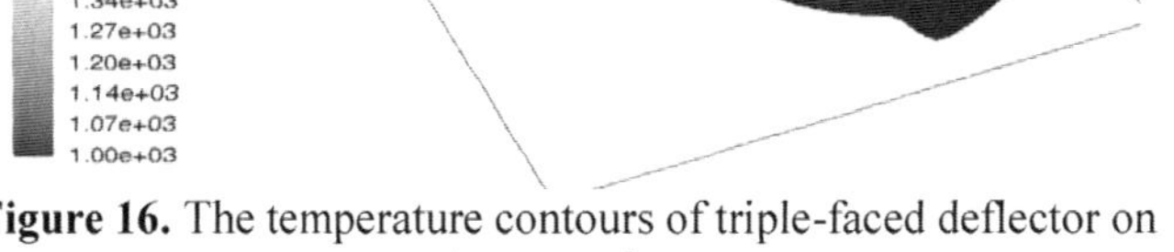

Figure 16. The temperature contours of triple-faced deflector on the ground

Generally, due to the deflector, region affected by jet flow primarily concentrated in the guiding side, the region influenced by jet flow of single-faced deflector is the minimum, and it is at the back of deflector. The region influenced by jet flow of double-faced deflector is at either side of the deflector, it could influence the vehicle. The region influenced by jet flow of triple-faced deflector is the maximum, and without baffle, some jet flow is exhausted to the front of deflector which is more likely to damage the vehicle and other devices around the deflector.

5. Conclusions

1) Single-faced deflector has a simplest structure, jet flow will mostly be deflecting to the backside of deflector, it will not affect the launching vehicle basically and however, it appears a little backwashing which will causes ablation of the bottom of missile, but the backwashing can be reduced by optimizing the shape of deflector plate.
2) Double-faced deflector has the best guiding performance, no backwashing, jet flow is deflecting to the side of deflector, and it could influence the launching vehicle. Nevertheless, its profile is a little difficult to processing.
3) As for triple-faced deflector, without baffle, the jet flow will be guided to the front of the deflector which has a bad effect on launching vehicle. At the same time, its profile is the most complicated. Baffles and optimization of the shape of deflector are needed if you want to use it.

6. References

[1] Miao Rui-sheng. *The Dynamics of the Jet Flow*. (National Defense Industry Press)
[2] Hu Xiao-lei, Sheng Wen-cheng.Study on Double-faced Deflector of Missile Vertical Launched System.*Journal of Fire control &Command Control*, Vol. **38**,No. 11 pp.53-55,2013
[3] Chen jing-song. Study on deflecting Features of Combustion-gas from Double-faced Deflector. *Journal of Missiles and Space Vehicles*,pp,11-15 2006 (2)
[4] Zhao Cheng-qing, JIANG Yi. *The Dynamics of the Jet Flow*. (The publishing company of the Beijing institute of the technology)
[5] LIC,JISM,TAN DP.Softness abrasive flow method oriented to tiny scale mould structural surface.*International Journal of Advanced Manufacturing Technology*,pp,975-987,61(9-12),2012.
[6] Zhang Fu-xiang.*The Dynamics Of The Rocket Jet Flow*.(The publishing company of the Harbin Engineering University)
[7] Yang Yong, Zhang Fu-xiang,Cao Cong-yong,Li Jun. Numerical Simulation Of Rocket Jet Flow Impinging On An Oblique Plane. *Journal of Ballistics*,vol.7 no.3 pp.52-55,1995.

IOP Conf. Series: Materials Science and Engineering **187** (2017) 012043 doi:10.1088/1757-899X/187/1/012043

Movement Characteristics Analysis and Dynamic Simulation of Collaborative Measuring Robot

MA guoqing, LIU li, YU zhenglin, CAO guohua and ZHENG yanbin

School of Mechatronical Engineering, Changchun University of Science and Technology, Changchun 130022

Abstract. Human-machine collaboration is becoming increasingly more necessary, and so collaborative robot applications are also in high demand. We selected a UR10 robot as our research subject for this study. First, we applied D-H coordinate transformation of the robot to establish a link system, and we then used inverse transformation to solve the robot's inverse kinematics and find all the joints. Use Lagrange method to analysis UR robot dynamics; use ADAMS multibody dynamics simulation software to dynamic simulation; verifying the correctness of the derived kinetic models.

1. Introduction

Since human-machine collaborations are currently in high demand, collaborative robot applications are being used with increasing frequency. UR robots have a force feedback function, meaning that when they collide, they automatically stop working. Therefore, in most applications a collaboration robot model does not require a security fence[1].

We selected a UR10 robot as our research subject for this study. First, we applied D-H coordinate transformation of the robot to establish a link system, and we then used inverse transformation to solve the robot's inverse kinematics and find all the joints. Using Lagrange method to analysis UR robot dynamics; using ADAMS multibody dynamics simulation software to dynamic simulation; verifying the correctness of the derived kinetic models.

2. Kinematics Solution

The traditional process to solve kinematics involves studying movement relationships, but does not consider the interaction force between each link[2]. To solve a forward kinematics problem we must solve the position and orientation of the end-effector in the base coordinate system, which is in turn based on the joint angle and dimensions of the known links. Once this has been done, the essential components are in place and it is possible to solve kinematic equations. With inverse kinematics, we strike each joint's angles when the end-effector reaches a location based on its known position and orientation the in base coordinate system.

2.1. Establishing the connecting coordinate system

UR10 is a series robot with six revolute joints. In this study, we used the D-H coordinate transformation method to solve the position and orientation relationship between each link of the robot.The steps are as follows[3]:

(1) Establish the coordinate system $x_0y_0z_0$ and confirm initial position and orientation;

(2) Establish a second joint coordinate system based on the DH coordinate principle;

(3) Establish a robot tool coordinate system $x_6 y_6 z_6$.

Fig. 1 below illustrates the DH coordinate system of a UR10 robot.

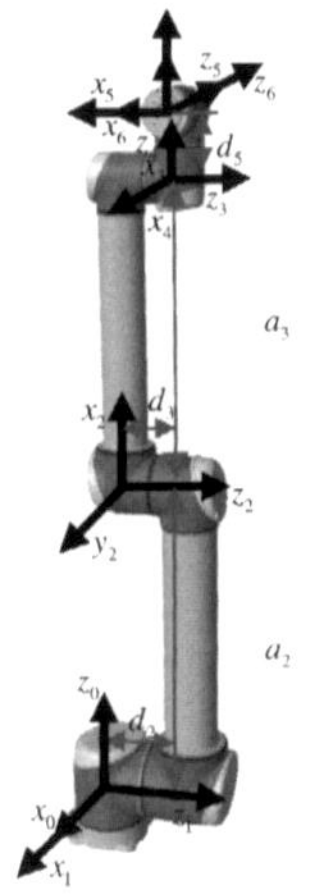

Figure 1. DH coordinate system of UR10

2.2. Forward kinematics solution

DH parameters can be determined based on the conversion relationship between adjacent joints and the links' coordinate system. All the joints of UR10 are revolute joints. This means that only the joint angle θ is variable, while joints with a twist angle α, link length a, and link distance d are constant. Table 1 shows the DH parameters for a UR10.

Table 1 DH parameters of UR10 robot

joint	α/(°)	a/(mm)	d/(mm)	θ/(°)	range(°)
1	-90°	0	0	θ_1	-360~360
2	0°	a_2	d_2	θ_2	-360~360
3	0°	a_3	d_3	θ_3	-360~360
4	90°	0	0	θ_4	-360~360
5	90°	0	d_5	θ_5	-360~360
6	0°	0	0	θ_6	-360~360

The transformation matrix between the robot's adjacent joint coordinates are shown in Formula (1) .

$$
\begin{aligned}
{}^{i-1}T_i &= Rot(z,\theta_i)\times Trans(0,0,d_i)\times Trans(a_i,0,0)\times Rot(x,\alpha_i) \\
&= \begin{bmatrix} c\theta_i & -s\theta_i c\alpha_i & s\theta_i s\alpha_i & a_i c\theta_i \\ s\theta_i & c\theta_i c\alpha_i & -c\theta_i s\alpha_i & a_i s\theta_i \\ 0 & s\alpha_i & c\alpha_i & d_i \\ 0 & 0 & 0 & 1 \end{bmatrix}
\end{aligned} \tag{1}
$$

In order to obtain the transformation matrix between the adjacent coordinate system, we must enter the DH parameters of Table 1 into Formula (1). Subsequently, we multiply the resulting transformation matrix, which provides positive solutions as shown in Formula (2). ($c_i = \cos\theta_i, s_i = \sin\theta_i, c_{ijk} = \cos(\theta_i + \theta_j + \theta_k), s_{ijk} = \sin(\theta_i + \theta_j + \theta_k)$ $i = 1\cdots 6, j = 1\cdots 6, k = 1\cdots 6$)。

$$
{}^0T_6 = {}^0T_1\,{}^1T_2\,{}^2T_3\,{}^3T_4\,{}^4T_5\,{}^5T_6 = \begin{bmatrix} n_x & o_x & a_x & p_x \\ n_y & o_y & a_y & p_y \\ n_z & o_z & a_z & p_z \\ 0 & 0 & 0 & 1 \end{bmatrix} \tag{2}
$$

In which

$$n_x = c_1 s_{234} s_6 - c_6 \left(s_1 s_5 - c_1 c_5 c_{234}\right); n_y = s_1 s_{234} s_6 + c_6 \left(c_1 s_5 + s_1 c_5 c_{234}\right)\ n_z = s_{234} s_6 - c_5 c_6 s_{234}; o_x = c_1 s_{234} c_6 + s_6 \left(s_1 s_5 - c_1 c_5 c_{234}\right)$$

$$o_y = s_1 s_{234} c_6 - s_6 \left(c_1 s_5 + s_1 c_5 c_{234}\right); o_z = c_5 s_{234} s_6 - c_6 c_{234}\ a_x = c_1 c_{234} s_5 + s_1 c_5; a_y = s_1 c_{234} s_5 - c_1 c_5; a_z = -s_5 s_{234}$$

$$p_x = d_5 c_1 s_{234} + c_1 a_3 c_{23} + c_1 a_3 c_{23} - s_1 d_2 - s_1 d_3\ p_y = d_5 s_1 s_{234} + c_2 a_2 s_1 + s_1 a_3 c_{23} + c_1 d_3 + c_1 d_2\ p_z = d_5 c_{234} - s_2 a_2 - s_3 s_{23}$$

2.3. Formatting author affiliations The inverse kinematics solution

According to the configuration characteristics of UR10, the second, third, and fourth joint axes are parallel. This meets the condition that three adjacent joint axes are parallel or intersecting at one point. Therefore, inverse kinematics for the UR robot exist.

1) Solve θ_1

Put ${}^0T_6 = {}^0T_1 {}^1T_2 {}^2T_3 {}^3T_4 {}^4T_5 {}^5T_6$ left and right side multiplied by $\left({}^0T_1\right)^{-1}$, then

$$\left({}^0T_1\right)^{-1} {}^0T_6 = \begin{bmatrix} c_1 n_x + s_1 n_y & c_1 o_x + s_1 o_y & c_1 a_x + s_1 a_y & c_1 p_x + s_1 p_y \\ -n_z & -o_z & -a_z & -p_z \\ -s_1 n_x + c_1 n_y & -s_1 o_x + c_1 o_y & -s_1 a_x + c_1 a_y & -s_1 p_x + c_1 p_y \\ 0 & 0 & 0 & 1 \end{bmatrix} \tag{3}$$

$$\text{make: } {}^1T_2 {}^2T_3 {}^3T_4 {}^4T_5 {}^5T_6 = \begin{bmatrix} A_{11} & A_{12} & A_{13} & A_{14} \\ A_{21} & A_{22} & A_{23} & A_{24} \\ A_{31} & A_{32} & A_{33} & A_{34} \\ A_{41} & A_{42} & A_{43} & A_{44} \end{bmatrix} \tag{4}$$

Take the third row and transpose into:

$$\left[A_{31} \quad A_{32} \quad A_{33} \quad A_{34}\right] = \left[s_5 c_6 \quad -s_5 s_6 \quad -c_5 \quad d_2 + d_3\right] \tag{5}$$

The third row and the fourth column of Formula (3) and Formula (4) correspond with:

$$-s_1 p_x + c_1 p_y = d_2 + d_3 \tag{6}$$

Perform the following triangle substitution for Formula (6)

$$\begin{cases} p_x = r\cos\varphi \\ p_y = r\sin\varphi \end{cases} \quad r = \sqrt{p_x^2 + p_y^2} \quad \varphi = \arctan\left(\frac{p_y}{p_x}\right) \text{So} \quad \theta_1 = \arctan\left(\frac{p_y}{p_x}\right) - \arctan\frac{\left(d_2 + d_3\right)^2}{\pm\sqrt{r^2 - \left(d_2 + d_3\right)^2}}.$$

2) Solve θ_5 The third row and third column of Formula (3) and Formula (4) correspond with:

$$-c_5 = c_1 a_y - s_1 a_x \tag{7}$$

So

$$\theta_5 = \arccos(s_1 a_x - c_1 a_y)$$

3) Solve θ_6 The third row and the first column of Formula (3) and Formula (4) correspond with:

$$c_6 s_5 = c_1 n_y - s_1 n_x \tag{8}$$

So $\theta_6 = \arccos(\frac{c_1 n_y - s_1 n_x}{s_5})$ Where $s_{234} = \frac{-a_z}{s_5}$ $c_{234} = \frac{c_1 a_x + s_1 a_y}{s_5}$

4) Solve θ_2 Put ${}^0T_6 = {}^0T_1 {}^1T_2 {}^2T_3 {}^3T_4 {}^4T_5 {}^5T_6$ left and right side multiplied by the $\left({}^0T_1 {}^1T_2\right)^{-1}$,then

$${}^2T_3 {}^3T_4 {}^4T_5 {}^5T_6 = \begin{bmatrix} c_{34} c_5 c_6 + s_{34} s_6 & s_{34} c_6 - c_{34} c_5 s_6 & c_{34} s_5 & c_3 a_3 + d_5 s_{34} \\ s_{34} c_5 c_6 - c_{34} s_6 & -c_{34} c_6 - s_{34} c_5 s_6 & s_{34} s_5 & s_3 a_3 - d_5 c_{34} \\ s_5 c_6 & s_5 s_6 & -c_5 & d_3 \\ 0 & 0 & 0 & 1 \end{bmatrix} \tag{9}$$

make:

$$\left({}^0T_1 {}^1T_2\right)^{-1} {}^0T_6 = \begin{bmatrix} B_{11} & B_{12} & B_{13} & B_{14} \\ B_{21} & B_{22} & B_{23} & B_{24} \\ B_{31} & B_{32} & B_{33} & B_{34} \\ B_{41} & B_{42} & B_{43} & B_{44} \end{bmatrix} \tag{10}$$

The second column, third column and fourth elements (respectively):

$$\begin{bmatrix} B_{12} \\ B_{22} \\ B_{32} \\ B_{42} \end{bmatrix} = \begin{bmatrix} c_1c_2o_x + s_1c_2o_y - s_2o_z \\ -c_1s_2o_x - s_1s_2o_y - c_2o_z \\ c_1o_y - o_xs_1 \\ 0 \end{bmatrix} \tag{11}$$

$$\begin{bmatrix} B_{13} \\ B_{23} \\ B_{33} \\ B_{43} \end{bmatrix} = \begin{bmatrix} c_1c_2a_x + s_1c_2a_y - s_2a_z \\ -c_1s_2a_x - s_1s_2a_y - c_2a_z \\ c_1a_y - s_1a_x \\ 0 \end{bmatrix} \tag{12}$$

$$\begin{bmatrix} B_{14} \\ B_{24} \\ B_{34} \\ B_{44} \end{bmatrix} = \begin{bmatrix} c_1c_2p_x + s_1c_2p_y - s_2p_z - a_2 \\ -c_1s_2p_x - s_1s_2p_y - c_2p_z \\ p_yc_1 - p_xs_1 - d_2 \\ 1 \end{bmatrix} \tag{13}$$

The first row and second column of Formula (9) and Formula (10) correspond with:

$$c_1c_2o_x + s_1c_2o_y - s_2o_z = s_{34}c_6 - c_{34}c_5s_6 \tag{14}$$

So
$$\theta_2 = \arctan \frac{c_1o_x + s_1o_y + c_5c_6\dfrac{c_1a_x + s_1a_y}{s_5} + c_6\dfrac{a_z}{s_5}}{o_z - c_6\dfrac{c_1a_x + s_1a_y}{s_5} + c_5c_6\dfrac{a_z}{s_5}}$$

5) Solve θ_3 The second row and third column, and the second row and fourth column of Formula (9) and Formula (10) correspond with:

$$c_{34}s_5 = -c_1s_2a_x - s_1s_2a_y - c_2a_z \tag{15}$$

$$s_3a_3 - d_5c_{34} = -c_1s_2p_x - s_1s_2p_y - c_2p_z \tag{16}$$

So
$$\theta_3 = \arcsin\left(\frac{-c_1s_2p_x - s_1s_2p_y - c_2p_z + d_5\dfrac{c_1c_2a_z + s_1c_2a_y - s_2a_z}{s_5}}{a_3}\right)$$

6) Solve θ_4 Based the above, joints θ_{234} θ_2 θ_3 are known, so $\theta_4 = \theta_{234} - \theta_2 - \theta_3$

$$\theta_4 = \arcsin\left(\frac{-a_z}{s_5}\right) - \arctan\frac{c_1o_x + s_1o_y + c_5c_6\dfrac{c_1a_x + s_1a_y}{s_5} + c_6\dfrac{a_z}{s_5}}{o_z - c_6\dfrac{c_1a_x + s_1a_y}{s_5} + c_5c_6\dfrac{a_z}{s_5}} - \arcsin\left(\frac{-c_1s_2p_x - s_1s_2p_y - c_2p_z + d_5\dfrac{c_1c_2a_z + s_1c_2a_y - s_2a_z}{s_5}}{a_3}\right)$$

3. Dynamics Equation of Measuring Robot

There is many methods to study robot dynamics, mainly Newton-Euler method (Newton-Euler)[4], Lagrangian (Lan-grange) method[5], Gauss (Gauss) law[6], Kane (Kane) method and Robertson-Weiden Fort (Robson-Wittenburg) method[7]. In this paper, creating dynamic equations of the robot used Lagrange way.Lagrange equation is:

$$\frac{d}{dt}\frac{\partial L}{\partial \dot{q}_i} - \frac{\partial L}{\partial q_i} = Q_i\left(i = 1, 2 \ldots 6\right) \tag{17}$$

Type, q_i, $\dot{q}_i$ is the generalized coordinates and generalized velocity; Q_i is the generalized force for the generalized coordinates.Seting the translational velocity vector is v_{ci}, the angular velocity vector is ω_i, for the quality is m_i, relative to the center of mass of inertia tensor is I_i in the base coordinate system, the kinetic energy of i th joints is:

$$T_i = \frac{1}{2}m_iv_{ci}{}^Tv_{ci} + \frac{1}{2}\omega_i{}^TI_iw_i \tag{18}$$

Among them, the first item is the kinetic energy of the object in the translational motion, the second is the kinetic energy of the object rotating around the center of mass. The i th joint velocity in basic coordinate system and the joints and before each joint can be expressed as the relationship between the speed

$$\begin{bmatrix} v_{ci} \\ w_i \end{bmatrix} = \begin{bmatrix} J_{L1}^{(i)} & \cdots & J_{Li}^{(i)} \\ J_{A1}^{(i)} & \cdots & J_{Ai}^{(i)} \end{bmatrix} \begin{bmatrix} \dot{q}_1 \\ \vdots \\ \dot{q}_i \end{bmatrix} = \begin{bmatrix} J_L^{(i)} \dot{q}_i \\ J_A^{(i)} \dot{q}_i \end{bmatrix}$$

Because joints for the rotational joint, so the total kinetic energy for the robot is:

$$T = \sum_{i=1}^{n} T_i = \frac{1}{2} \sum_{i=1}^{n} \dot{q}^T M \dot{q} = \frac{1}{2} \sum_{i=1}^{n} \sum_{j=1}^{n} M_{ij} \dot{q}_i \dot{q}_j \quad (19)$$

M_{ij} is the n square matrix of M element

$$M = \sum_{i=1}^{n} \left(m_i J_L^{(i)T} J_L^{(i)} + J_A^{(i)T} I_i J_A^{(i)} \right) \quad (20)$$

The robot's potential energy $U = \sum_{i=1}^{n} m_i g^T O_{P_{ci}}$ Among them, g is acceleration of gravity vector on the basis of coordinates system, $O_{P_{ci}}$ is the vector of origin of coordinates to the center of mass of the vector on the basis of the origin of coordinates. Generalized force is: $Q = \tau + J^T F$

Among them, τ and F respectively joint force vector and the force vector of end of actuator contact with the external environment.Put the robot's kinetic energy, potential energy and generalized force generation into the Lagrange equation, available the robot dynamics equation.

$$\sum_{j=1}^{n} M_{ij} \ddot{q}_j + \sum_{j=1}^{n} \sum_{k=1}^{n} \left(\frac{\partial M_{ij}}{\partial q_k} - \frac{1}{2} \frac{\partial M_{ij}}{\partial q_i} \right) \dot{q}_j \dot{q}_k + \sum_{j=1}^{n} m_i g^T J_{Li}^{(j)} = Q_i \quad (21)$$

As can be seen by the above formula is derived to calculate the theoretical advantage of solving the dynamic equations of the robot more complex, so the general multi-body dynamics simulation method to accomplish each robot joint force or torque solving.

4. Dynamics Simulation of Measuring Robot

We us ADAMS to simulate the dynamics of the measuring robot, the simulation time is set to 5s, step number is set to 500. add drivers in each of the joints as follow setting.

1)The first joint rotation control function
STEP(time,0,0,0.5,72d)+STEP(time,0.5,0,5.5,0)+STEP(time,5.5,0,6,-72d),
2)The second joint rotation control function
STEP(time,0,0,1,40d)+STEP(time,1,0,9,0)+STEP(time,9,0,10,-40d)
3)The third joint rotation control function
STEP(time,0,0,1,40d)+STEP(time,1,0,9,0)+STEP(time,9,0,10,-40d)
4)The forth joint rotation control function
STEP(time,0,0,1,50d)+STEP(time,1,0,7,0)+STEP(time,7,0,8,-50d)
5)The fifth joint rotation control function
STEP(time,0,0,1,60d)+STEP(time,1,0,6,0)+STEP(time,6,0,7,-60d)
6)The sixth joint rotation control function
STEP(time,0,0,1,80d)+STEP(time,1,0,5,0)+STEP(time,5,0,6,-80d)

4.1. Angular Velocity of each joint

Each joint angular velocity curve shown in figure 2, which play a very important role in the dynamics of the robot's overall design, therefore needs to measure each joint angular velocity changes during the simulation.

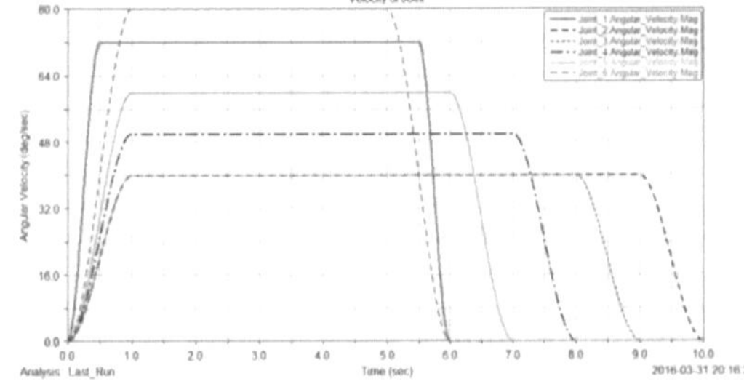

Figure 2 angular velocity curve of each joint robot

As we can see from the joint angular velocity graph, during the period of the robot dynamics simulation, the angular velocity of each joint in line with the changes of setted, uniformly accelerated start - uniform run - even retarding the working conditions, and in the acceleration phase and deceleration phase transition is relatively stable, there is no mutation phenomenon, each robot joint operation is in accordance with the actual situation, will not affect the overall performance of the robot.

4.2. Force and Moment of each joint

The force and moment of each joint in the robot dynamics simulation are an important parameter. By measuring the torque of each joint suffered,can verify whether the drive motor selected can meet the requirements, but also can verify whether the reduction ratio and torque harmonic reducer chosed can meet the design requirements, so dynamics simulation has play a very important role in the design of the robot ,joint forces and torques suffered is hown in fig 3 each joint of UR10 robot can withstand the torque value is shown in table 2.

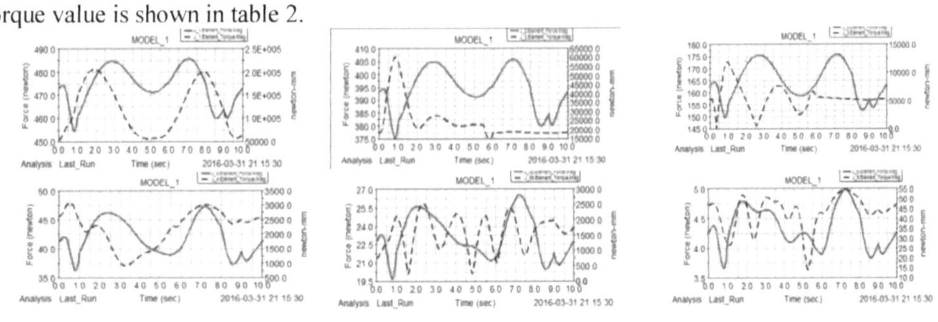

Figure 3 the station of stress and torque of each joint

Table 2 withstand torque value of UR robot

Joint	Joint 1	Joint 2	Joint 3	Joint 4	Joint 5	Joint 6
torque	330N m	330N m	150N m	56N m	56N m	56N m

We can see from figure 3, the force of each joint and moment of force during robot movement are withstand the torque range of UR10 robot can suffer as shown in table 2 withstand the torque range, meets the needs of the movement. moment of force.

5. Conclusions

In this study, we selected a UR10 robot as our research subject. We first used D-H coordinate transformation of the robot to establish a link system. Then, we used inverse transformation to solve inverse kinematics for the robot and locate all the joints. Use Lagrange method to analysis the UR robot dynamics, use ADAMS multibody dynamics simulation software to realize dynamic simulation.

The results show that the measuring robot UR10 has a good static balance of performance, stability and dynamic response of the motion.

6. References

[1] LIU Kun, LI Shi-zhong, WANG Bao-xiang. Research of the Direct Teaching System Based on Universal Robot[J]. Science Technology and Engineering, 2015,10(28):22-26.
[2] Gao Yi,Ma Guoqing,Yu Zhenglin,etal. Kinematics Analysis of an 6-DOF Industrial Robot and Its Visualization Simulation[J]. China Mechanical Engineering.2016.27(13):1726-1731.
[3] Huang Xiaochen,Zhang Minglu,Zhang Xiaojun,etal. Improved DH Method to Build Robot Coordinate System[J]. Transactions of the Chinese Society for Agricultural Machinery, 2014,45(10):313-318.
[4] ZHAO Tieshi,GENG Mingchao,CHEN Yuhang,etal. Kinematics and Dynamics Hessian Matrices of Manipulators Based on Screw Theory[J]. Chinese Journal of Mechanical Engineering,2015,02:226-235.
[5] Zhao Junwei,Li Xuefeng,Chen Guoqiang. The Dynamics Equation of A 3- PRS Parallel Manipulator Based on Lagrange Method[J], Machine Design and Research.2015,31(2):1-5.
[6] LIU Yingjie,ZHAO Youqun,XU Jianxiong,etal. Vehicle Handling Inverse Dynamics Based on Gauss Pseudospectral Method while Encountering Emergency Collision Avoidance[J], Journal of Mechanical Engineering.2012, 48 (22):127-132.
[7] YU Changjuan,ZHANG Minglu,ZHANG Jianhua,etal. Dynamics modeling and simulation of multi-motion mode robot[J]. Journal of Machine Design. 2015,3(3):68-71.

IOP Conf. Series: Materials Science and Engineering **187** (2017) 012044 doi:10.1088/1757-899X/187/1/012044

Experimental Research on Micro-nozzle Applied on Micro-propulsion Systems based on MEMS

Zhang Bao-jun[1], Li Xing-chen[1], Huang Yi-yong[1,*] and Xu Xiang-ming[2]

[1] Micro Satellite Engineering Centre, National University of Defence Technology, China

[2] College of Mechatronic Engineering and Automation, National University of Defence Technology, China

E-mail: *yiyong_h@sina.com; chambj@sina.com; lixingchen_hit@163.com

Abstract. In order to study the influence of the structural parameters of micro thruster applied in micro satellite attitude adjustment and orbital maneuver on its propulsion performance, this paper considers the factors influencing the performance of the thruster, and utilizes the orthogonal test design to obtain nine groups of micro-nozzles with different structural parameters. We processed this series of micro nozzles through MEMS (Micro-Electro-Mechanical Systems) technology. The micro-nozzles are made of single crystal silicon and glass through the anode bonding, and the electric heating wire is creatively processed through MEMS in the thrust chamber to improve the performance of the micro thruster. Experiments were carried out in a vacuum chamber. Finally, we analyse the experimental results by analysis of variance and analysis of range. The experimental results show that the performance of the micro nozzle is optimal when the semi-shrinking angle is 30 degrees, the semi-expansion angle is 15 degrees and the area ratio is 6.22. Meantime, the experiment verifies that it is feasible to improve the propulsive performance of micro-propulsion system through electronic heater strip.

1. Introduction

With the launch of various satellites, the sustainable development of aerospace technology has not only promoted the continuous improvement of modern science and technology, but also brought earth-shaking changes for people's lives. Small satellite technology emerging in 1980s has attracted increasing attention from major aerospace countries by virtue of its small size, light weight, low cost, short development cycle, and excellent performance as well as the functions of rapid launch and spatial network and to achieve the functions of large satellites[1-4].In order to realize the attitude control, gravitational compensation, formation flying and other functions of microsatellite, the demands for propulsion system are continuously enhanced[5, 6].The existing propulsion systems and thrusters cannot meet the demands of micro-satellites because of their size, mass, and thruster requirements , etc. [7]. Therefore, it is necessary to develop a new type of thruster suitable for micro-satellites, which has light mass, small size, precise thrust and low power consumption.

Traditional manufacturing methods cannot produce the appropriate micro-propulsion system, thus the micro-propulsion system based on MEMS (Micro-Electro-Mechanical-Systems) emerged cater to the demands and was developed subsequently [8-11]. The existing micro-thrusters applied in the spacecraft include: high pressure cool gas micro-thruster, chemical micro-thruster, electric micro-thruster, and so on. High pressure cool gas micro-thruster is relatively mature and the mostly widely utilized at present [12-15].

In this paper, we use MEMS technology to design and process a series of micro-thrusters with different aspect ratios and shrinkage ratios. The detailed processing flow is given. In addition, a special form of heating wire is processed in the micro-thruster chamber. The micro-thruster uses nitrogen as propellant and the designed thrust is 10mN. Based on the principle of torsion pendulum measurement [16], the thrust force is measured indirectly. Finally, we analyze the experimental results and put forward some suggestions for the improvement.

2. Design of cool gas micro-thruster

The cool gas micro-thruster utilizes a classic Laval nozzle configuration (as shown in Figure1). Under the condition of high Reynolds number and low Knudsen number, the following design hypotheses about the ideal engine are satisfied:

(1) Propellant is homogeneous (mainly gaseous, the amount of any aggregate phase can be ignored);

(2) Propulsive working medium obeys the ideal-gas law;

(3) The entire flow process is adiabatic, and there is no heat transfer through the engine wall;

(4) In the whole gas flow process, there is no obvious friction, and all of the boundary layer effects are ignored;

(5) No shock waves or discontinuity occurs in the nozzle flow;

(6) The flow of the propellant is a steady state. The expansion of the working medium is uniform and stable. The transient effect time is short and can be ignored.

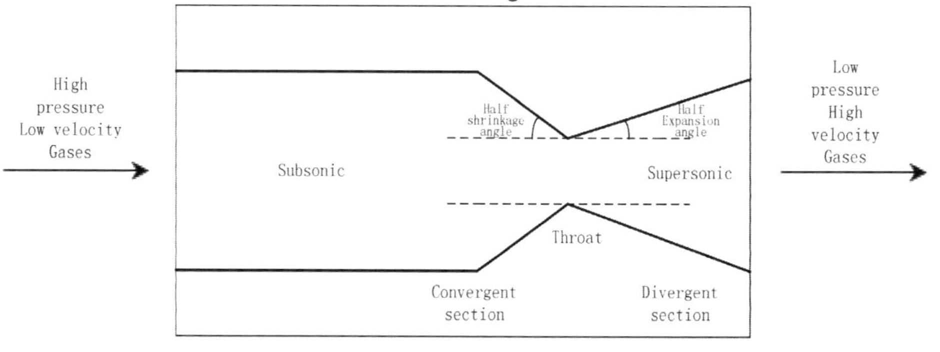

Figure 1. Typical configuration of Laval nozzle

2.1. Structural design of Micro-thruster

The cool gas micro-thruster is designed as a composite layer. That is, a single crystal silicon wafer is utilized as the lower substrate, and Pyrex glass is utilized as the upper substrate, both of which are combined by anodic bonding process to form a MEMS micro-nozzle. Subject to the MEMS processing technology [17], the cross-section of the driving nozzle is rectangular. In order to improve the jet velocity and efficiency, the heating wire is processed by sputtering on the substrate in the thrust chamber.

In this paper, we design three sets of micro-nozzle chips through the L_9 (27) orthogonal experimental design method [18]. Each set has several thrusters with different semi-shrinkage angles, semi-expansion angles and different area ratios, namely, nine groups. The channel depths of the three sets of micro-nozzle chips are 100μm, 200μm and 400μm, respectively. The corresponding parameters are shown in Table1.

Table 1. The structural parameters of nozzles chips

Number	Semi-shrinkage angle /°	Semi-expansion angle /°	Area ratio
1	30	20	10
2	45	30	10
3	30	15	6.22
4	45	15	4
5	60	15	10
6	60	30	6.22
7	30	30	4
8	60	20	4
9	45	20	6.22

2.2. Specific impulse and thrust calculation of thruster based on adiabatic process

The gas flow process of the micro-thruster can be calculated by the nozzle flow equation (1),

$$v_e = \left\{ \frac{2\gamma}{\gamma - 1} RT_0 \left[1 - \left(\frac{p_e}{p_0} \right)^{\frac{\gamma-1}{\gamma}} \right] \right\}^{1/2} \tag{1}$$

Where v_e is the nozzle outlet velocity, p_0 is the total inlet pressure. T_0 is the total inlet temperature. A_t is the cross-sectional area of the nozzle throat, $R=R_0/M$ is a gas constant. M is the molecular weight of the gas, and γ is the specific heat ratio of the gas.

We can obtain the specific impulse of the micro-thruster by equation (2)[19],

$$I_{sp} = c^* C_f / g_0 \tag{2}$$

where $c^* = \sqrt{\gamma R U_0} \gamma^{-1} \left(\frac{2}{\gamma+1} \right)^{-\frac{\gamma+1}{2(\gamma-1)}}$, $C_f = \left\{ \frac{2\gamma^2}{\gamma-1} \left[\frac{2}{\gamma+1} \right]^{\frac{\gamma+1}{\gamma-1}} \left[1 - \left(\frac{p_e}{p_0} \right)^{\frac{\gamma-1}{\gamma}} \right] \right\}^{1/2} + \varepsilon \frac{p_e - p_a}{p_0}$

The thrust of the micro-thruster can be obtained through the following equation,

$$F = C_f p_0 A_t \tag{3}$$

Where I_{sp} is a gas specific impulse. F is the thrust. ε is the area ratio of the nozzle, p_e and p_a are the nozzle outlet pressure and the ambient pressure, respectively.

In space, since the external environment of the thruster is vacuum microgravity, the ambient pressure p_a=0. Assuming that the expansion ratio of the nozzle is infinite, p_e=p_a=0. We can get the expression of ideal specific impulse of the cool gas thruster in equation (4),

$$I_s = \left(\frac{2\gamma R T_0 g_0}{\gamma - 1} \right)^{1/2} \tag{4}$$

Equation (4) shows that the ideal specific impulse of the cool gas thruster is only related to the molecular weight of the gas M, the total inlet temperature T_0, and the specific heat ratio γ.

For the cool gas micro-propulsion system, the whole process of gas flow from the cylinder into the micro-thruster is adiabatic. There is no loss of heat during the gas flow process. Therefore, the inlet temperature of the thruster is equal to the gas temperature in the cylinder, namely, $T=T_0$.

Then, the calculation formula of ideal thrust of the micro-thruster is equation (5),

$$F = C_f p_0 A_t = p_0 A_t \left[\frac{2\gamma^2}{\gamma-1} \left(\frac{2}{\gamma+1} \right)^{\frac{\gamma+1}{\gamma-1}} \right]^{1/2} \tag{5}$$

For the selected propellant, it can be observed that the ideal thrust depends mainly on the inlet pressure p_0 and the nozzle throat area. Therefore, we shall adjust the inlet pressure and nozzle throat area to achieve the ideal thrust.

During the actual operation of the micro-thruster, since the nozzle expansion ratio is not infinite, the nozzle outlet pressure of the thruster is not zero, i.e.$p_e \neq 0$. As we can see from equation (3), the thrust of the thruster will be slightly smaller than the ideal case.

The mass flow of the micro-thruster can be calculated through equation (6):

$$\dot{m} = \frac{F}{I_{sp} g_0} = \frac{A_t p_0 \gamma}{\sqrt{\gamma R T_0}} \left[\left(\frac{2}{\gamma+1} \right)^{\frac{\gamma+1}{\gamma-1}} \right]^{1/2} \tag{6}$$

According to equation (6), if the theoretical thruster F, micro-nozzle expansion ratio ε and total inlet pressure are known, we can calculate the specific impulse and mass flow of the micro-thruster.

3. Manufacturing process of cool gas micro-thruster

The cool gas micro-thruster is manufactured through the standard MEMS technology. As a micro-electro-mechanical system, MEMS is the miniaturized revolutionary technology developed and expanded from microelectronic technology [20, 21]. It integrates physical, chemical and biological sensors, information processing and storage units as a micro-integrated systems, which refers to the micro-devices and system manufactured in large batch through micro-machining technology [22].

The micro-fabrication technology of micro-electro mechanical system is an important part of MEMS, which is the technological extension and relevant application in three aspects of theoretical basis, technical basis and application research. Modern micro-fabrication technology can be divided into four categories: first, the separation process, such as evaporation, sputtering, cutting, etc.; second, bonding processing, such as deposition, evaporation, etc.; third, deformation processing of materials, such as moulding deformation processing, and fluid deformation processing, etc.; fourth, the material processing and surface modification processing, such as diffusion, mixing and reaction incentives[23].

3.1. Processing of the micro-nozzle

At present, MEMS micro-processing technologies mainly include: lithography masking, high-energy beam etching (laser etching) and LIGA process [24]. In this paper, the processing mode utilized to process the micro-nozzle is lithography masking and high-energy beam etching (laser etching). The specific processes include oxidation of silicon wafer, deep reactive ion etching, laser drilling, removal of surface oxide layer, corrosion of Al material, etc. [25].

After the above-described processing flow, the micro nozzle with the thrust chamber and the heating function has been manufactured. The specific form of each component is as shown in Figure 2.

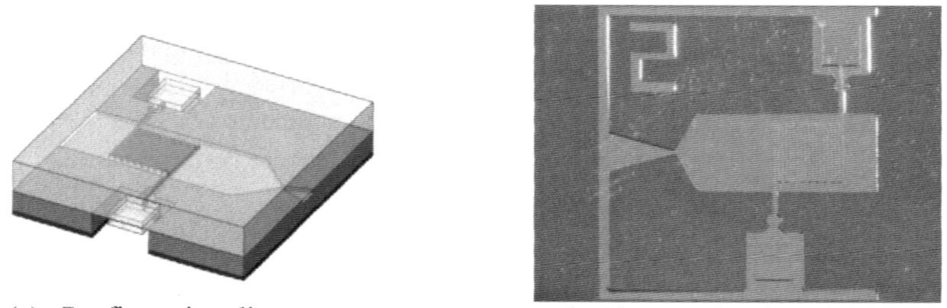

(a) Configuration diagram (b) Picture of real product

Figure 2. Micro-thruster with heating filaments

3.2. The connection of the heating wire and the fixing of the micro-nozzle

After the fabrication of the micro-nozzles, the copper wire is bonded to the Al electrodes through the conductive adhesive, as shown in Figure 3. The conductive adhesive can withstand temperature as high as 350 degrees which meets the heating requirements.

IOP Conf. Series: Materials Science and Engineering **187** (2017) 012044 doi:10.1088/1757-899X/187/1/012044

Figure 3. Micro-nozzle with heating wire

In order to verify the conductive properties and bond strength of conductive adhesive, we measured the resistance of each experimental part. The experimental results in Table2 show that the properties of the conductive adhesive are good and the resistance of the heating wire is not greatly changed, which can guarantee the stability of the heating power. However, the heating power is too large during the heating process, which leads to burnout of heating wire, which is as shown in Table 2.

Table 2. Heating system resistance change

Number	The resistance before bonding wire	The resistance after bonding wire
1	329 Ω	\
2	325 Ω	296.4 Ω
3	276 Ω	\
4	264 Ω	266.1 Ω
5	273 Ω	281.2 Ω
6	271 Ω	262.9 Ω
7	323 Ω	290.6 Ω
8	316 Ω	339.3 Ω
9	312 Ω	\

Figure 4 shows the components used to connect the air supply tube to the micro-nozzle, and figure 5 shows the cool gas micro-thruster manufactured after above steps.

Figure 4. Connectors between the gas line and the nozzle

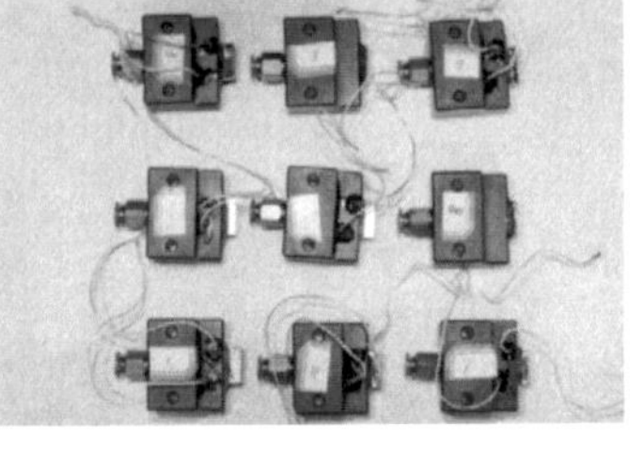

Figure 5. The finished products of micro-thruster

4. Experimental devices and thrust measurement

4.1. Thrust test gas path

Micro thrust test gas path is composed of cylinders, the corresponding valves, pressure sensors, flow sensors and control system. Figure 6 shows the schematic diagram of the gas path.

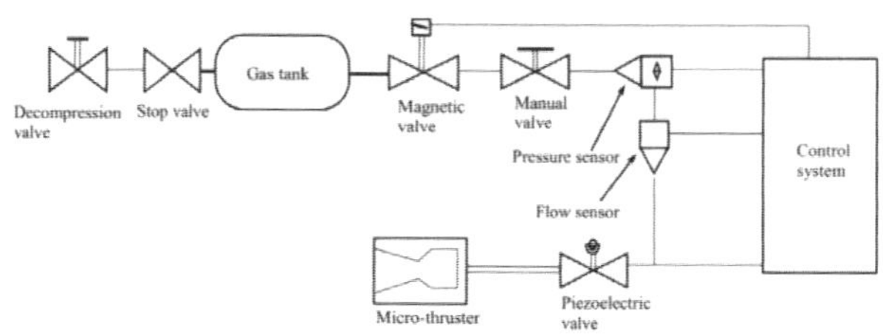

Figure 6. The schematic diagram of the gas path

The physical map of micro test gas path is shown in Figure7. Before placing the experimental system into the vacuum chamber, close the manual valve and charge 2 L of nitrogen gas into the cylinder through the filter relief-pressure valve. Close the stop valve after the inflation and remove the filter relief-pressure valve. Finally, open the control system, close the magnetic valve, open the manual valve, and the entire gas path equipment can be placed in the vacuum chamber together with the thrust test bench.

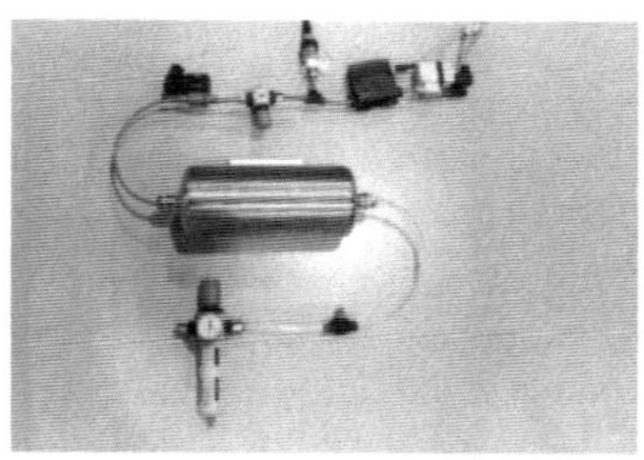

Figure 7. The physical map of micro test gas path

4.2. Thrust test bench

The micro thrust test is carried out based on the steady-state thrust measurement principle. Figure 8 shows the thrust test bench. The micro-thruster is attached to one end of the torsion bar and connected with the gas path through the silicone hose. First, thrust is exerted on the plane of the torsion bar, and converts into the torsion angle. Second, with the increase of the torsion angle, the thrust and the restoring force tend to balance. Finally, we can measure the steady-state voltage through the photoelectric sensor, and then calculate the thrust by the functional relationship between the voltage and the torsion angle.

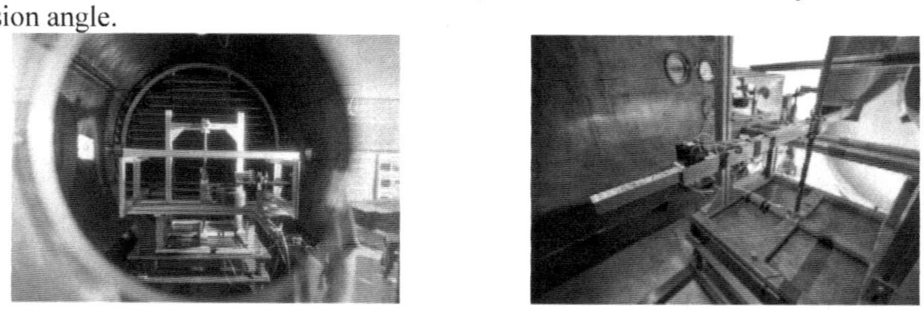

(a) Thrust test bench (b) Thruster is fixing on the torsion bar

Figure 8. Thrust test bench in vacuum chamber

In order to reduce the influence of environmental and human factors on the thrust test process and the thrust calibration process, we have taken the following measures. Firstly, both the thrust test and the thrust calibration are carried out in the vacuum chamber, ensuring that the thrust test bench will not move throughout the testing process. Secondly, the gas supply pipe is fixed to the torsion bar to prevent the test result from being affected by the swing of the pipe during the thrust test.

IOP Conf. Series: Materials Science and Engineering **187** (2017) 012044 doi:10.1088/1757-899X/187/1/012044

4.3. Thrust test process

4.3.1 Calibration of thrust test system. In order to guarantee the accuracy of the thrust test and calibration, the entire thrust test bench, especially the torsion bar, shall be in the horizontal state. Both thrust test and calibration are carried out in the vacuum chamber.

Several weights with different masses are selected, and the weights are equal to the tested thrust. Tie a filament to the torsion bar which fixes the nozzle, and hang the weight on the filament. Pull up the filament so that the weight is hung from the torsion bar plane, and the angle between the filaments on both sides of the weight is maintained at 135 degrees. According to the decomposition of the force, it can be known that the simulated thrust is equal to the mass of the weight, namely, $F_1=G$.

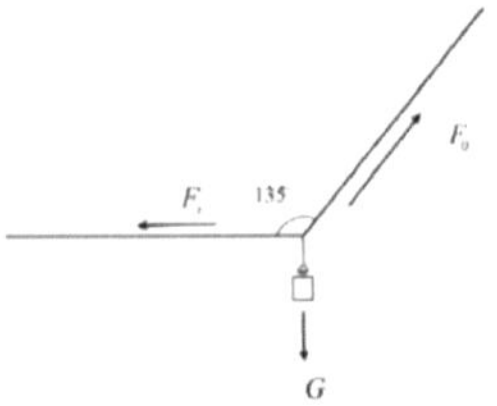

Figure 9. Calibration force diagram

The mass of the weight utilized to simulate micro-thrust is close to the measured thrust so as avoid the inconformity between the calibration value and the range of actual measured thrust. Change the weights with different masses, and record the sensor charge signal. The two signals on the display card are gradient. The difference between two stairstep signals is the mass of the weight, namely the simulated continuous thrust.

Each weight is calibrated for twelve times in different position of the filament. Assume that the calibration data at each point are randomly distributed, and we can obtain the precision of the measurement by calculating the variance of the data distribution at each point.

$$\left.\begin{aligned} \bar{x} &= \frac{1}{12}\sum_{i=1}^{12} x_i \\ v_i &= x_i - \bar{x} \\ \sigma &= \sqrt{\frac{\sum_{i=1}^{n} v_i^2}{n-1}} \end{aligned}\right\} \tag{7}$$

Where x_i is the measurement value of each time, v_i is the measurement residual, $\bar{x}$ is the average measurement value, σ is the standard deviation of the measurement, and n is the times of measurements.

According to Figure 10, the calibration curve approximates a straight line. There is a high degree of linearization between the output voltage signal (horizontal axis) and the simulated thrust (vertical axis) of the calibration system. The fitting curve can be set as $y=a+b*x$, where a and b are the fitting curve coefficient.

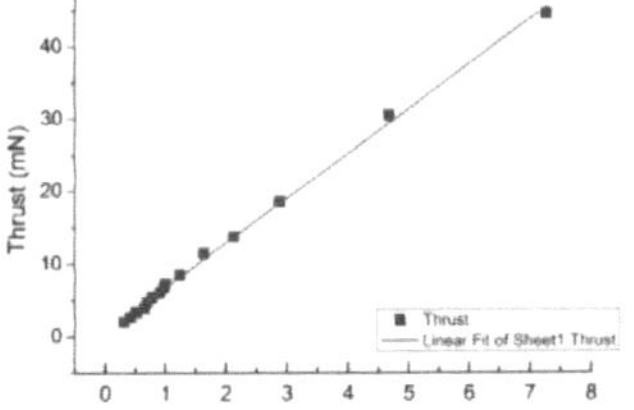

Figure 10. The calibration curve

IOP Conf. Series: Materials Science and Engineering **187** (2017) 012044 doi:10.1088/1757-899X/187/1/012044

4.3.2 Measurement of micro thrust. Before measuring the micro thrust, stabilize the test system and evacuate the vacuum chamber to 10 Pa. Select the measurement range on the digital charge meter, and reset the meter to carry out the experiment for data acquisition. Open the valve to jet, and start to collect the signal when it comes to stable, after a period of time to close the valve; continue to gather the signal for a period of time. The measured stairstep signal is shown in Figure11. The magnitude of the thrust is expressed as the height difference between the two steps in the signal data.

The measured voltage has a lot of ambient noise. After the signal is filtered, the signal-to-noise ratio is obviously improved, and the measurement accuracy is increased.

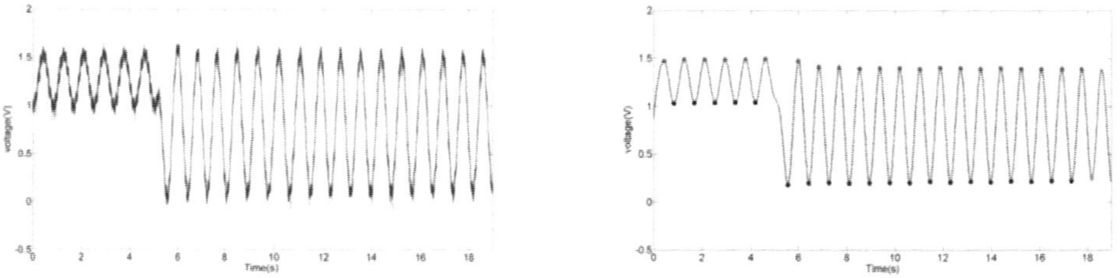

(a) Original signal (b) Processed signal

Figure 11. Original and processed signals of thrust

4.4. Error in measurement

There are three main sources of error in this experiment. One is the influence of the environment, the other is the design defect of the experimental system and processing error and the third is the error of the signal acquisition system.

5. Experimental results and discussion

5.1. Orthogonal test analysis of thrust measurement results

The range and variance of the experimental results of the nine groups of micro-thrusters at the inlet pressure of 1.29 bars are analysed. The optimized parameters are initially screened through the significant level order of the factors obtained from the range analysis. Afterwards the significance level of each factor is evaluated by the F test of variance analysis to get the influence of each factor on the thrust.

The range analysis results indicate that semi-shrinkage angle is the main factor affecting the test index, that is, the change of semi-shrinkage angle has the greatest influence on the thrust value.

In order to the influence of various factors on the test index trends and laws more intuitively reflect, the factor level is taken as the abscissa, and the average of the test indicators as the vertical axis to draw factors and index change trend chart as shown in figure 12.

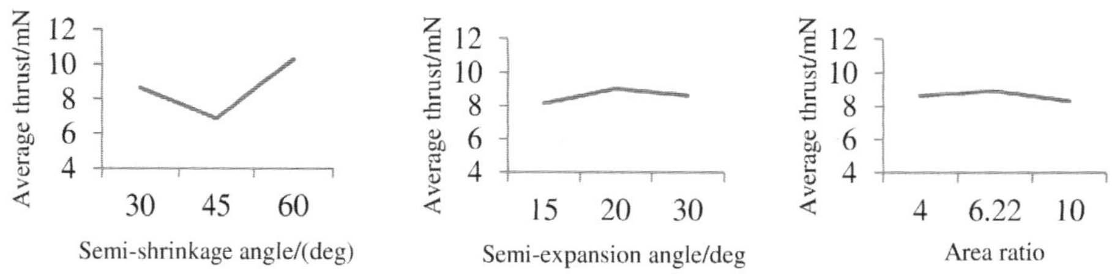

(a) (b) (c)

Figure 12. Factors and index change trend

As shown in Figure 12, we can see that in the case of inlet pressure of 1.29 bars, the optimal structure parameters of the micro-nozzles can be determined as semi-expansion angle of 30 degrees, semi-expansion angle of 20 degrees and area ratio of 6.22.

Due to the poor precision of range analysis, we perform a variance analysis on the experimental results so as to get a quantitative standard to judge the influence of various factors on the thrust, and the results are shown in Table 3.

Table 3. Analysis of variance

Factor	DevSq	DOF	F	$F_{0.05}(2,2)$	Significance level
A[a]	17.944	2	36.251	19.000	*
B[b]	1.178	2	2.380	19.000	
C[c]	0.562	2	1.135	19.000	
Error	0.49	2			

[a] Semi- shrinkage angle.
[b] Semi-expansion angle.
[c] Area ratio.

$F_{0.05}$ (2, 2) = 19.000, we can see in table 3 that $F_A > F_{0.05}$ (2, 2). Therefore, factor A is significant, that is semi-shrinkage angle has a greater impact on micro thrust, which is consistent with the previous range analysis.

5.2. Thrust and Mass flow rate at different heating power

Figure 13 shows the change in micro thrust at different heating powers. At the same inlet pressure, the thrust increases with the heating wire heating power. Moreover, and the effect of heating on the thrust is relatively large. This is because the gas in the thrust chamber absorbs heat, and the temperature gradually rises, which leads to expansion, and increases the gas inlet pressure. According to equation (1), the outlet gas flow rate is increasing. Formula (5) indicates that thrust is also increasing.

In addition, as shown in Figure 13, with the changes of inlet pressure changes, thrust also increases accordingly, which has been verified by formula (5).

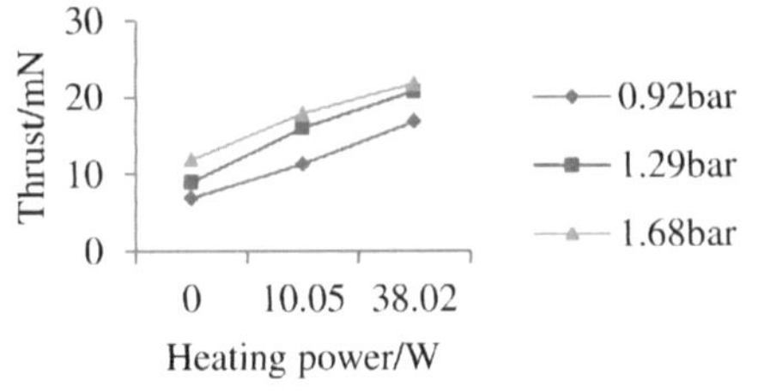

Figure 13. Thrust change under different heating power

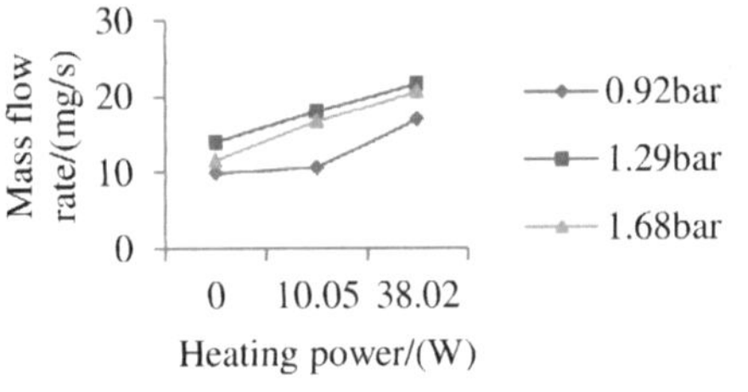

Figure 14. Mass flow rate change under different heating power

According to Figure 14, as the heating power of the heating wire increases, the mass flow rate of the gas enhances. This is for that the gas is heated to accelerate its flow velocity, thereby taking more gas in the same time. However, larger inlet pressure does not necessarily result in the greater mass flow rate. As can be seen from Figure 14, when the inlet pressure is 1.29bar, the mass flow rate of the gas is the largest. This may be for that the air flow cannot be fully expanded due to the large inlet pressure, which aggravates the friction loss.

6. Conclusion

Nine groups of thrusters with different structure parameters designed by the orthogonal experimental design method were tested in a vacuum chamber. Through the experimental results, range analysis and variance analysis were utilized to get a better set of structural parameters, i.e., the semi-shrinkage angle is 30 degrees, the semi-expansion angle is 20 degrees, and the area ratio is 6.22.For whether the result is the optimal, it remains to be further explored in the late test. In addition, we utilized the heating wire in thrust chamber to heat the gas. The results show that the thrust of the propeller increases with the heating power, and the change of the thrust is obvious. This indicates that the heating wire set in the thrust chamber of the micro propeller is conductive to improving the thrust.

7. References

[1] Tang L G. 1999 Aerospace Shanghai **6** pp.47-49
[2] Lin L X. 2006 Aerospace China **11** pp.43-47
[3] Shi R, Li X, Deng K. 2016 Aerospace Electronic Warefare **32**(1) pp.8-13
[4] V Bakanauskas, A Domeika. 2015 Solid State Phenomena. **220** pp.561-564
[5] Henry Helvajian, Siegfried W and Janson. 2009 *The Future of Small Satellites. Small Satellites: Past, Present, and Future*(California: The Aerospace Press)
[6] M Keidar, T Zhuang, A Shashurin, G Teel, et al. 2015 Plasma Physics &Controlled Fusion **57**(1) pp.14005-14014
[7] Hang G R, Hong X and Kang X L. 2013 Journal of Rocket Propulsion **39(**5) pp.7-15
[8] You Z, Lin Y and Zhang G F.2005 Aerospace China **6** pp.12-15
[9] C K Kuan, G B Chen, Y C Chao. 2015 Journal of Propulsion & Power **23**(6) pp.1313-1320
[10] Q P Wang, J F Bao, Y Ling, et al. 2015 Microsystem Technologies **21**(8) pp.1805-1810
[11] Lin L X. 2010 Spacecraft Engineering **19**(6) pp.13-20
[12] Schmuland D, Carpenter C, Masse R, et al. 2013 27th Annual AIAA/USU Conference On Small Satellites pp.1-12
[13] B Y Yao, H Li, H Y Xu, et al. 2016 Space International **447**(3) pp.23-26
[14] X Jin, Y J Hong, W J Zhou, H Chang. 2015 Journal of Propulsion Technology **10** pp.1554-1559
[15] C Ru, J Dai, J Xu, Y Ye,& P Zhu. 2016 European Physical Journal Applied Physics **74**(3)pp.4-9
[16] Anthony, R., Mathuna, C. O., & Rohan, J. 2016 Microelectronic Engineering **155**
[17] C J Zhang. 2015 *Experimental Design and Data Processing*(BeiJing: Chemical Industry Press)
[18] Dieter K, David H, Huang. 1992 *Modern Engineering for Design of Liquid Propellant Rocket Engines*(AIAA)
[19] L Z Yang, Y M Wei, X H Liu. 2016 Aerospace Control and Application **42**(1) pp.13-19
[20] C B Ru, Y H Ye, C L Wang, et al. 2014 Applied Mechanics & Materials **490** pp.1042-1046
[21] Watanabe, K., Shimooka, Y., Saito, T., & Ikehashi, T. 2016 European Heart Journal **32**(5) pp.537-544
[22] Zhang W, Zhang D C and Wang Y Y. 2002 Micronanoelectronic Technology **39**(1) pp.22-27
[23] Y C Dai, J Z Zhou, J Wang, et al. 2006 Machine Tool & Hydraulics **5** pp.15-19
[24] Y Q Xu, Y J Yang. 2010 MEMS Device & Technology **47**(7) pp.425-431
[25] C J Qiu. 2016 *MEMS Technology Foundation and Application*(Harbin: Harbin Institute of Technology Press)

A Dummy Scan Flip-Flop Insertion Algorithm based on Driving Vertex

H L Liu[1,a], L Li[1], Z X Zhang[2] and W T Zhou[1]

[1]Research Institute of Electronic Science and Technology, University of Electronic Science and Technology of China
No.2006, Xiyuan Avenue, High-Tech West Zone, Chengdu, Sichuan, China
[2]Taiwan Semiconductor Manufacturing Company(Nanjing)
No.16, Zifeng Road, Pukou Economic Development Zone, Nanjing, Jiangsu, China

[a]hailiangliu@std.uestc.edu.cn

Abstract. Commonly termed as Hardware Trojans, is an emerging issue for global hardware security. The research on Hardware Trojan detection is urgent and significant. Dummy Scan Flip-Flop(DSFF) structure could be used to improve the probability of hardware Trojan activation, which is significant to hardware Trojan detection, especially during the design phase. In this express, an algorithm for inserting the DSFF structure based on driving vertex is proposed. According to the experimental results, under the same transition probability threshold(P_{th}), compared to the state-of-art, the proposed algorithm can reduce both the inserting complexity and the induced area overhead of the DSFF insertion. The maximum area optimization rate can reach 44.8%. The simulation results on S386 and S38584 benchmark circuits indicate that the proposed algorithm can significantly reduce Trojan authentication time by increasing activation probability of hardware Trojan circuits.

1. Introduction

With the global expansion of the semiconductor industry, it has become a major challenge to ensure the trustworthiness of a given Integrated Circuit(IC). Commonly termed as Hardware Trojans, is an emerging issue for global hardware security in special application, such as military, economics, et al [1]. Hardware Trojans are kinds of extra and malicious circuits integrated in ICs spitefully, which could leakage information or change the original function, etc, and be a crucial threat for global information security [2].

As shown in figure 1, the basic structure of hardware Trojan is composed of two parts: the trigger and payload [3]. Generally hardware Trojan selects input signals from internal nets of main circuit to trigger the Trojan and modifies circuit function through Trojan payload. The significant characteristic of hardware Trojan is the low transition probabilities so that they can be dormant and have low influence on side channel signals such as power or delay. However, so long as the Trojan is trigged (fully or partially), the corresponding information can be used for Trojan detection using side channel analysis[4,5] or some other methods during the chip design phase.

The authentication time is the time consumed by confirming which a chip is attacked by hardware Trojan or not. In order to have an efficient detection and reduce Trojan authentication time, it's critical to:

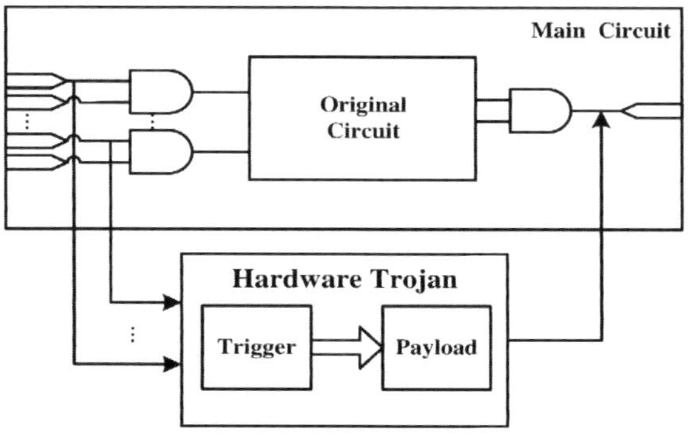

Figure 1. The basic structure of Hardware Trojan

(i)Increase the probability of generating a transition in hardware Trojan;
(ii)Seek for effective detection methods to reduce the detection time.

To increase the activation probability of hardware Trojan circuit (i), hardware Trojan transition probability is modelled with geometric distribution [6] and a Dummy Scan Flip-Flop (DSFF) structure is proposed [8]. Inserting DSFF structure, it's not only propitious to design test vectors to activate hardware Trojan in design phase but also conducive to generate test vectors to activate Trojan to increase the related side channel information during chip test phase [9, 10]. In a word, it is significant to increase the hardware Trojan activation probability.

This paper proposed a DSFF insertion algorithm based on driving vertex, which is shown in section 3. The proposed algorithm greatly reduces the insertion complexity and has lower effect on the circuit area. The rest of the paper is organized as follows: Section 2 presents the background of DSFF. The proposed algorithm, the new insertion procedure and the simulation results are described in Section 3. Section 4 concludes this paper.

2. Background of Dummy Scan Flip-Flop

This section describes the transition probability of the net modelled with geometric distribution [6] and the background of DSFF structure.

2.1. *The transition probability and activation probability*

In one clock cycle, assuming a net, the probability of which appears a logic 0 is P_0 and logic 1 is P_1. The transition probability of the net is defined as P_t and modelled with geometric distribution[6], which is :

$$P_t = P_0 \times P_1 \tag{1}$$

As a net is either logic 0 or logic 1, we have the following equation:

$$P_0 = 1 - P_1 \tag{2}$$

According to the mathematical formula, it's easy to know that only when $P_0 = P_1 = 0.5$, we can achieve the maximum probability of a net's transition. When the probability of a net of logic 0 or logic 1 is close to 1, that is, $P_0 << P_1$ or $P_1 << P_0$, the transition probability of the net will be reduced dramatically.

A hardware Trojan may have the following conditions: 1) with low transition probability of nets; 2) rare combinations. Assuming the trigger circuit has q input ports, the probability of generating a specific trigger vector is:

$$P_{trigger-vector} = \prod_{i=1}^{q} P_i \tag{3}$$

where

$$P_i = \begin{cases} P_i0 & \text{for trigger input } Net_i \text{ to be logic 0} \\ P_i1 & \text{for trigger input } Net_i \text{ to be logic 1} \end{cases} \quad (4)$$

From equation (3), we know that a trigger circuit could have several input ports and if P_i is low, then the activation probability of the trigger circuit $P_{trigger-vector}$ will be low. Therefore, to improve the activation of Trojan circuits, for each net of Trojan circuits, we should increase the transition probability of net(P_i). During IC design phase, in order to improve the transition probability of nets, we need to insert DSFF structure at nets with transition probability lower than threshold P_{th}.

2.2. The background of Dummy Scan Flip-Flop structure
The DSFF structure was proposed in paper [7] at first. Figure 2 shows two kinds of DSFF structures. DSFF is mainly include a scan chain which is made of one MUX and one Flip-flop, a gate (AND gate, OR gate) circuit with two input ports. SI is a test signal for MUX and TE is an enable signal.

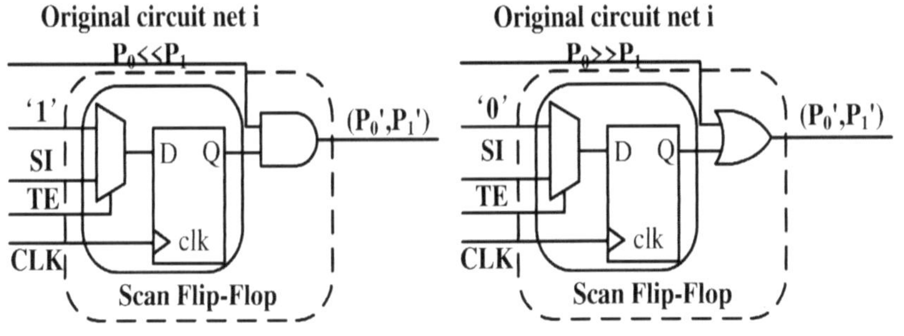

Figure 2. Two kinds of DSFF structures [7]

If the probability of original circuit net logic value 1 is much higher than the logic value 0. selects the left DSFF structure, whereas selects the right DSFF structure. When the circuit works at normal mode, TE is logic 0, the value of nets don't change. When the circuit works at test mode, TE is logic 1, SI is a test input signal, the probability of logic 0 or logic 1 is 0.5. Then the probability of logic 0 or logic 1 in Q port of flip-flop is also 0.5. If $P_0 << P_1$ in the original circuit net, by inserting DSFF AND gate structure, we can improve the probability of logic 0; if $P_1 << P_0$, we can also improve the probability of logic 1 by inserting DSFF OR gate structure. Therefore, DSFF structure could make a net's P_0 and P_1 become closer, which could improve the transition probability of a net.

3. The proposed algorithm

The inserted objects in the algorithm in the paper[8] are nets(stored in array M)with transition probability higher than P_{th}.Then nets in M are sorted by transition probability in an increasing order and net with lowest probability is selected as target net which is to be inserted with DSFF structure. Next the number of low transition nets is judged. If the number is decreased, the inserted DSFF is kept, otherwise the DSFF would be ignored. As the quantity of M is N_M, the insertion complexity is $O(N_M)$[8].

This express proposes a new DSFF insertion algorithm based on driving vertex. The driving vertex is defined as net located at the top of drive with transition probability lower than P_{th}. Figure 3 shows the diagrammatic sketch of driving vertex.

From net M1 of the circuit topology, the transition probability of nets is lower than P_{th}. So M1 is a driving vertex. M2 and M3 are drivers of M1.

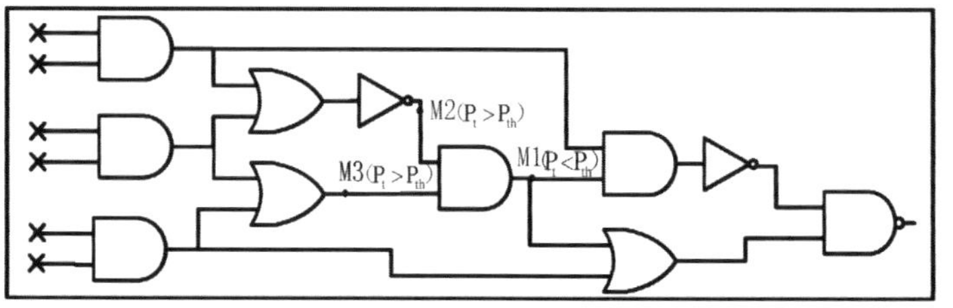

Figure 3. The diagrammatic sketch of driving vertex

Shown as in figure 4, First, nets with transition probability lower than P_{th} are stored in array L and all driving vertexes in L are stored in array T. Second, all drivers of driving vertexes are selected from nets with transition probability higher than P_{th} and stored in array B. Third, all nets in array B are sorted by transition probability in an increasing order and the net with lowest probability is selected as target net and inserted with DSFF structure. Fourth, next target net will be selected and inserted with DSFF structure until the transition probability of all nets in the design are higher than P_{th}. The quantity of array B is N_B, so the insertion complexity of proposed algorithm is $O(N_B)$. Compared to paper[8], as $N_B << N_M$, $O(N_B) < O(N_M)$. The proposed algorithm reduces the insertion complexity. Meanwhile, the proposed algorithm has a higher insertion efficiency and decreases the inserted quantity of DSFF, reduces the induced area overhead of DSFF insertion.

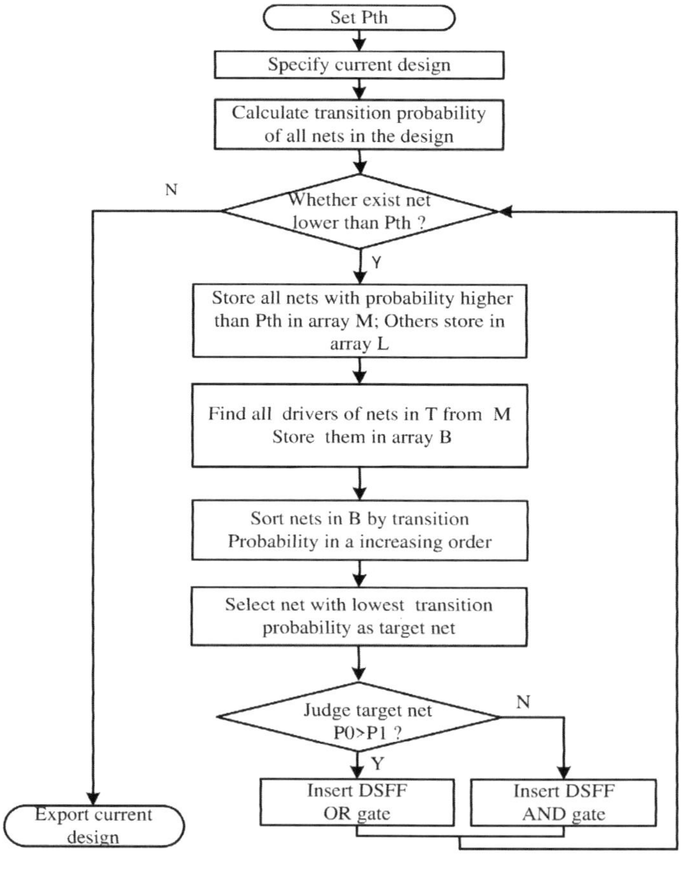

Figure 4. The new process of inserting DSFF structure

We choose S386 of ISCAS'89 benchmark circuit to verify the proposed algorithm. Table I lists the effect on circuit area of two insertion algorithms under different P_{th}. The proposed algorithm has a higher insertion efficiency and decreases the inserted quantity of DSFF, reduces the induced area overhead of the DSFF insertion.

As shown in table 1, We define a parameterη(η= ((B-A)/B) * 100%), which is to show the optimization on inserted DSFF circuit area that proposed algorithm makes relative to algorithm[8]. The maximum area optimization rate can reach 44.8%.

Table 1. Effect on circuit area of two insertion algorithms

P_{th}	area percentage increase in proposed algorithm (A)	area percentage increase in[8] (B)	Area optimization rate η=((B-A)/B)*100%
10^{-5}	7%	9.3%	24.8%
10^{-4}	7%	12.7%	44.8%
10^{-3}	14%	19.7%	28.9%

To show the influence of inserted DSFF structure, the n32 and n42 nets with the lowest transition probability of the S386 benchmark circuit are selected for analysis. As shown in table 2, if n32 and n42 are attacked as input ports of a hardware Trojan, the transition probability of the net before inserting DSFF structure is 0.99999e-5. After inserting DSFF structure, the transition probability increases to 0.174, which is 4 orders of magnitude better than before.

Table 2. Comparison on the transition probability of nets before and after insertion of DSFF

net	No DSFF			Inserted with DSFF		
	P0	P1	$P_{n32}1xP_{n42}0$	P0	P1	$P_{n32}1xP_{n42}0$
n32	0.99999	e-05	0.99999e-05	0.824	0.176	0.174
n42	0.99999	e-05		0.991	0.009	

We select S38584 circuit of ISCAS'89 benchmark circuits as primary circuit. Under the condition of 10^5 groups random test vectors, figure 5 and figure 6 show the transition in Trojan 1 circuit before and after inserting DSFF structure with the threshold value of $P_{th} = 10^{-5}$. n3937 and n3717 are internal nets of S38584 benchmark circuit with the transition probability less than 10^{-5}, which are proper to be selected as access points of hardware Trojan. m1 is internal net of Trojan circuit and g29212 is the output of Trojan. Before the insertion of DSFF structure, the whole circuit keeps dormant until n3717 generates a transition at about $8*10^9$ps. After inserting DSFF, we can see in figure 6, g29212 is triggered several times and n3937, n3717, m1 generate a number of transitions within $2*10^9$ps. The transition probability of Trojan circuit nets is increased after the insertion of DSFF structure. So the activation probability of Trojan circuit is also increased according equation (1) and (3).

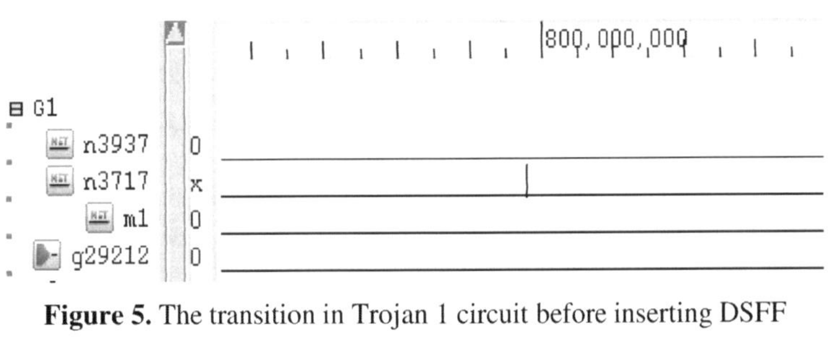

Figure 5. The transition in Trojan 1 circuit before inserting DSFF

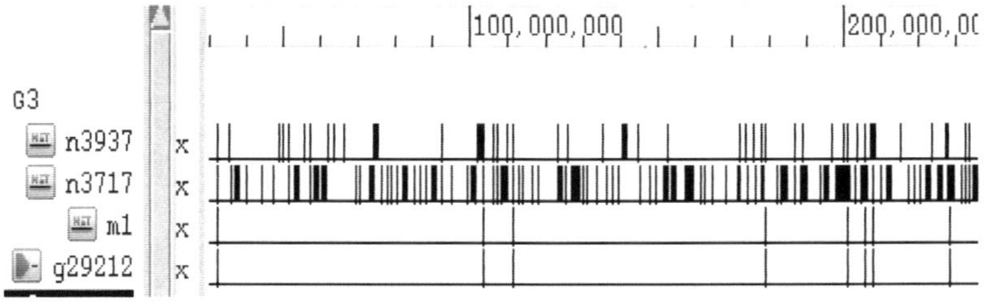

Figure 6. The transition in Trojan 1 circuit after inserting DSFF

4. Conclusion

Hardware Trojan threat is becoming serious global security issues and research on hardware Trojan detection is urgent and significant. In this express, we proposed a DSFF insertion algorithm based on driving vertex. Compared to paper[8], the proposed algorithm can reduce both the DSFF structure inserting complexity and the induced area overhead of the DSFF insertion. The maximum area optimization rate can reach 44.8%. The simulation results on S386 and S38584 benchmark circuits indicate that the proposed algorithm can significantly increase activation probability of hardware Trojan circuits.

5. Reference

[1] Tony F. W, Karthik G, Yunqing Alexander H, H. -S. Philip W, Simon W and Subhasish M 2014 *J. IEEE Trans. on Computer-Aided Design of Integrated Circuits and Systems* **35** 521-534
[2] Mingfu X, Wei L, Aiqun H and Youdong W 2014 *J. IEICE Trans. Info* **E97-D** 606
[3] Francis W, Chris P, Swarup B and Rajat S.C 2008 C. *Int. Conf. on Design, Automation and Test in Europe(Munich)* (IEEE Conference Publications) p 1362-1365
[4] Seetharam N, Dongdong D, Rajat Subhra C, Somnath P, Francis G. W, Christos A. P, Kaushik R and Swarup B 2012 *J. IEEE Trans. on Computers* **62** 2183-2195
[5] Chenxu W, Jinghu L, Mingyan Y and Jinxiang W 2013 *J. IEICE Electronics Express* **10** 20130602
[6] Dennis D. W, William M and Richard L. S 2008 *B. Thomson Higher Education(Belmont,CA 94002-3098,USA)7th* **Chapter 3** 140-149
[7] Hassan S, Mohammad T and Jim P 2009 *Proc. Int. Workshop on Hardware-Oriented Security and Trust* (San Francisco) vol 9 (IEEE Press) p 66-73
[8] Hassan S, Mohammad T and Jim P 2012 *J. IEEE Trans. on Very Large Scale Integration Systems* vol 20 (IEEE Press) p 112-125
[9] M. R, G. H, B V Bhargav R and Siva Y 2015 *Proc. Int. Sym. on VLSI Design and Test (*Hsinchu*)* vol 19 (IEEE Press) p 1-6
[10] M. R, G. H and Siva Y 2015 *Proc. Int. Conf. on Smart Technologies and Management for Computing, Communication, Controls, Energy and Materials(*Chennai*)* vol 1 (IEEE Conference Publications) p 439-442

Acknowledgments

We are grateful for the relative research work done by related researchers and this work was supported in part by the National Nature Science Foundation of China (Grant No.: U1630133).

IOP Conf. Series: Materials Science and Engineering **187** (2017) 012046 doi:10.1088/1757-899X/187/1/012046

Operational Plan Ontology Model for Interconnection and Interoperability

F Long, Y K Sun and H Q Shi

Operation Software and Simulation Institute, Dalian Naval Academy, 667 Jiefang Road, Dalian, China

doubao.0123@aliyun.com

Abstract. Aiming at the assistant decision-making system's bottleneck of processing the operational plan data and information, this paper starts from the analysis of the problem of traditional expression and the technical advantage of ontology, and then it defines the elements of the operational plan ontology model and determines the basis of construction. Later, it builds up a semi-knowledge-level operational plan ontology model. Finally, it probes into the operational plan expression based on the operational plan ontology model and the usage of the application software. Thus, this paper has the theoretical significance and application value in the improvement of interconnection and interoperability of the operational plan among assistant decision-making systems.

1. Introduction

The theory research and equipment construction of the system operations based on the information system are developing rapidly. The assistant decision-making system's demand for the data and Knowledge processing of the operational plan is on the increase. The traditional expression method of the operational plan cannot adapt to the sharing and interaction among systems, which has severely impeded the system combat capability [1]. Ontology theory, extensively studied by civil information systems, has been proved effective in solving similar problems [2], so it has a great potential in the application to military information systems [3]. Therefore, revolving around the bottleneck of the system combat capability, this paper will make a deep research on the operational plan ontology model with the intention of promoting the settlement of the problems of interconnection and interoperability of the assistant decision-making system.

2. Analysis of the method of operational plan expression

2.1. Traditional expression method and its problems

Among the same type of assistant decision-making systems, operational modes of specific domains correspond to several fixed description templates which are used for the expression of different types of operational plan, while the fixed description templates for the different type of assistant decision-making systems vary [4-5]. In order to deal with the problems of interconnection and interoperability, some standards or data models with the description format of distributed static data structure appear to support the fixed description templates to express [6].

The following problems exist in the above expression method: First, fixed description templates cannot cope with the flexible descriptions of operational plan in practical operations. Second, standards or data models have difficulties in realizing the interconnection and interoperability of the different type of heterogeneous assistant decision-making systems. Third, adapters are necessary when format conversion of the heterogeneous operational plan between two different types is under conduction. The installation of adapters will immensely add to the workload and decrease the reliability.

2.2. Prominent advantages of the ontology model

Operational plan ontology model, equivalent to the meta-model of the expression of operational plan, plays a dual role: First, it provides a unified framework of the operational plan model, which guides the development of operational plan models in various professional domains of the different type of assistant decision-making systems. Second, it regulates expression of the elements and relationships of the operational plan, so as to guarantee the consistency of the general description of operational plan among the different type of assistant decision-making systems.

Ontology model can effectively cope with the problems of operational plan expression. First, the development of multiple types of operational plan models of diverse scales will meet the need of flexible description. Second, operational plan elements are unified and the reliance on fixed templates is eliminated, which ensure the interconnection and interoperability. Third, in the process of format conversion of the operational plan, the complexity and workload of are reduced, and the reliability increases in both newly developed and existing systems.

3. Elements of operational plan ontology model

Operational plan ontology model consists of the following five elements in the operational plan domain knowledge: category, category attribute, relationship, axiom and instance, which can be defined as:

$$\text{Onto\_ plan}=\langle C, A, R, X, I\rangle$$

Wherein: C, expressing the category, is the set of objects with similar features, and it describes the important concepts of operational plan domain knowledge. A, expressing the category attribute, is a set of properties of the category, and it describes the inner structure of the operational plan ontology model category. R, expressing the relationship, is the interrelation between categories (category and attribute), and it describes the interaction between the categories (categories and attributes) of operational plan ontology model. X, expressing the axiom, is the statement of permanent validity, and it describes the valid assertion in any case in the operational plan domain knowledge. I, expressing the instance, is the concrete object corresponding to the category, and it describes the specific element of operational plan domain knowledge.

4. The basis of construction of the operational plan ontology model

4.1. Fundamental basis of the data hierarchy

Two aspects are included in the basis of the construction of the data hierarchy. First, the content of the operational plan that needs to be exchanged among assistant decision-making systems. Second, user's demand for data description and expression while processing the operational plan. Thus the following three elements are determined. (1) Category. All the categories necessary for the description are included. (2) Category attribute. The demand of data exchange is met. (3) Instance. All the elements necessary for description are included.

4.2. Advanced basis of the knowledge hierarchy

Three aspects are included in the basis of the construction of the knowledge hierarchy. First, the huge amounts of relationships among the internal elements of the operational plan. Second, the logical test of the operational plan. Third, the intelligent reasoning of the operational plan [7]. The following two elements are determined in this way. (1) Relationship. All the relationships among the various factors are included. (2) Axiom. The inspection and judgment rules and professional operation rule are included.

5. Construction of operational plan ontology model

According to the basis of construction, the operational plan ontology model for interconnection and interoperability, including category, attribute and relationship, is at a semi-knowledge level, and the other two elements, including axiom and instance, can be expanded according to the application demand.

5.1. Category

The operational plan ontology model is composed of 14 categories (1+7+6), as is shown in figure 1. There are one general category and seven core categories, in which time category and space category are generic categories, basing on relative references [8-9]. There are six subordinate categories. The solid lines between the core and subordinate categories in figure 1 symbolize their affiliated associations. The definition of category is shown in table 1[10].

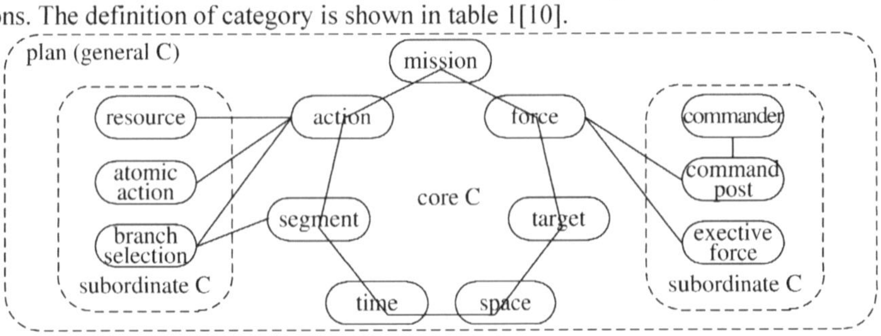

Figure 1. Composition of categories of operational plan ontology model

Table 1. Definition of categories of operational plan ontology model

No.	C name	C definition
1	plan	the plan of readiness and implementation for the implementing mission
2	mission	the task of the force for achieving the predetermined objective
3	segment	the general designation of the schedule and phase, consisting of a series of complicated processes and actions
4	action	attacks or defenses carried out by forces in order to accomplish the mission
5	force	the general designation of various organizations, staff and armory for the implementing mission
6	target	the object of attack, capture, or defense in actions
7	atomic action	the tactical actions using resources, and the basic units of the action
8	resource	the general designation of all kinds of resources used by the forces in the action
9	branch selection	according to the constraints, selecting the corresponding segment or action from a segment set or action set in which the branch relationships are mutually exclusive
10	command post	temporary agency established for the command of the operations and non-combat military operations
11	executive force	also called the object of command , referring to the general designation of the units and staff under direction in the commanding relationship
12	commander	staff who have been entrusted with and carry out the right of command

5.2. Category attribute

Category attribute can be determined according to the construction basis and the definition and concept of category of operational plan ontology model, as shown in figure 2. (abbreviation: Sg= segment, Ct= constraint, Ex= executive, Sup= superior, Cm= command, Cmdr= commander)

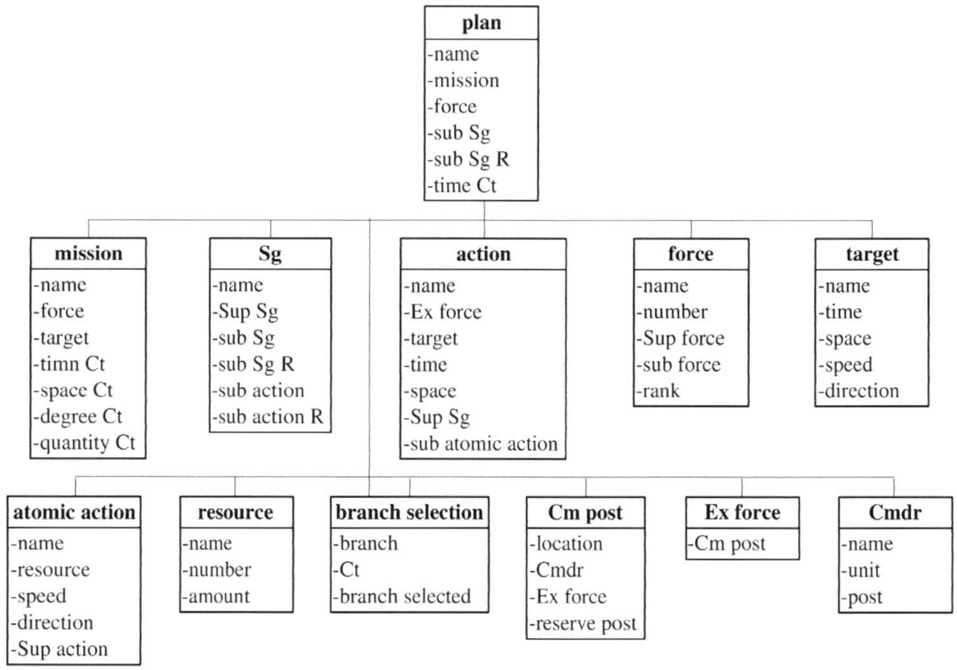

Figure 2. Category attribute of operational plan ontology model

5.3. Relationship

5.3.1. Definition of relationship. The relationship of operational plan ontology model can be defined using the following form:

$$R\ (C_1, C_2),\ \text{or}\ R\ (C_1, A_1)$$

Wherein: R is relationship. C_1 is associated with C_2, or A_1 with the way R. C_1 is category, and C_2 or A_1 is either category or attribute.

5.3.2. Relationship description. There are 19 types of relationships between operational plan ontology models, and each type of relationship consists of several specific relationships, with a total number of 74. These relationships are classified as general relationship and special relationship according to the degree of combination with domain knowledge. There are 10 types of general relationships, which are attribute, parent-child, decomposition/aggregation, selection, serial, parallel, mutually exclusive, subordinate/superior, ownership, and constraint relationship. There are 9 types of special relationships, which are completion, execution, targeting, undertaking, in (space), at(time), interaction, usage and command relationship.

Figure 3 describes the relationships of all the 14 composition categories with conceptual graph. The relationships between categories form a complicated direct network [11], in which the relationships of the operational action categories, involving 8 types, are the most complex. The relationships of the categories are obviously a knowledge network with operational action as its core. The complexity of relationships of operational forces category comes next, involving 5 types. (abbreviation: decomposition/aggregation= De/Ag)

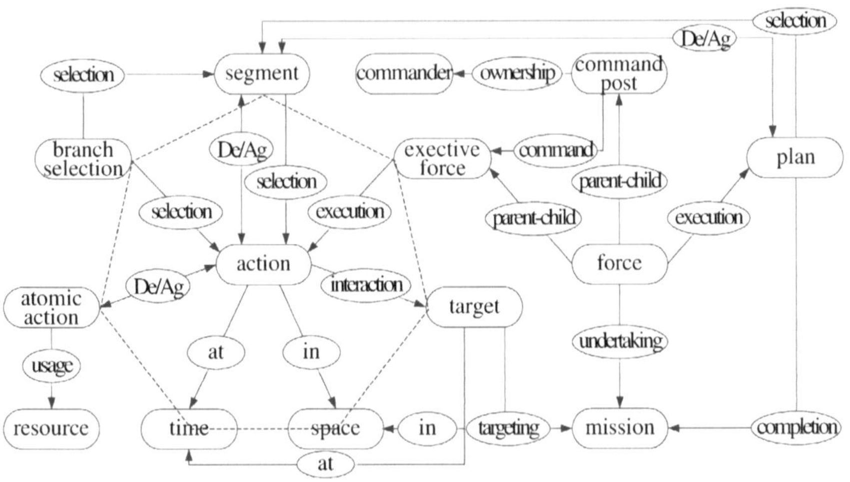

Figure 3. Relationships among the operational plan ontology model categories

Due to space limitation, only the main specific relationships are given below. Table 2 and 3 are the decomposition/aggregation and constraint relationships respectively belonging to general relationship. Table 4 and 5 are the execution and interaction relationship respectively belonging to specific relationship.

Table 2. Decomposition/aggregation relationships

No.	R (C_1, C_2), or R (C_1, A)
1	De/Ag (plan, sub segment)
2	De/Ag (sibling segment, sub segment)
3	De/Ag (sibling segment, sub action)
4	De/Ag (sibling action, sub atomic action)
5	De/Ag (sibling Ex force, sub Ex force)

Table 3. Constraint relationships

No.	R (C_1, C_2), or R (C_1, A)
1	Constraint (time constraint, plan)
2	Constraint (space constraint, plan)
3	Constraint (degree constraint, mission)
4	Constraint (condition, branch selection)

Table 4. Execution relationships

No.	R (C_1, C_2), or R (C_1, A)
1	Execution (force, plan)
2	Execution (Ex force, action)

Table 5. Interaction relationships

No.	R (C_1, C_2), or R (C_1, A)
1	Strike (action, target)
2	Seize (action, target)
3	Defense (action, target)

6. Application of operational plan ontology model

The core application of operational plan ontology model is the expression of operational plan. First, the basic applications, which is to describe specific operational plan completely with the model independently. Second, advanced applications, which is to develop the application software to realize the basic applications and other more advanced extensions.

6.1. Expression of the operational plan based on the ontology model

By using the category, category attribute and relation of the operational plan ontology model, specific operational plans can be described according to the process in a unified and standardized way. The steps are as follows: The first step is building the tree of the operational forces. The second step is building the tree of the operational actions, which is the core work. The third step is expressing other elements, such as operational plan, operational mission, operational target, and commander, etc..

6.2. Application software of the operational plan ontology model

The operational plan ontology model can really play the value only through the development of application software. In the assistant decision-making system, the application software is in an important position, so it is called the core layer. It is closely related to the application layer and user layer. Application layer refers to the operational plan software modules in different professional fields. User layer refers to the human-computer interaction part of the assistant decision-making system. The localization and application method of application software is shown in figure 4.

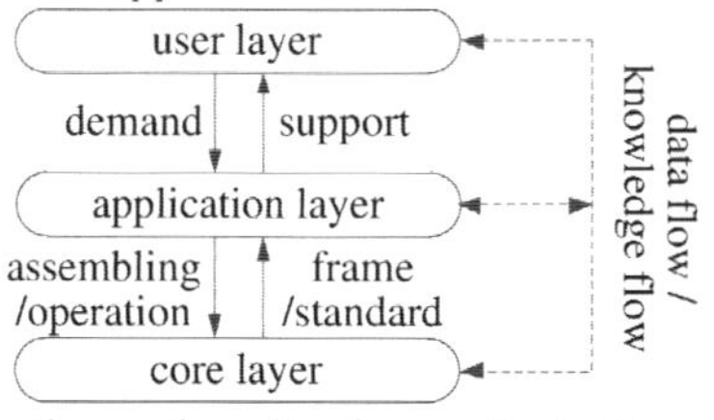

Figure 4. Localization and application method of application software

7. Conclusion

Faced with the interconnection and interoperability of operational plan among assistant decision-making systems, ontology technology has significant advantages. The article focuses on the theoretical exploration of the construction of the semi-knowledge-level operational plan ontology model, and opens up a new solution to the bottleneck problems. To actively promote the application of the results of the research, follow-up study in two aspects needs to be carried out: First, constantly improve and perfect the operational plan ontology model through demonstration and trial projects. Second, the application of operational plan ontology model, especially the design and development of software. To realize the core value of the ontology model and promote the leapfrog development of the equipment with assistant decision-making system.

8. References

[1] Zhang W P 2011 *Research on System Operation Architecture of Command Information System*(Beijing: National Defense University Press)

[2] He K Q He Y F and Wang C 2008 *Theory, Method and Application of Ontological Metamodeling*(Beijing: Science Press)

[3] Wang C 2016 *Artificial Intelligence Technology and its Military Application*(Beijing: National Defense Industry Press)

[4] Fan Q Y Liu D B and Ruan Q M 2009 *J. Computer Technology and Development* **19** 204-6

[5] Chen X J Qi H 2012 *J. Fire Control & Command Control* **37** 90-3

[6] DMWG 2009 *The Joint C3 Information Exchange Data Model 3.0.2*(Greding: MIP-NATO Management Board)

[7] Wickler G Potter S and Tate A 2007 *Proc. the IEEE/WIC/ACM Int. Conf. on Intelligent Agent Technology* pp 222-8

[8] Hu X 2011 *The Research on Ontology Based Expression and Validation of Joint Operation Plan*(Changsha: National University of Defense Technology)

[9] Qian M Liu Z and Yao L 2010 *J. Systems Engineering and Electronics* **32** 994-1000

[10] Long F and Zhang L 2013 *Proc. 4th Global Congress on Intelligent Systems* pp 89-93

[11] Long F Zhang L Deng K 2014 *Proc. 7th Int. Symposium on Computational Intelligence and Design* pp 494-7

Acknowledgments

The work is supported by the China Postdoctoral Science Foundation under Grant No. 2016M592919.

Institute of Physics
Dirac House, Temple Back
Bristol BS1 6BE UK

ISSN: 1757-8981
ISBN 978-1-5108-3842-0